APOLLO

VII – XVII

APOLLO

VII – XVII

Floris Heyne
Joel Meter
Simon Phillipson
Delano Steenmeijer

Edited by
Neil Pearson

With a special foreword by Apollo 9 astronaut
Rusty Schweickart

Featuring the exclusive essay *"The Good Old Days"*
A personal reflection on Apollo 7 photography by astronaut
Walt Cunningham

teNeues

"When you get back... you will be a national hero. But your photographs... they will live forever. Your only key to immortality is the quality of your photography."

Richard W. Underwood
NASA Chief of Photography for Mercury, Gemini and Apollo

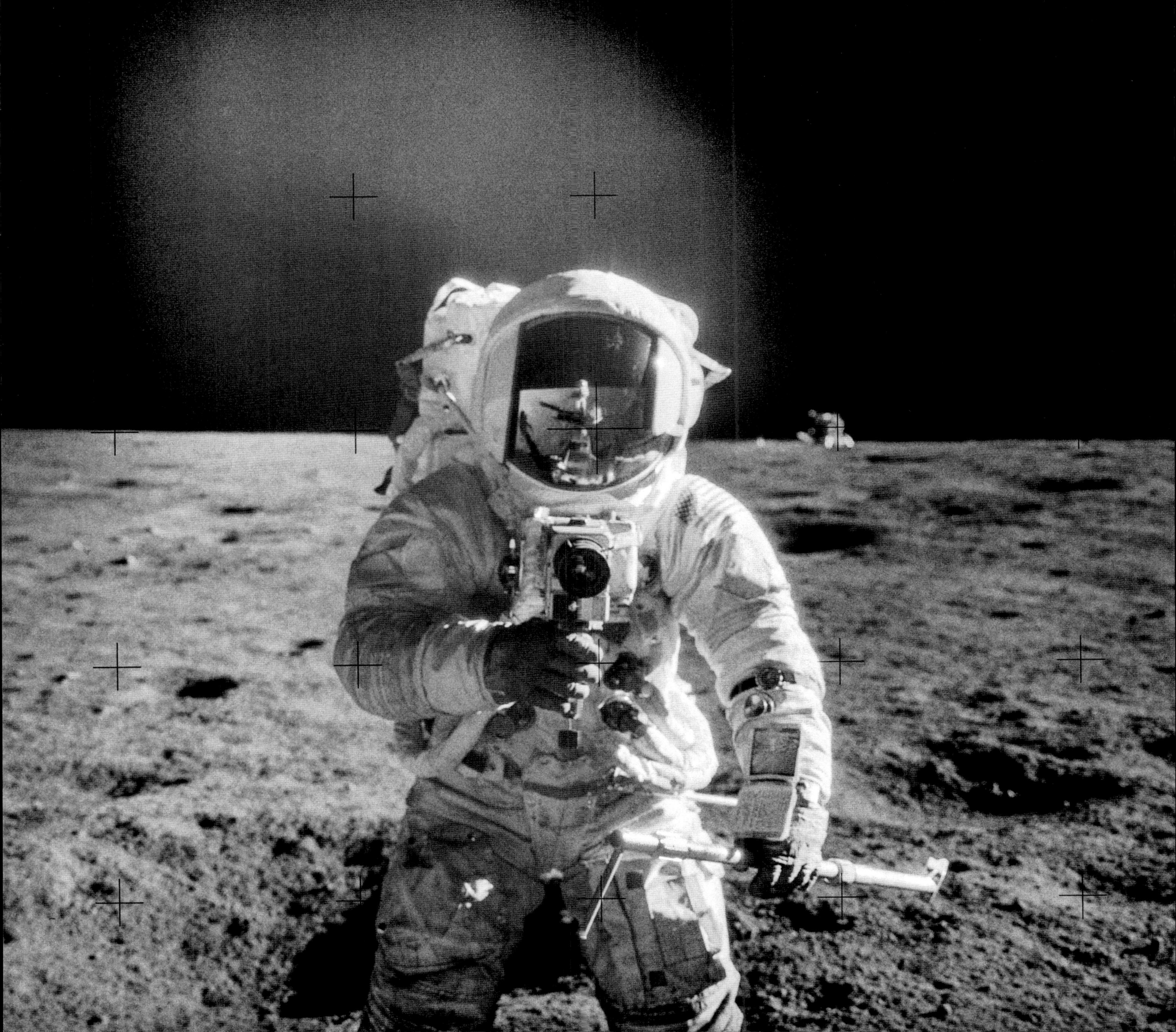

Images

My mental map says that we use images the way dogs and other animals use smell: they give us an astonishing amount of information at a glance. That happens because our brains allow us to project ourselves into the scene. Whether it's a portrait of a friend, a historic figure, or a dear relative. We find ourselves there with them — if we allow it.

If the image is an incredibly beautiful landscape, we almost automatically place ourselves in the position of the photographer and marvel at the captured beauty. And the phenomenon repeats, invited or not — for pain, happiness, sadness, death, and birth. Through photographs, we share lives and emotions we would otherwise have missed.

The shot I took of Dave Scott — partially out the Command Module hatch taking a picture of me — was cute. But what really inspired me to take that shot was that incredibly thin, iridescent blue margin between the Earth and the infinite blackness of the universe: our atmosphere.

It is so very thin, and it seems so tenuous, especially from that perspective. Yet we owe our very existence to that life-enabling membrane. It is so beautiful.

At an infinitely larger scale are the amazing images taken by astronomers. I've been interested in astronomy since I was a kid. I consider myself an astronomer of sorts — amateur, but deep into it.

A few years ago, I suddenly stopped looking at those astonishing images of distant galaxies and stellar nurseries as simply fantastic pictures. I had one of those eureka moments when I realized I was looking at real physical locations — actual environments where life doubtlessly thrives. Life, as precious as ours. Life of all kinds — less advanced, more advanced — shaped by emotions and experiences beyond our most creative imagining.

I have never looked at an astronomical image the same way since that day. We are life. We are consciousness. We are the lovers of the great universe.

We — humans, Earthlife — are at the very beginning of an extraordinary process, something I've come to call Cosmic Birth.

The Apollo 8 crew were the first to set eyes directly on "Mom" from the far end of the birth canal. I had the unbelievable good fortune to see and reflect on that iconic *Earthrise* image, taken by my good friend Bill Anders, while preparing for my own EVA on Apollo 9.

After processing that image — and my personal experience — for several years after my flight, I came to understand that we, we Earthlife, will collectively determine whether we continue to contribute to this amazing universal experiment: Life.

If we care — if we love and respect life enough — we will.

I invite you to look deeply at these images that I, and my dear friends and fellow Apollo astronauts, were able to capture — images of this evolutionary moment in our collective experiment of life.

Think.
Enjoy.

Rusty Schweickart
April 2025

Right:
Lunar Module Pilot Russell 'Rusty' Schweickart taking a photograph during his Extravehicular Activity (EVA), testing the new spacesuit during the Apollo 9 mission. (AS09-19-2982)

Previous page:
Apollo 12 Commander Charles 'Pete' Conrad Jr., holding a core sample tube, photographs Lunar Module Pilot Alan L. Bean during an EVA. The mission, which landed near the Surveyor 3 probe, collected valuable geological samples from the Ocean of Storms. (AS12-48-7071)

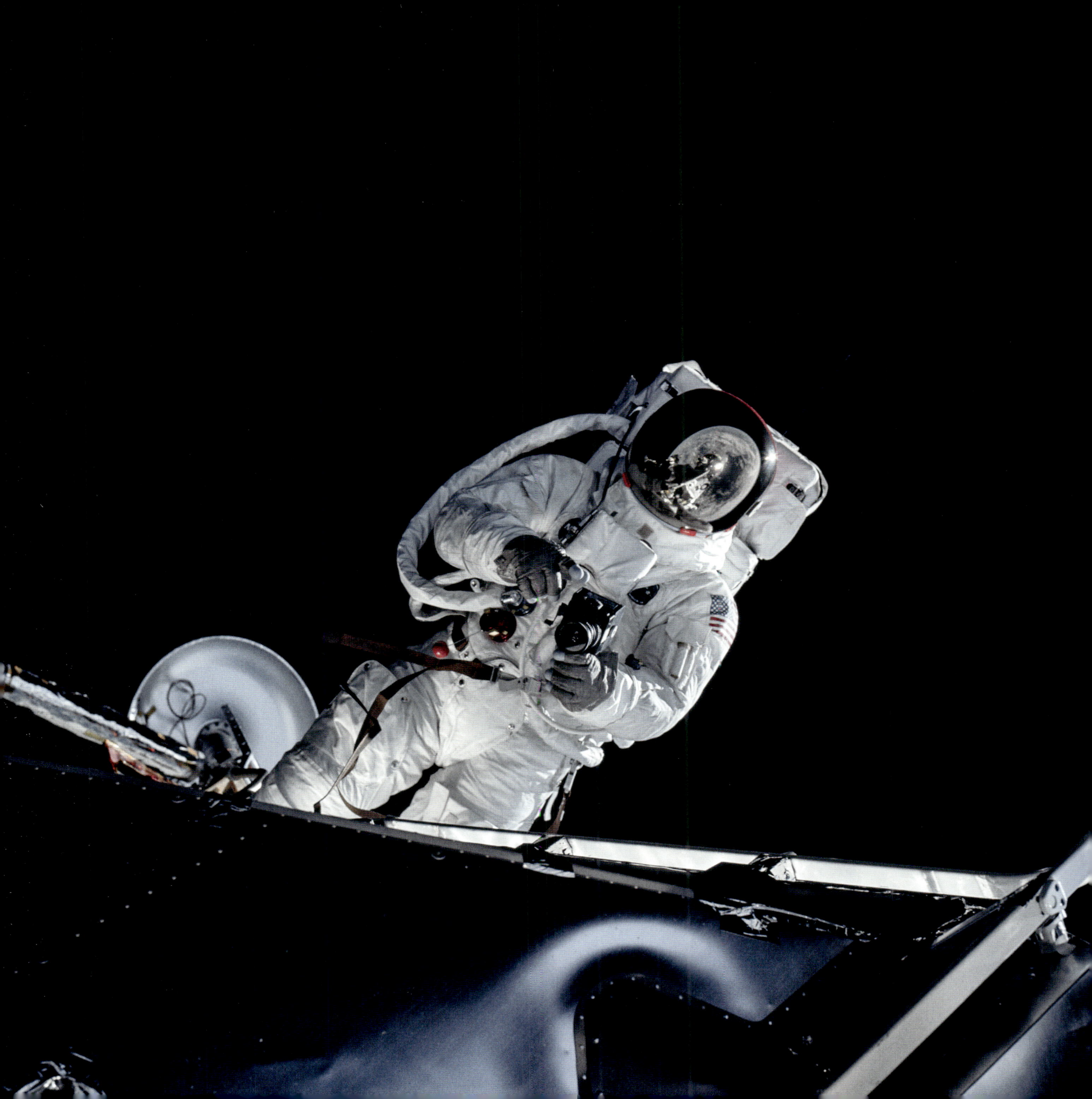

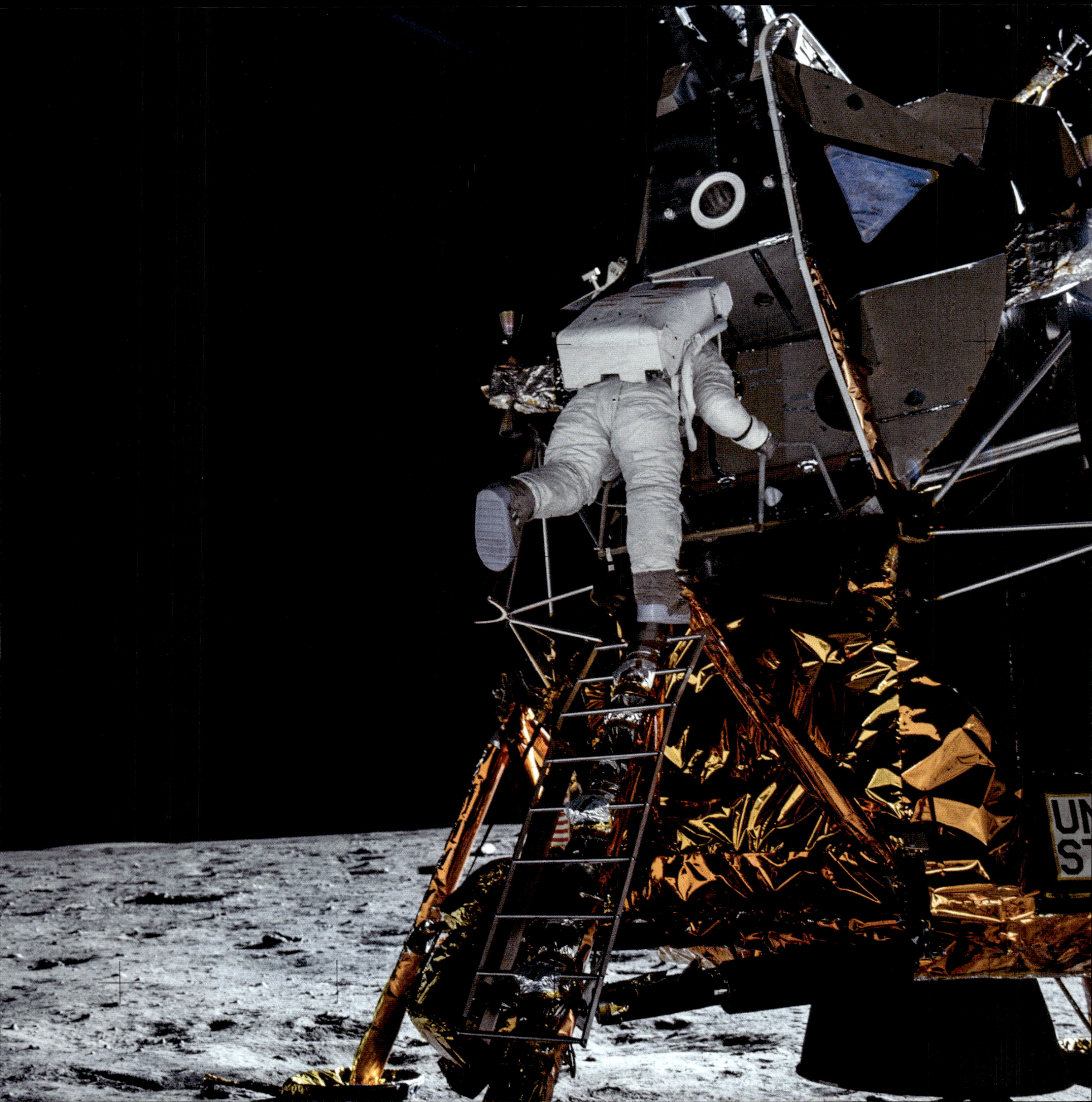

Small steps. Giant leaps.

The English word 'photograph' is made up of the ancient Greek words 'photos' and 'graphos' which literally mean 'light writing'. At the time of the Apollo program, that meant exposing a chemically-treated film to patterns of light and it was this process which made it possible to seize moments in time and share them with the world.

A single photograph can tell a story to billions of people. It transcends language barriers, physical barriers and requires no prior knowledge of the subject. This is the beauty of photography: it reaches out to all people and everybody can intuitively understand its form and content.

Thus photography became a crucial technique for documenting the history of mankind. Our ancestors depended on the words of the Greek historian Thucydides to recount the war between Athens and Sparta. They relied on the woven illustrations of the Bayeux Tapestry to see scenes from the Norman invasion of Britain. Only in the last two centuries has photography enabled us to see actual light captured during such historical events.

The early exploration of space is one such historical event, unrivaled among humanity's achievements. Fortunately space travelers were able to bring back beautiful, moving and instantly recognizable images to depict it. Those images add a new understanding to what it means to be human, what it means to live on a delicate little orb circling the Sun since time immemorial.

Not only did Apollo bring us this photographic testimony, but many major advancements in photographic technology date back to the extensive research and engineering that was part of the hugely complex project.

Now, as NASA's Artemis program prepares to return astronauts to the Moon, this visual legacy takes on new relevance. The photographs in this book are more than documentation — they are a foundation. They remind us of what was achieved, what was seen, and how it was shared. And just as Apollo marked the beginning of our photographic presence beyond Earth, Artemis will extend that story — through new eyes, new lenses, and new ways of seeing our place in the universe.

In sumptuous detail and color, this book presents some of the most significant and stunning photographs ever taken. It also celebrates the skill and daring of a very particular group of photographers — the Apollo astronauts.

Left:
Lunar Module Pilot Edwin 'Buzz' Aldrin descending the steps of the Apollo 11 Lunar Module to become the second man to stand on the Moon. (AS11-40-5866)

The Apollo program

Project Mercury put the first American, Alan Shepard, into space on May 5th, 1961. Less than three weeks later on May 25th, at a special joint session of Congress, U.S. President John F. Kennedy announced that "this nation should commit itself to achieving the goal, before this decade is out, of landing a man on the Moon and returning him safely to the Earth". After a further five Mercury solo missions, Project Gemini was seen as the stepping stone to the Moon where two-man crews flew ten manned missions during 1965 and 1966 with three main objectives: to fly extended-duration missions (the time it would take to get to the Moon and back), to develop EVA (extravehicular activity) techniques and to develop the procedures necessary to rendezvous and dock. With those three goals achieved, NASA was well on its way to manned missions to the Moon with the three-man spacecraft of the Apollo program.

After the tragic loss of the Apollo 1 crew during a launch pad test in 1967, Apollo 7 became the first successful crewed mission of the program, launching into Earth orbit on October 11th, 1968. A further ten Apollo missions were launched and many landmark and historical moments were achieved during the program.

Apollo 8 was the first manned spacecraft to orbit another celestial body and the crew of Apollo 13 traveled the farthest away from the Earth, something which has never been repeated to date. The most famous of all moments was when the goal of the Apollo program was realized. On July 20th, 1969, Neil Armstrong and Buzz Aldrin touched down on the lunar surface, becoming the first humans to have set foot on another celestial body. Apollo 17 was the last mission to fly, drawing the program to a close soon after returning to Earth on December 19th, 1972.

During the 11 Apollo missions, 29 men lifted off from Earth with 12 of them touching the surface of the Moon. It is these men that took the photographs that still give humans, over four decades later, the opportunity to see things which had never been seen by the human eye before. Views such as the first *Earthrise* over the lunar surface, one of history's most-seen images, or *The Blue Marble*, capturing the whole of the Earth in one delicate frame.

Apollo Mission Insignias

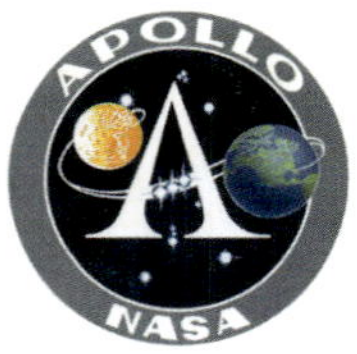

Apollo Program

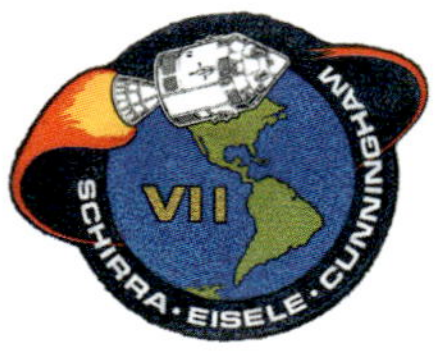

Apollo 7

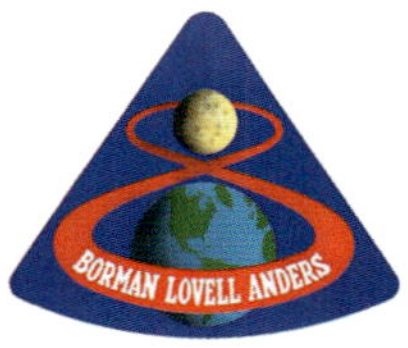

Apollo 8

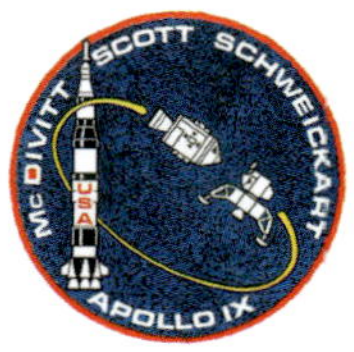

Apollo 9

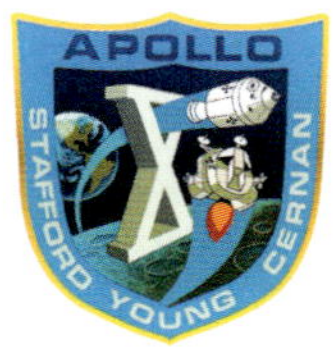

Apollo 10

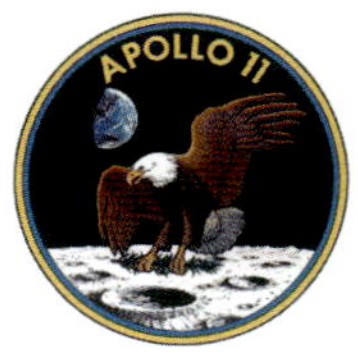

Apollo 11

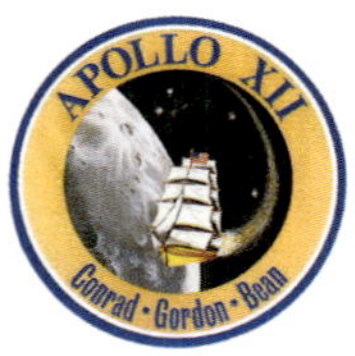

Apollo 12

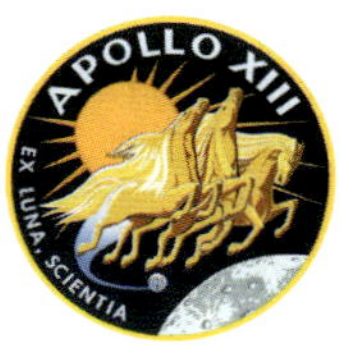

Apollo 13

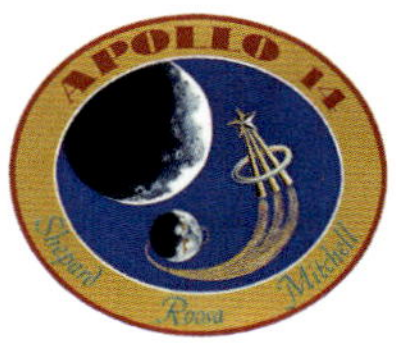
Apollo 14

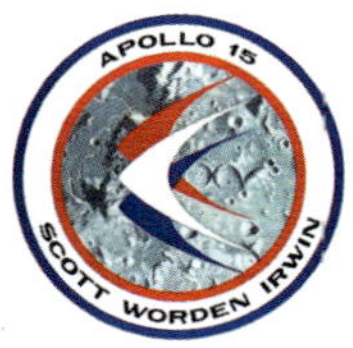

Apollo 15

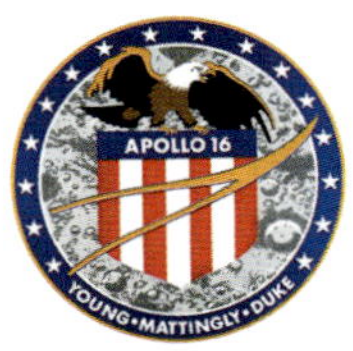

Apollo 16

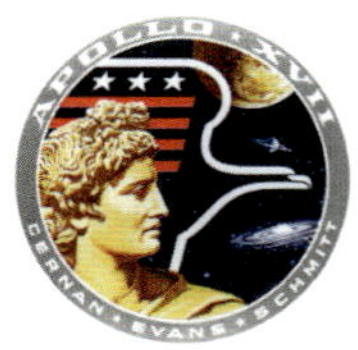
Apollo 17

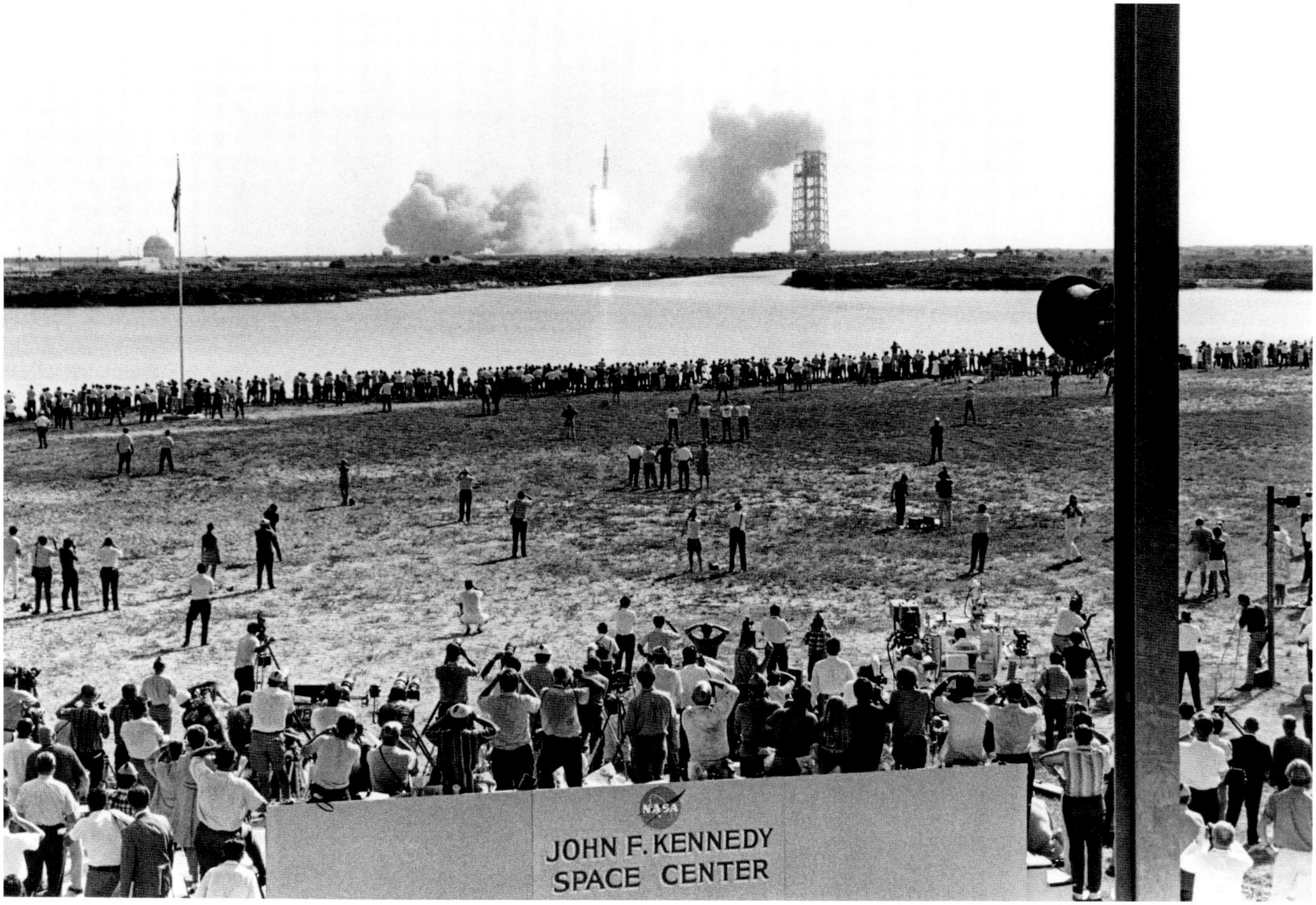

Crowds from the launch complex press site watch as Apollo 11 launches from the Kennedy Space Center on July 16th, 1969.

'The good old days'

An introduction by Apollo 7 astronaut Walt Cunningham

Apollo 7 was the first mission from the Apollo program. The mission's aim was to ensure the spacecraft and launch vehicle operated as designed. Since this was the first time a crew had flown on the Command/Service Module, the astronauts had to test the multitude of systems and subsystems, recording all aspects of the flight in order to improve their knowledge of the craft and how it handled in space.

With no Lunar Module to fly on the mission, along with his other tasks Walt Cunningham took nearly 70 percent of the photographs brought back from this pioneering mission. The lessons he learned had a significant impact on the future of space photography in the Apollo project. Although Walt said he thought of himself as a physicist and fighter pilot, rather than a photographer, he graciously agreed to write this introduction about his personal experiences and the photographic challenges and problems faced during the Apollo 7 mission.

Over the last 50 years, we have enjoyed magnificent photographs of our planet taken from space. Today, the International Space Station (ISS) covers the Earth's surface between plus and minus 51.6 degrees of latitude covering about 75 percent of our planet's land area, where about 95 percent of the Earth's population lives. The ISS has excellent windows, with some always offering a view of the Earth. The space station has a number of high resolution digital cameras enabling them to eventually take pictures of most things near the station's ground tracks.

Photography is but one area of activity that has shown amazing improvements as space exploration has evolved over the past 50 years. In the 'good old days' of the Apollo program, we started improving the capability of doing space photography. We had the pleasure of carrying a Hasselblad 500C camera modified for use in space with an 80mm lens and nine magazines holding about 500 frames of 70mm film.

We trained with the Hasselblad camera for two years prior to our first Apollo launch. We used it at home and in our office but we also carried it with us when flying our T-38s around the country for training and engineering activities. It was not a particularly conveniently sized camera for our small T-38 cockpits but we did enjoy this part of our training activities.

Today, most of the public interested in space thinks we took beautiful pictures of nearly everything on our planet back in the 1960s Unfortunately, in the 1960s, space missions were vastly different than they are today. We were still learning how to survive in space and only two of our Apollo crews remained in Earth orbit – Apollo 7 and Apollo 9. It gave us the opportunity to enjoy the beauty of our planet as seen from space. The other nine Apollo missions traveled to the Moon. Then why did we not get more than a few hundred pictures of our planet's surface?

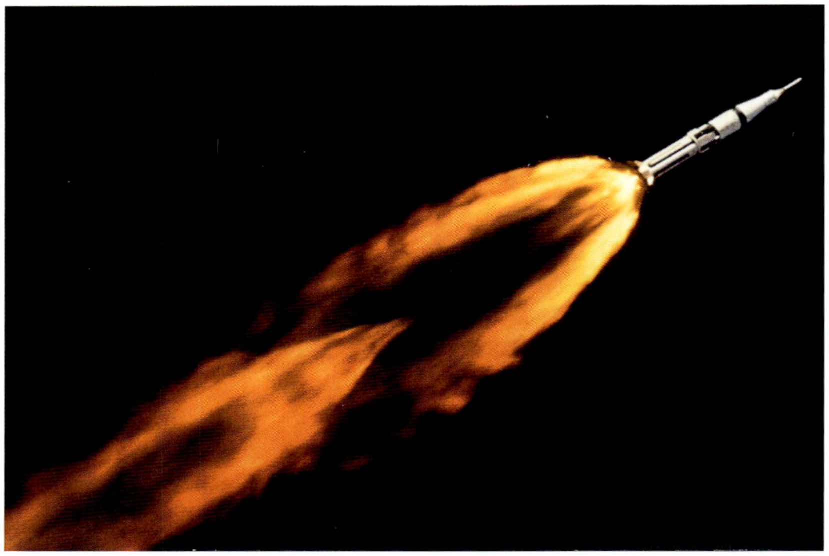

The launch of Apollo 7 Saturn IB on October 11th, 1968.

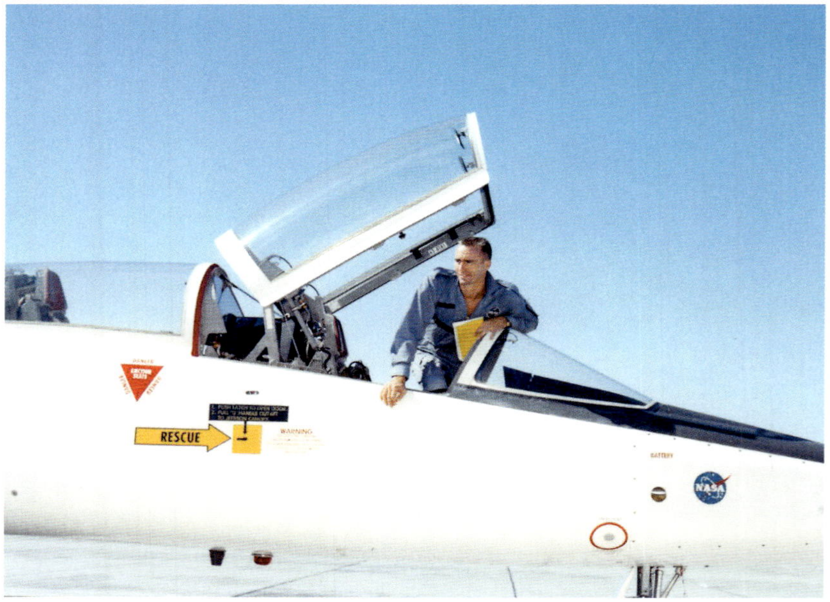

Walt Cunningham about to enter the cockpit of a Northrop T-38 Talon.

Opposite:
Walt looking out of the Command Module window during their 11-day orbit around the Earth. (AS07-04-1584)

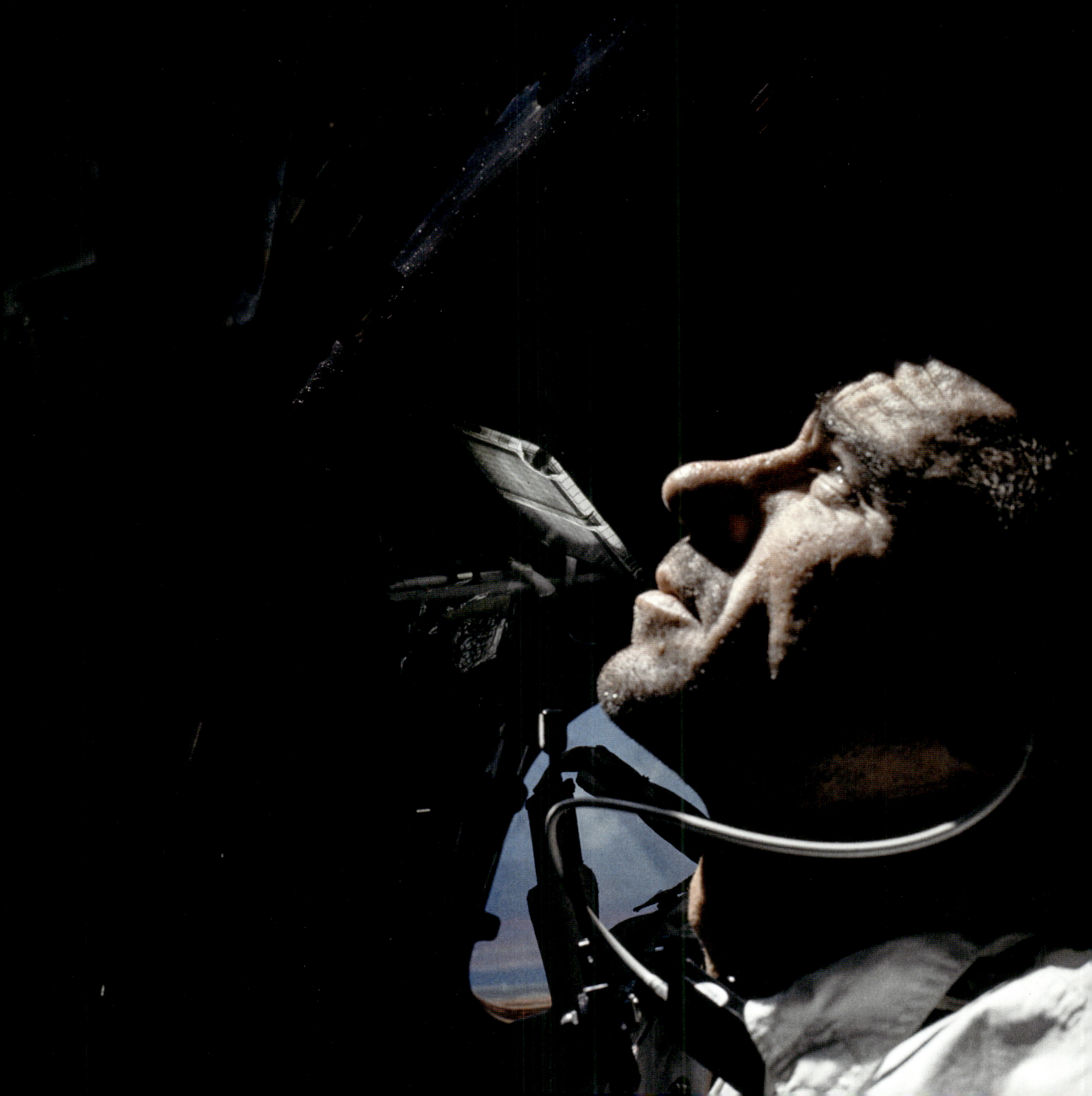

There were vast differences between the hardware and the cabin environment for the Apollo missions in the 1960s and the Space Shuttle and ISS expeditions of the modern era. For example, the ISS has, virtually, 100 percent air to ground communications. On our Apollo earth orbital missions, we had all of four percent air to ground communications. Looking back on it, that may have been an advantage!

Our Apollo 7 crew felt particularly fortunate to be there and have the opportunity to occasionally photograph the surface of the Earth. Our mission was the first manned Apollo mission with a focus on testing all of the spacecraft systems and helping to develop many of the operational procedures necessary to land a man on the Moon. Some of our operational testing and development work required surface photos to evaluate the procedures. This included terrain and weather photography.

Our first principal camera assignment was to document our Command Module separation from the S-IVB (the second stage of the Saturn IB rocket) and the rendezvous and near docking with it the following day. While Schirra was circling our S-IVB booster, I was focusing the Hasselblad on the booster and taking photos with the Earth's surface in the background. We learned after returning and the film was developed that I had taken a surprising number of good pictures of the surface geography in the background behind the S-IVB.

How did the design characteristics of our spacecraft and our orbital tracks across the surface of the earth impact what we were able to photograph on our beautiful planet?

A good many of our orbital objectives required us to use our Reaction Control System (RCS) rockets to control the spacecraft attitude. This was to determine the best techniques and to minimize the fuel required. Also, on a test flight, the mission objectives are 'front end' loaded in case the mission has to be aborted before completing the full duration. We experienced such a flight plan for our 11-day mission.

Even from the beginning, we couldn't take pictures of anything we wanted to. Random photographs could only be taken when we had the spare time and were not involved in test objectives. The last two or three days of our mission left us with a little more time to use our Hasselblad if we had a window pointing at the ground. Unfortunately, we were forced to restrict ourselves to only a few photos a day because of the small number of frames we had left.

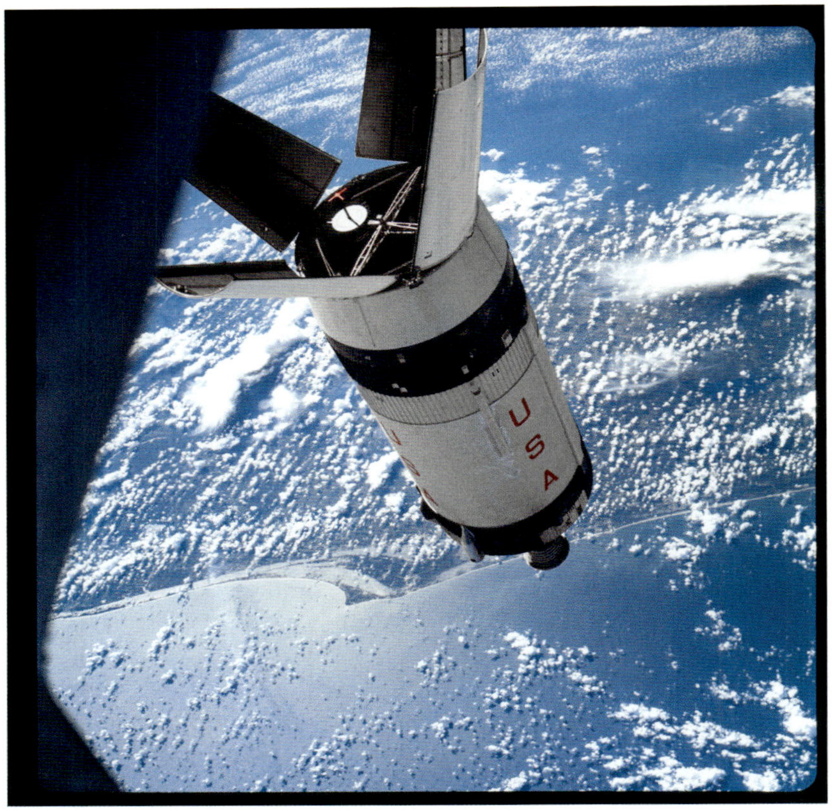

Two of Walt Cunningham's favorite photographs that he took during the mission. The photograph above shows the S-IVB stage just after separating from the Command/Service Module. Florida and the Kennedy Space Center can be seen below (AS07-03-1545). Opposite is a unique photograph in which Walt captured the 13 highest peaks of the Himalayas all within a single frame. (AS07-07-1748)

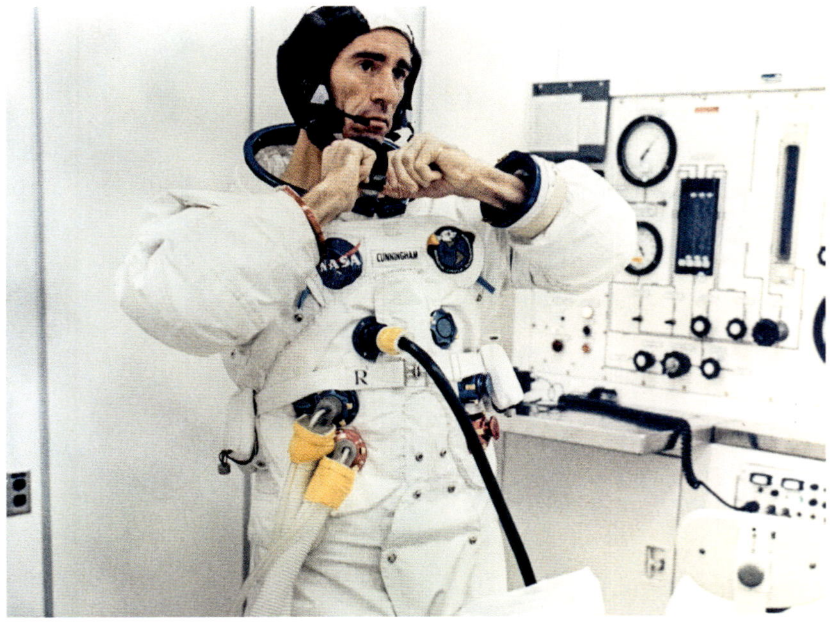

Walt Cunningham fitting his pressure suit on the morning of the Apollo 7 launch.

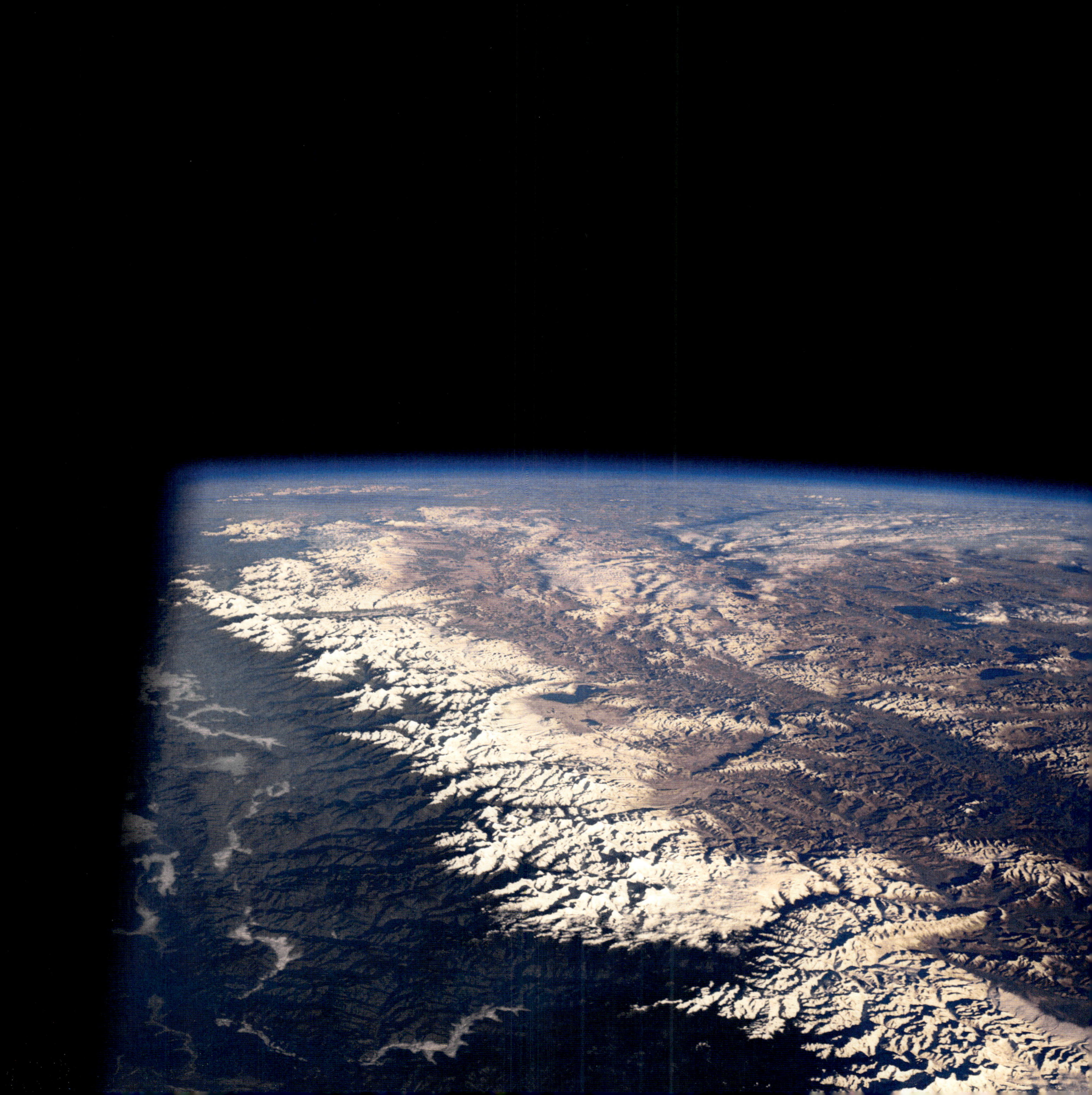

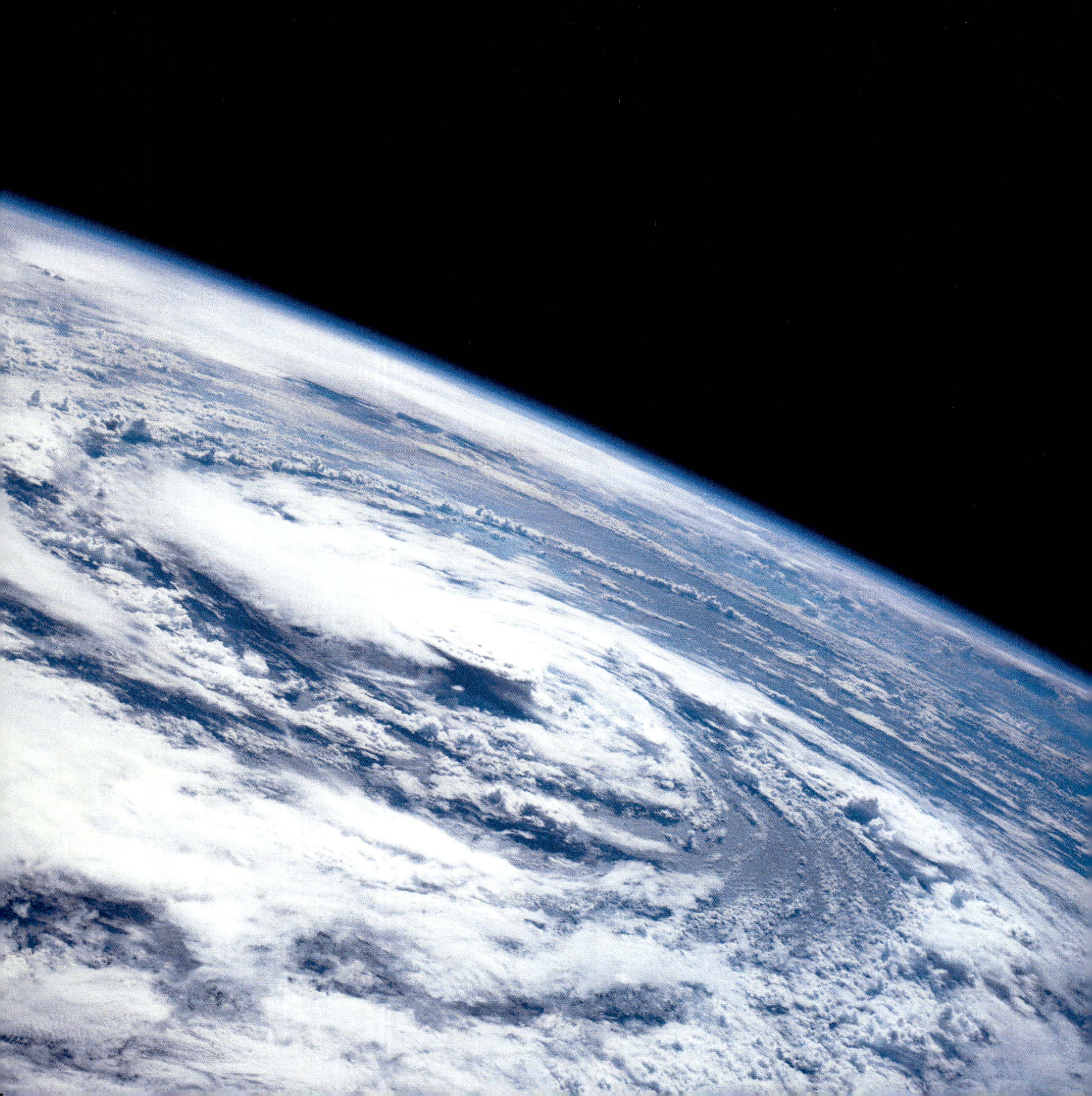

The photographic challenges of capturing Earth from space

Many factors had a negative impact on our ability to take pictures, even if we had planned for them in advance. A number of these were 'hard' constraints that were impossible to overcome. The public was pretty much unaware of them.

Geographical factors:
The orbital inclination of Apollo 7 was 31.6 degrees. That meant that all of our orbits carried us North and South of the equator as far as 31.6 degrees of latitude. We could fly as far North as San Diego, Houston and Orlando in the Northern Hemisphere and almost down to Uruguay in South America. This limited what we could fly over that we might find of interest to photograph.

Ground track:
When it takes only 90 minutes to travel around the world, you will make 16 orbits every 24 hours. Then we began repeating our trips over the same surface tracks. Each of those tracks were separated by 22.5 degrees of longitude. At the equator those tracks were separated by about 1500mi (2414km) and were a little closer north and south of the Equator. That left most of our planet's surface uncovered.

Day/night cycle:
When you circle the earth every 90 minutes, you are moving from night to day or from day to night every 45 minutes. That meant we only had enough light to take photographs during half of each orbit.

Cloud cover:
On average, 55 percent of the Earth's surface is covered by clouds. That provided more opportunity for cloud photos than for the Earth's surface. About 45 percent of the surface was out from under clouds for the photographs we were more interested in taking.

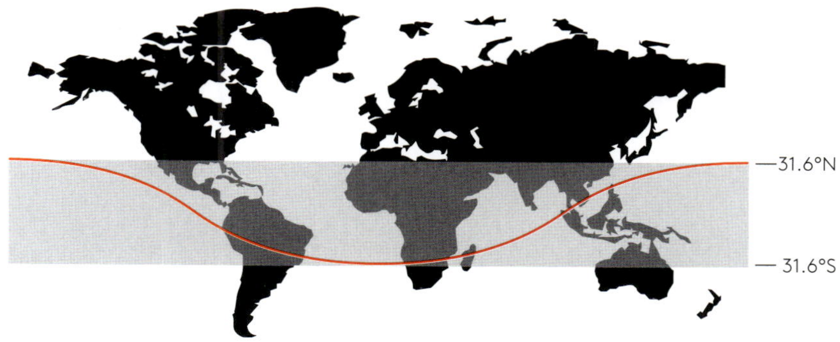

Chart showing the approximate area covered with an orbital inclination of 31.6 degrees.
— Orbital revisit time: every three weeks
— Revisit time with the same lighting conditions: every three months

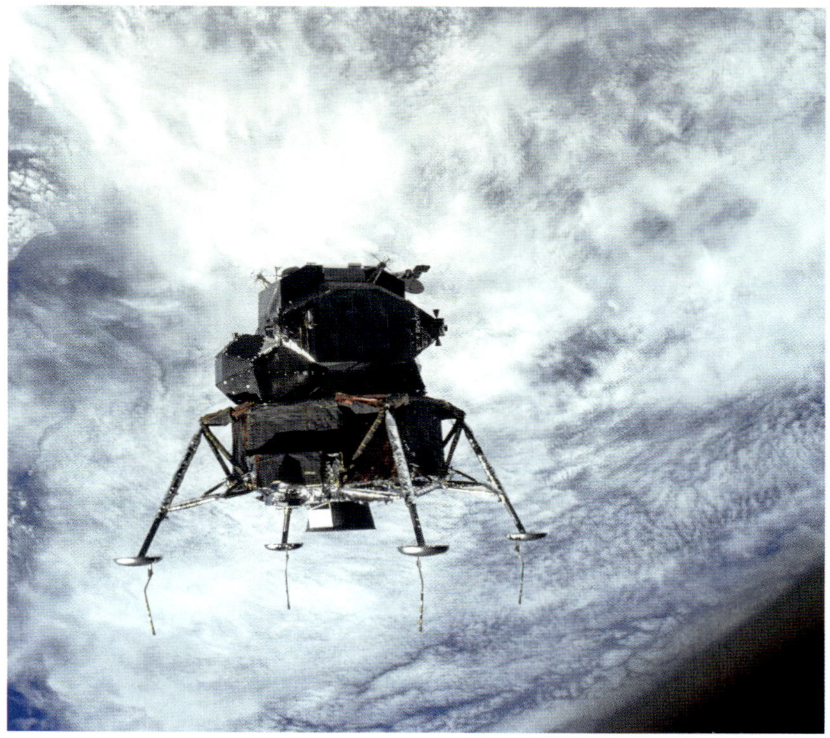

The cloud cover seen behind the Apollo 9 Lunar Module in the photograph above gives a good example of the dense and obscuring cloud cover that was often encountered. (AS09-21-3199)

Opposite:
Another of Walt's favorite photographs, showing Hurricane Gladys forming over the Gulf of Mexico. (AS07-07-1877)

Window clarity:
The Apollo Command Module had five windows, all pretty much in the same plane. This encompassed an arc of about 150 degrees and provided us a relatively narrow view outside the capsule. The largest window of the five, and the one we expected to provide our best view outside, was a 10in (25.4cm), round window located in the spacecraft hatch.

Drifting spacecraft:
A number of mission objectives required a particular spacecraft attitude or changing the attitude. That required controlling the spacecraft attitude utilizing the small rockets of our reaction control system and consuming RCS fuel, of which we had a very limited amount. When our operations did not require maintaining attitude control, we were drifting. In a random drift, most of the time our windows were pointing out into space rather than down toward the Earth's surface.

Window degradation:
One of our major viewing problems on Apollo 7 was one that we never anticipated. Each window consisted of two 0.75in (19mm) thick quartz panes, with a very tiny gap separating the two panes. Unfortunately, in zero gravity, the 'glue' securing the window panes in place began to out-gas. This slowly contaminated the space between the two panes. The first to deteriorate was the hatch window with the largest view. The windows slowly degraded one by one. Fortunately for me, the largest window on my side was still clear at the end of our mission. This design problem was corrected from Apollo 9 onwards.

All of these factors were essentially hard constraints, impacting our opportunities to take pictures outside of the spacecraft. We also had a couple of 'soft' constraints. These were suggestions that our photography experts suggested we try to follow. They suggested that we avoid taking pictures until an hour after sunrise and stop an hour before sunset. That was for 24 hour days. For our 90 minute days in orbit, that would reduce our photo time from 45 minutes to about 38 minutes.

The photographic experts also suggested that we improve the quality of our photos by pointing the camera no more than 30 degrees off of the local vertical because of the increasing atmospheric thickness. Beyond that arbitrary choice, they felt it would have too big an impact on the quality of the photos. In reality, if it was daylight on the ground, we had a good window pointing toward the surface of the earth, there was little cloud cover and we found time to look out, we paid very little attention to these two soft constraints. Our favorite window was whichever one was pointing toward the surface and was still clear with no out-gassing.

There was one more management constraint for our mission that we followed perfectly: "Take no pictures of China." That was an easy rule to follow. In 11 days and many passes over China, we never saw the surface because of the cloud cover and smog in their atmosphere.

Over the next 11 chapters you will find a carefully chosen selection from the many thousands of stunning photographs that were taken by me and my fellow Apollo astronauts. I think it's great to see a book that has processed and presented this unique collection of photographs in such a clear and beautiful way. It was a real privilege and honor to be able to go up there and bring back some of these images that I can share with you and the rest of the world.

Walt Cunningham
October 2016

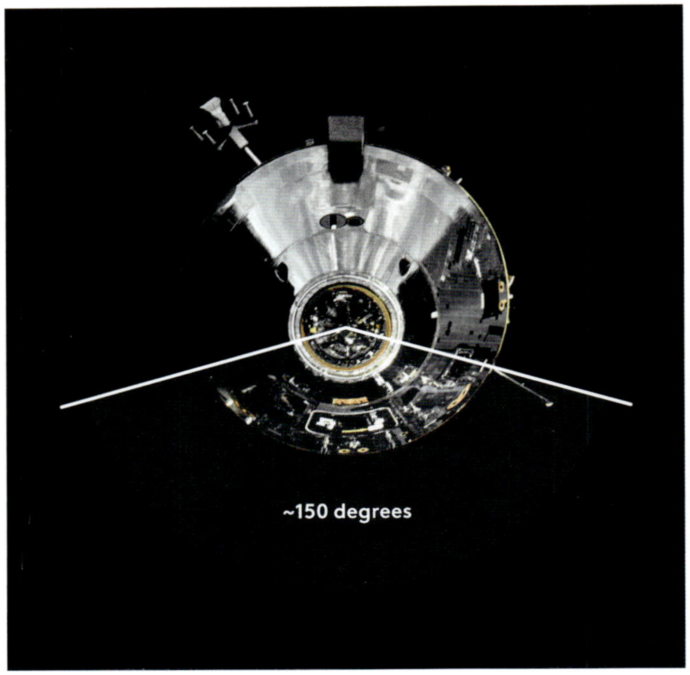

The lines give an indication of the 150 degree views available to the crew from the five windows fitted to the Command Module. (AS09-24-3652)

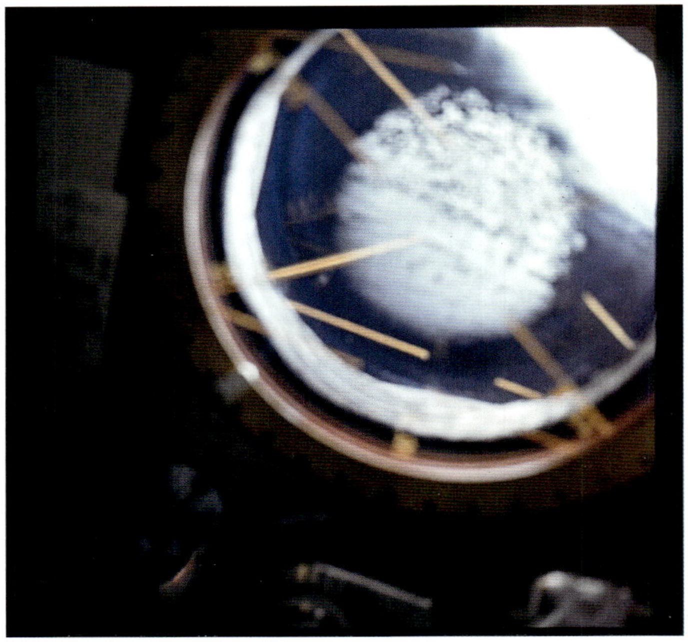

The 'out-gassing' of the window sealant can be clearly seen in the middle of the round window in this photograph taken from within the Command/Service Module. (AS07-04-1581)

Opposite:
Another favorite photograph of Walt Cunningham, taken by crew member Wally Schirra, of Princess Charlotte Bay and the Great Barrier Reef, Australia. (AS07-08-1902)

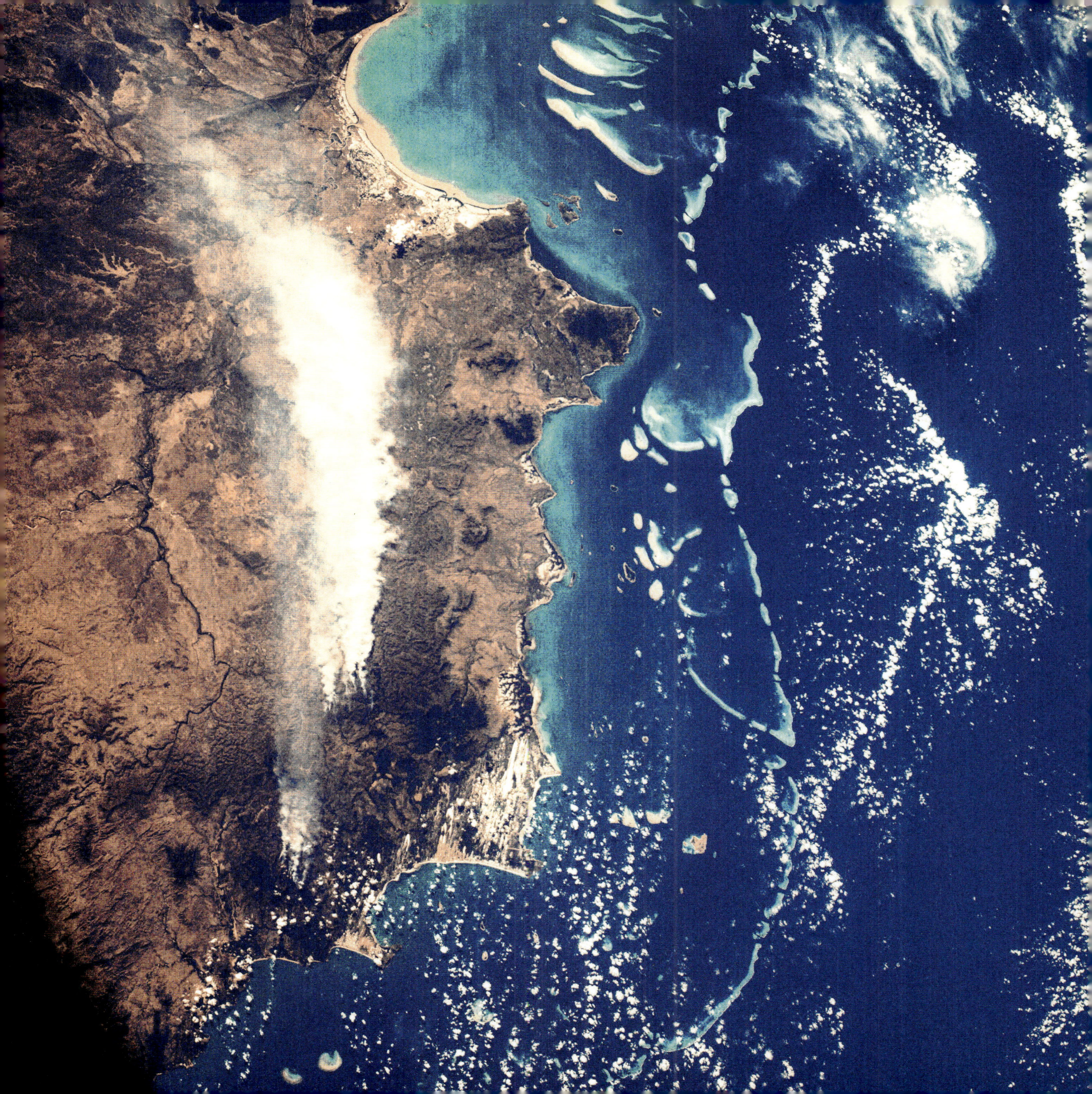

Apollo 7

VII

Walter M. Schirra

Donn F. Eisele

R. Walter Cunningham

Apollo 7

Launch: October 11th, 1968
Splashdown: October 22nd, 1968

It was in 1968 that the United States of America launched the first crew into space under the Apollo program, the first mission to carry astronauts since Gemini 12 in November 1966. The program had been delayed by the tragic circumstances of the Command Module fire, in which the crew of Apollo 1 lost their lives during a launch rehearsal in January 1967. During the stand-down of the accident investigation, several unmanned test flights of the Saturn V launch vehicle and the Apollo spacecraft were made. The improved spacecraft and safety procedures helped ensure that Apollo 7 could fulfill Apollo 1's objective to test the Apollo Command/Service Module in low Earth orbit.

Apollo 7 completed 163 low Earth orbits during which numerous tests were undertaken, such as practising a simulated Lunar Module rendezvous and docking.

Photography was focused on synoptic terrain and weather targets. NASA used the term 'target' to instruct the Apollo crews to photograph a particular area or flight procedure, allowing scientists and engineers to observe and evaluate it.

The images captured during Apollo 7 had a huge scientific impact, advancing geological and oceanographic understanding. Modern 'space-to-ground' photography was very much in its infancy since there were only a few satellites in orbit capable of photographing and returning images back to Earth. Astounding photographs were taken, capturing many sights never seen before, such as the world's 13 tallest peaks (the Himalayan mountain range) all framed within one picture, or the complete storm and eye of Hurricane Gladys that formed over the Gulf of Mexico during the mission.

Walter M. Schirra
Commander

Born on March 12th, 1923 in Hackensack, New Jersey. In 1962, he was Backup Pilot of Mercury-Atlas 7, piloted the six orbit Mercury-Atlas 8 flight, served as Backup Command Pilot for the Gemini 3 mission, and was the Command Pilot of Gemini 6A. He was the only astronaut to have flown Mercury, Gemini and Apollo.

Donn F. Eisele
Command Module Pilot

Born on June 23rd, 1930 in Columbus, Ohio. He was one of the third group of astronauts selected by NASA in October 1963. In addition to flying on the Apollo 7 mission, he served as Backup Command Module Pilot of Apollo 10.

R. Walter Cunningham
Lunar Module Pilot

Born on March 16th, 1932 in Creston, Iowa. He joined NASA as one of 14 astronauts named in October 1963. After Apollo he served as NASA Chief of the Skylab Branch of the Astronaut Office from 1968 to 1971.

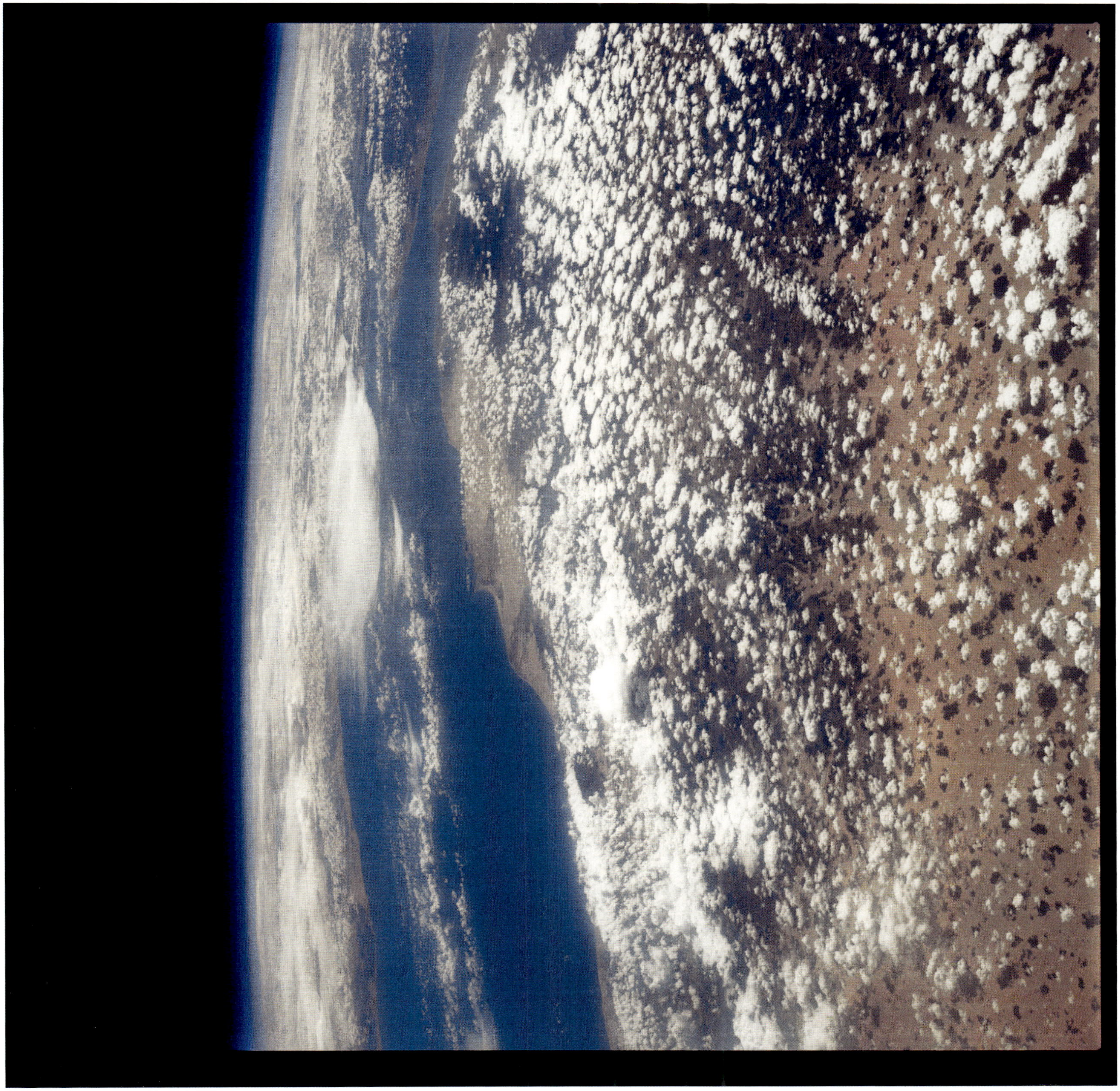

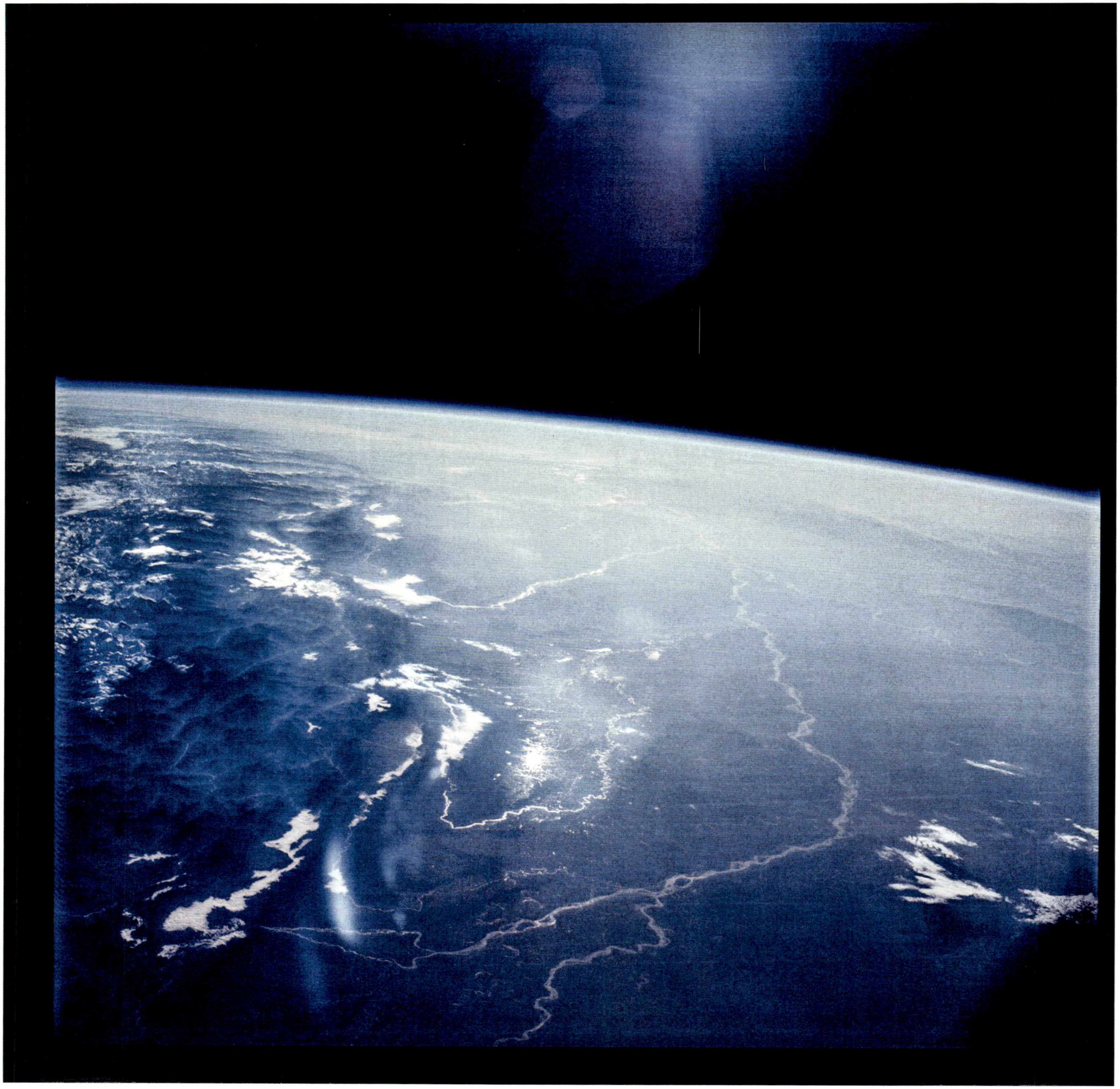

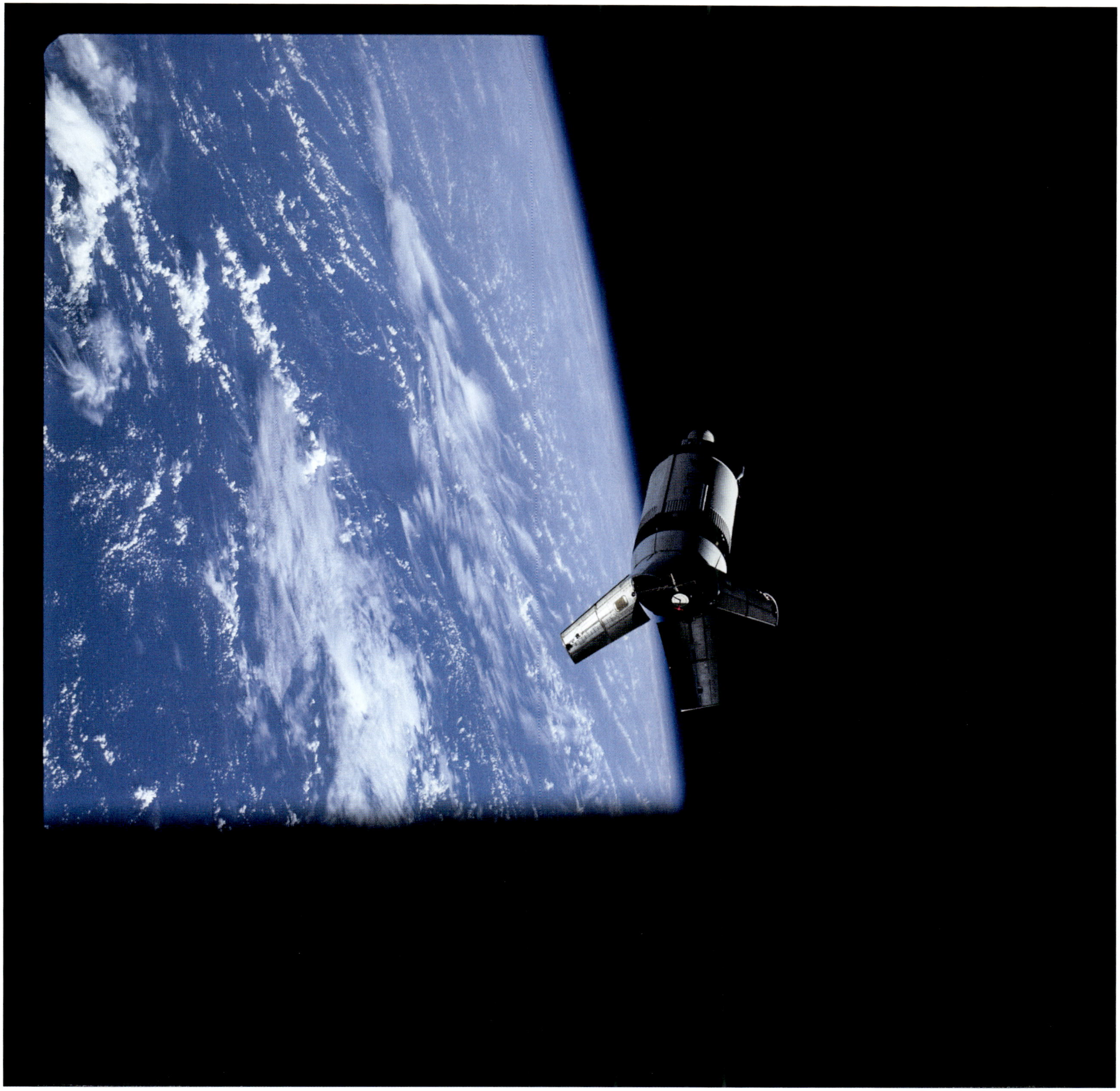

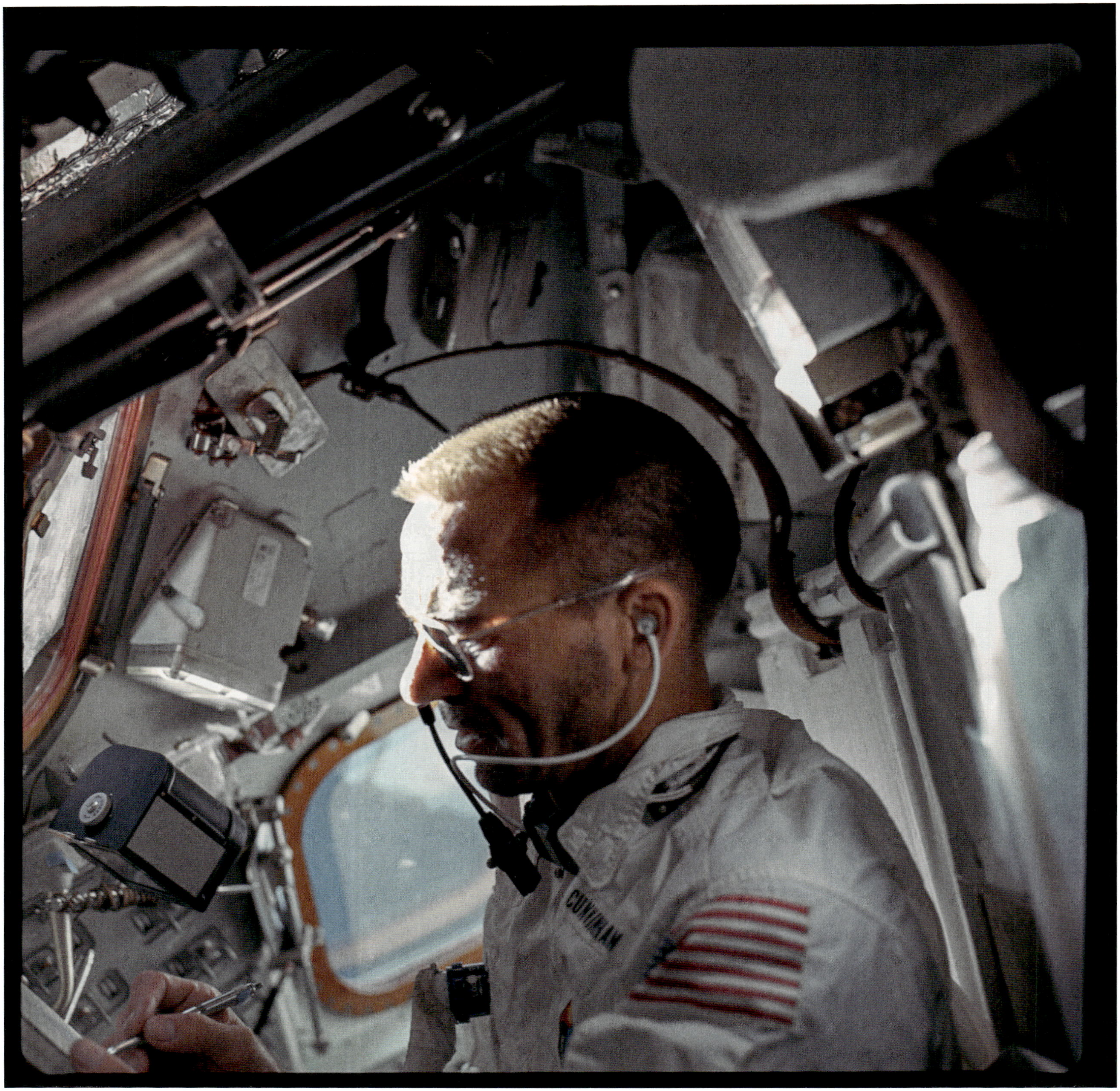

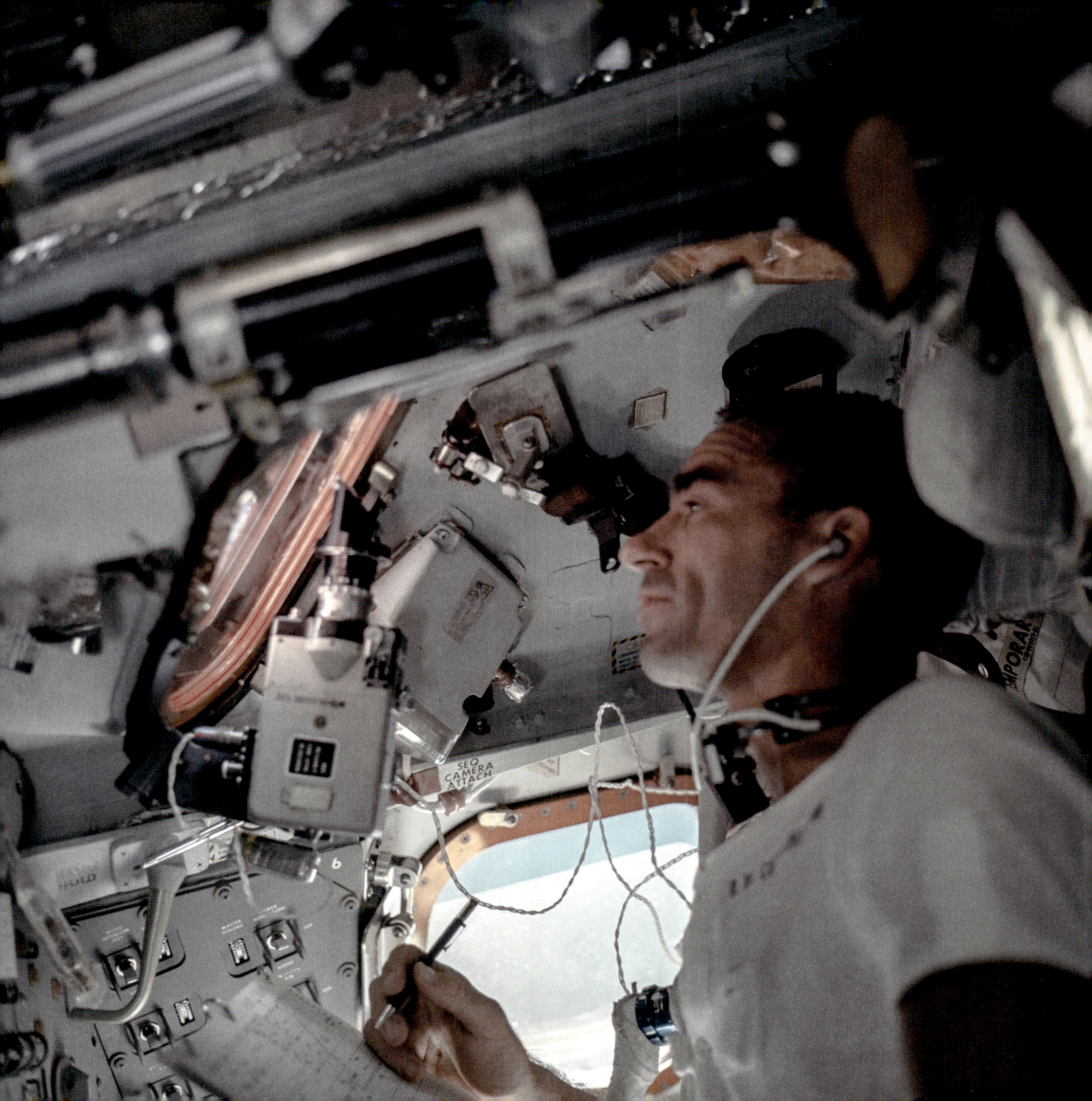
SEQ
CAMERA
ATTACH

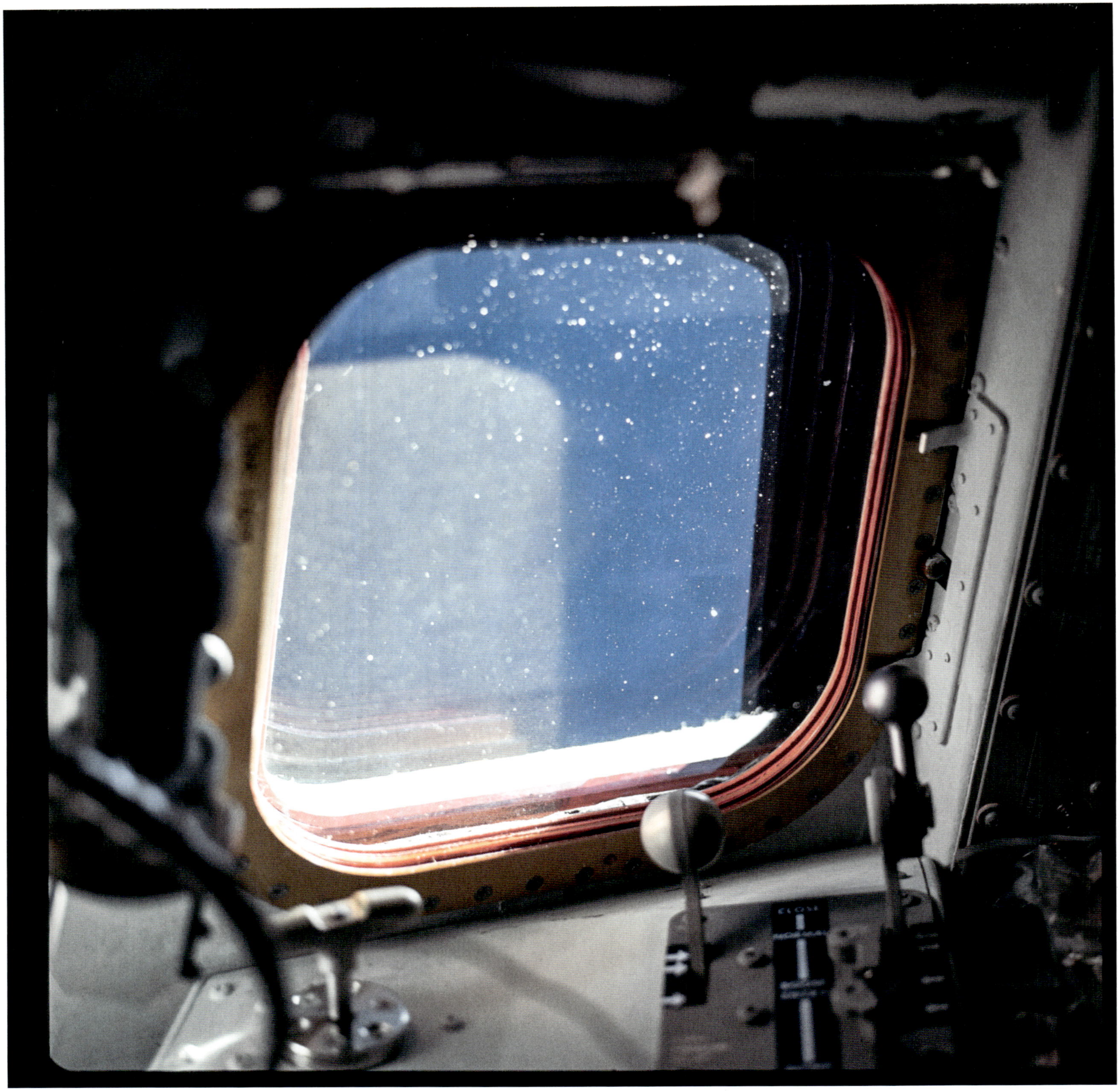

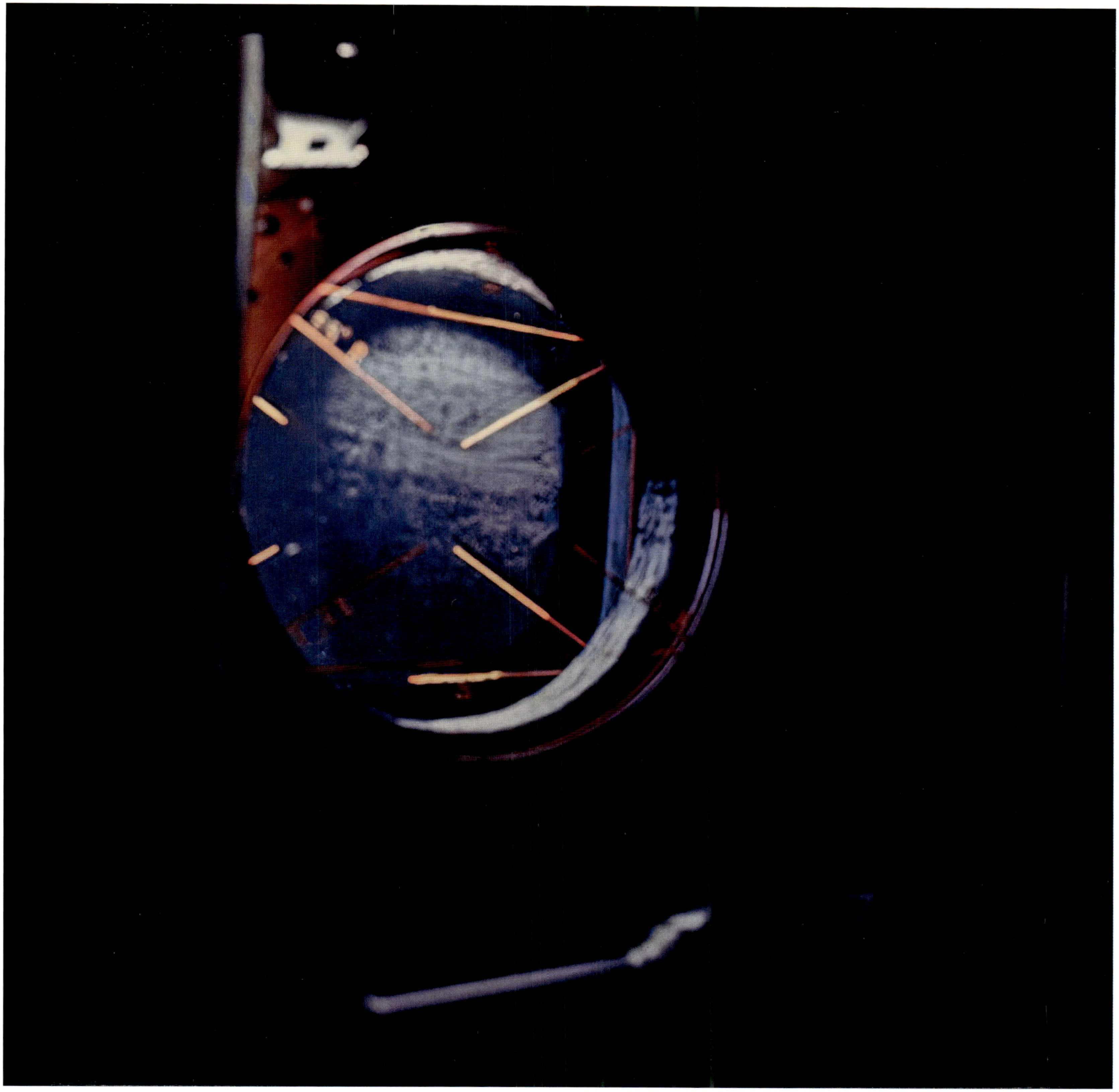

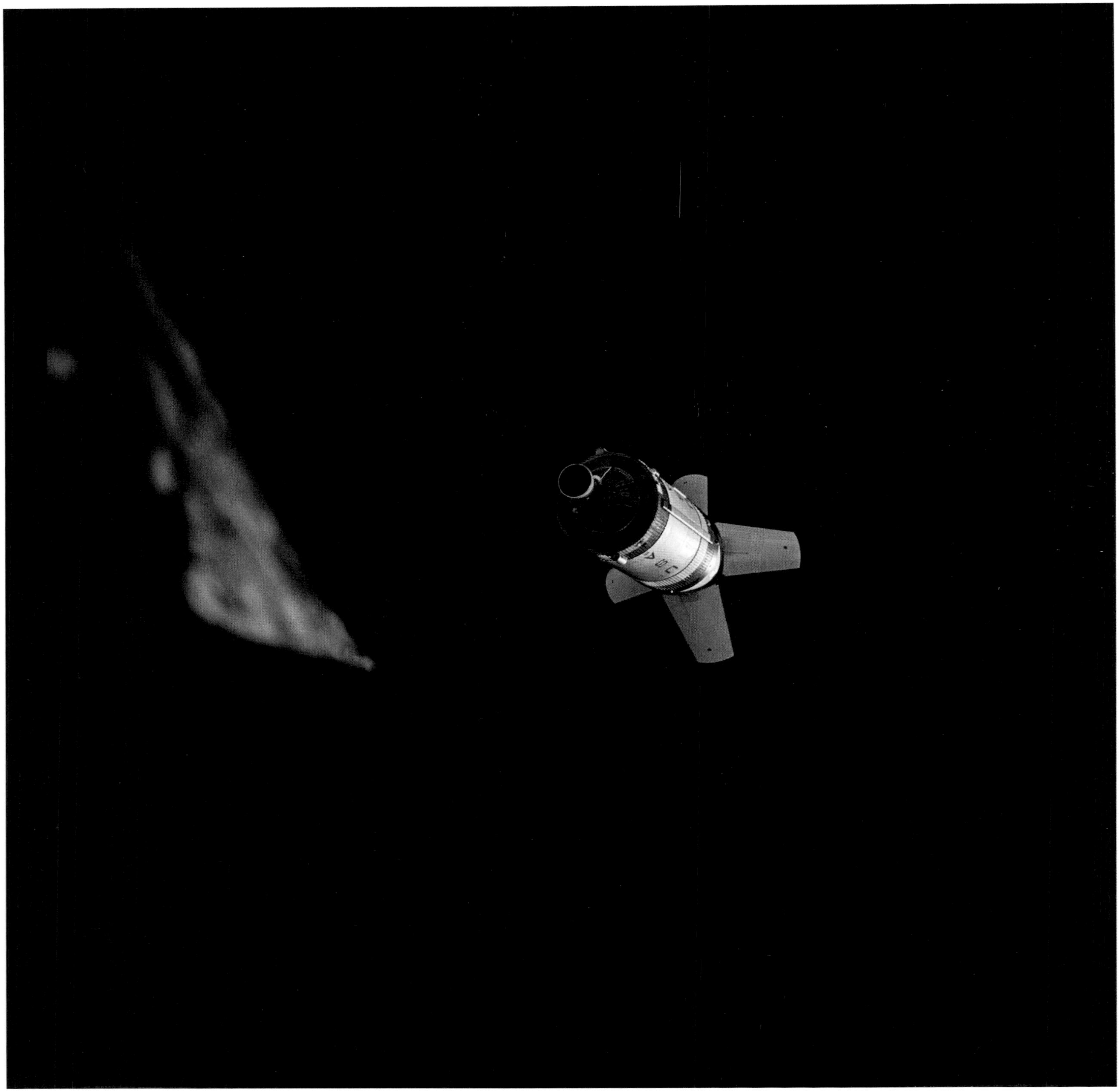

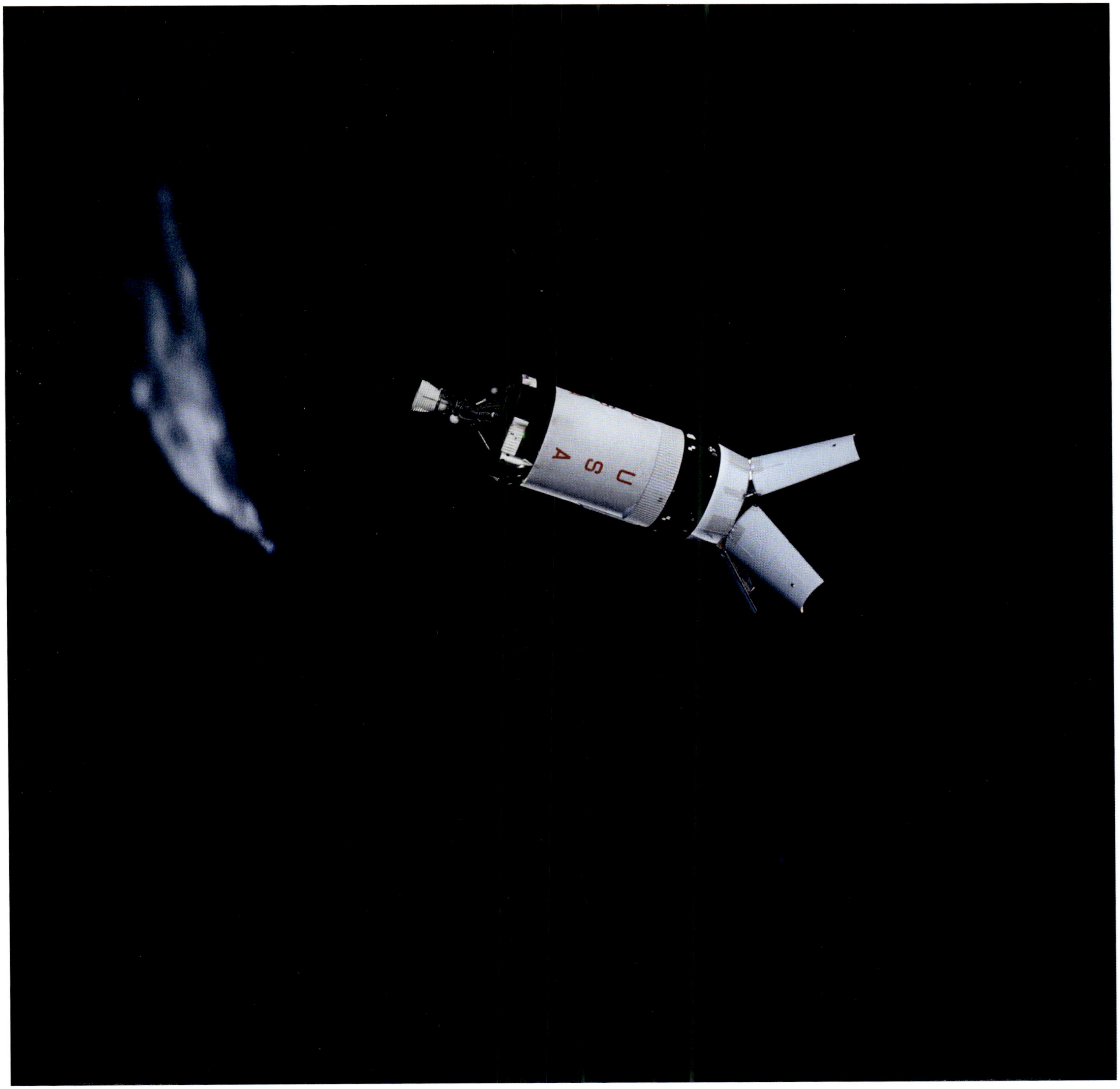
U
S
A

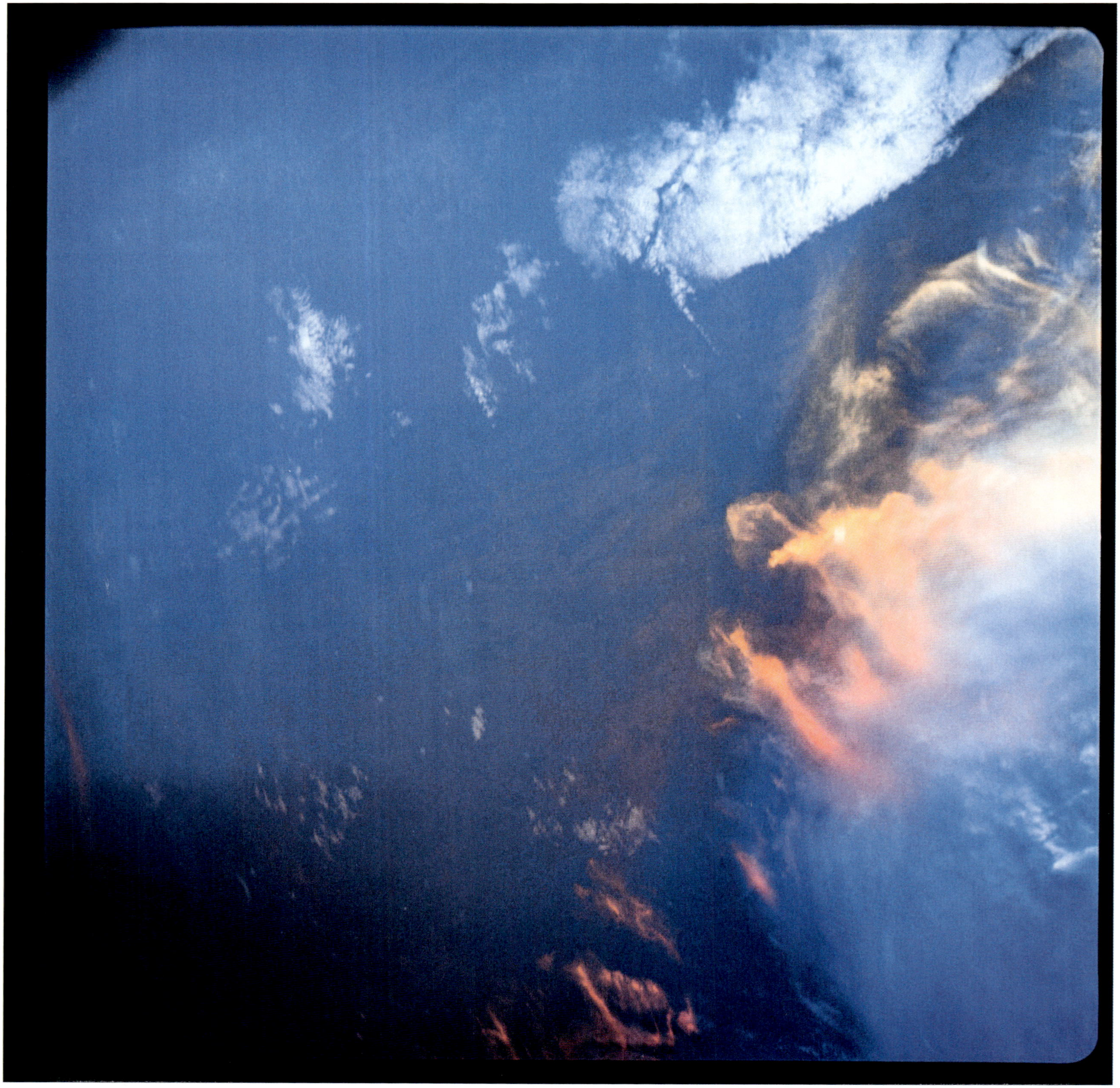

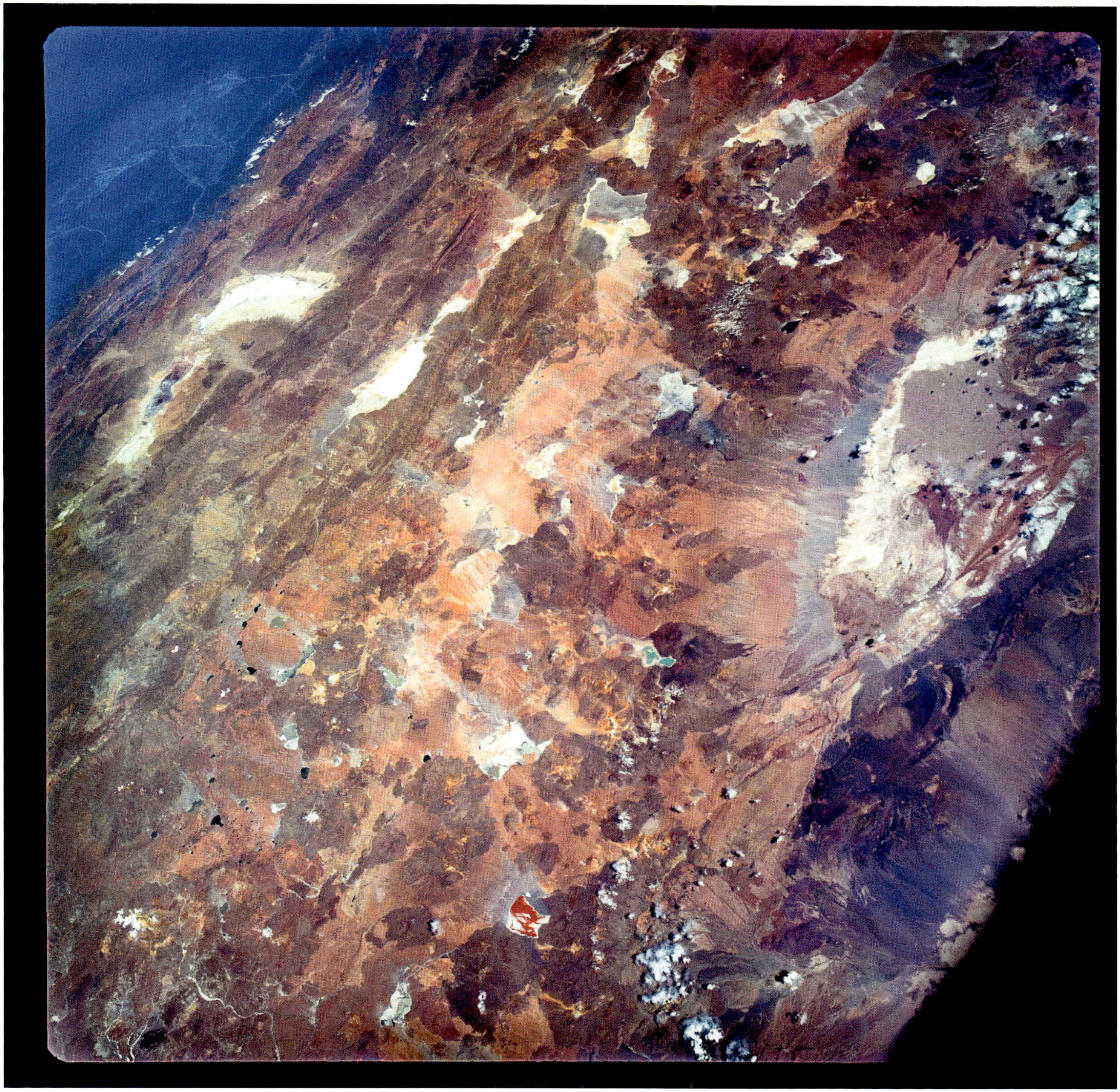

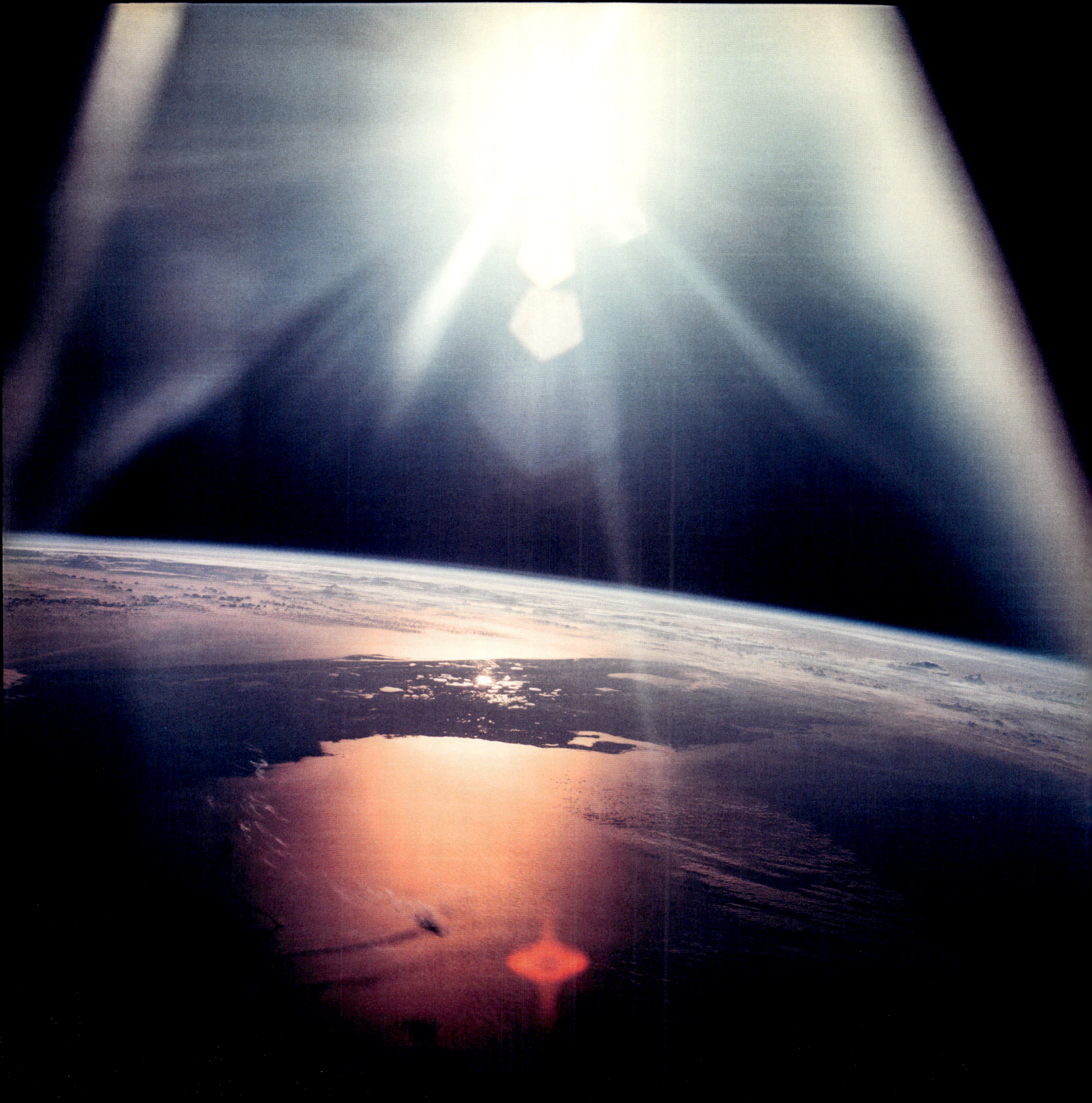

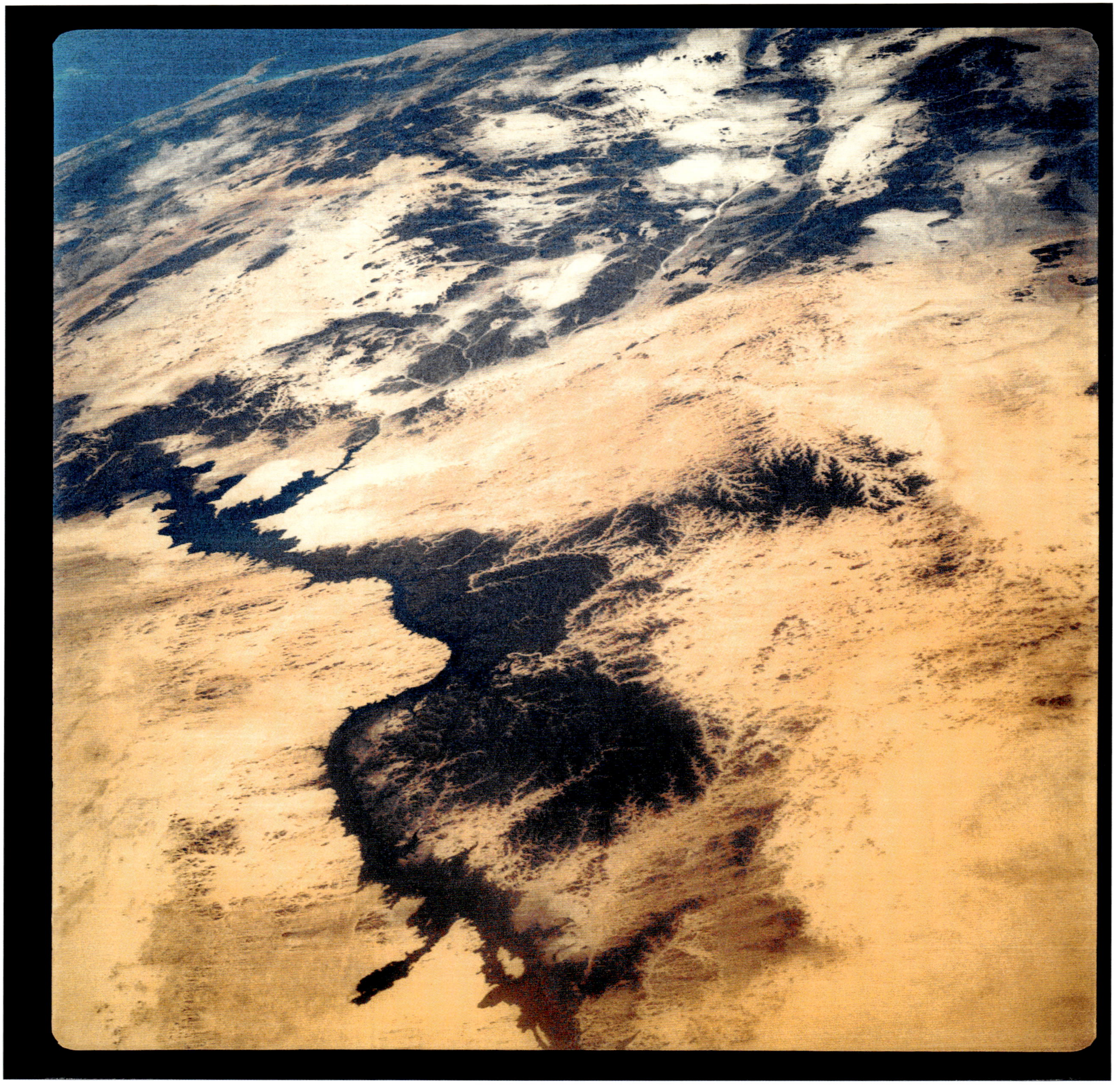

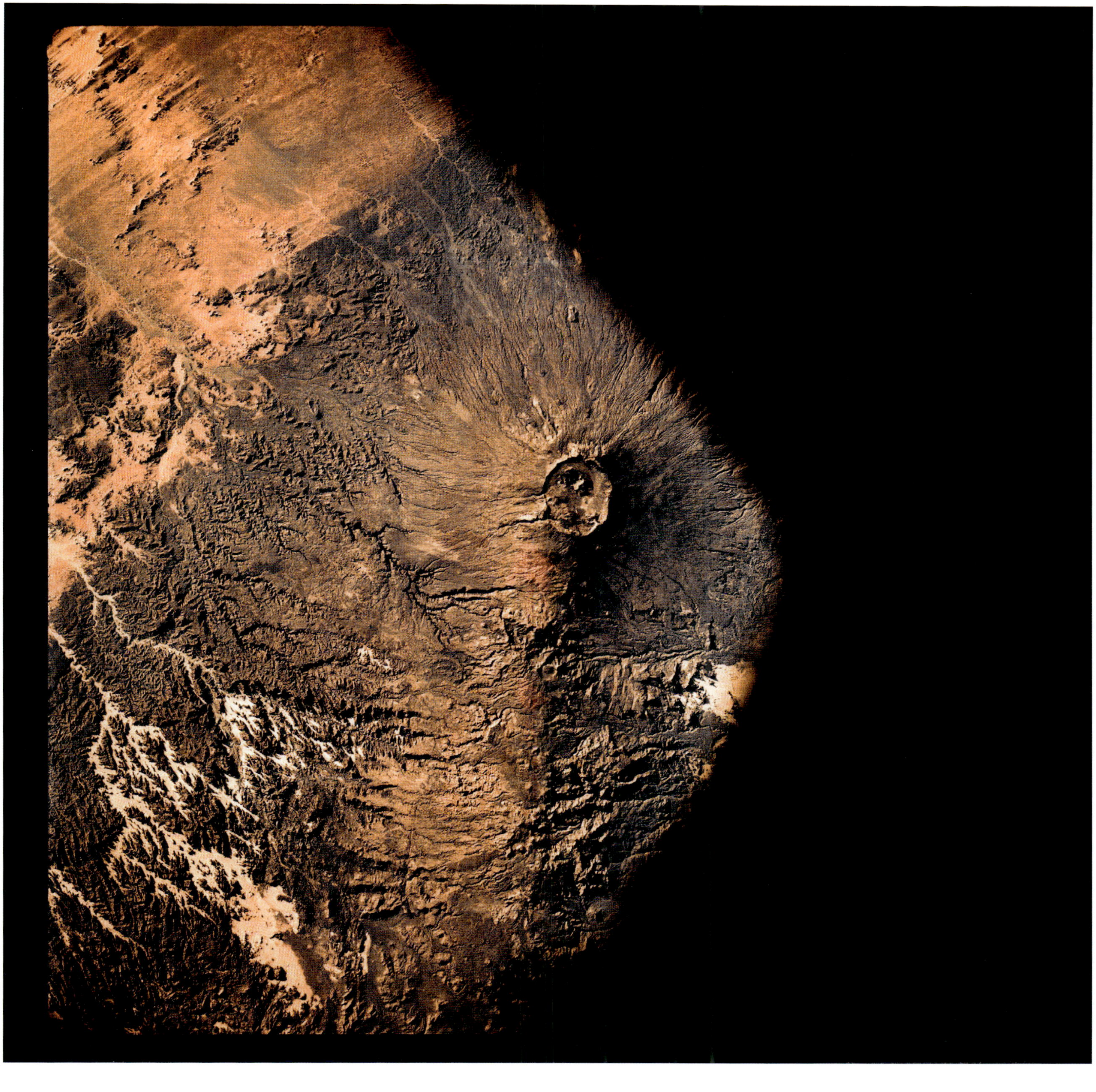

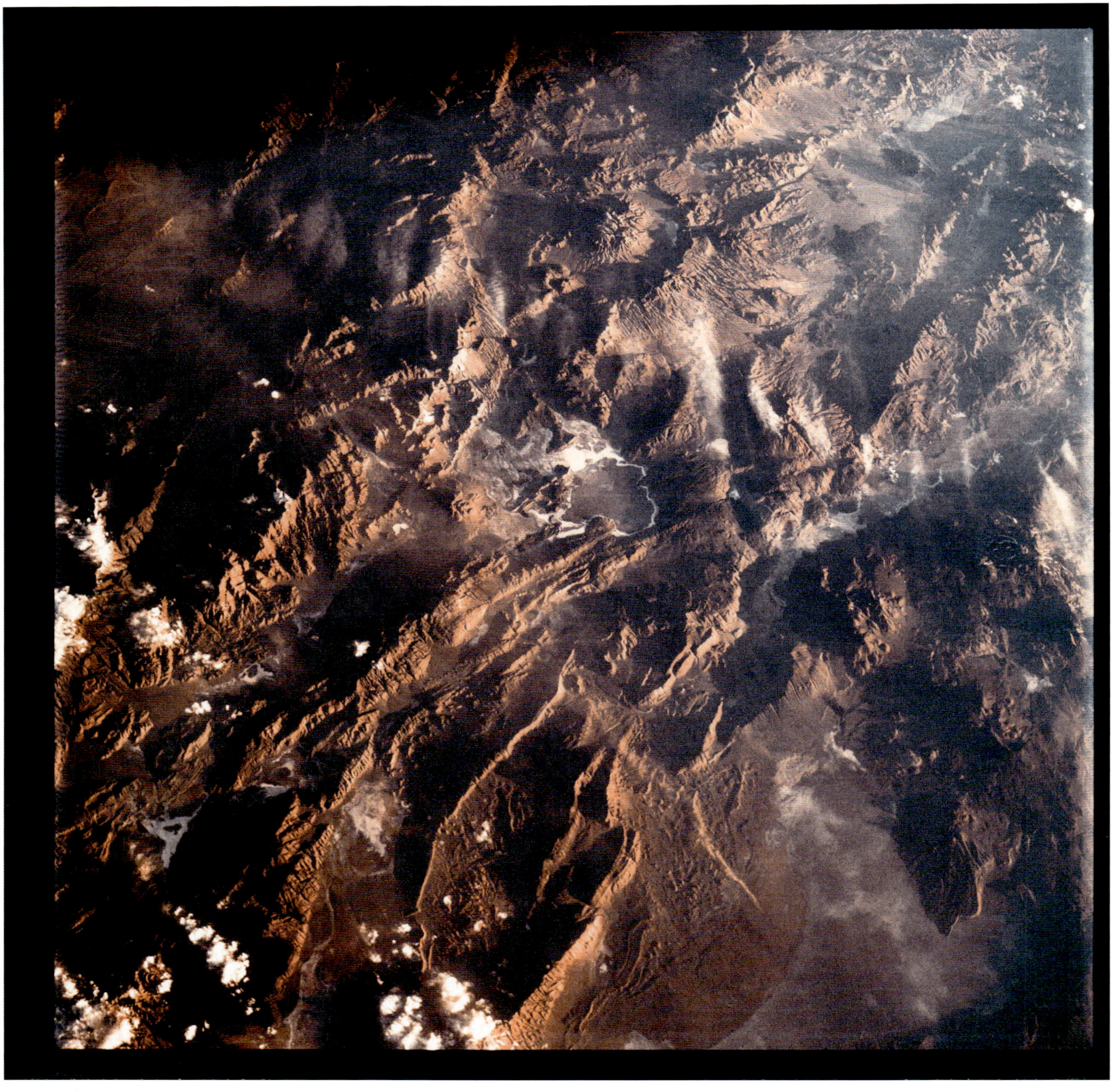

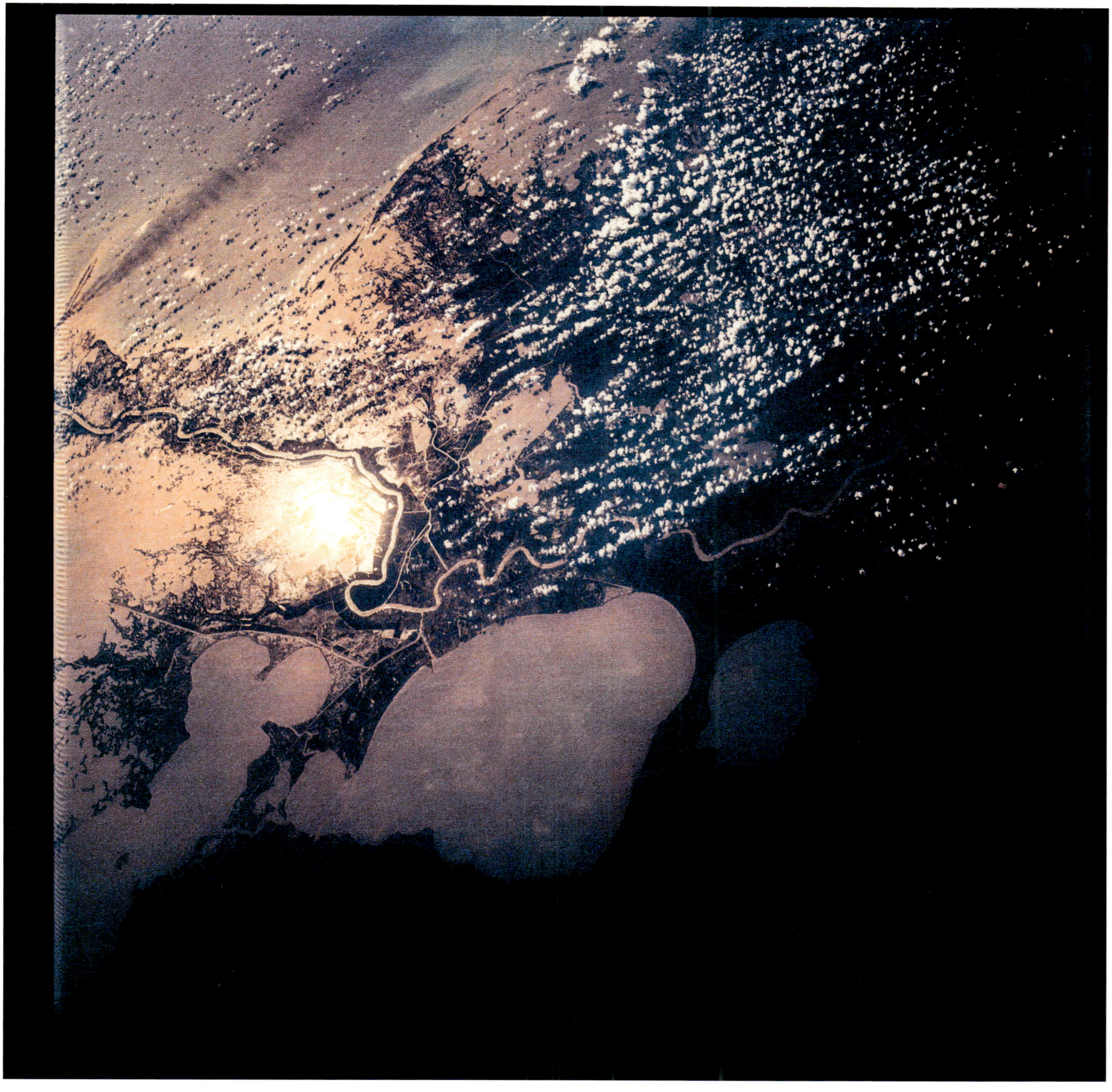

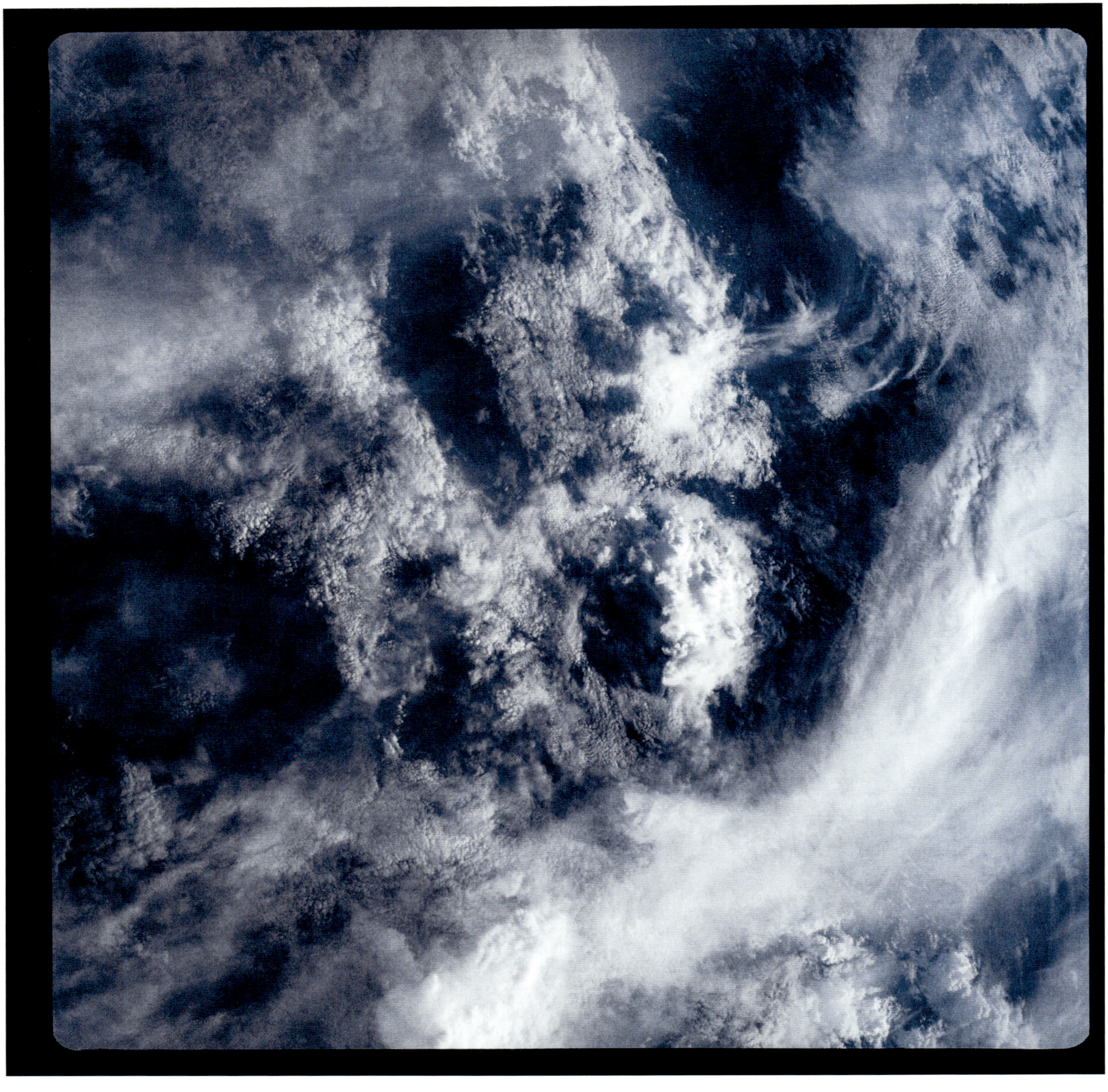

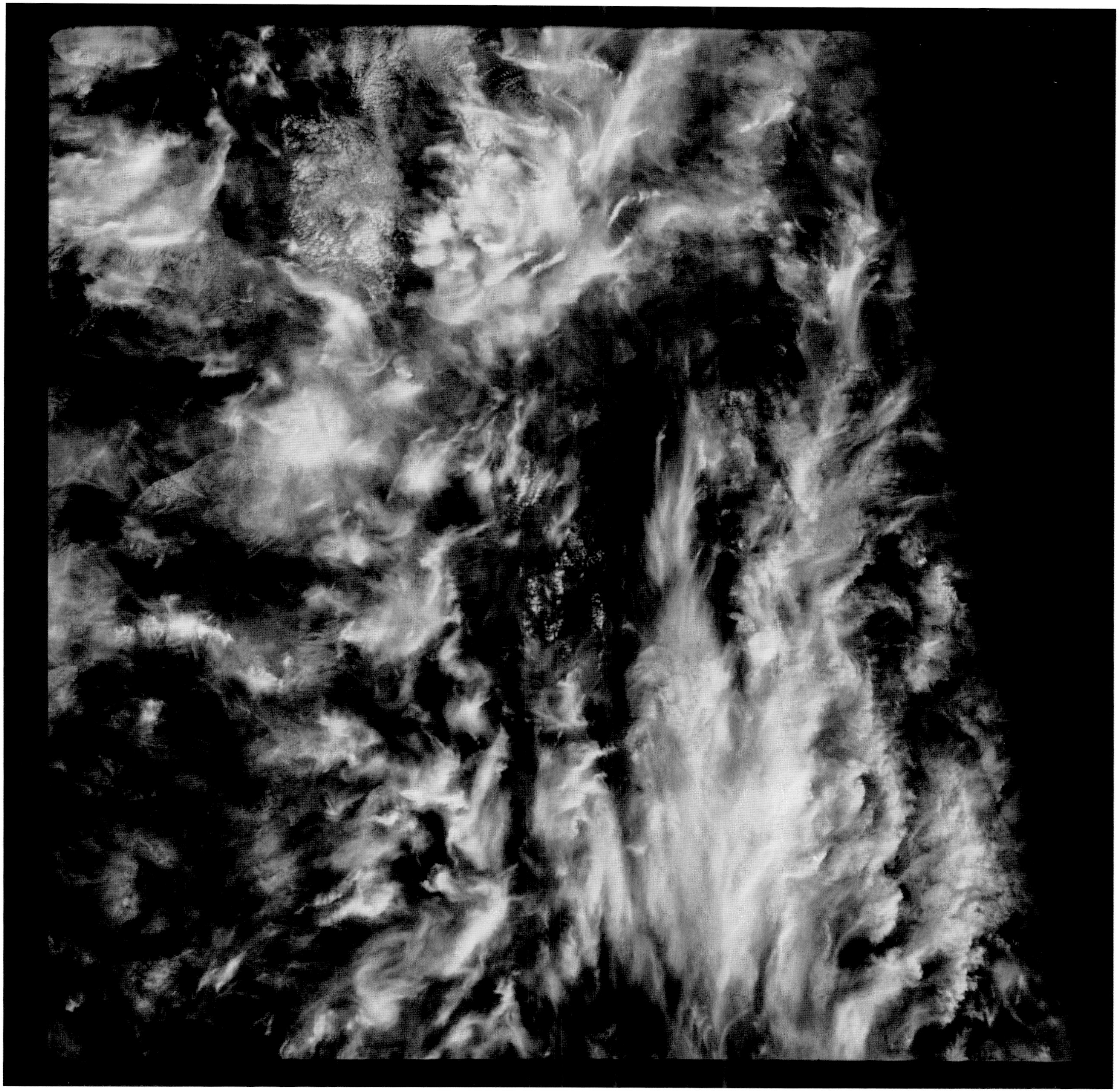

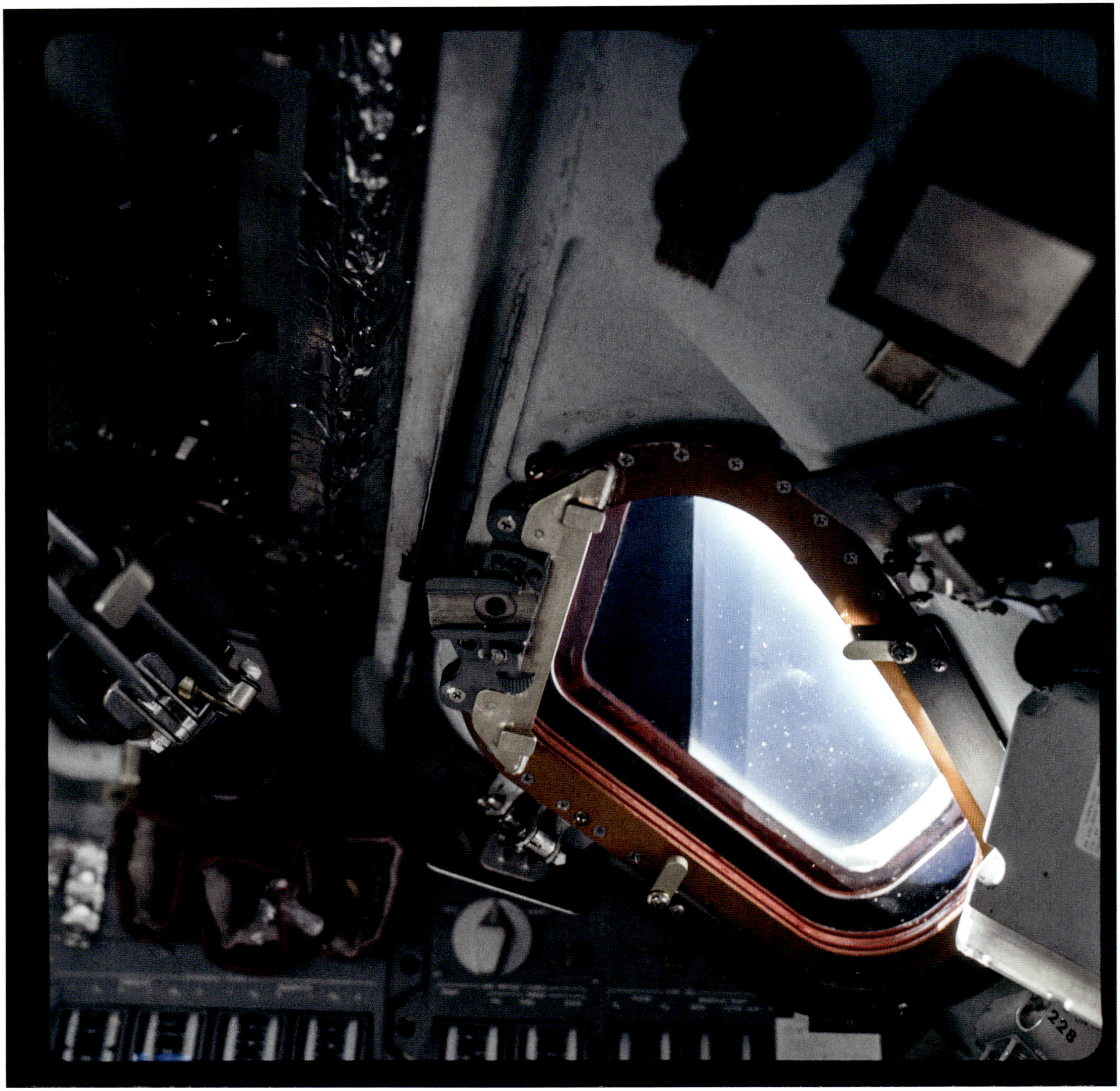

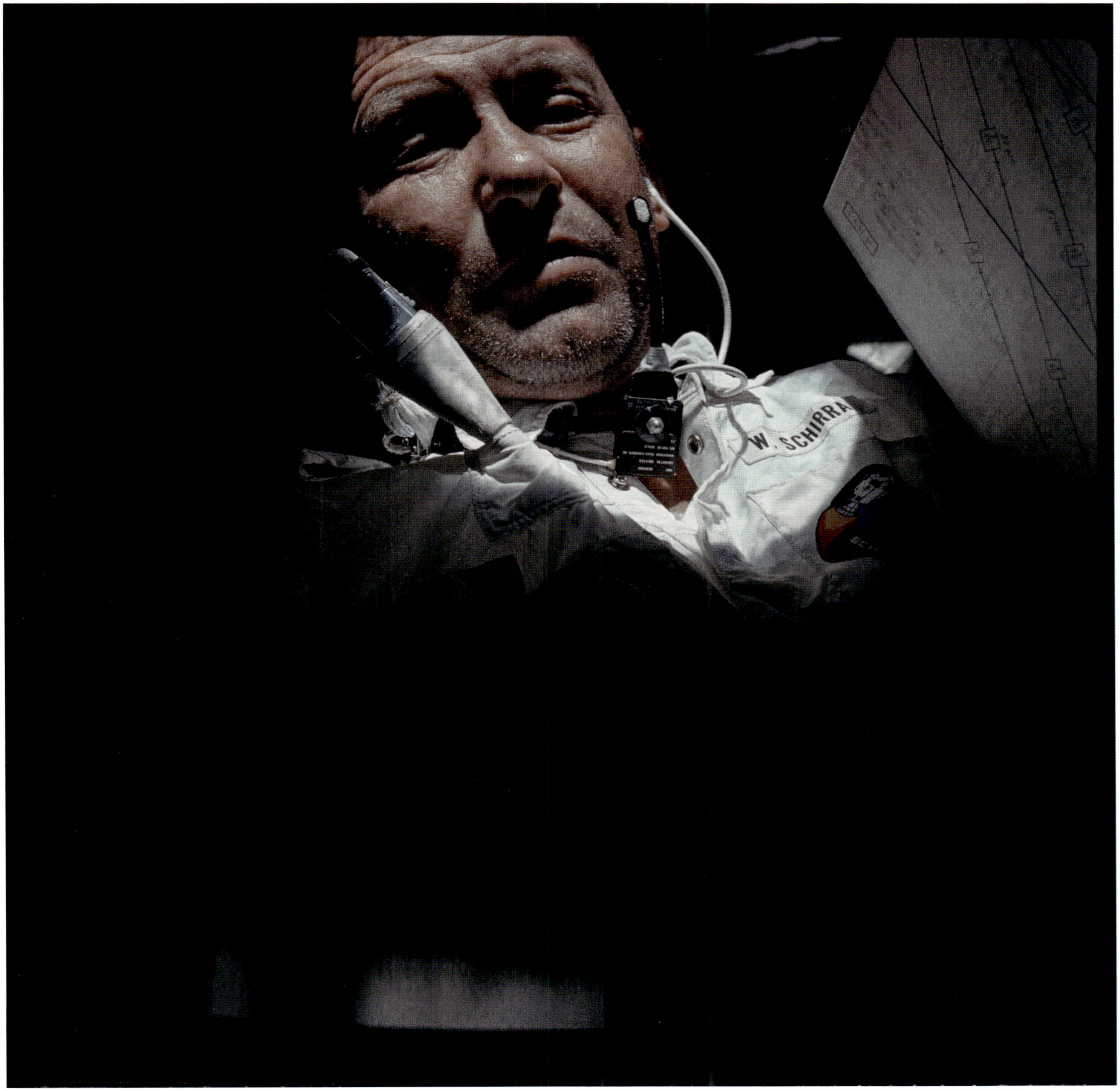
W. SCHIRRA

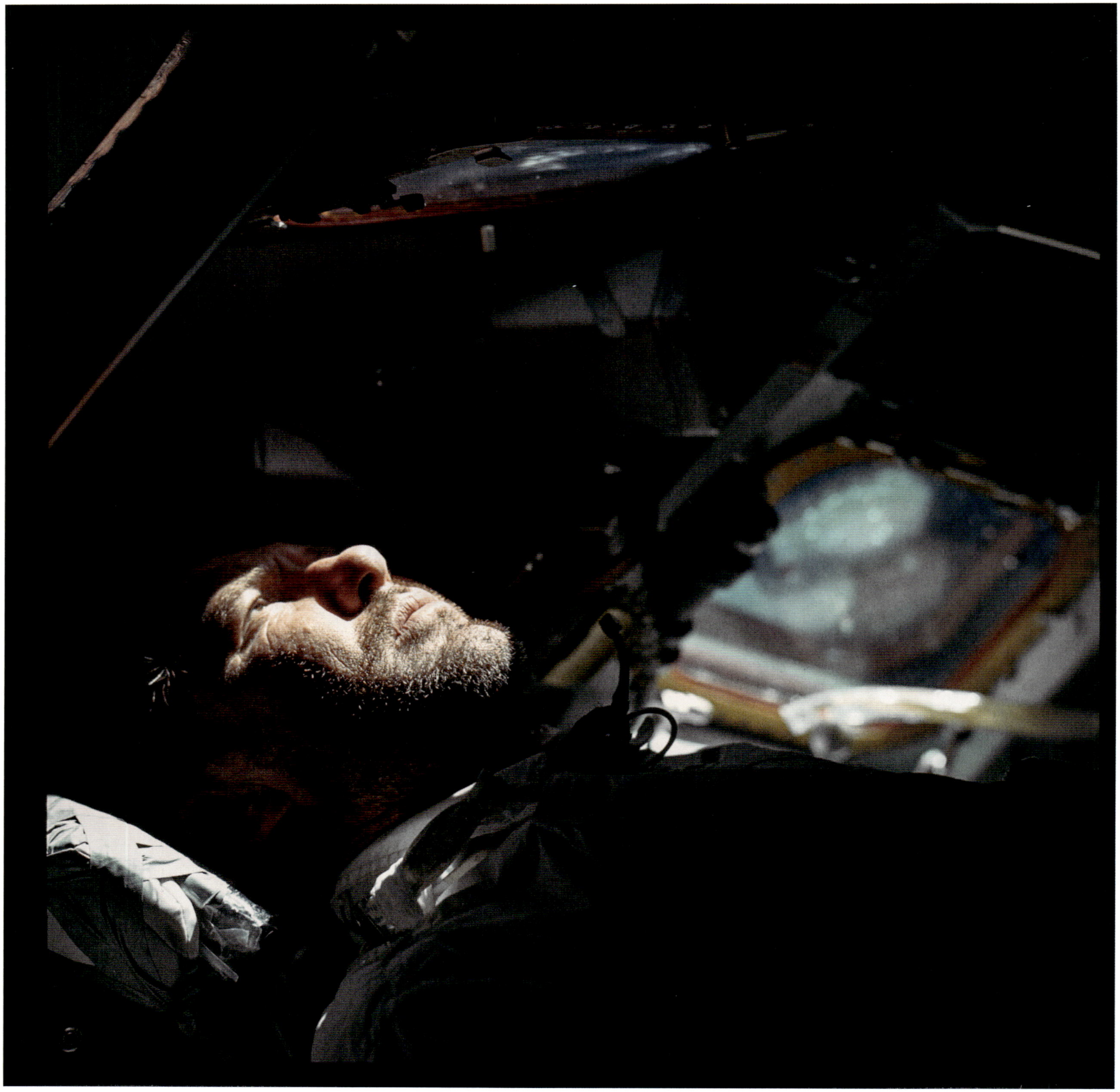

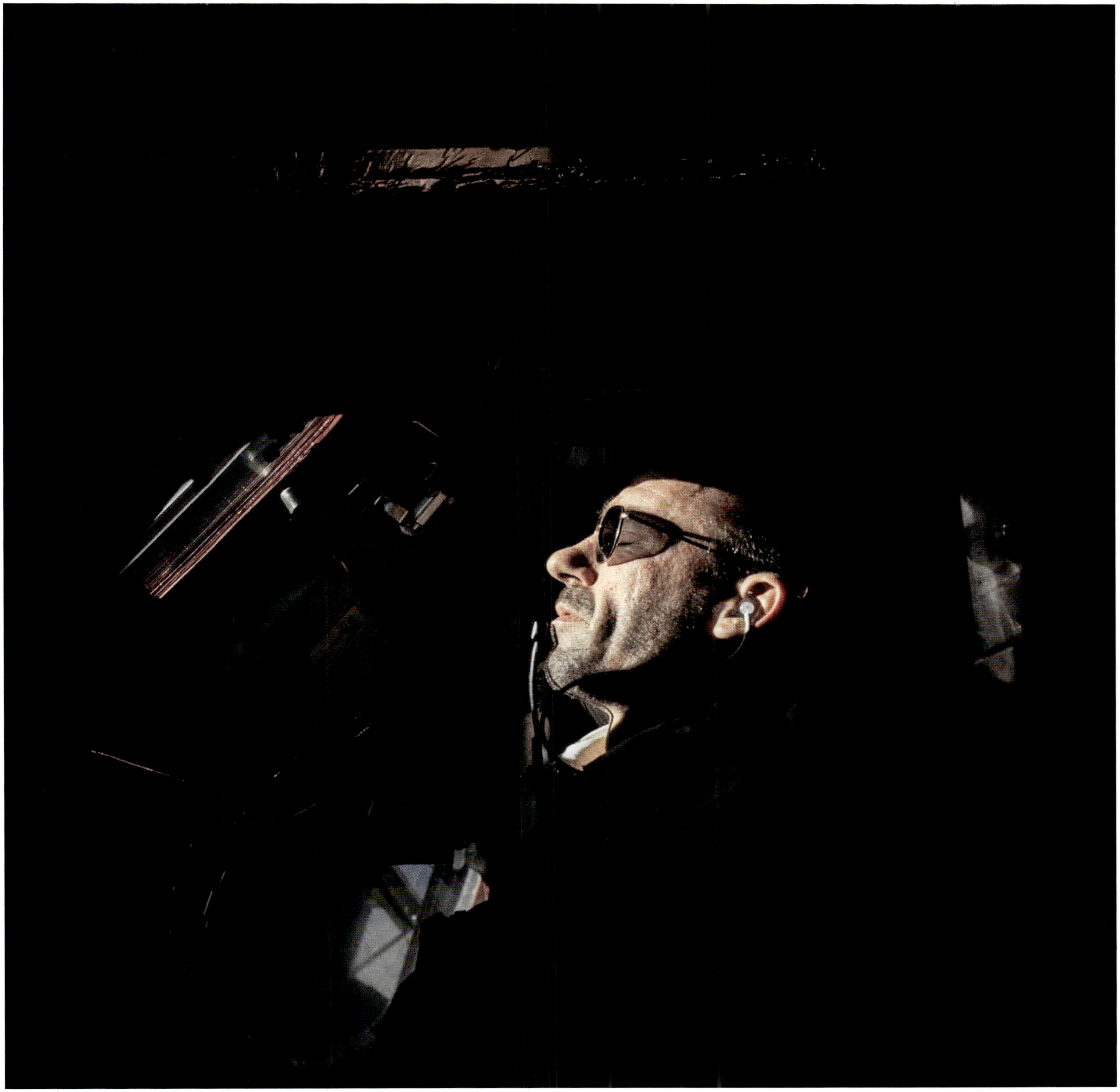

23 (AS07-07-1776)
The Gulf of Aden and the Empty Quarter (Rub' al Khali), the world's largest continuous sand desert. The contrast between the golden dunes and deep blue waters highlights Earth's stunning diversity from space.

24 (AS07-07-1809)
The Ganges River Plain, stretching across India, Nepal, and Tibet, with the Himalayas, Annapurna, and Dhaulagiri peaks in view. This fertile plain, carved by the Ganges River, is one of Earth's most vital agricultural regions.

25 (AS07-03-1550)
The S-IVB stage, seen after separation in Earth orbit. One of the Spacecraft-Lunar Module Adapter (SLA) panels failed to fully open, leading to redesigns on later missions, where they were jettisoned after Command/Service Module (CSM) separation. Apollo 7 did not carry a Lunar Module (LM). Instead, the crew tested a dummy docking target during a rendezvous exercise to evaluate the CSM's maneuvering for future lunar missions.

26 (AS07-04-1586)
Lunar Module Pilot R. Walter Cunningham inside the Command Module (CM), wearing sunglasses due to intense sunlight flooding the cabin, as seen in this photo. Apollo 7 did not carry an LM, focusing instead on testing the CSM in Earth orbit.

27 (AS07-04-1559)
Cunningham, writing on a chart while observing Earth through a spacecraft window. The mission tested CSM navigation, reaction control thrusters, and life support systems to ensure the success of future lunar missions.

28 (AS07-03-1557)
A view through the CSM window during Apollo 7. The spacecraft's five windows were vital for navigation, inspections, and Earth observations.

29 (AS07-11-2043)
The hatch window of Apollo 7's CM, the largest of the spacecraft's five windows at 10in (25.4cm) in diameter. Designed for navigation, docking, and Earth observation, it was expected to provide the best external view. It is often overlooked that all Apollo photographs taken, excluding spacewalks and lunar surface operations, were captured through these small windows.

30 (AS07-04-1576)
The thruster section of the S-IVB stage is seen in this photo. During this series of images, the crew performed a simulated LM rendezvous, testing navigation and docking procedures. Over 11 days and 163 orbits, Apollo 7 successfully demonstrated spacecraft systems and conducted Earth photography.

31 (AS07-04-1575)
Another angle of the Apollo 7 S-IVB stage, with "USA" clearly marked on the side. It was 58.4ft (17.8m) long and 21.7ft (6.6m) in diameter. In later Saturn V missions, the S-IVB fired a second time to send astronauts toward the Moon after Earth orbit insertion.

32 (AS07-03-1511)
The Democratic Republic of the Congo, showcasing the Congo River Basin's lush rainforest. The Congo River, visible here, is Africa's second-longest and the world's deepest river. Early space photography presented significant challenges, including limited film, motion blur, and exposure issues. These orbital views of Earth had rarely been captured before, offering new perspectives on our planet's landscapes and weather patterns.

33 (AS07-04-1593)
Salar de Atacama, Chile, one of the driest places on Earth. The salt flat features a rich variety of colors, with mineral-rich lagoons and salt deposits creating red, green, and blue hues.

35 (AS07-08-1933)
The morning sun reflects off the Florida coastline and the Atlantic Ocean. Captured from Apollo 7 at 120mi (193km) altitude. Most of the Florida peninsula appears as a dark silhouette. Captured during the spacecraft's 134th orbit, this photo was taken 213 hours and 19 minutes after liftoff.

36 (AS07-08-1905)
The Nile River, showing the fertile delta contrasting with the surrounding arid landscape. The Nile, the longest river in the world, has sustained Egyptian civilization for millennia. As part of Apollo 7's Earth observation program, the crew captured large-scale landscape and weather features to study geography from orbit.

37 (AS07-05-1621)
Emi Koussi, the highest peak in the Tibesti Mountains of Chad, towering 11,204ft (3,415m) above the Sahara.

38 (AS07-07-1829)
Salar del Hombre Muerto, Argentina, a high-altitude salt flat over 13,000ft (4,000m) above sea level. Its salt crusts and mineral-rich lakes make it a visually striking landscape.

39 (AS07-07-1789)
New Orleans, Louisiana, with the Mississippi River's crescent-shaped bend visible. Built near the river's delta, the city's location has shaped its history around the water. In this image, you can also see Lake Pontchartrain to the north and the Mississippi River Delta.

40 (AS07-06-1683)
Extensive cloud cover over Earth, captured during Apollo 7. Cunningham noted that 55% of Earth's surface was covered by clouds, often obstructing the crew's photographic targets. Despite this, Apollo 7 provided valuable early space-based meteorological data.

41 (AS07-10-1956)
A black-and-white image of cloud cover over Earth, transforming atmospheric patterns into a dramatic, high-contrast view. The monochrome format highlights the textures and movement of Earth's weather systems in a way that color photography does not. Early Apollo images like this helped refine space-based Earth observation.

42 (AS07-04-1602)
A view through the CM window, framed by the spacecraft's interior. At the time, the CM was cutting-edge technology, but today it appears remarkably analog, with dials, switches, and gauges instead of modern digital screens.

43 (AS07-04-1582)
A striking close-up portrait of Apollo 7 Commander Walter M. Schirra, captured from a low angle as he focuses on his work. The interplay of light and shadow makes this photograph a captivating and dramatic work of art.

44 (AS07-04-1596)
Another dramatic close-up of Schirra, bathed in intense light from the spacecraft window as he concentrates on his work. The high-contrast lighting and low-angle shot create a cinematic portrait.

45 (AS07-04-1588)
Cunningham, eyes closed, resting inside the CSM. Despite the high-pressure environment of spaceflight, he appears peaceful and at ease in this intimate portrait. The Apollo 7 crew followed a strict sleep schedule but found that microgravity and cabin noise made rest difficult.

Apollo 8

VIII

Frank Borman

James A. Lovell, Jr.

William A. Anders

Apollo 8

Launch: December 21st, 1968
Splashdown: December 27th, 1968

Apollo 8 launched towards the end of 1968 and carried the first ever crew to leave the gravitational influence of the Earth and orbit another celestial body. The crew themselves only gave the mission a 50-50 chance of fully succeeding. The mission, however, was an outstanding success, and they became the first three humans to see the Earth in its entirety and first to see the dark side of the Moon.

As it was the first time a manned spacecraft had reached the vicinity of the Moon, a major objective of this mission was to return high-resolution photography of the proposed landing areas and other locations of interest, specifically around the Sea of Tranquility which was envisaged as the first lunar landing site. The launch time of Apollo 8 had been especially selected so the crew would be able to photograph the Sea of Tranquility in the best possible lighting conditions. Over 700 images of the Moon were recorded along with another 150 of Earth.

During the fourth orbit of the Moon, Frank Borman and his crew witnessed the first ever 'Earthrise'; the Earth emerging from behind the lunar horizon. Life Magazine later picked William Anders' photograph of the event as one of '100 photographs that changed the world'. Anders himself described how *Earthrise* gave humanity a new perspective on its home planet: "...despite all the training and preparation for an exploration of the Moon, the astronauts ended up discovering Earth."

Six television broadcasts were conducted during the mission: two while flying to the Moon, two during lunar orbit and two while returning to Earth. These transmissions were broadcast worldwide and in real time. During the transmissions, the astronauts described the scenes that nobody before them had seen such as how the Earth looked from their perspective and the range of colors visible on both Earth and the Moon. During the broadcast on Christmas Eve, the crew read from the Book of Genesis and wished viewers a "good night, good luck, a Merry Christmas and God bless all of you — all of you on the good Earth."

Apollo 8 circled the Moon ten times after which the mission objectives were met and it was time to return. On the morning of Christmas Day, the spacecraft had to attempt the first-ever escape from lunar orbit with a single, well-calculated engine burn. Mission Control anxiously awaited Lovell's confirmation, until he finally radioed: "Roger, please be informed there is a Santa Claus."

Frank Borman
Commander

Born on March 14th, 1928 in Gary, Indiana. He was chosen with the second group of astronauts in 1962. He was Backup Command Pilot of Gemini 4 and Command Pilot of Gemini 7. In May 1969 he became Field Director of the Long-Term Space Station Program.

James A. Lovell, Jr.
Command Module Pilot

Born on March 25th, 1928 in Cleveland, Ohio. He was Backup Pilot of Gemini 4, Pilot of Gemini 7, Backup Command Pilot of Gemini 9A, Command Pilot of Gemini 12, Backup Commander of Apollo 11 and Commander of Apollo 13.

William A. Anders
Lunar Module Pilot

Born on October 17th, 1933 in Hong Kong. He was chosen with the third group of astronauts in 1963. He was the Backup Pilot of Gemini 11 and Backup Command Module pilot of Apollo 11. He resigned from NASA and the Air Force in September 1969.

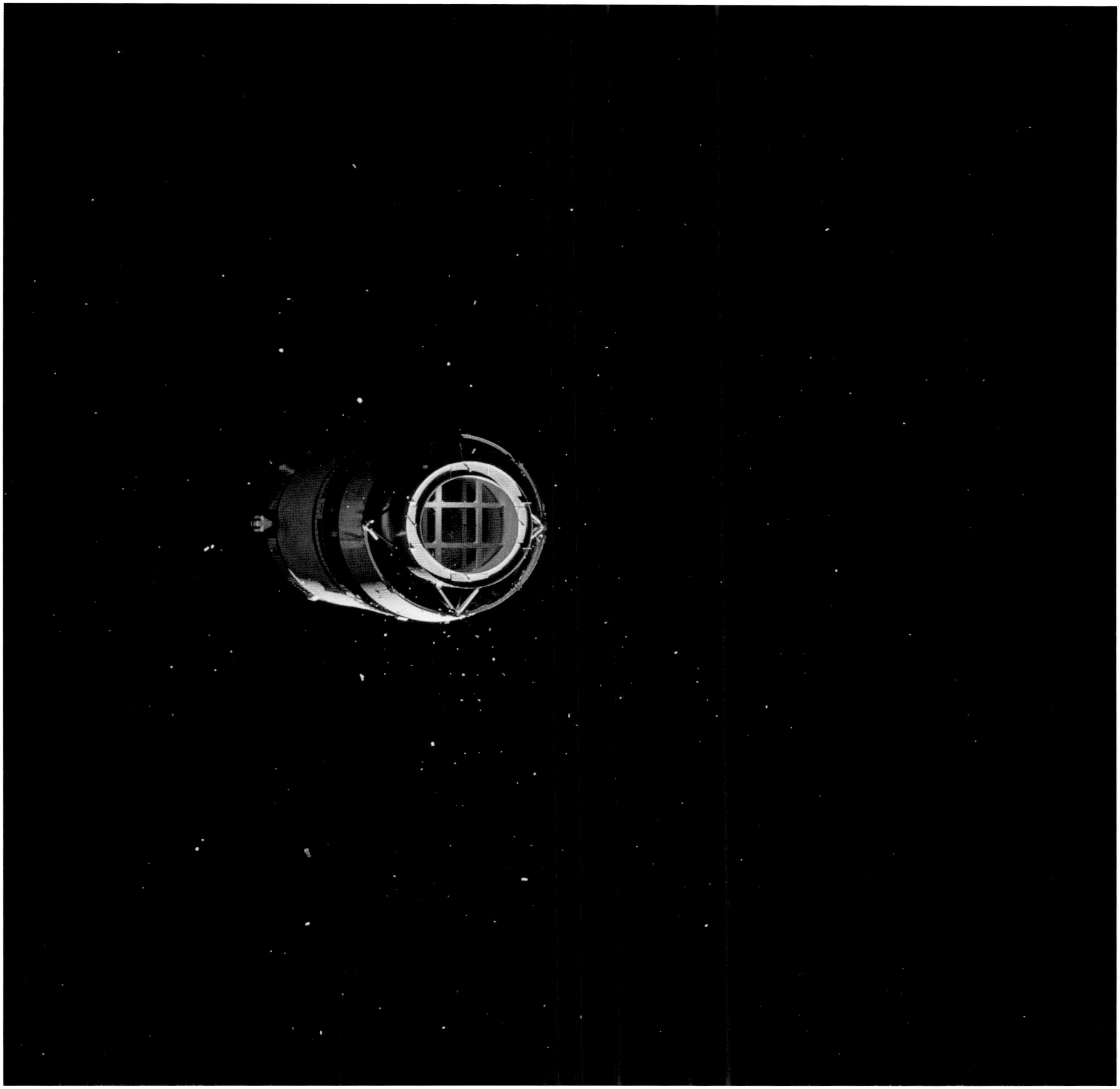

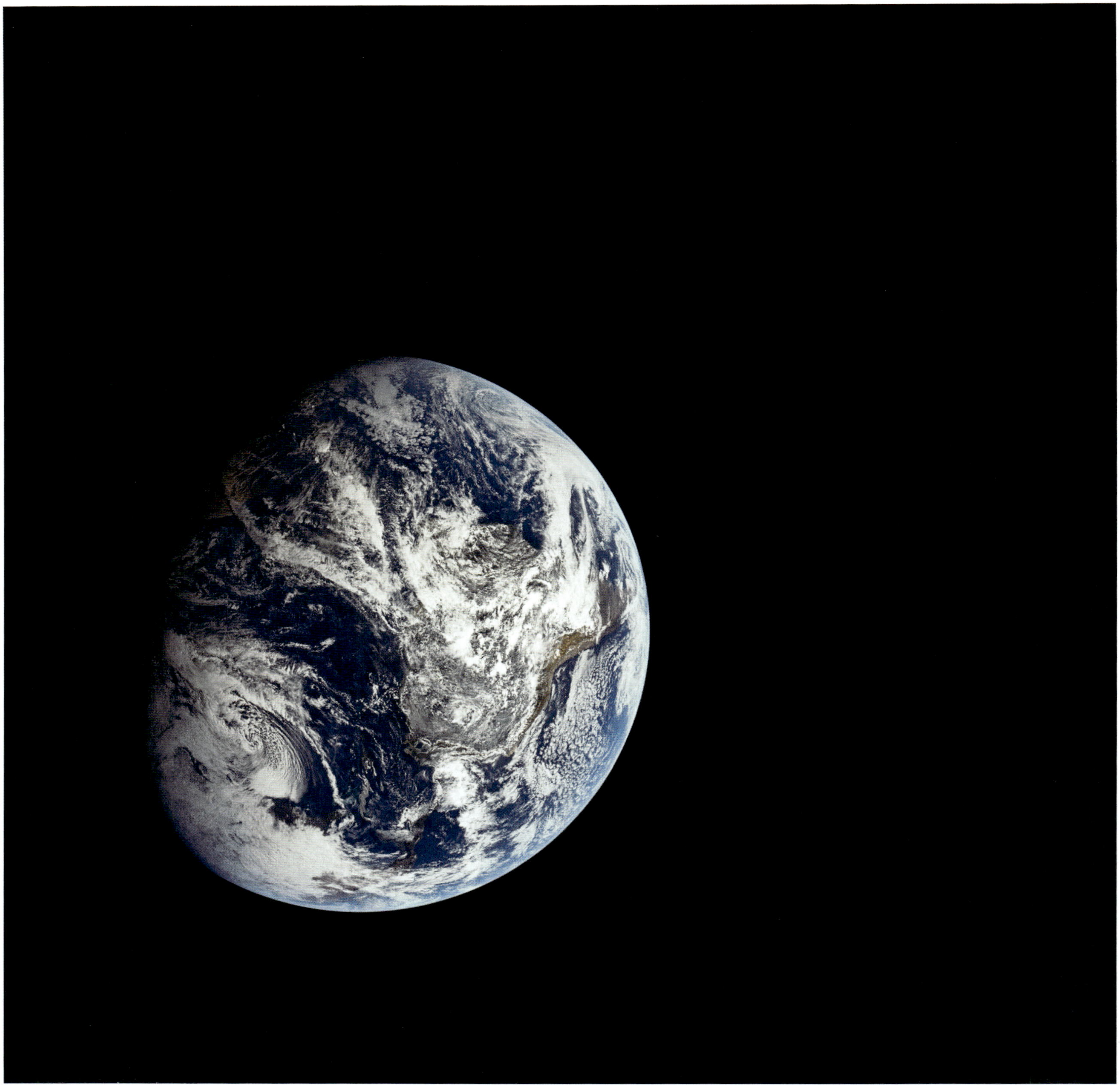

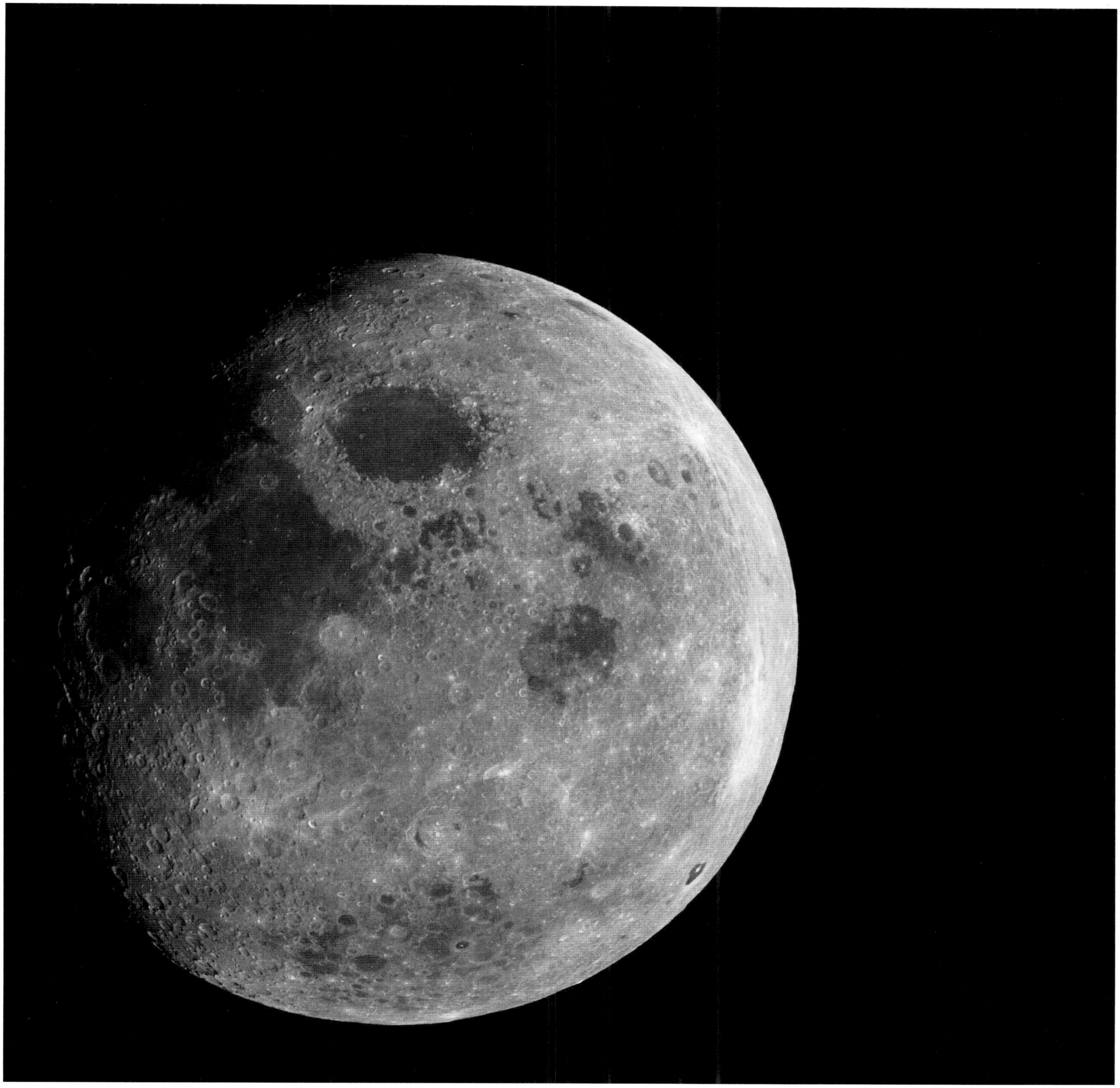

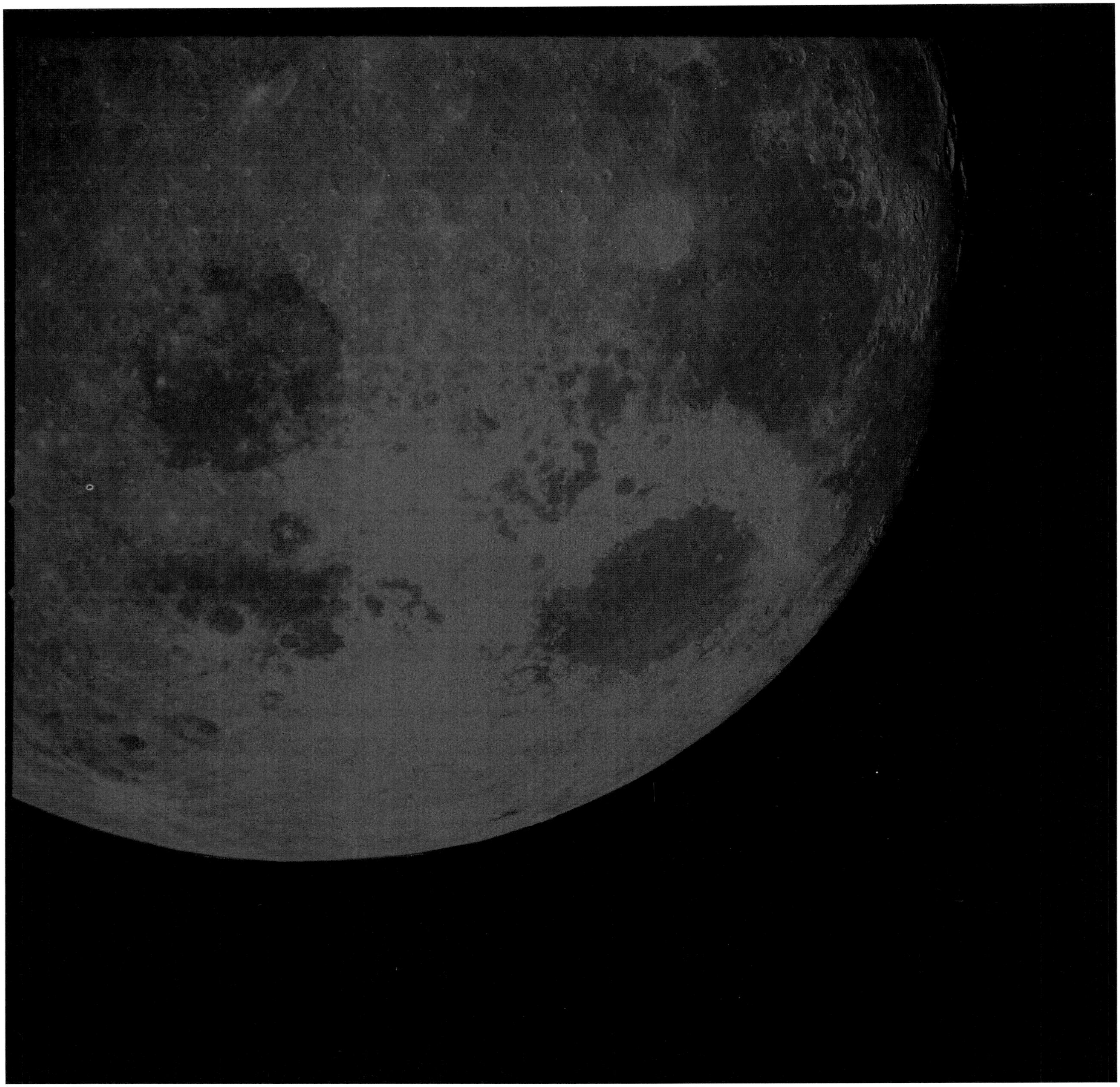

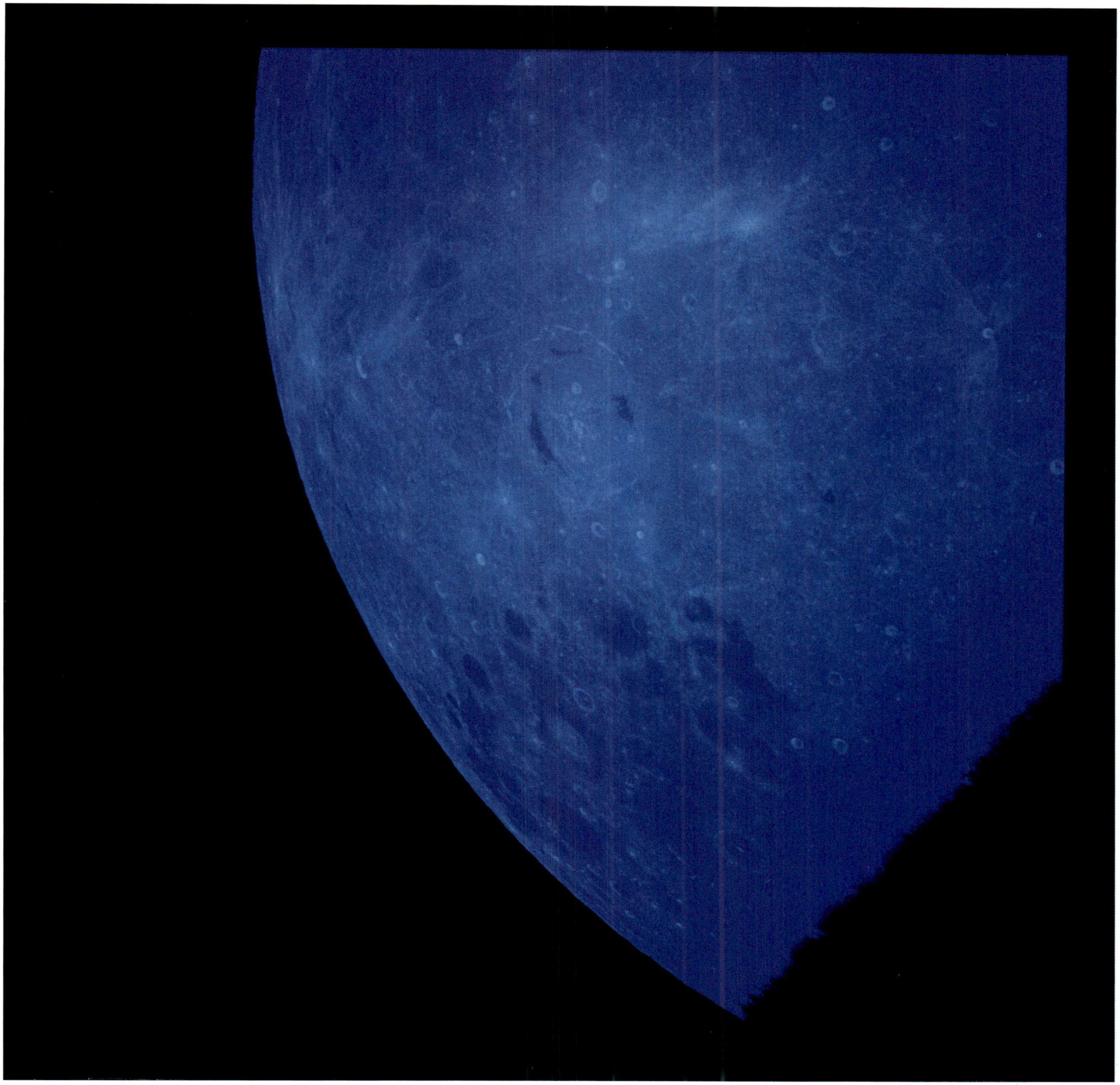

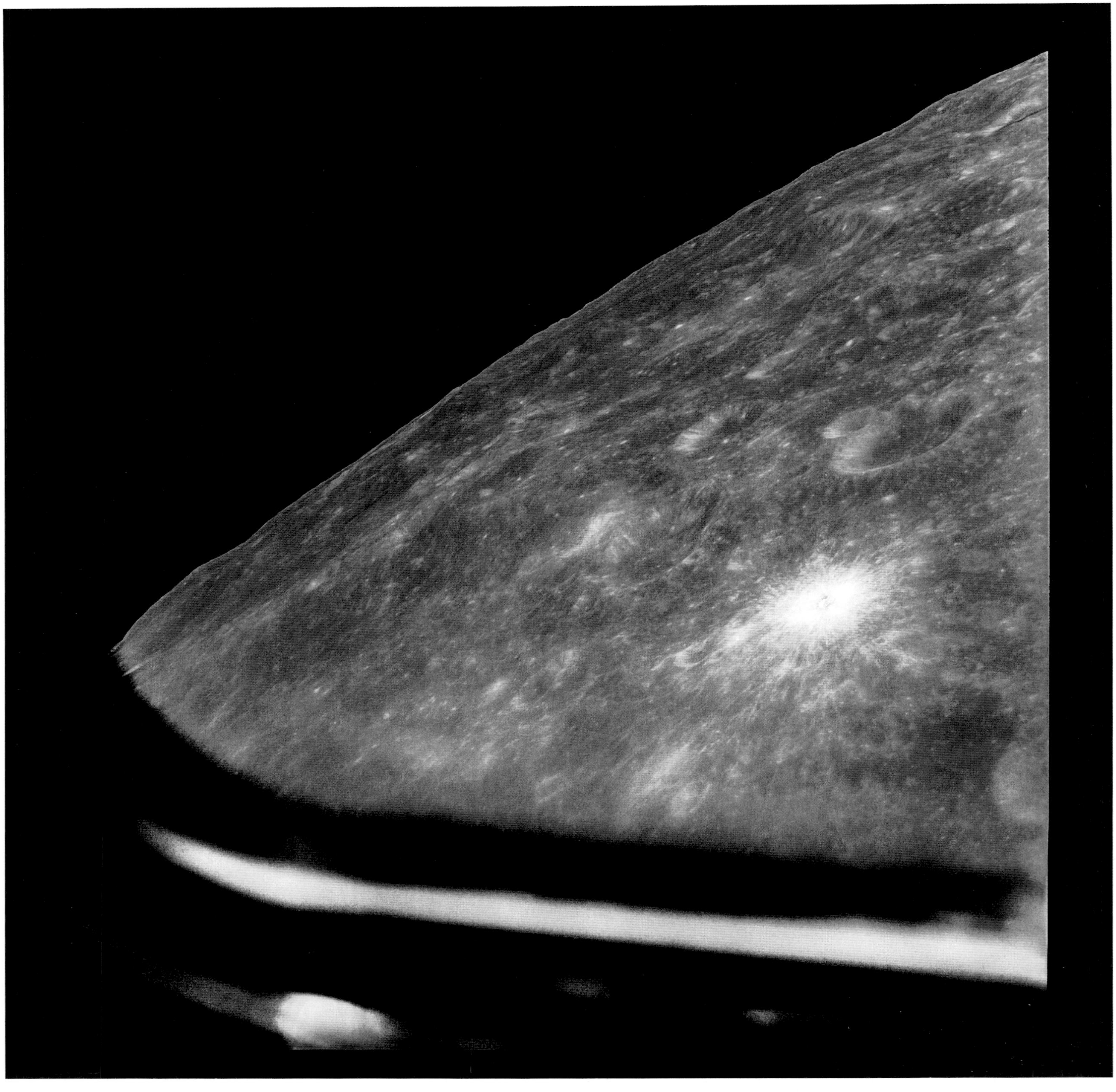

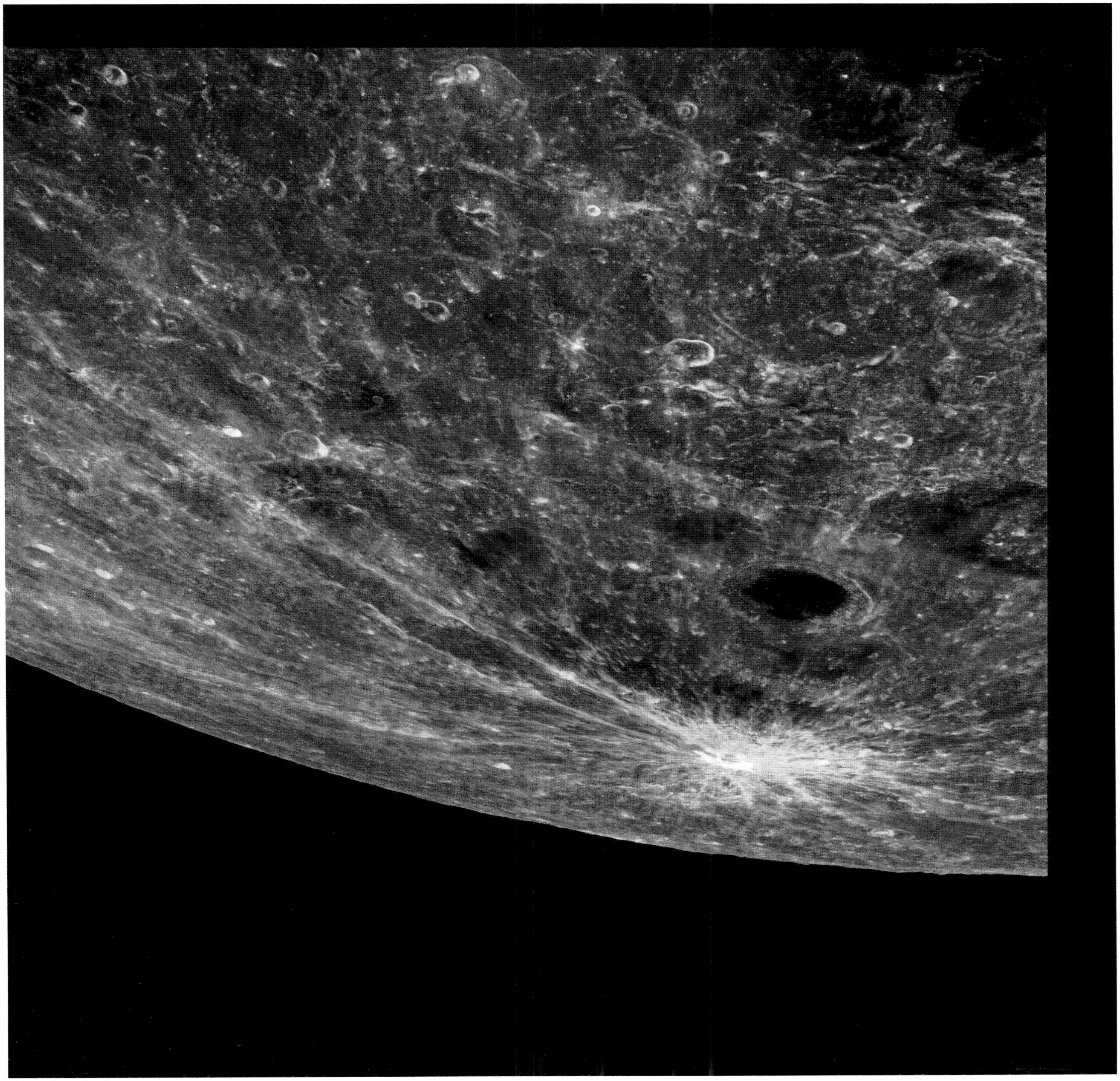

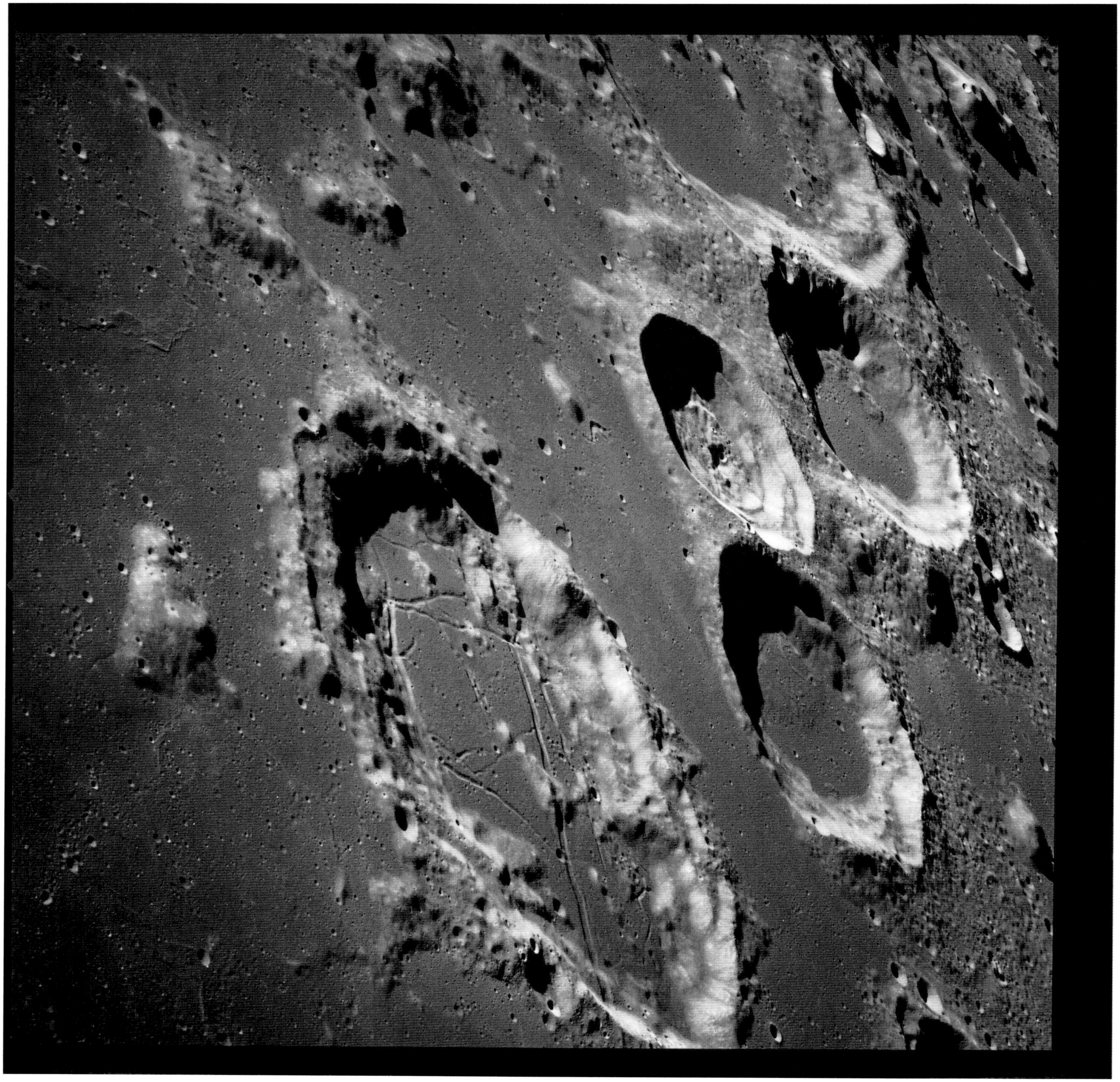

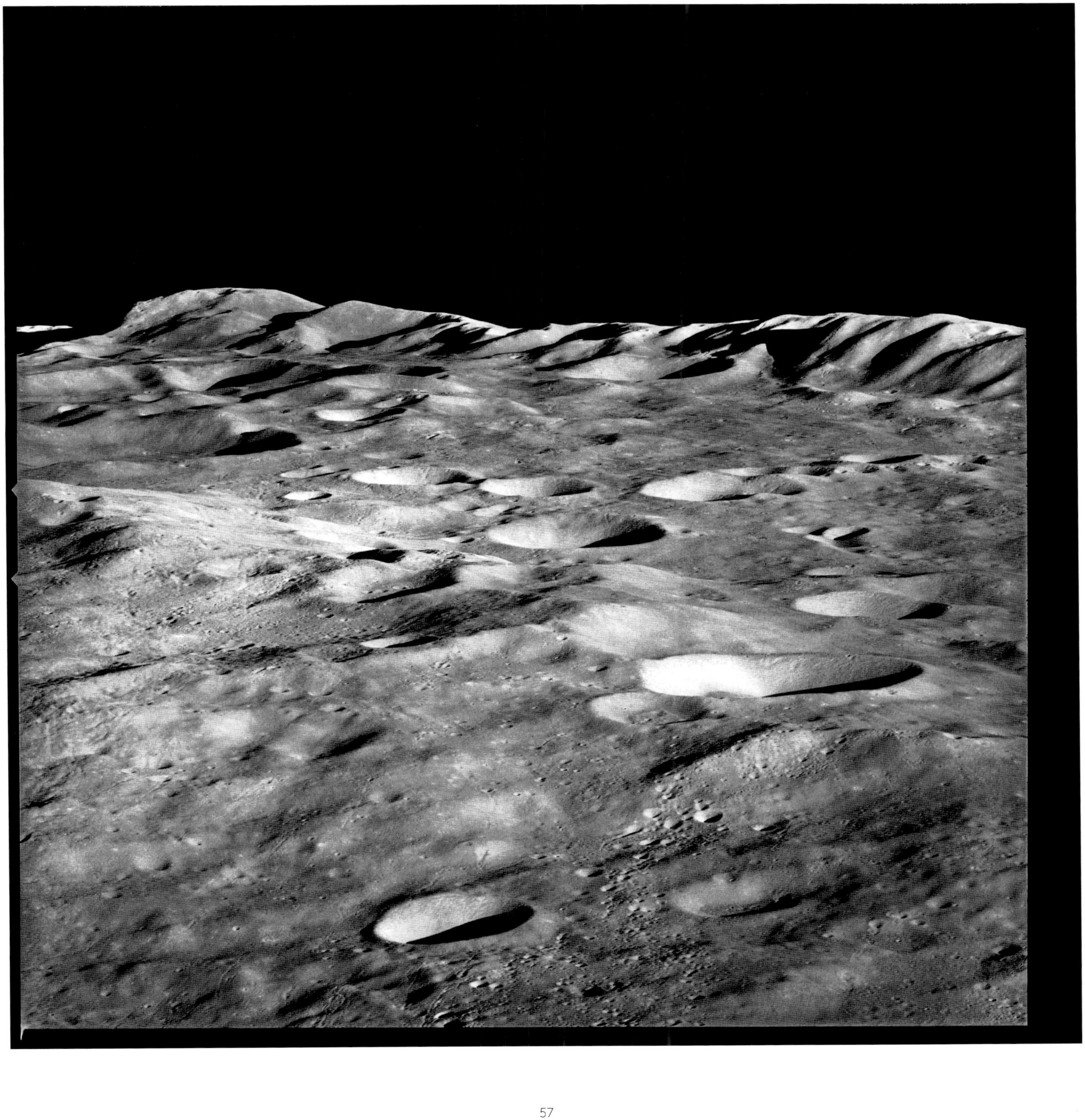

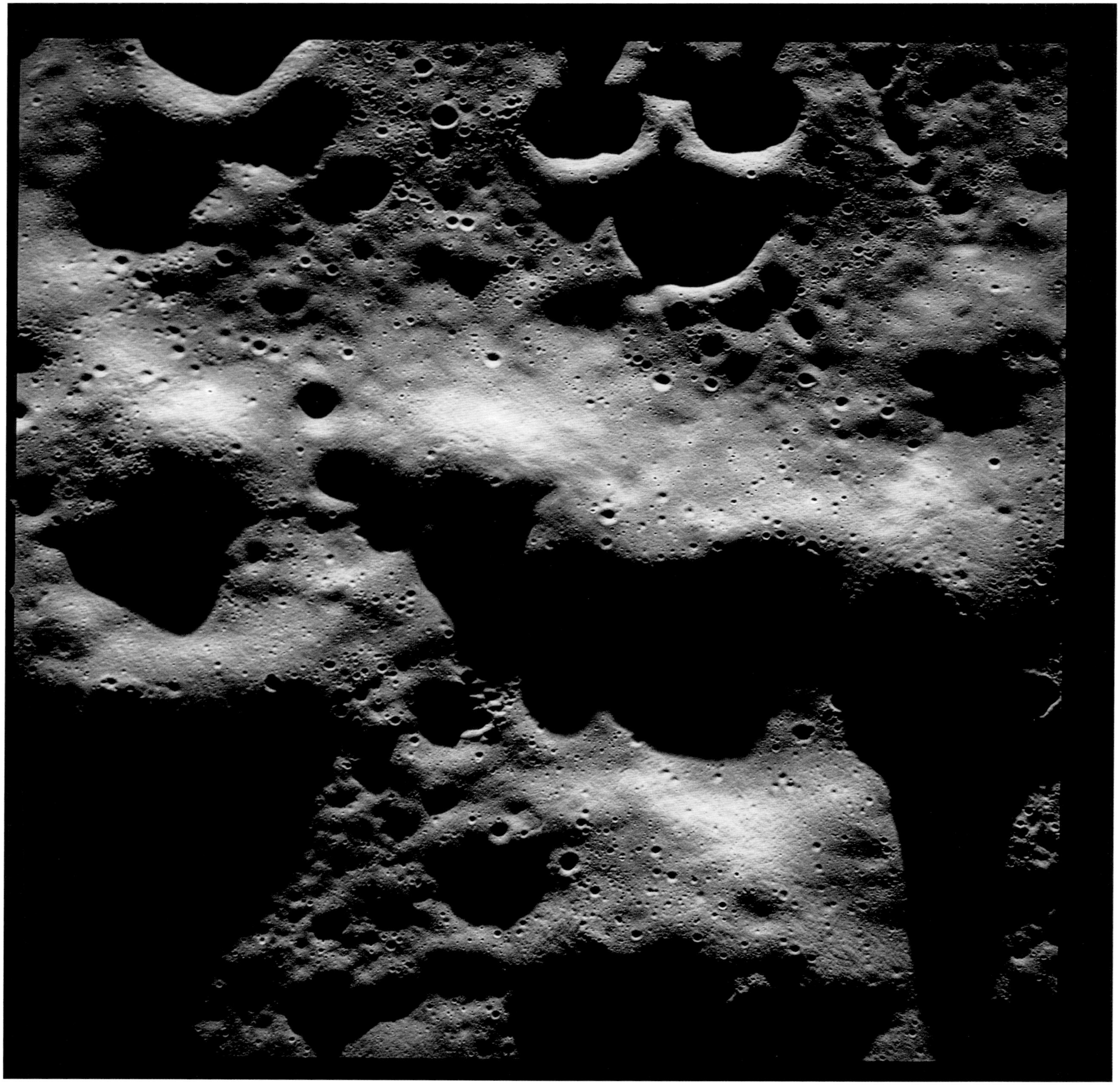

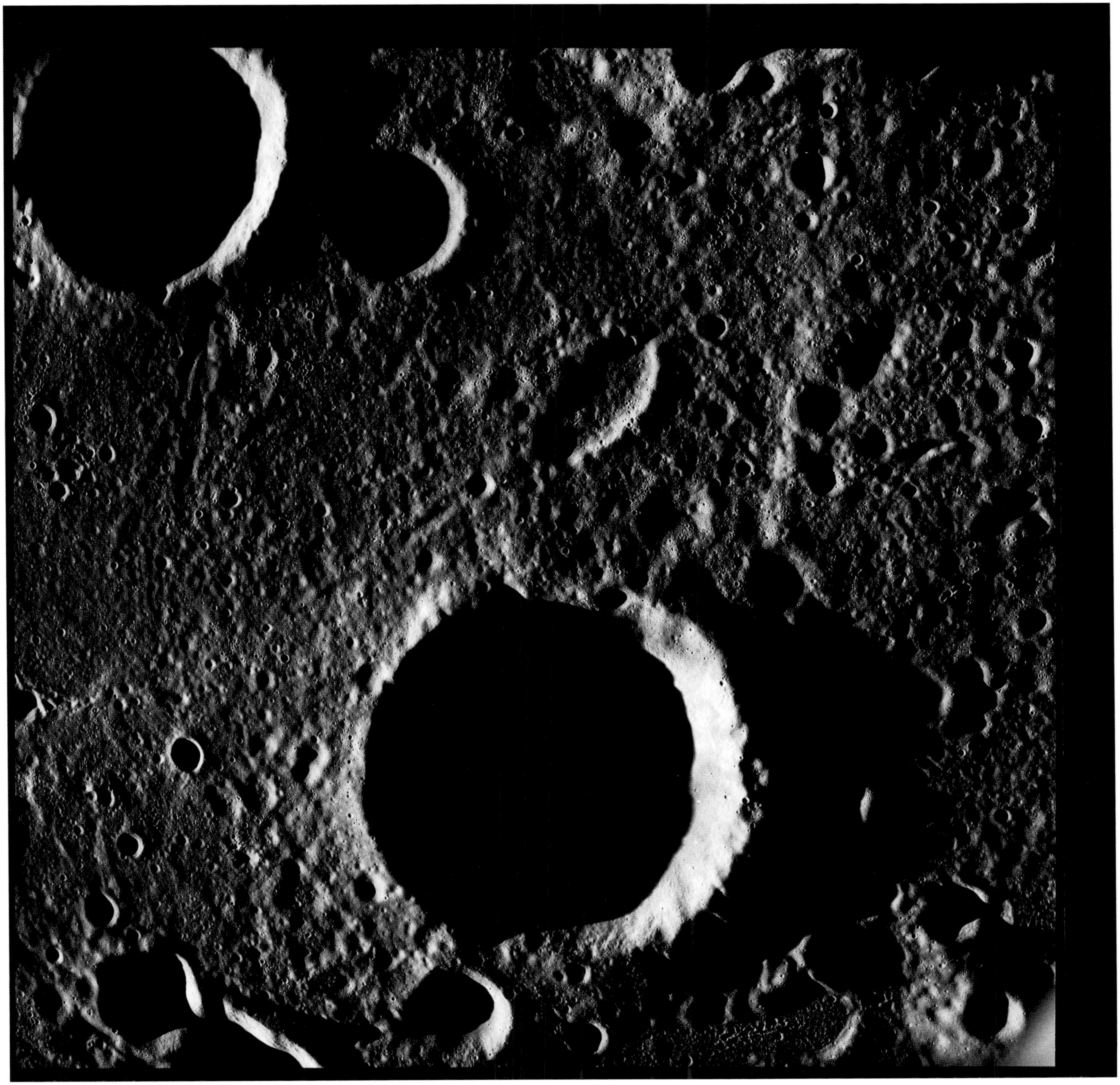

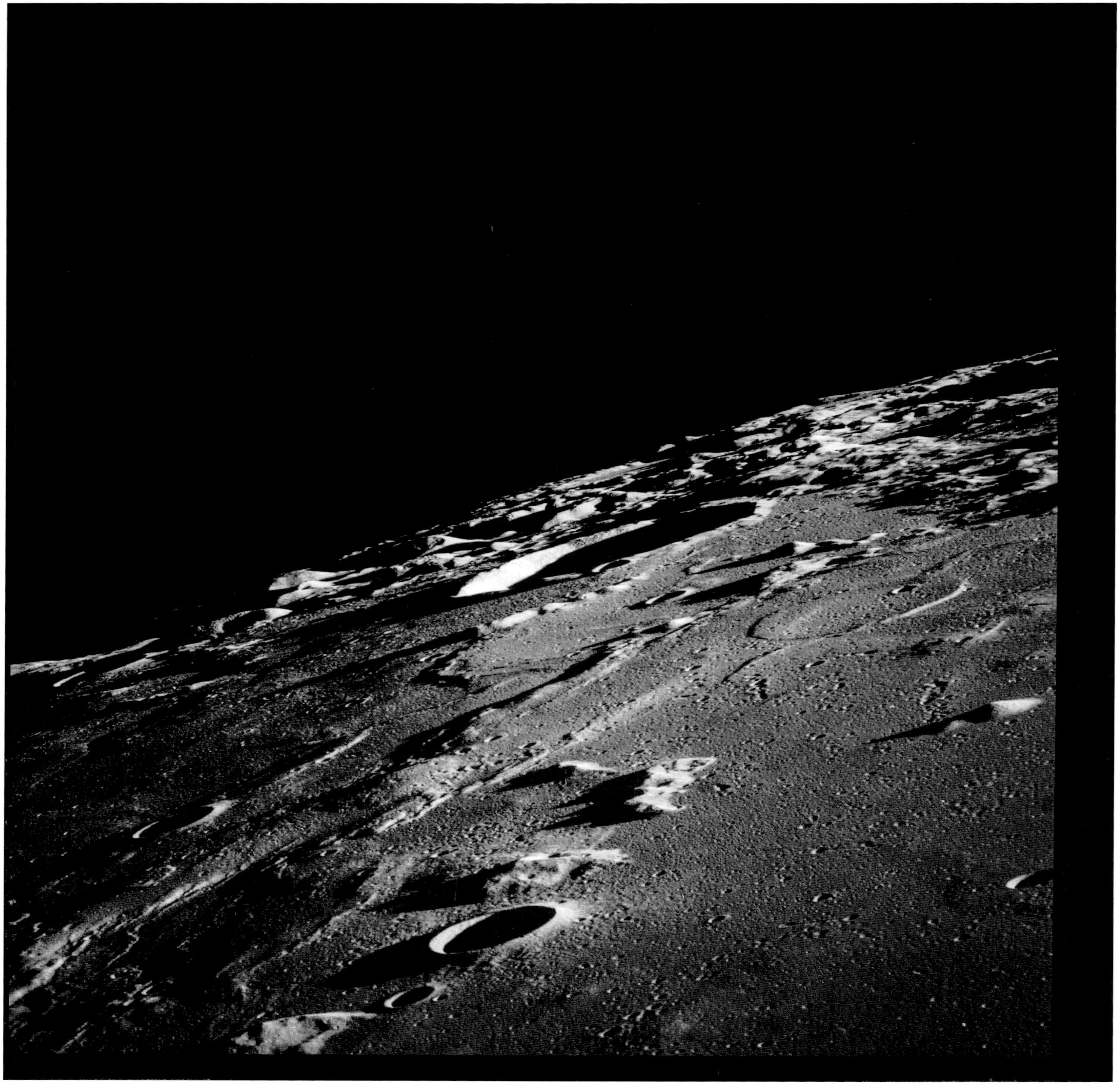

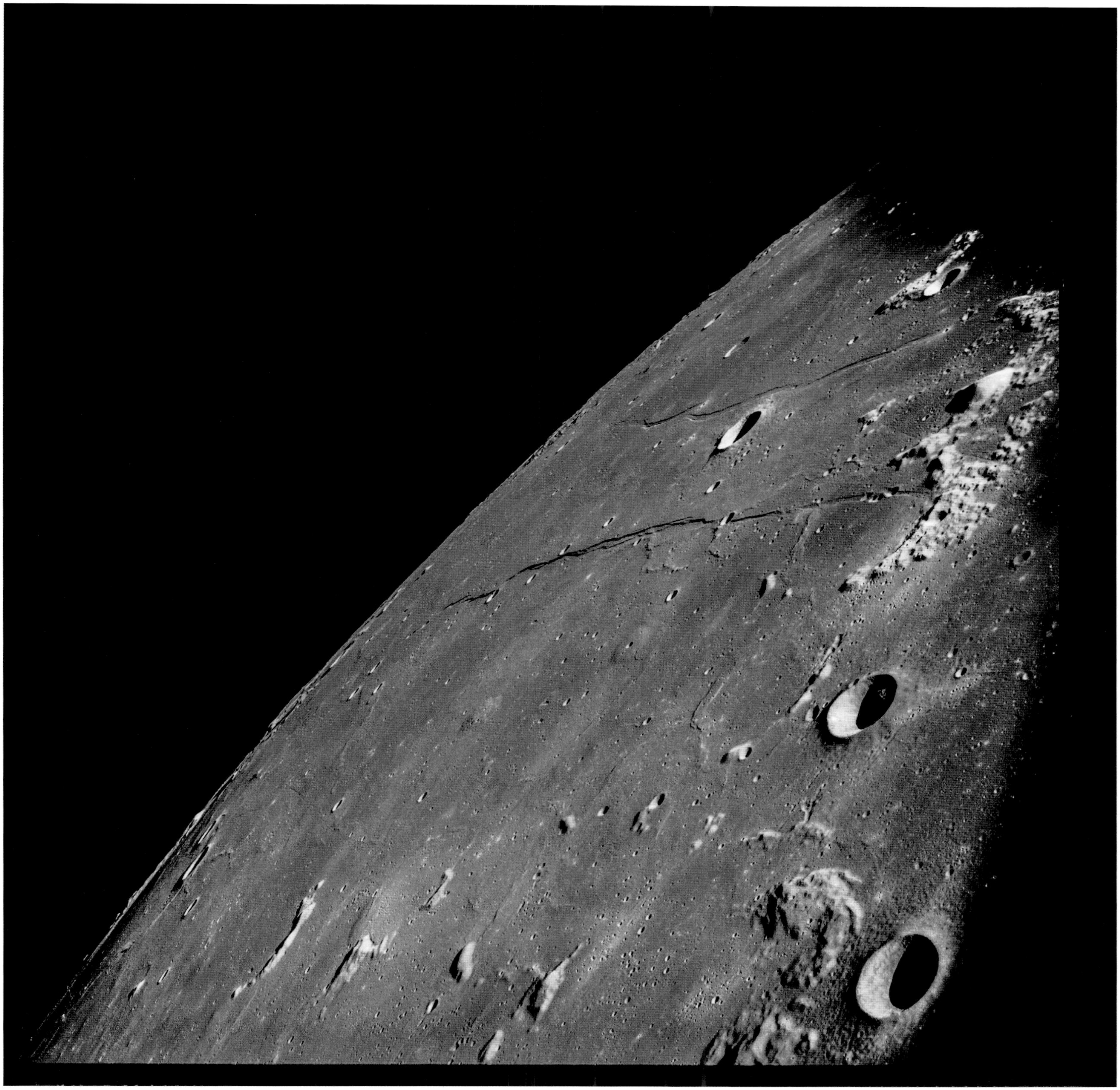

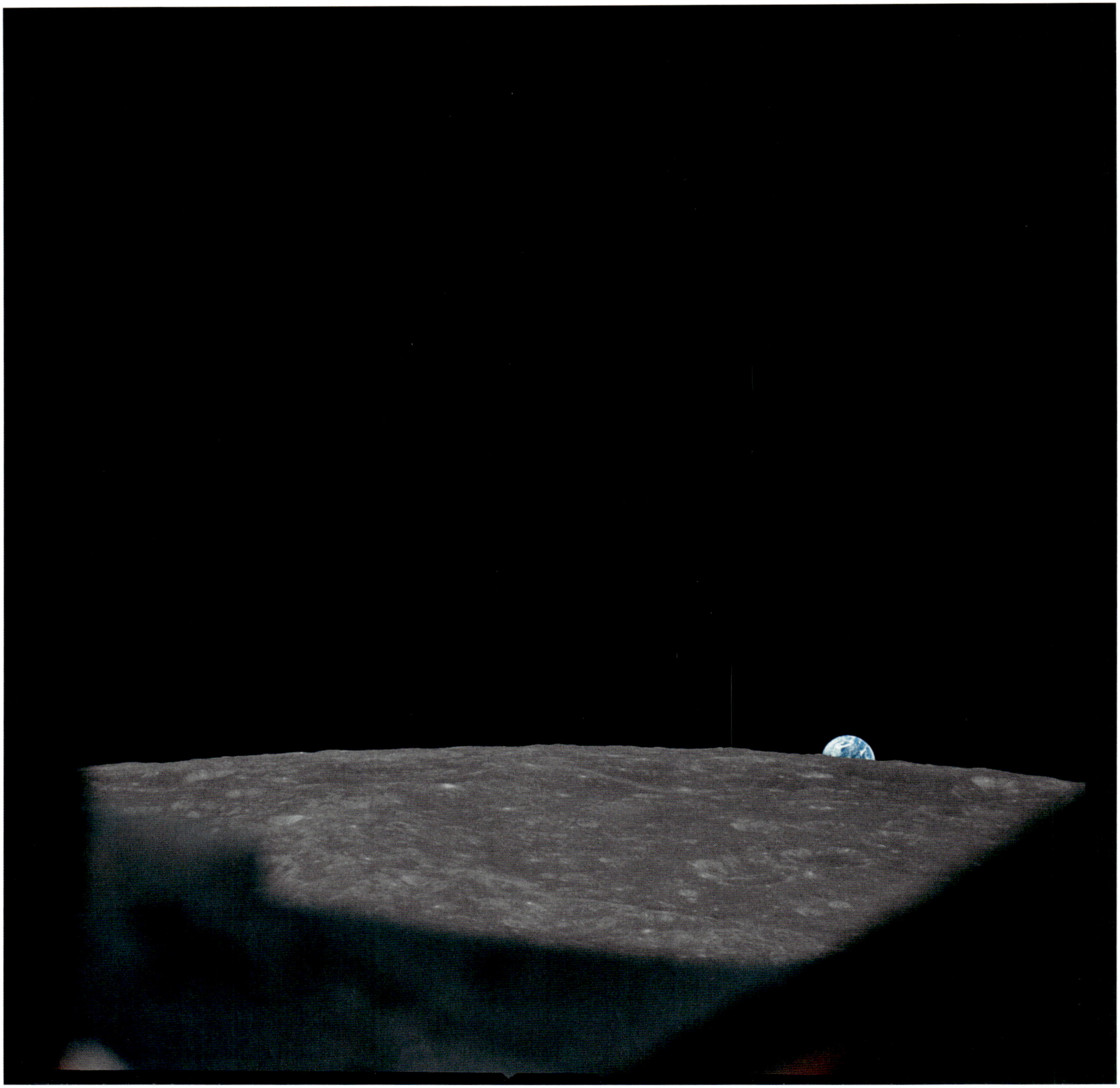

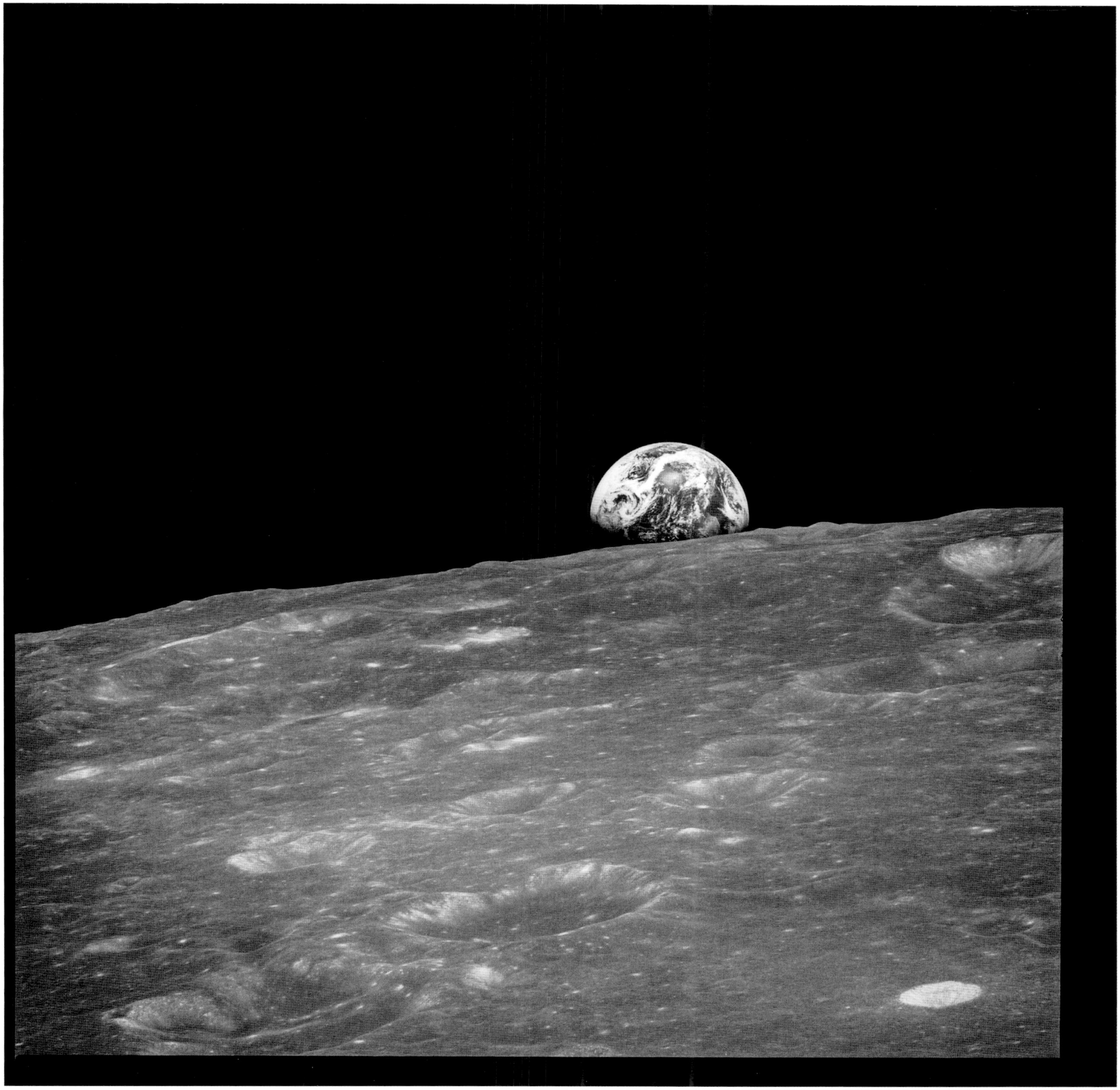

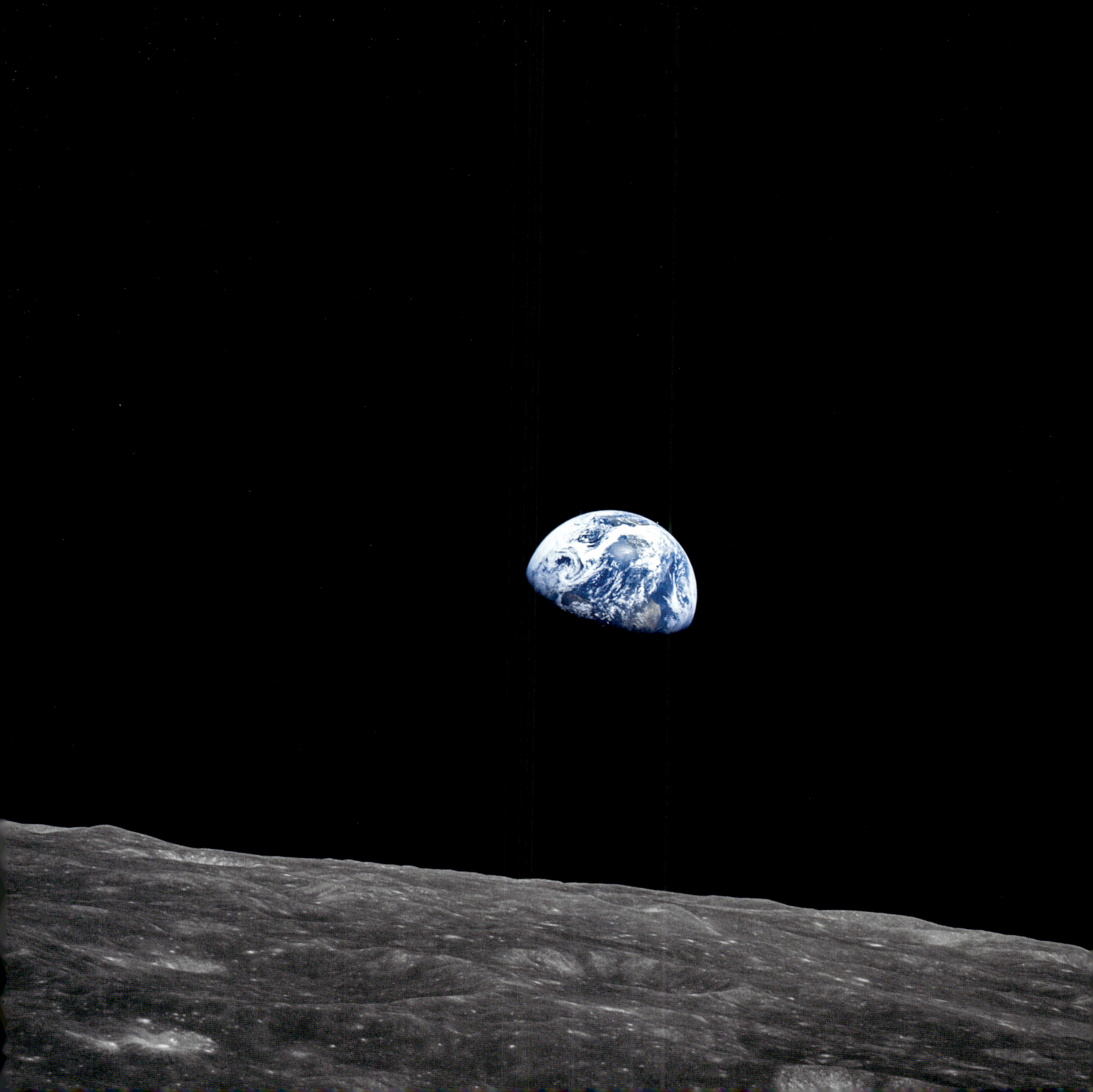

49 (AS08-16-2583)
The S-IVB stage, illuminated by sunlight, with the 'firefly' phenomenon — glowing ice crystals or debris reflecting light as they drift away from the spacecraft.

50 (AS08-16-2593)
The first full photograph of Earth, taken by Apollo 8 on December 22nd, 1968, from 18,641mi (30,000km) away. The image captures South America, North America, and Africa, with the terminator line marking the transition from day to night.

51 (AS08-14-2506)
After ten lunar orbits, Apollo 8 performed Trans-Earth Injection (TEI) on December 25th, 1968. This image, captured during their journey back to Earth, shows Mare Crisium, Mare Tranquillitatis, and Tsiolkovsky Crater, along with far-side features rarely seen from Earth.

52 (AS08-14-2475)
Captured during Apollo 8's return to Earth, this image of the receding Moon was mistakenly taken with a red filter on color film, giving it an unnatural tint. The red filter was typically used in black-and-white photography to enhance contrast by reducing glare and emphasizing surface features, helping scientists analyze lunar craters, maria, and terrain composition.

53 (AS08-14-2465)
Another filter mistake, this time with a blue filter on color film, captured during Apollo 8's return to Earth. Blue filters were used in black-and-white photography to enhance surface texture and highlight albedo differences, but on color film, they created an unintended color shift.

54 (AS08-12-2187)
A detailed view of the rugged terrain south of Tsiolkovsky Crater on the Moon's far side. The crater's lava-flooded floor and towering central peak contrast with the surrounding cratered highlands.

55 (AS08-12-2209)
A northeastward view of the lunar surface, captured on December 24th, 1968, featuring Joliot-Curie Crater, 98mi (158km) wide, on the Moon's far side.

56 (AS08-13-2224)
This view of southwest Mare Fecunditatis, taken on December 24th, 1968, features Goclenius Crater. The rilles within the crater indicate past lunar tectonic activity, while the surrounding basaltic plains formed through ancient volcanic eruptions.

57 (AS08-13-2319)
The South Pole–Aitken Basin's rim, photographed on December 24th, 1968. One of the largest impact structures in the solar system, it spans 1,553mi (2,500km) and may contain material from the Moon's mantle, revealed by the ancient impact.

58 (AS08-14-2399)
A far-side lunar landscape, captured on December 24th, 1968, near the terminator, where shadows highlight craters and mountains. These images were essential for mapping the Moon, analyzing its geology, and preparing for future Apollo landings.

59 (AS08-17-2664)
A far-side view of the Korolev Basin, featuring Korolev L, Korolev T, and the Keyhole feature. A small crater on the northern rim of Keyhole was Command Module Pilot James A. Lovell, Jr.'s control point for orbital navigation exercises.

60 (AS08-13-2347)
This image highlights the Sea of Tranquility (Mare Tranquillitatis), a dark basaltic plain formed by ancient volcanic activity. This region would later become famous as the Apollo 11 landing site, chosen for its smooth surface, which reduced the risk of hazards for the Lunar Module (LM).

61 (AS08-13-2344)
This image of the Sea of Tranquility provided crucial data for selecting the Apollo 11 landing site. Its flat, dark basaltic plains made it an ideal location for the first Moon landing.

62 (AS08-14-2390)
Part of the *Earthrise* series, this image shows Earth just beginning to emerge from behind the Moon, unlike the more well-known photo where our planet appears fully risen.

63 (AS08-13-2329)
On December 24th, 1968, this was the first *Earthrise* photo, taken before switching to color film. Captured by Lunar Module Pilot William A. Anders using a 250mm telephoto lens, it looks west across Crater Pasteur. This sharp image remains one of the best-focused early views of Earth from lunar orbit.

65 (AS08-14-2383)
As Apollo 8 orbited the Moon, the crew was focused on photographing the lunar surface when Anders suddenly noticed Earth coming into view. Excited, he quickly grabbed a Hasselblad camera, switched to color film, and captured one of history's most iconic images — *Earthrise*.

Apollo 9

IX

James A. McDivitt

David R. Scott

Russell L. 'Rusty' Schweickart

Apollo 9

Launch: March 3rd, 1969
Splashdown: March 13th, 1969

Command Module: Gumdrop
Lunar Module: Spider

Apollo 9, the third crewed Apollo mission, was the first space test of the complete Apollo spacecraft, including its Lunar Module. Apollo 9 demonstrated that the crew could successfully undock and re-dock the Lunar Module and Command Module, exactly as they would later do in lunar orbit. This critical mission laid essential groundwork for the Apollo 11 Moon landing, proving the spacecraft's design and procedures were ready for the ultimate goal — landing astronauts on the lunar surface.

During the Apollo 9 spacewalk, Rusty Schweickart had to pause midway while transferring along the Lunar Module's external handrails when Dave Scott discovered the handheld film camera had jammed. Mission Commander Jim McDivitt radioed: "OK Dave, you've got five minutes to try and fix it — Rusty, stay right where you are." Those unexpected five minutes became, in Schweickart's own words, "the most amazing five minutes of my life," as he floated silently, gazing at Earth suspended in the infinite blackness of space.

Such a quiet moment, rare amidst NASA's precisely choreographed missions, profoundly impacted Schweickart and became one of Apollo 9's most personally meaningful events.

Schweickart's powerful experience — often cited in discussions of the transformative 'overview effect' — highlights the deeply human side of space exploration, reminding us that even within tightly structured missions, there are moments when awe and wonder unexpectedly break through.

Over 1,400 images were captured during multispectral terrain photography experiments and general photographic documentation, significantly enriching scientific understanding of our planet.

James A. McDivitt
Commander

Born on June 10th, 1929 in Chicago, Illinois. McDivitt was Command Pilot for Gemini 4 in 1965 and was Backup Command Pilot in Apollo 1's first backup crew. In August 1969, he became manager of the Apollo Spacecraft program and was the Program Manager of Apollo 12 through to 16.

David R. Scott
Command Module Pilot

Born on June 6th, 1932 in San Antonio, Texas. Scott flew with Neil Armstrong on the Gemini 8 mission and later served as Backup Senior Pilot in Apollo 1's first backup crew, Backup Commander of Apollo 12 and Commander of Apollo 15. He retired from the Air Force in March 1975 with the rank of Colonel and over 5,600 hours of flying time.

Russell L. 'Rusty' Schweickart
Lunar Module Pilot

Born on October 25th, 1935 in Neptune Township, New Jersey. Schweickart was Backup Pilot in Apollo 1's first backup crew and was named Backup Commander for the first Skylab mission in 1973. He later went to NASA Headquarters in Washington, D.C. to serve as Director of User Affairs in the Office of Applications.

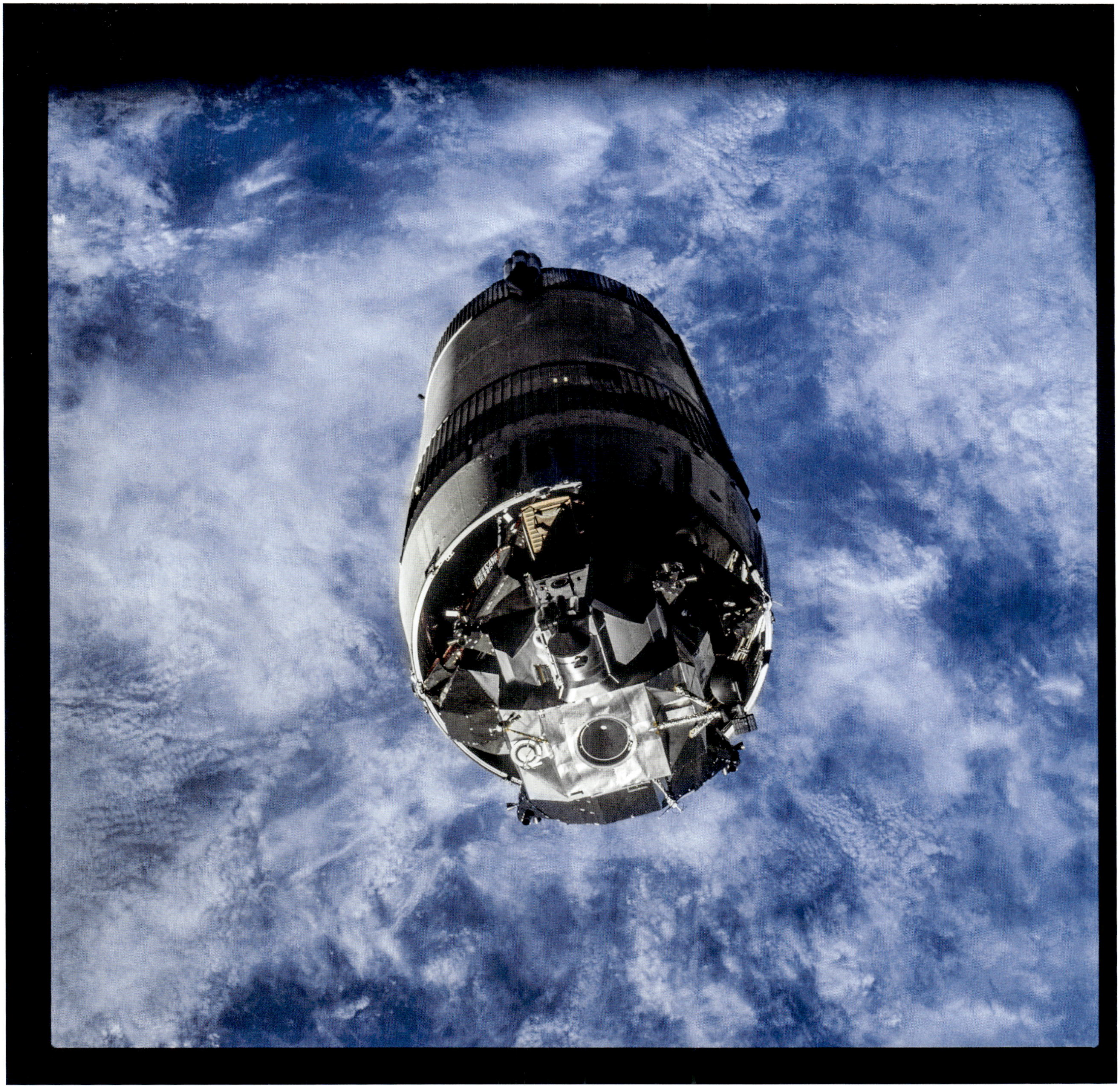

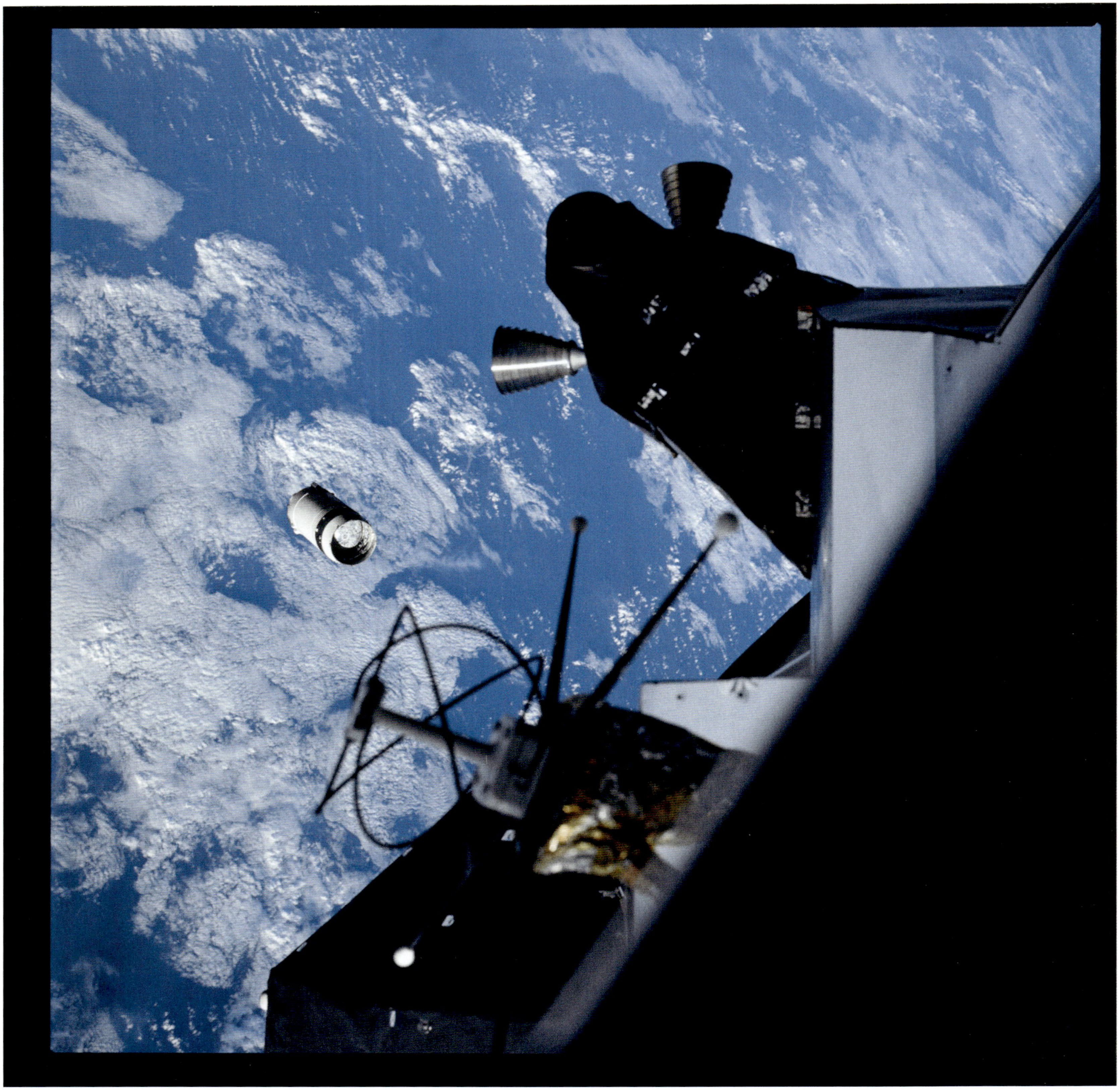

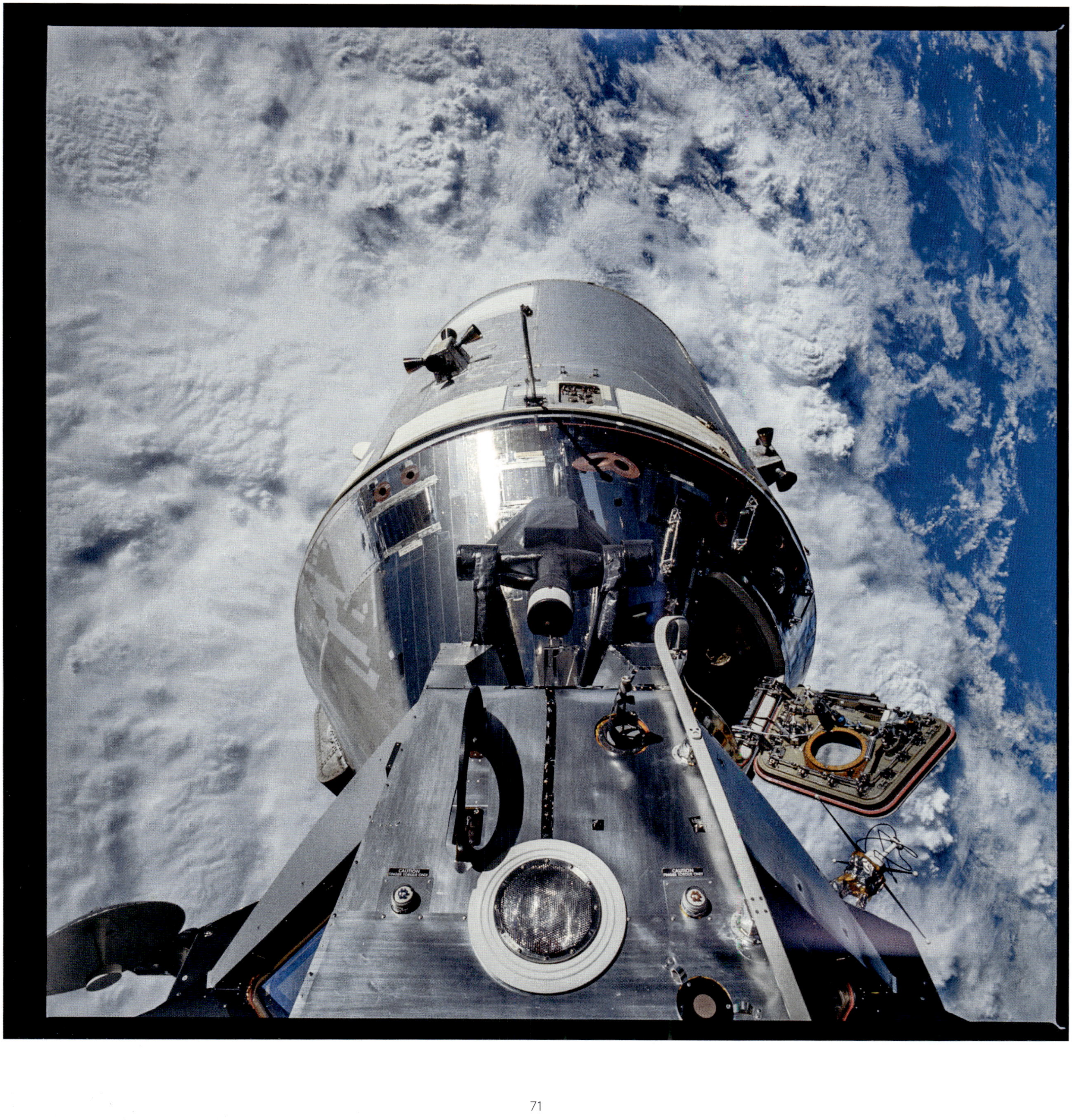

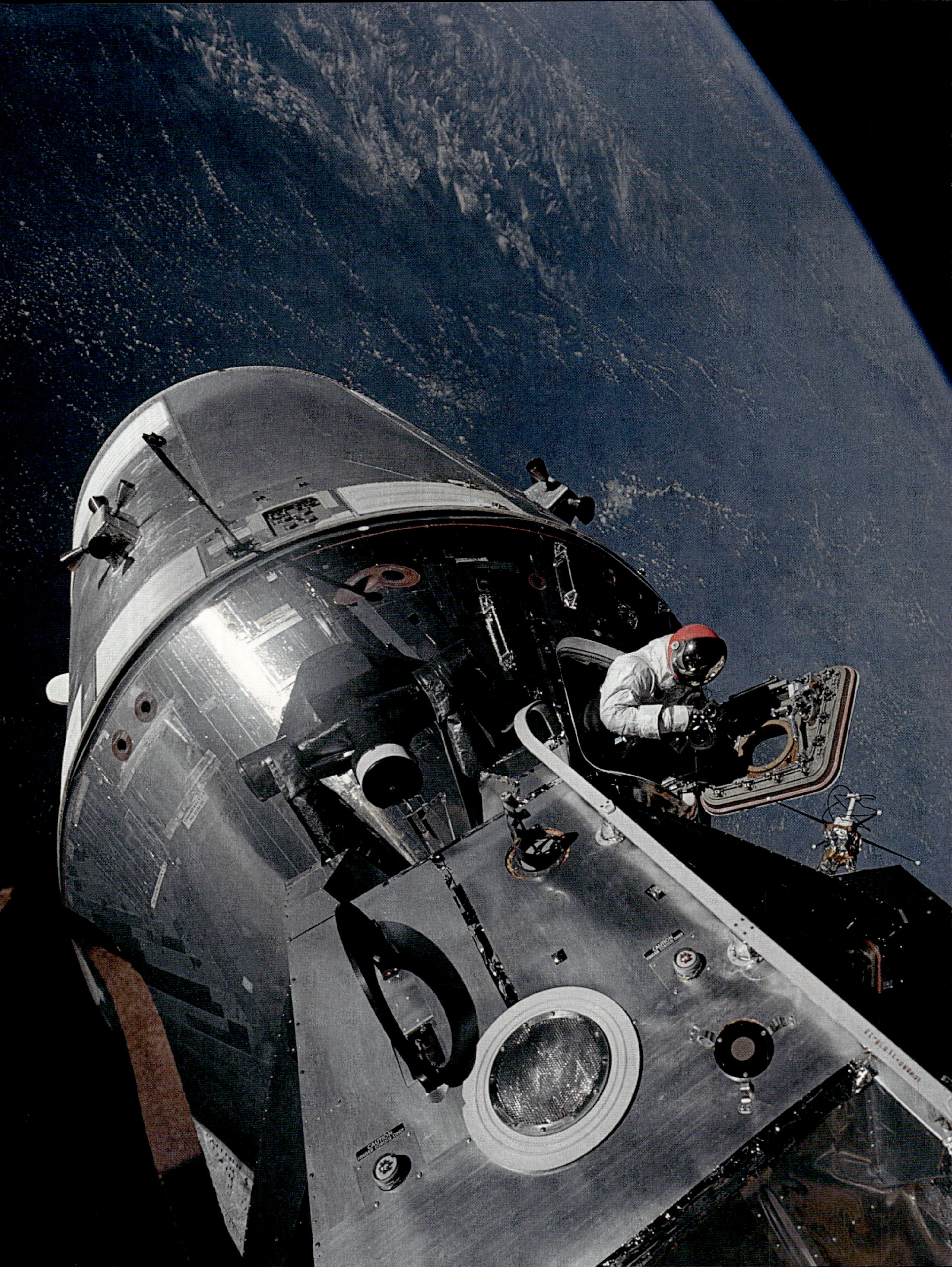

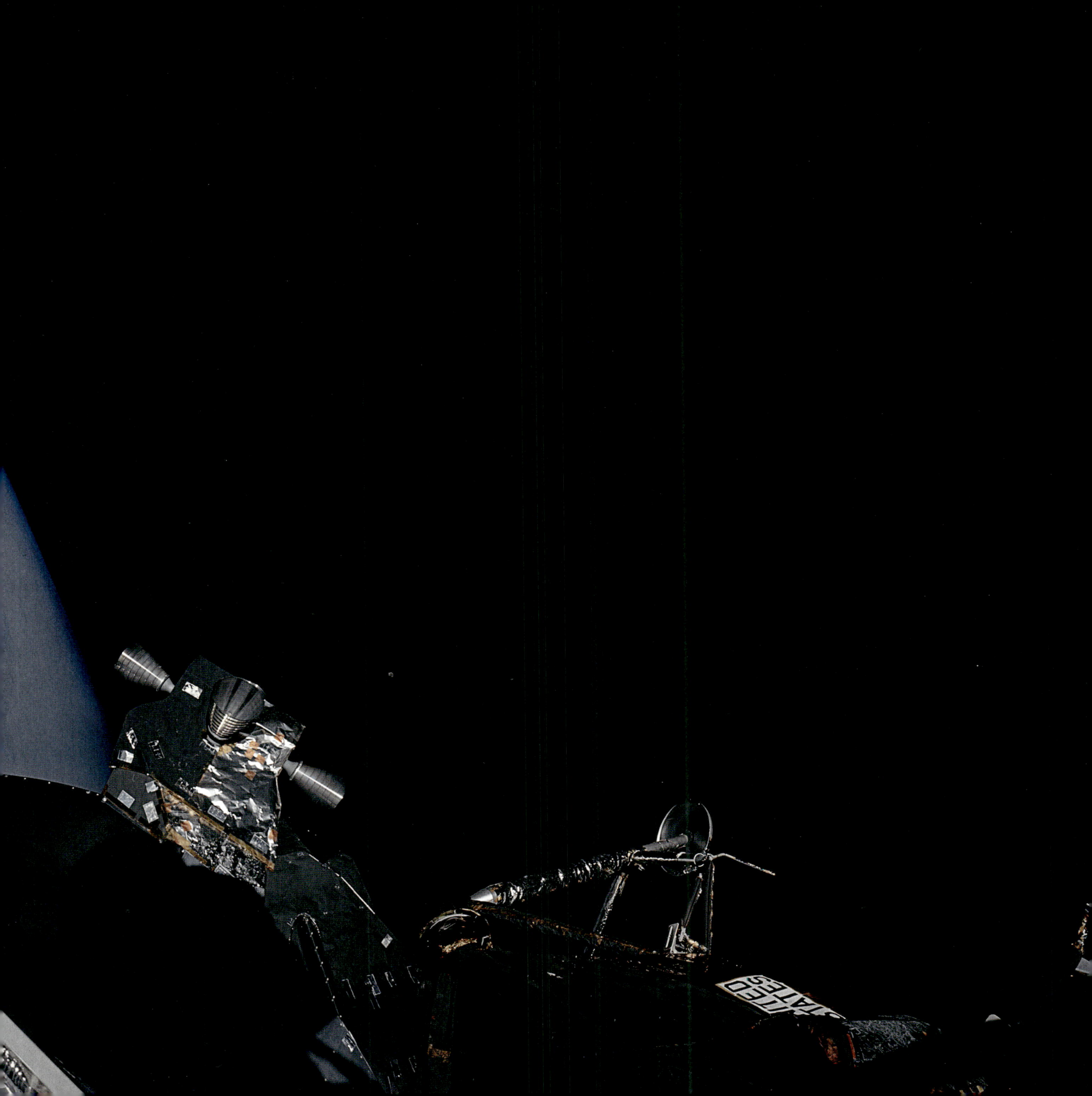

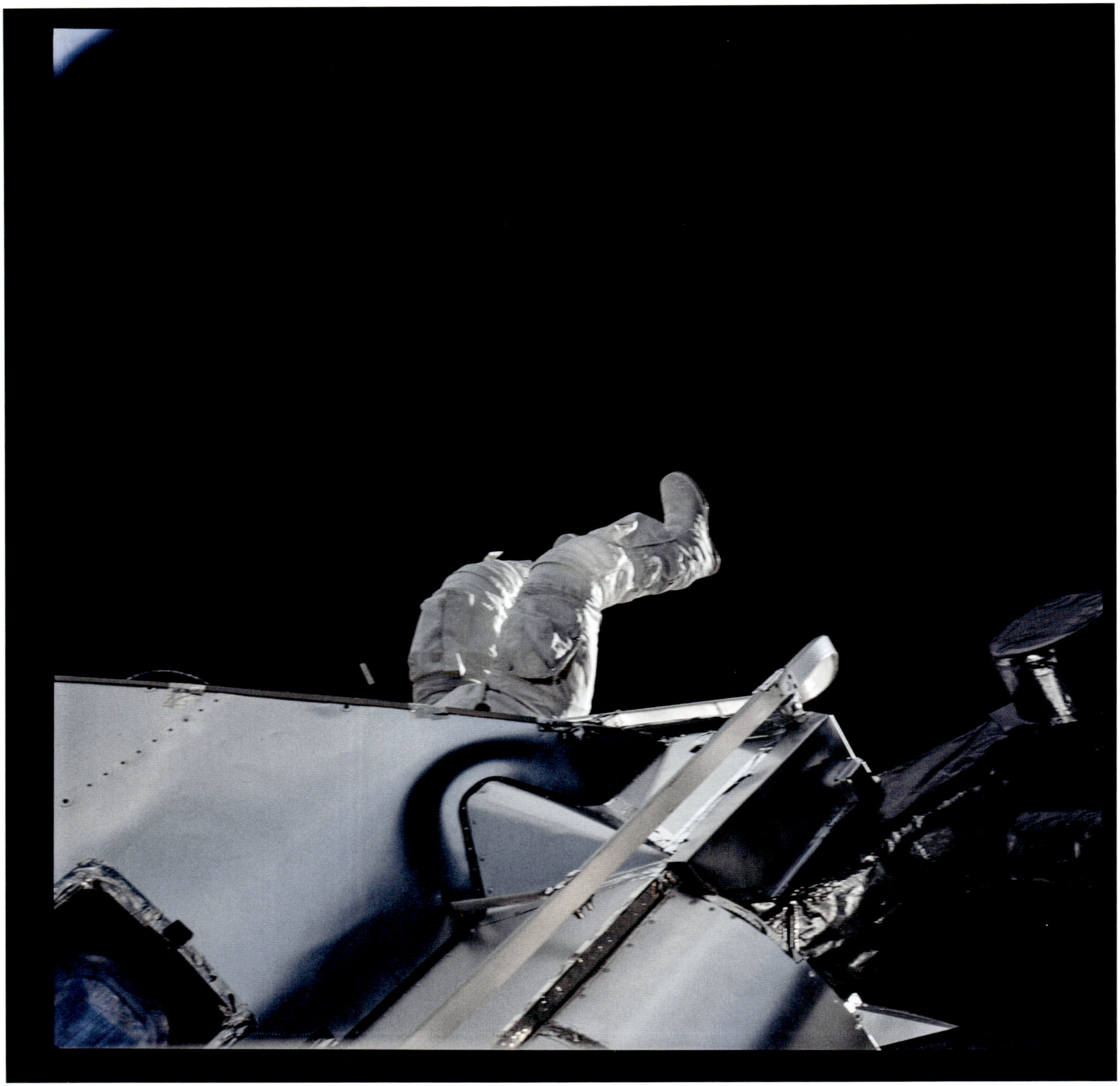

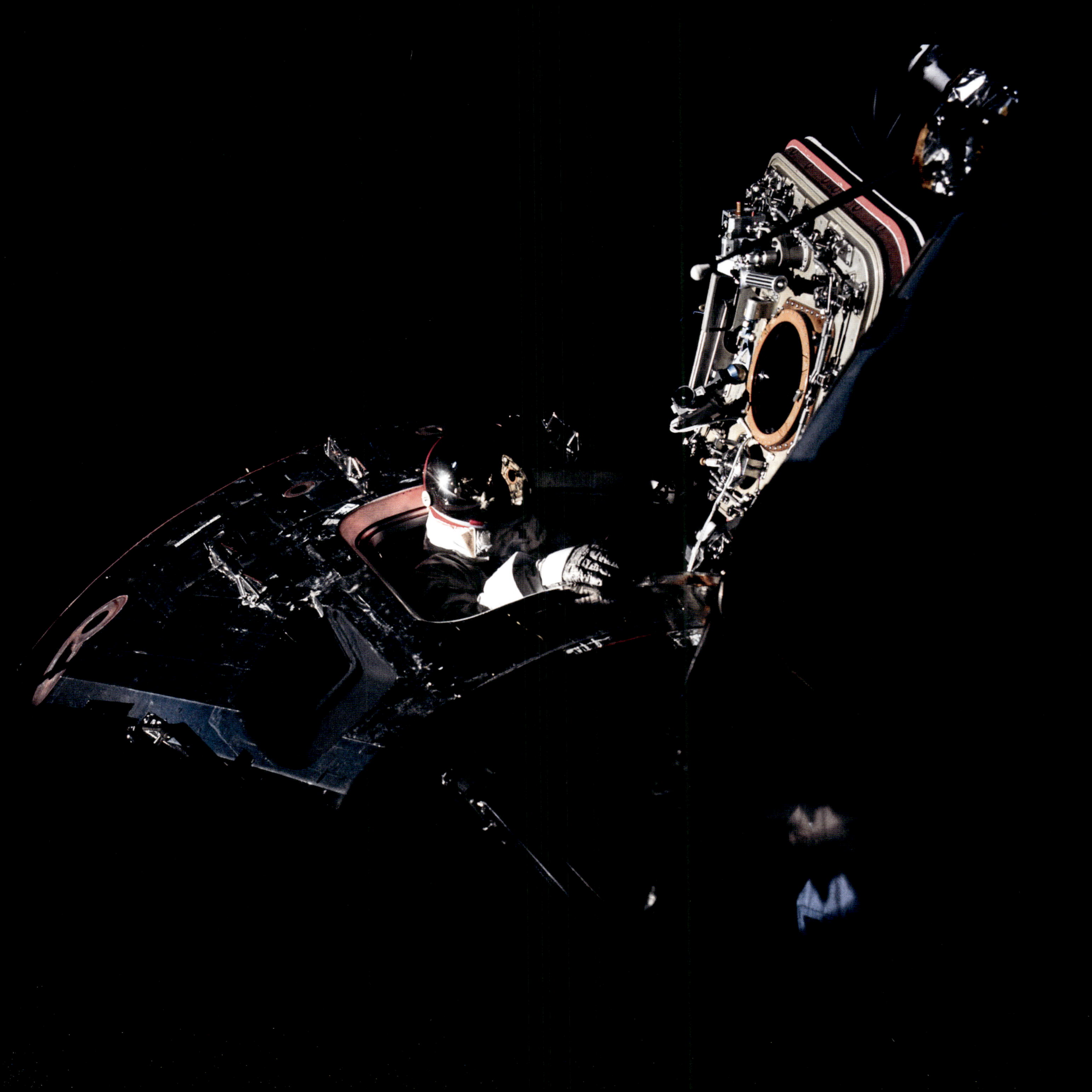

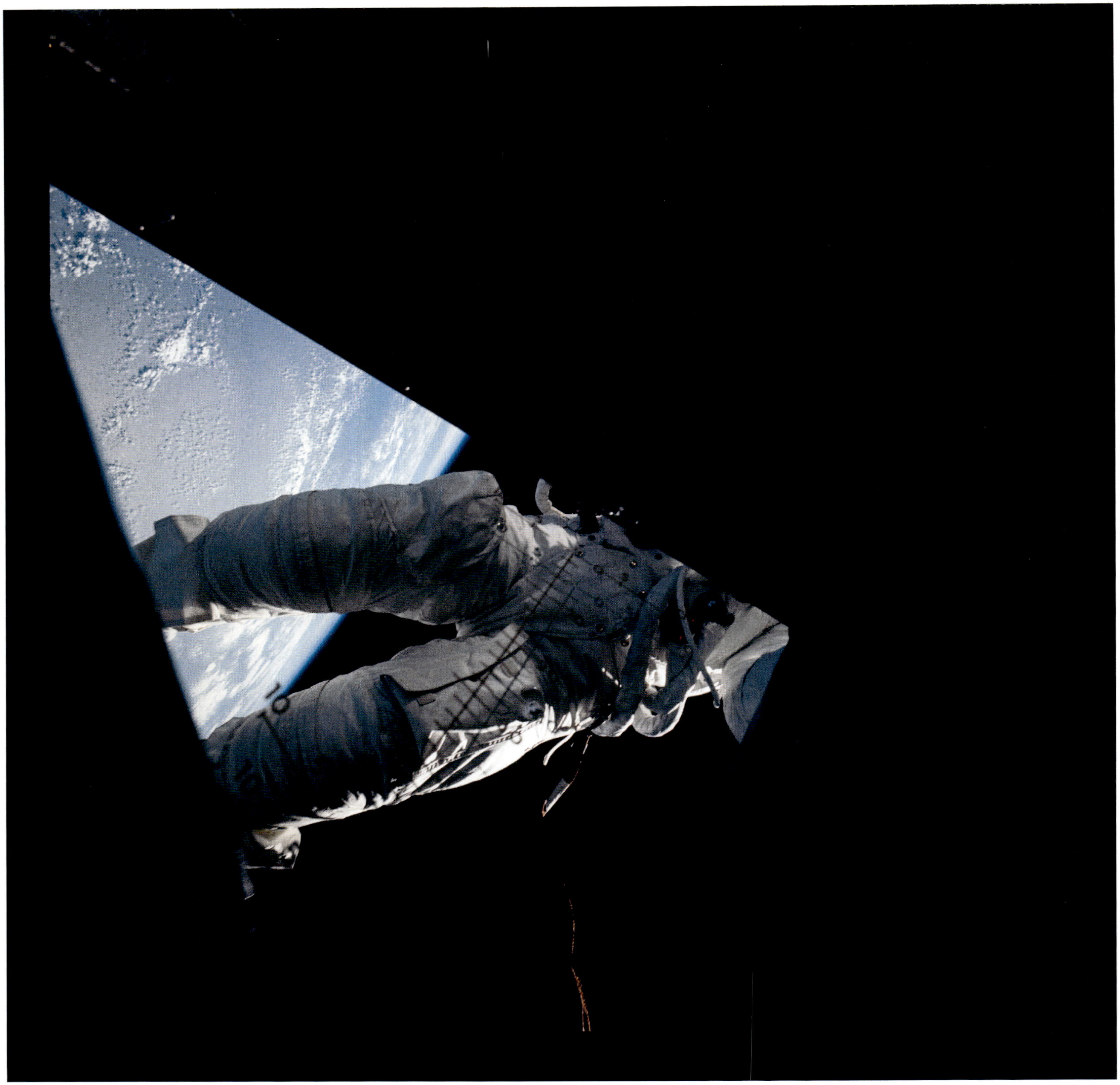

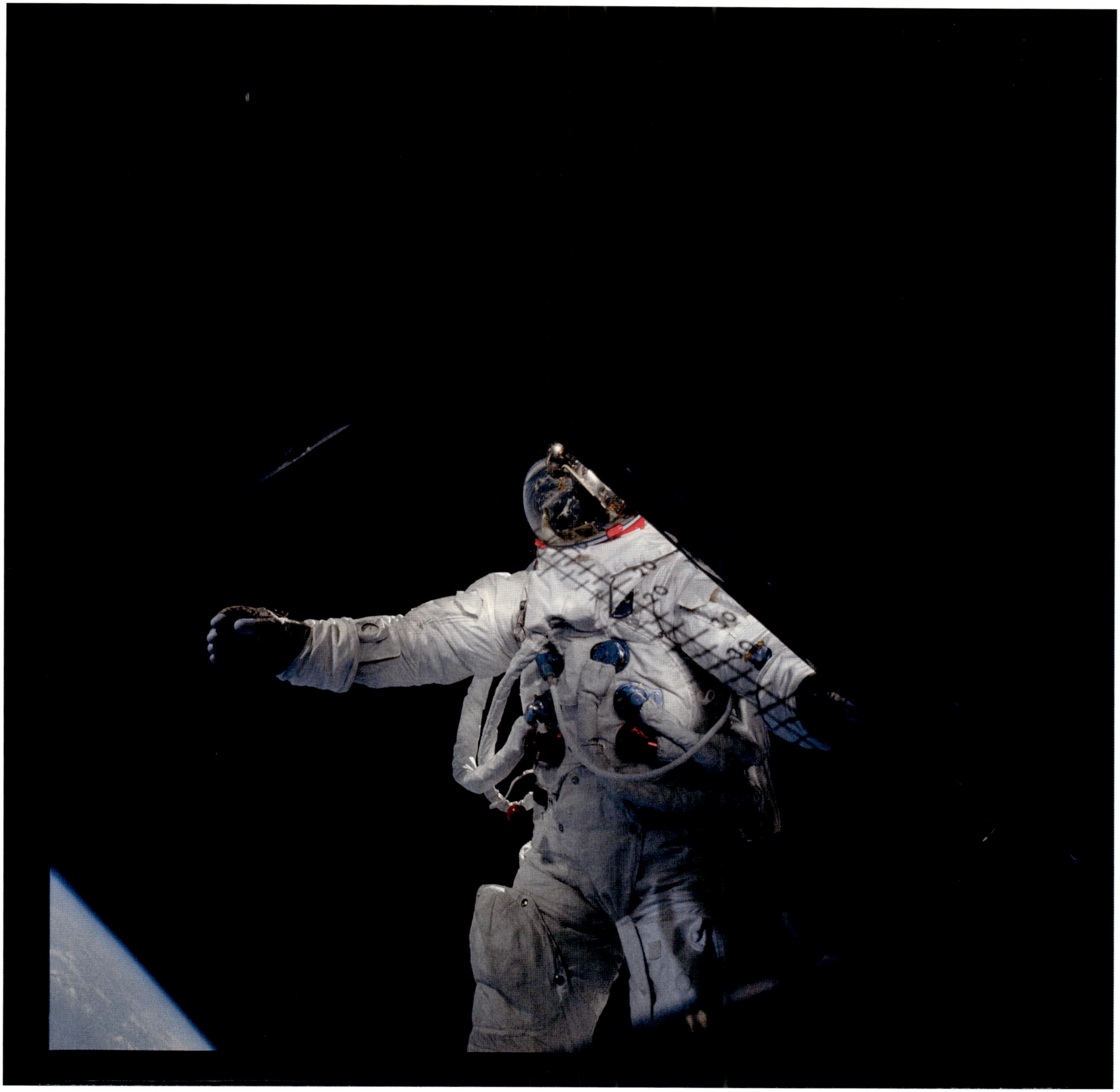

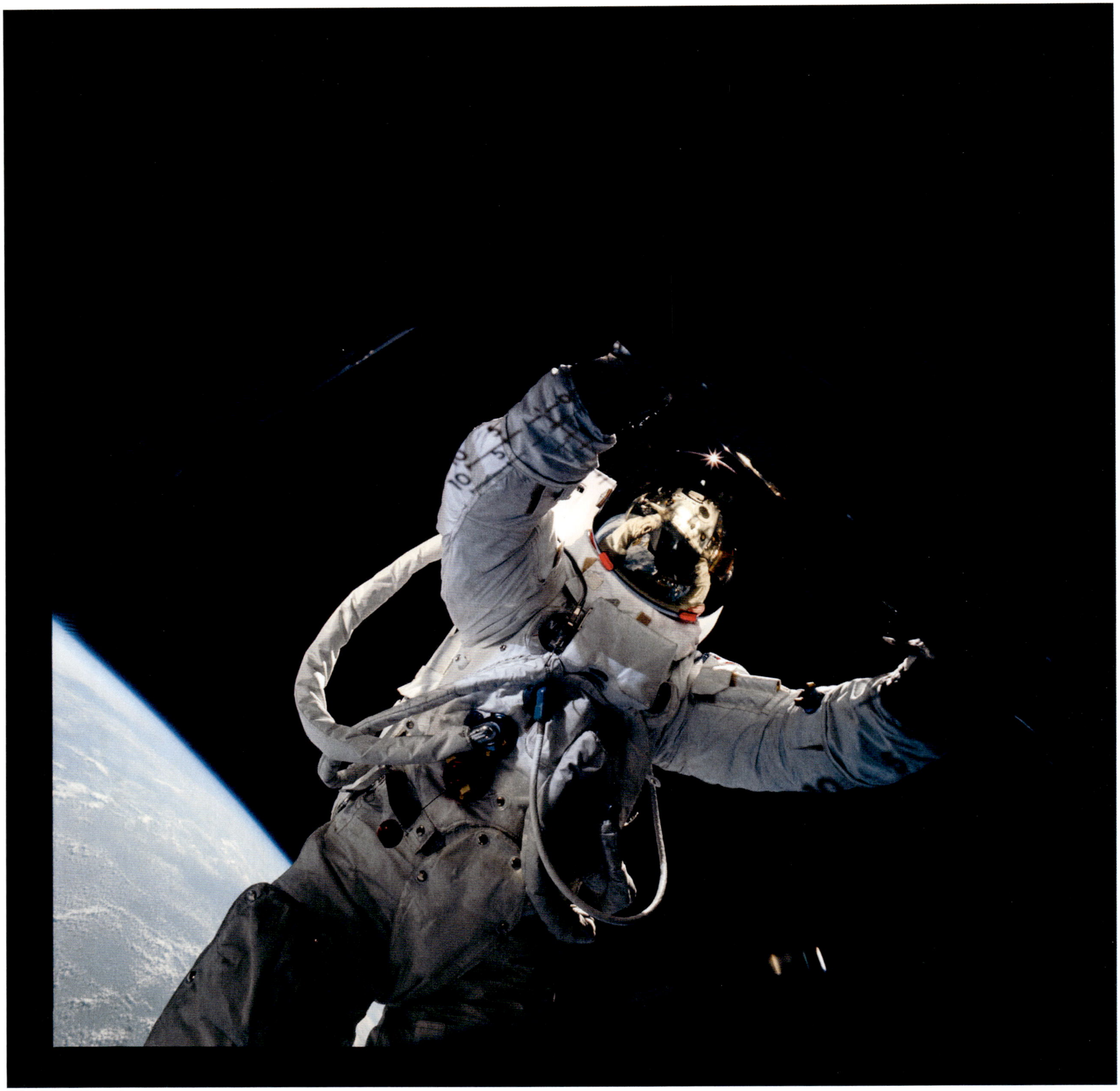

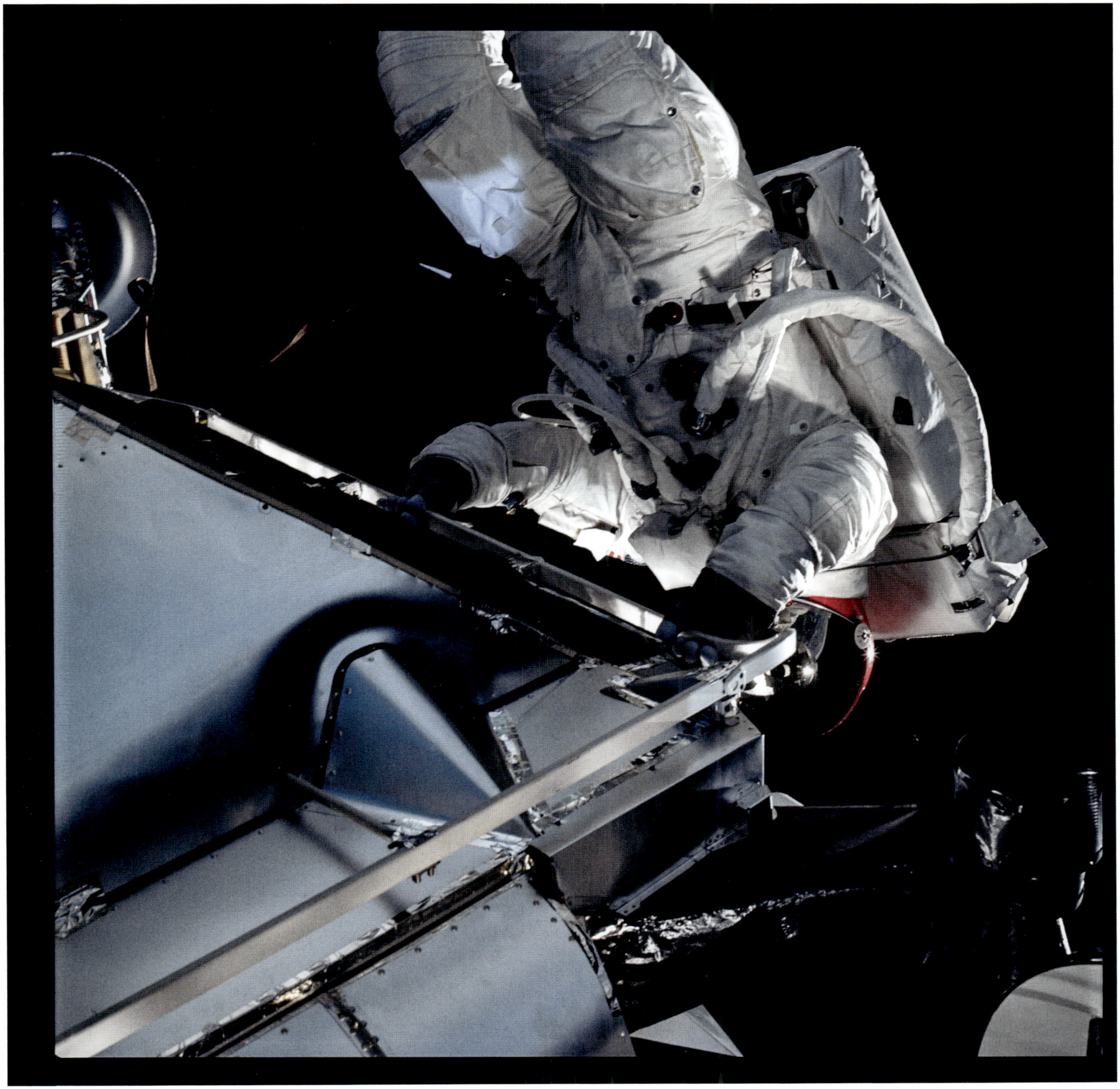

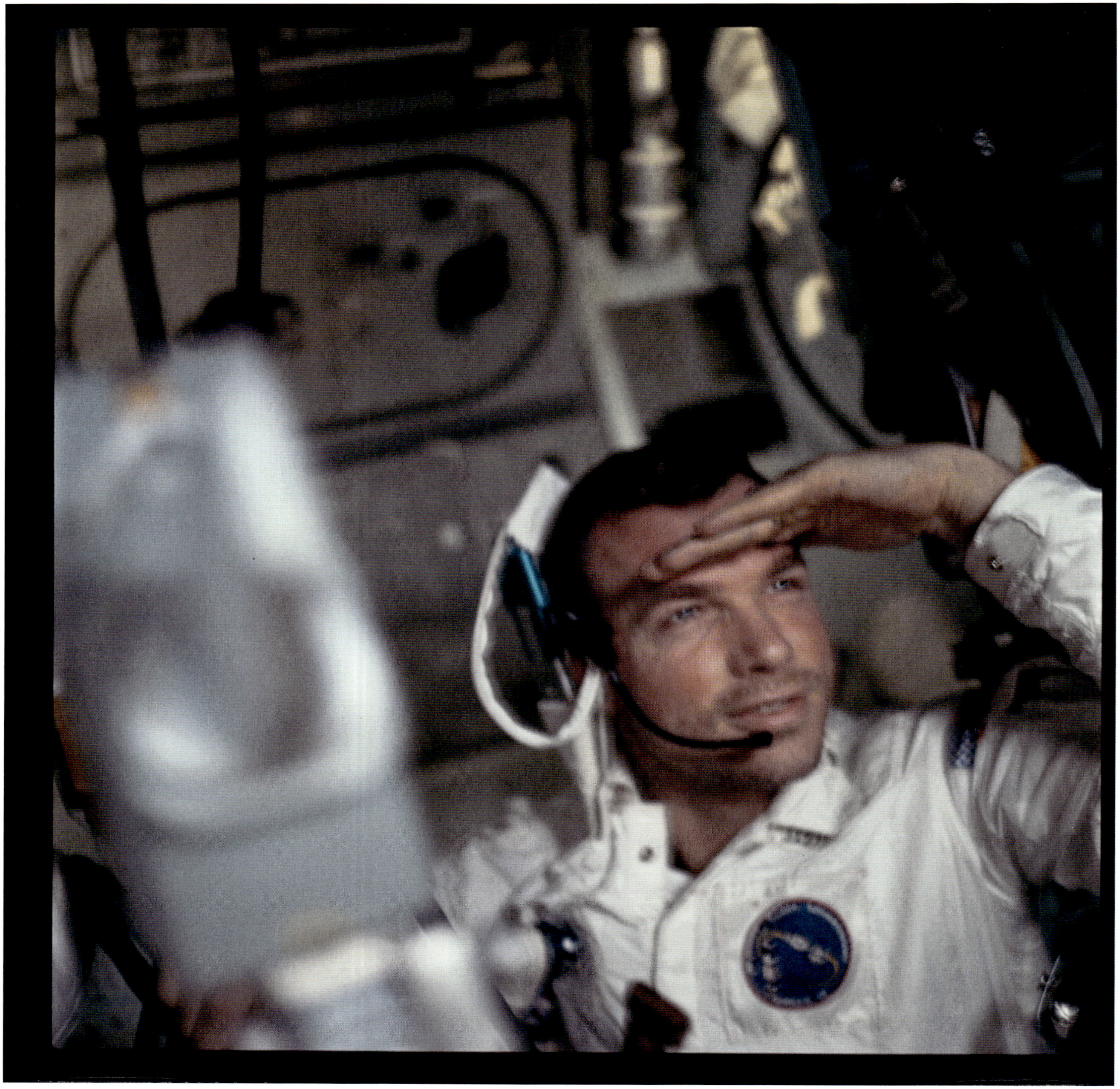

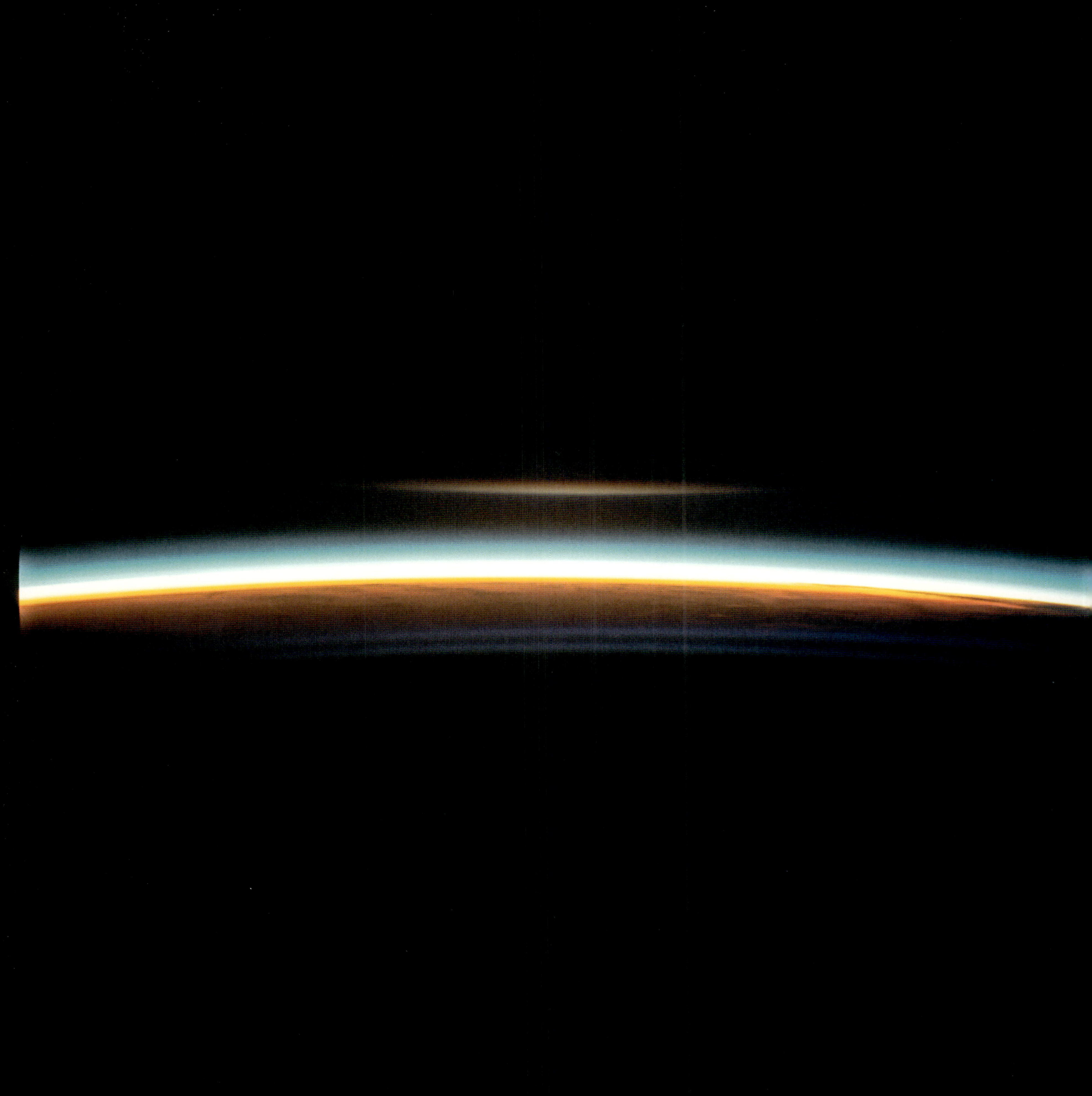

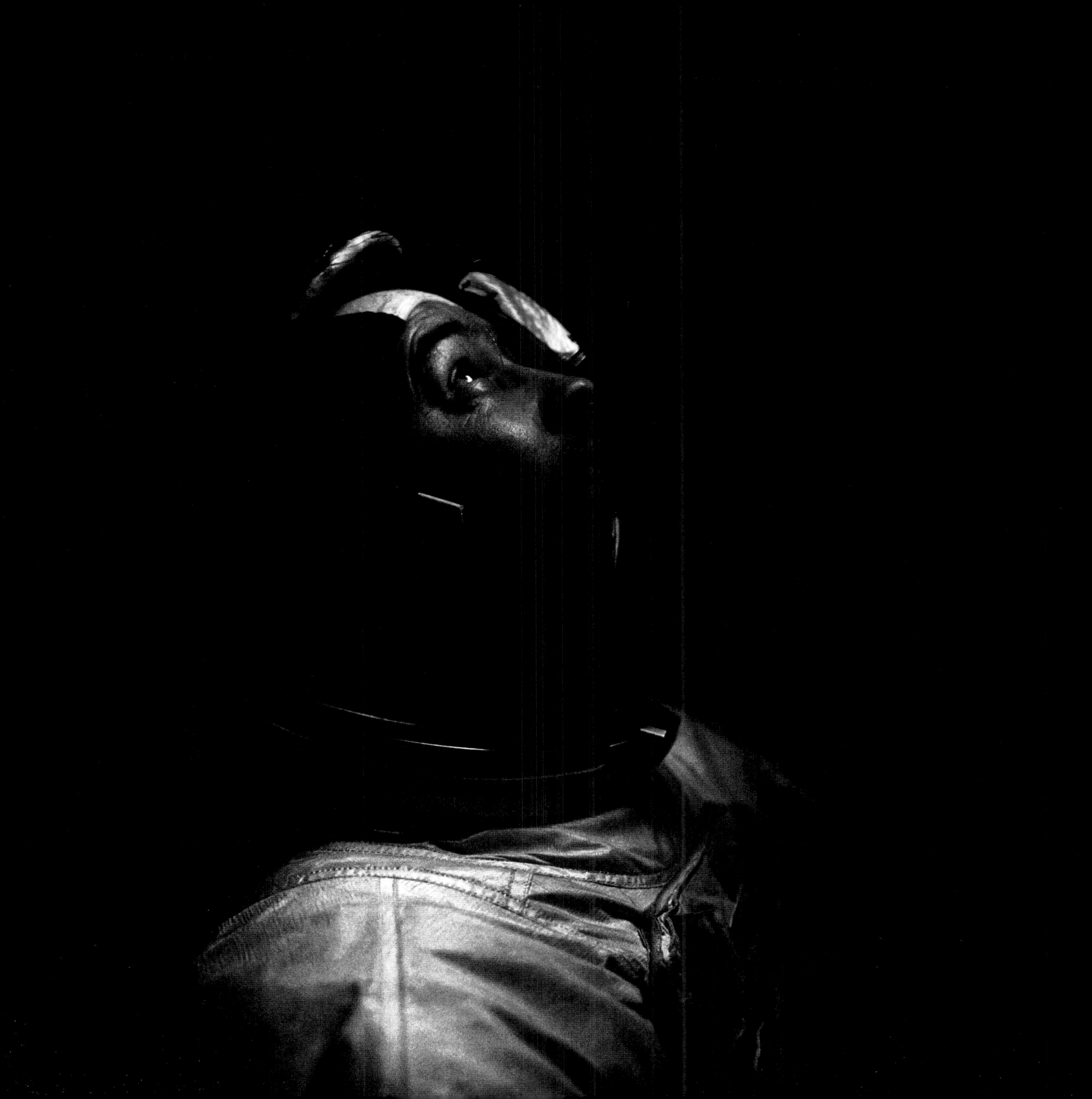

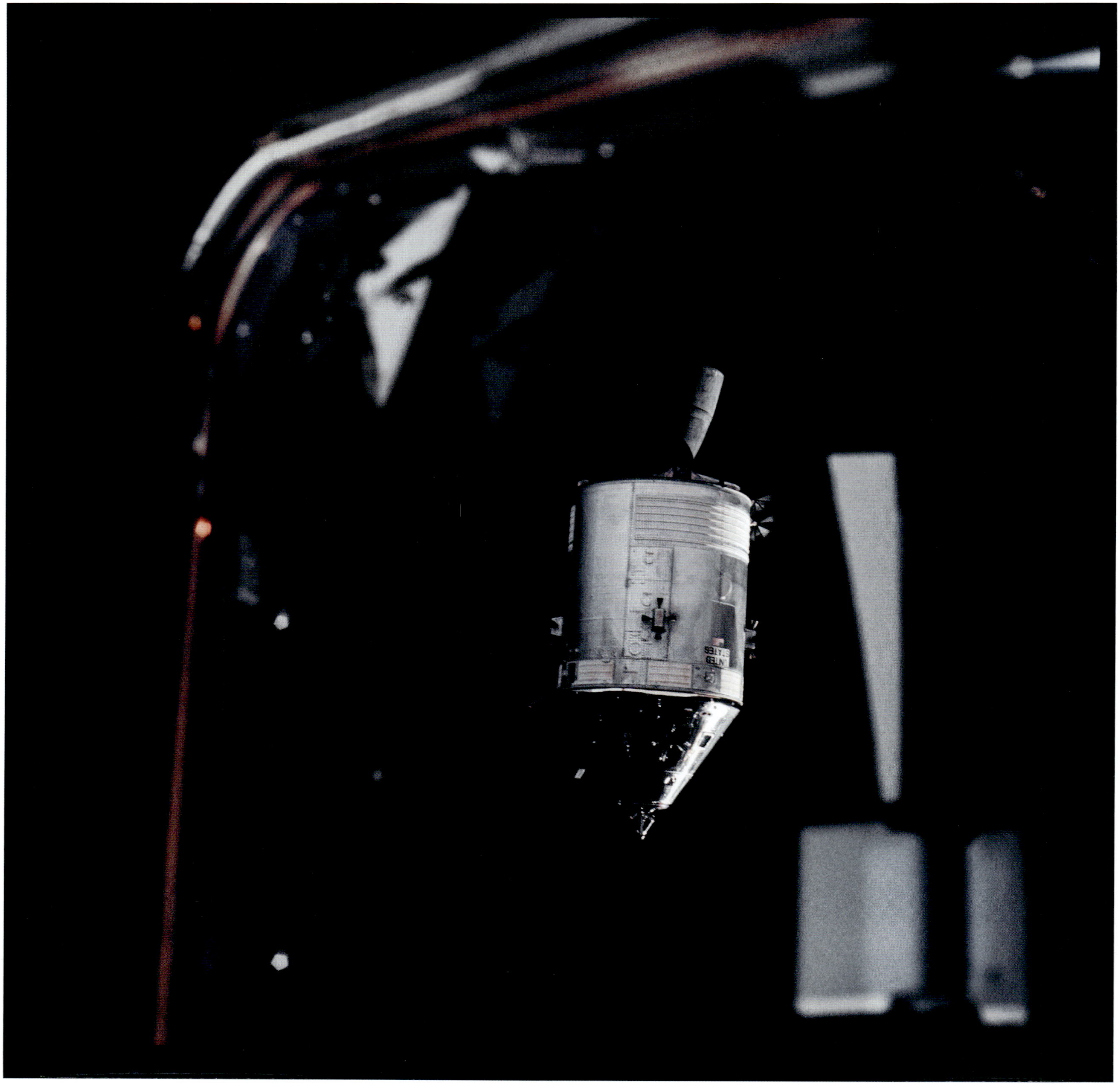

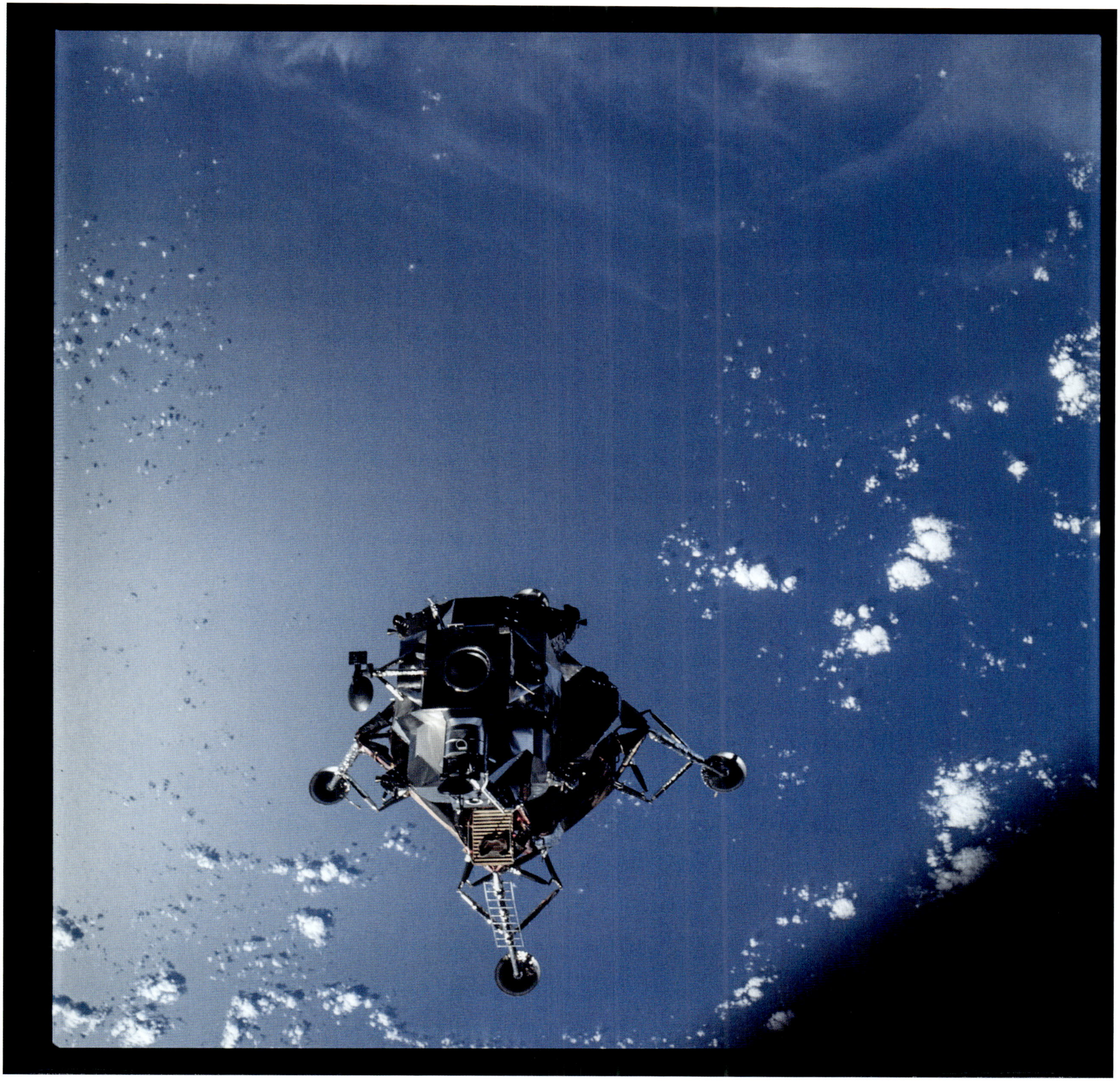

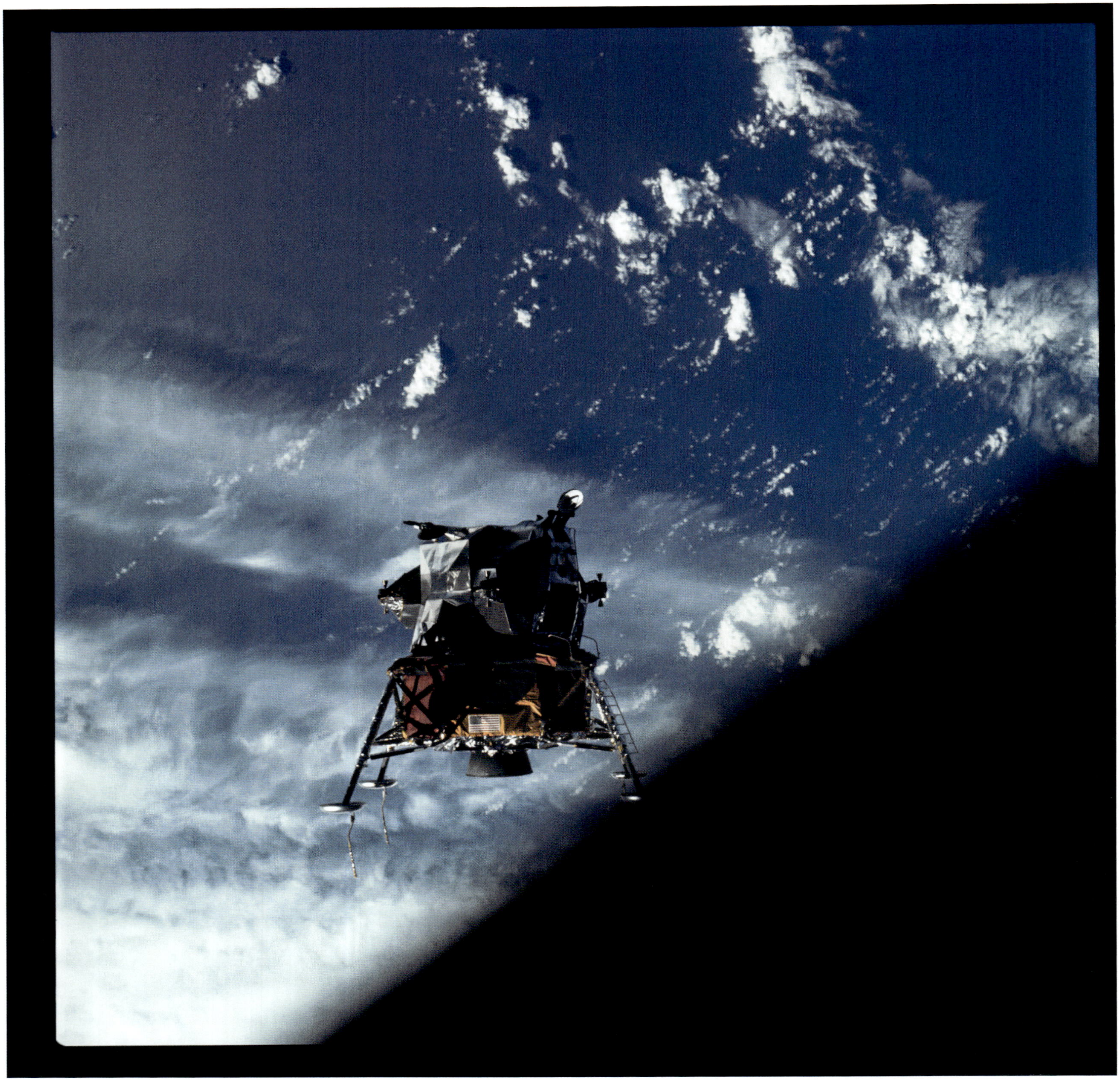

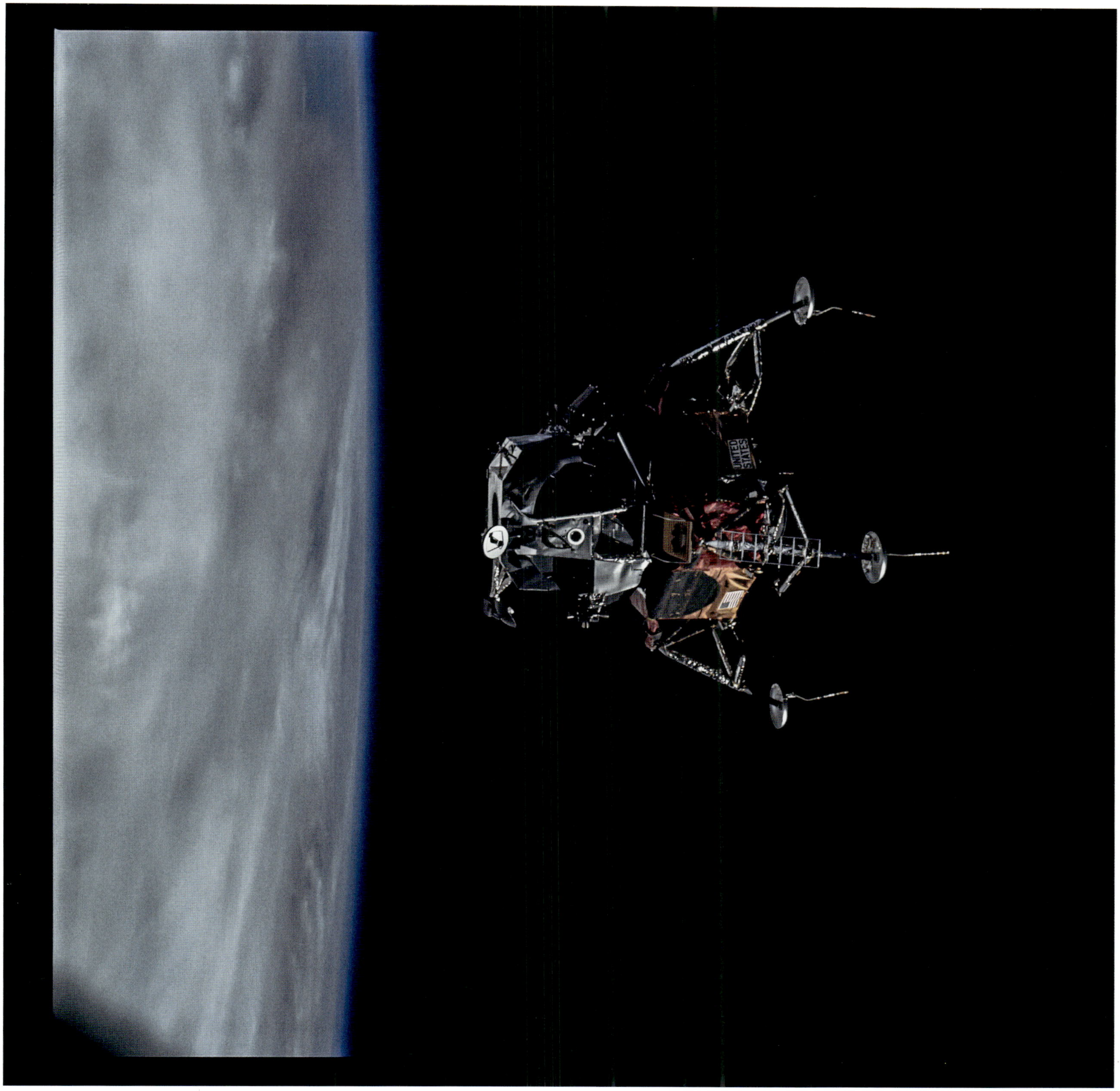
UNITED
STATES

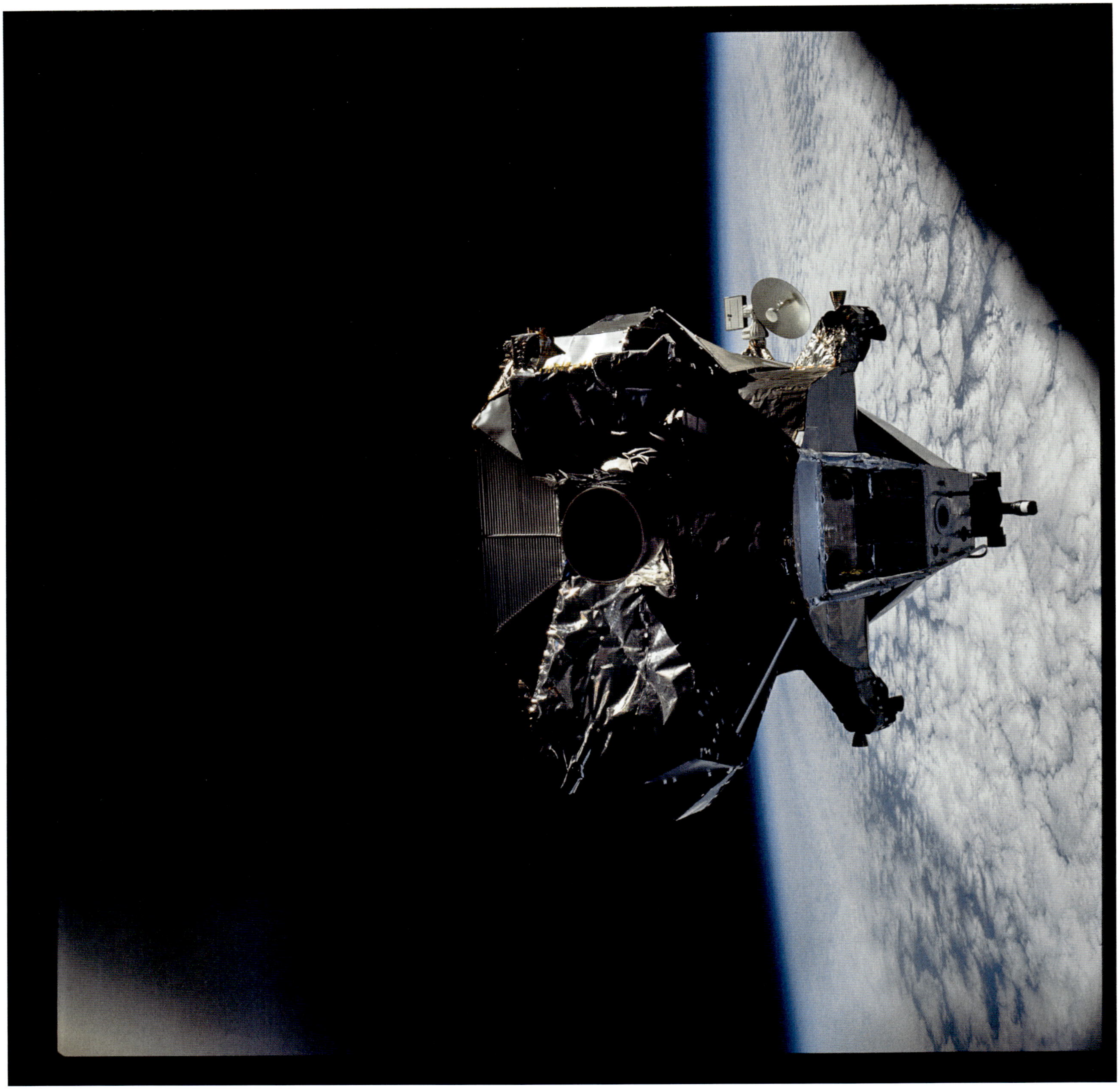

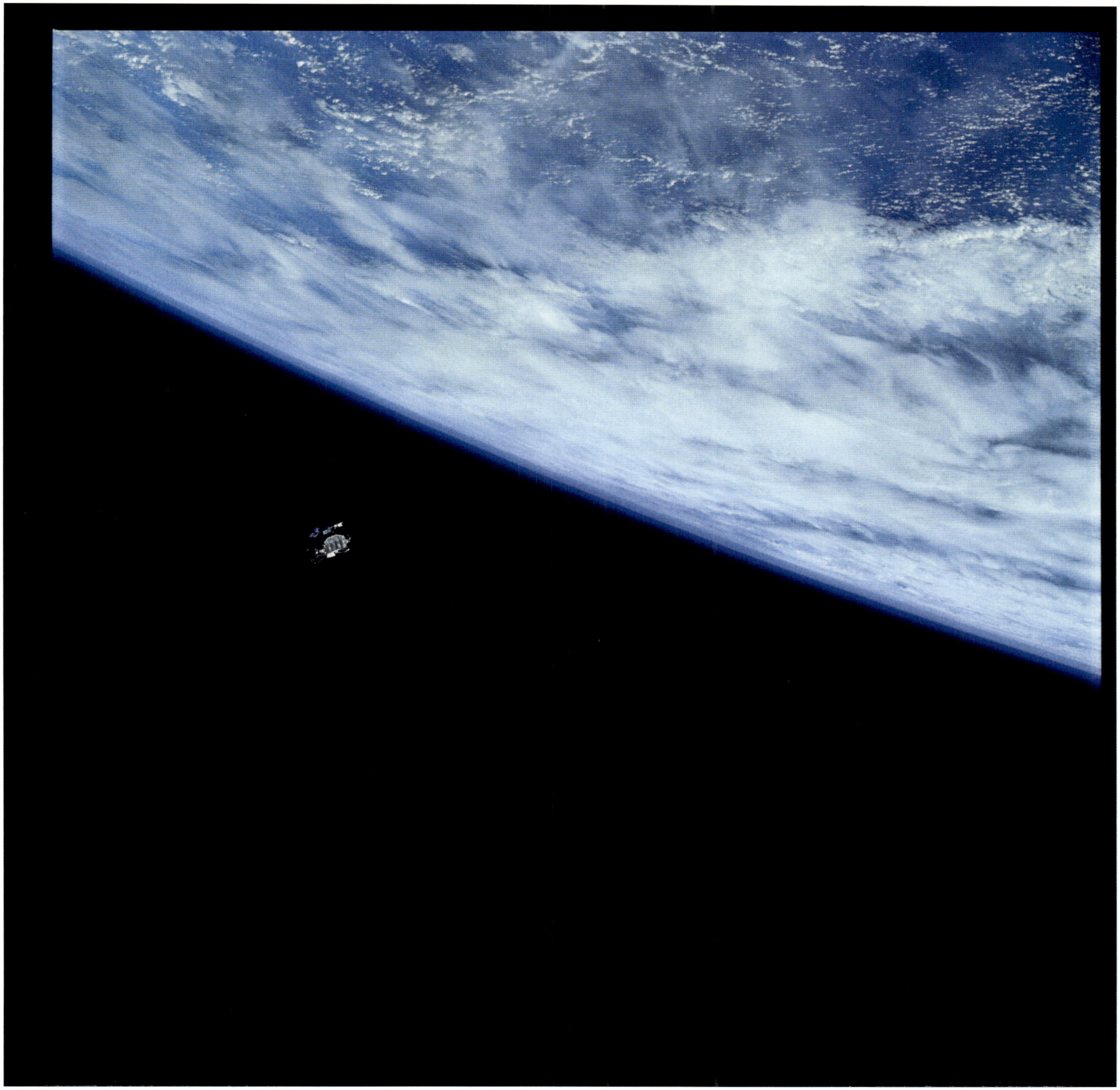

69 (AS09-19-2921)
A top-down view of the Lunar Module (LM) *Spider*, still inside the S-IVB stage, captured from the Command/Service Module (CSM) *Gumdrop* before the Transposition, Docking & Extraction (TD&E) maneuver.

70 (AS09-19-2941)
Spider, now docked to the CSM, drifts in space as the S-IVB third stage falls back toward Earth. After a successful TD&E maneuver, Apollo 9's S-IVB burned up in Earth's atmosphere.

71 (AS09-20-3071)
Captured from outside the LM, this image shows Command Module Pilot David R. Scott inside the CSM, his visor just visible in the darkness. Taken by Lunar Module Pilot Russell L. "Rusty" Schweickart during his Extravehicular Activity (EVA), while testing the Portable Life Support System (PLSS), the same life-support system later used for moonwalks.

72-73
(AS09-20-3064–AS09-20-3069)
This breathtaking panorama, taken by Schweickart, shows Scott standing fully upright in the CM hatch during Apollo 9's EVA. In the background, Earth appears as a stunning backdrop, along with parts of the LM connected to the CSM.

74 (AS09-19-3000)
A striking image — only Schweickart's legs are visible, floating in space behind the LM. Part of a series documenting Apollo 9's first spacewalk, the EVA tested the PLSS, proving astronauts could move and perform tasks in their bulky spacesuits.

75 (AS09-20-3054)
Scott stands in the CM hatch, observing Schweickart's historic EVA. His role was to monitor Schweickart's movements and test hatch operations for future Apollo missions.

76 (AS09-20-3087)
Seen from inside the LM, this image captures Schweickart outside during Apollo 9's EVA, testing the PLSS. His EVA also verified the oxygen supply, temperature regulation, and communications system, ensuring the spacesuit was safe for future Moon landings.

77 (AS09-20-3094)
Another view from inside the LM, this image of Schweickart's EVA was taken by Commander James A. McDivitt, who remained inside to monitor the spacewalk and test the docking hatch. This test ensured astronauts could safely transfer between spacecraft — an essential step for the Moon landings.

78 (AS09-20-3097)
A close-up photo detailing Schweickart's life-support system, including the Remote Control Unit (RCU), which controlled communications and suit functions.

79 (AS09-19-2998)
Schweickart is seen outside, grasping the sides of the LM during Apollo 9's EVA. This part of the mission tested the transfer between the CSM and LM in case the docking hatch developed a fault.

80 (AS09-23-3596)
Scott inside the CM, shielding his eyes as he looks out the window.

81 (AS09-20-3175)
A captivating photo of sunrise over Earth's atmosphere, where light bends through the thin blue air, creating a breathtaking gradient of colors. Orbiting at 17,000mph (27,400km/h), astronauts saw a sunrise and sunset every 90 minutes.

83 (AS09-24-3665)
McDivitt performs the first docking of two crewed spacecraft with internal transfer. The stunning lighting adds a dramatic effect, while his helmet visor reflects the docking window, Earth, and the Crew Optical Alignment Sight (COAS).

84 (AS09-24-3640)
CSM *Gumdrop*, framed by the LM window. This perspective, tested in Apollo 9, would later be seen by the moonwalkers as they departed for the lunar surface and docked back with the CSM before returning to Earth.

85 (AS09-21-3182)
LM *Spider*, floating with Earth as a backdrop — a rare scene, unlike future LM images, since Apollo 9 never left Earth orbit.

86 (AS09-21-3188)
A clear view of LM *Spider*, showing its landing gear, footpads, and Lunar Surface Sensing Probes (LSSP). These thin rods extended from the footpads to detect surface contact before landing.

87 (AS09-21-3216)
A stunning view of LM *Spider*, dramatically lit against Earth's swirling clouds. This composition, unique to Apollo 9, would never be seen again in future lunar missions.

88 (AS09-21-3238)
An unfamiliar view of LM *Spider*, showing the bottom of the ascent stage. The Ascent Propulsion System (APS), the engine that would later power Moon astronauts back to the CSM, is clearly visible.

89 (AS09-21-3254)
After the Apollo 9 docking tests were completed, the LM *Spider* was discarded into Earth's atmosphere, where it re-entered and burned up.

Apollo 10

X

Thomas P. Stafford

John W. Young

Eugene A. Cernan

Apollo 10

Launch: May 18th, 1969
Splashdown: May 26th, 1969

Command Module: Charlie Brown
Lunar Module: Snoopy

Apollo 10 was the 'dress-rehearsal' for the first Moon landing, following the exact flight plan of Apollo 11 on the same lunar day and time, except for landing. On the fifth day of the mission, Stafford and Cernan boarded the Lunar Module and separated from the Command Module. They descended from lunar orbit towards the Sea of Tranquility. This phase was planned to check all the procedures and systems in the Lunar Module. The astronauts brought the Lunar Module to within 9mi (14.5km) of the lunar surface, stopping at the point where the final powered descent would have to begin for an actual Moon landing.

John Young remained aboard the Command Module, documenting the Lunar Module descent and ascent in a series of photographs, some of which appear in this chapter.

As a precautionary measure, NASA short-fueled the Ascent Stage so that it would never be able to get off the surface of the Moon. This was to prevent the astronauts from actually trying to land the Lunar Module on the Moon, as NASA knew the crew might be tempted. Cernan himself even said that "a lot of people thought about the kind of people we (Stafford and Cernan) were: 'Don't give those guys an opportunity to land, 'cause they might!'"

The mission was one of the most successful Apollo missions, giving NASA the confidence to land a man on the Moon a mere eight weeks after Apollo 10's splashdown.

Nearly every photographic objective was completed, with nine magazines of 70mm film exposed. The main photographic objective was to take a series of space-to-lunar-surface shots to map out the entire landing area. This, along with landmark and tracking data, allowed NASA to reduce the size of the landing ellipse, one of the most important objectives of the mission.

Thomas P. Stafford
Commander

Born on September 17th, 1930 in Weatherford, Oklahoma. He was Backup Pilot of Gemini 3, Pilot of Gemini 6A, Command Pilot of Gemini 9A, Backup Commander of Apollo 7 and Commander of the Apollo-Soyuz Test Project.

John W. Young
Command Module Pilot

Born on September 24th, 1930 in San Francisco, California. He was Pilot of Gemini 3, Backup Pilot of Gemini 6A, Command Pilot of Gemini 10, Backup Command Module Pilot of Apollo 7, Commander of Apollo 16 and Backup Commander of Apollo 13 and 17. He also commanded the first and ninth Space Shuttle mission (STS-1 & STS-9).

Eugene A. Cernan
Lunar Module Pilot

Born on March 14th, 1934 in Chicago, Illinois. He was the Pilot of Gemini 9A, Backup Pilot of Gemini 12, Backup Lunar Module Pilot of Apollo 7 and Backup Commander of Apollo 14. As Commander of Apollo 17, he was the eleventh, and currently last, man to walk on the Moon.

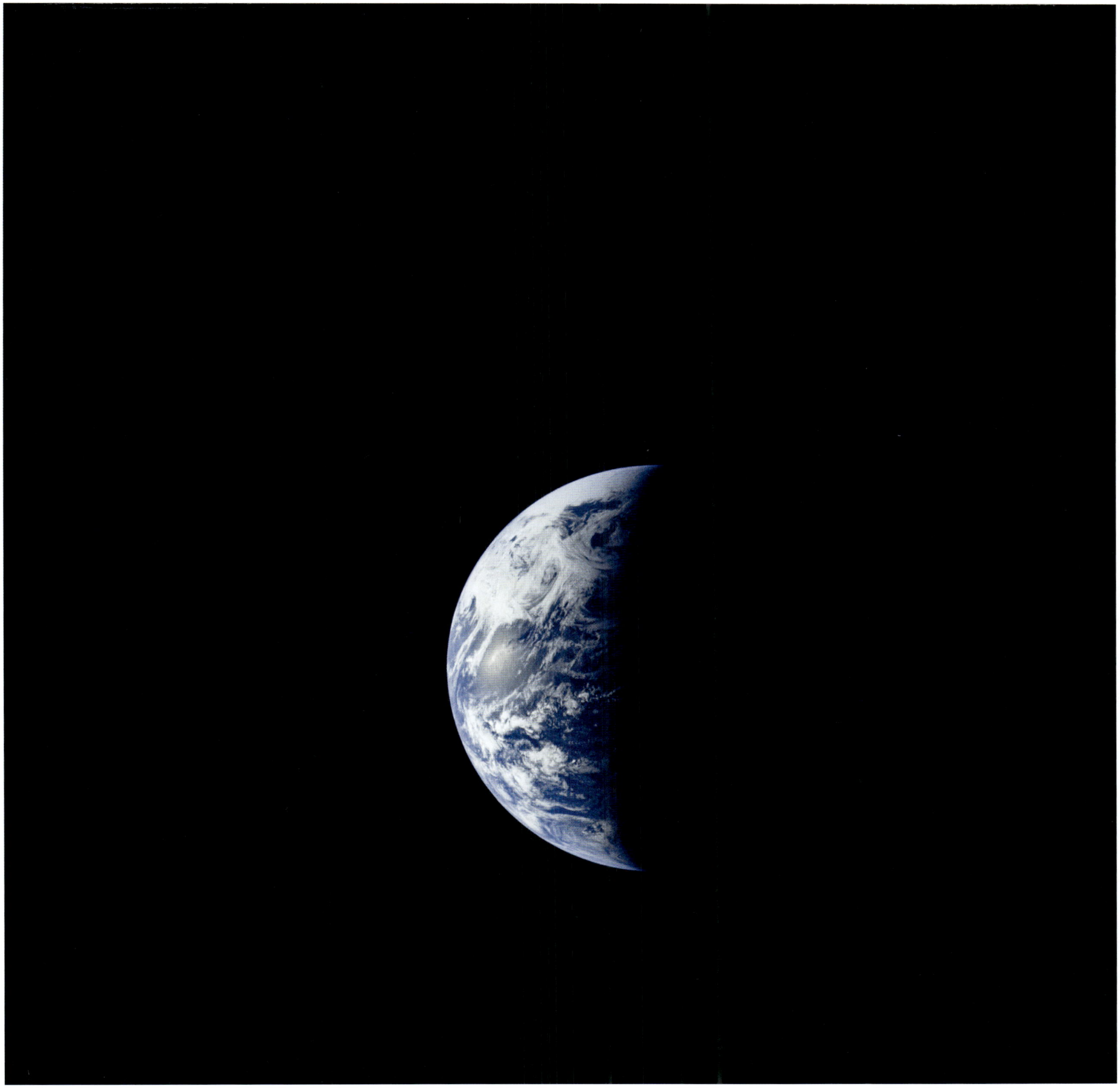

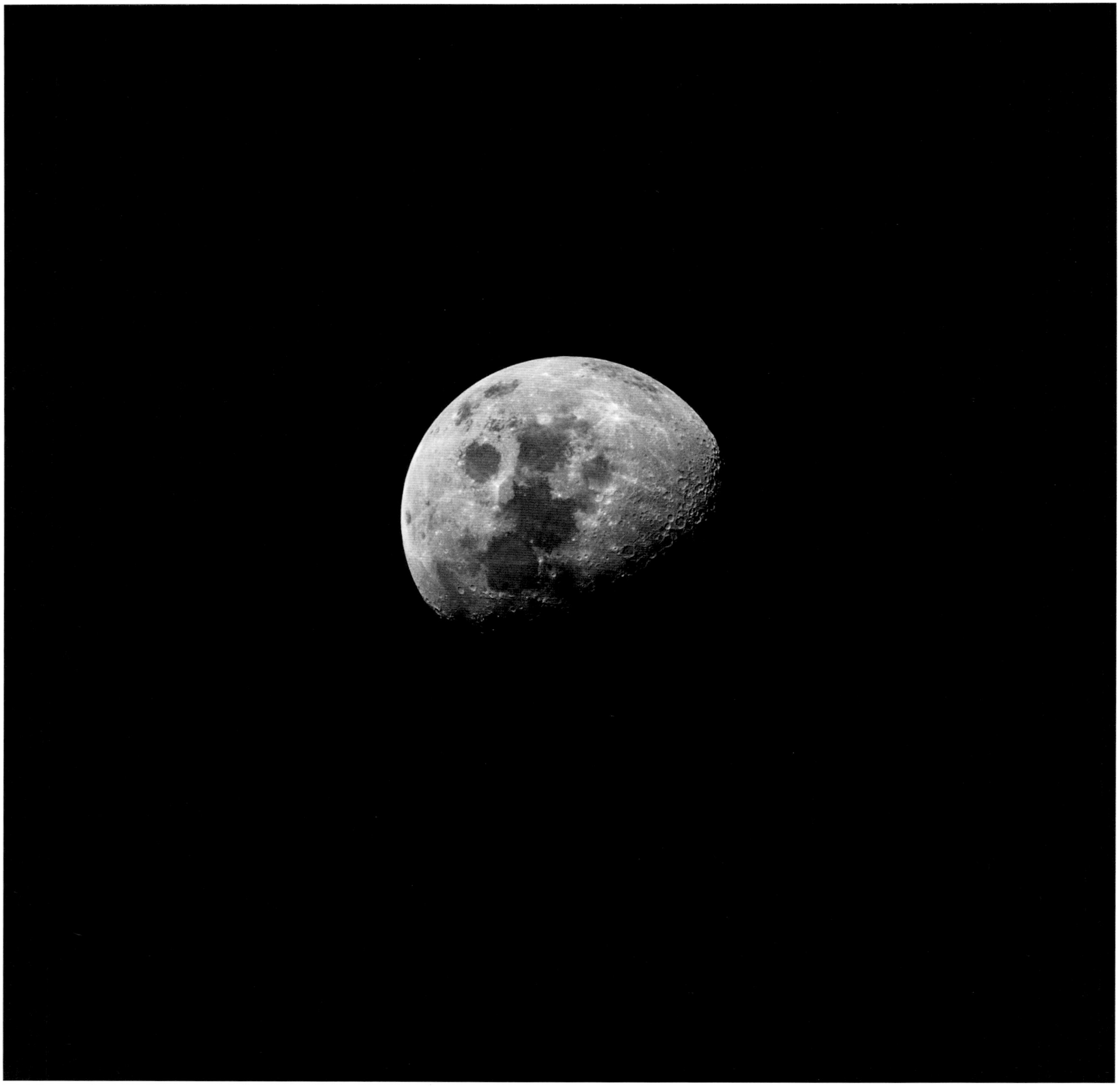

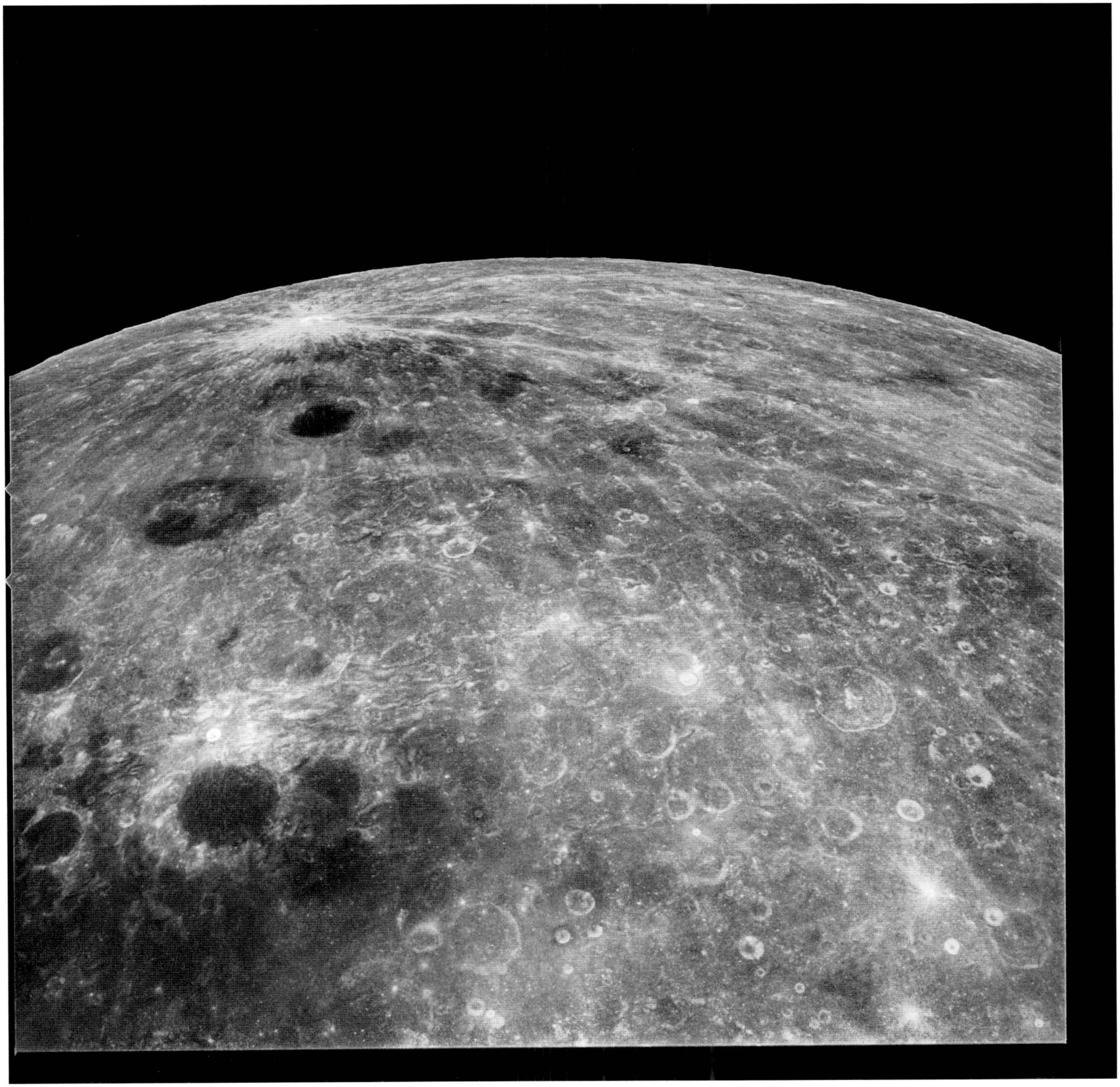

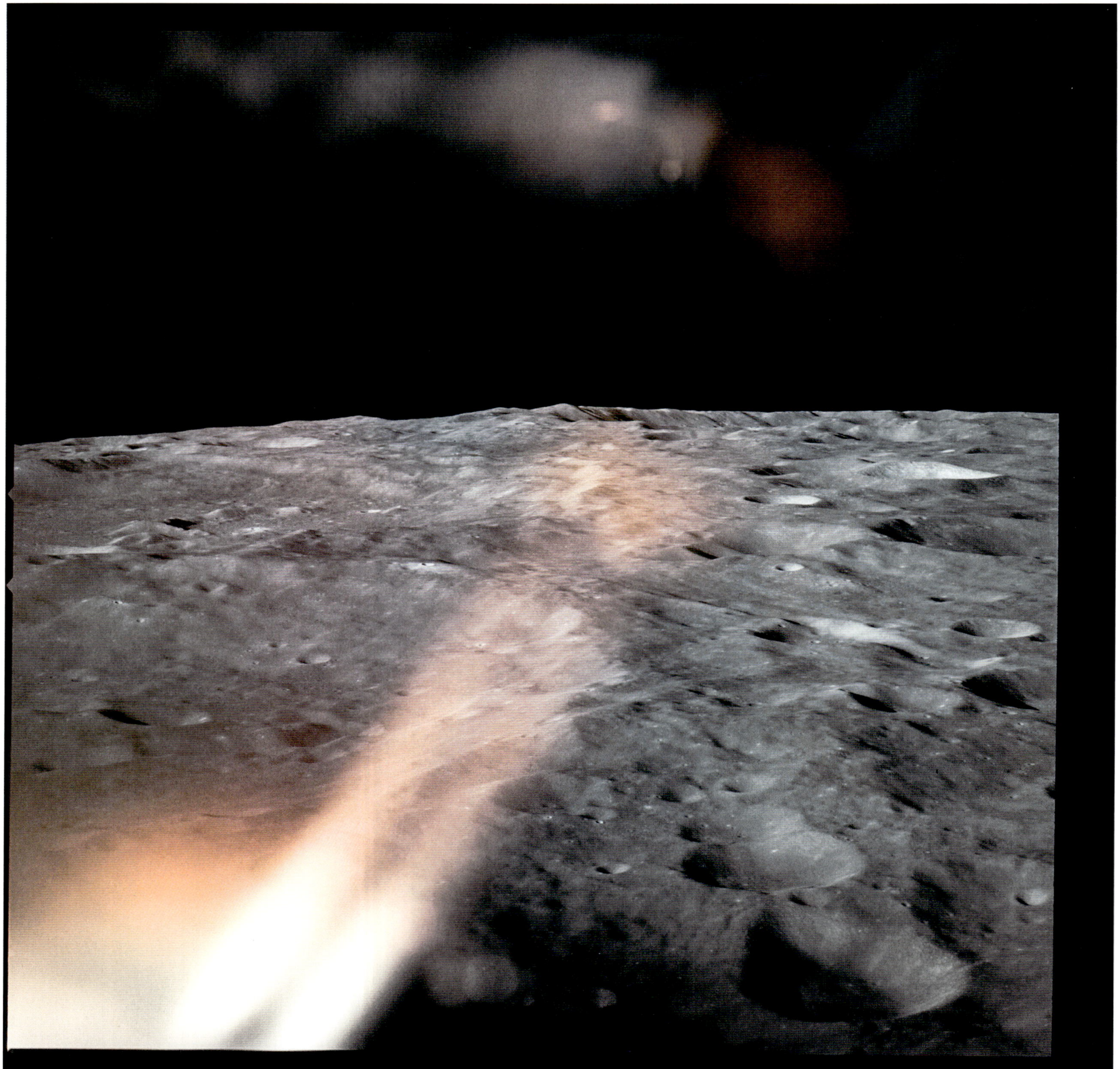

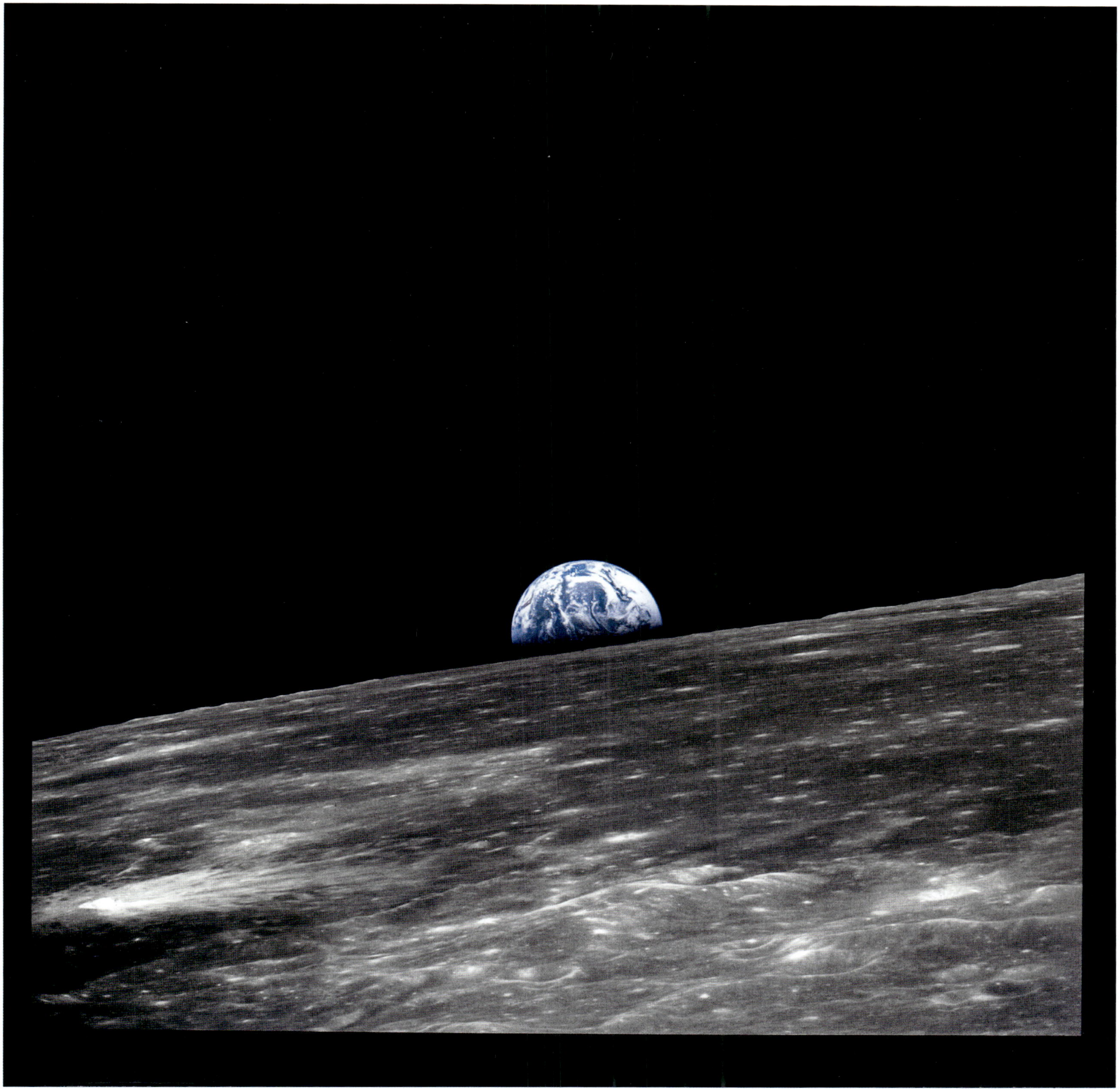

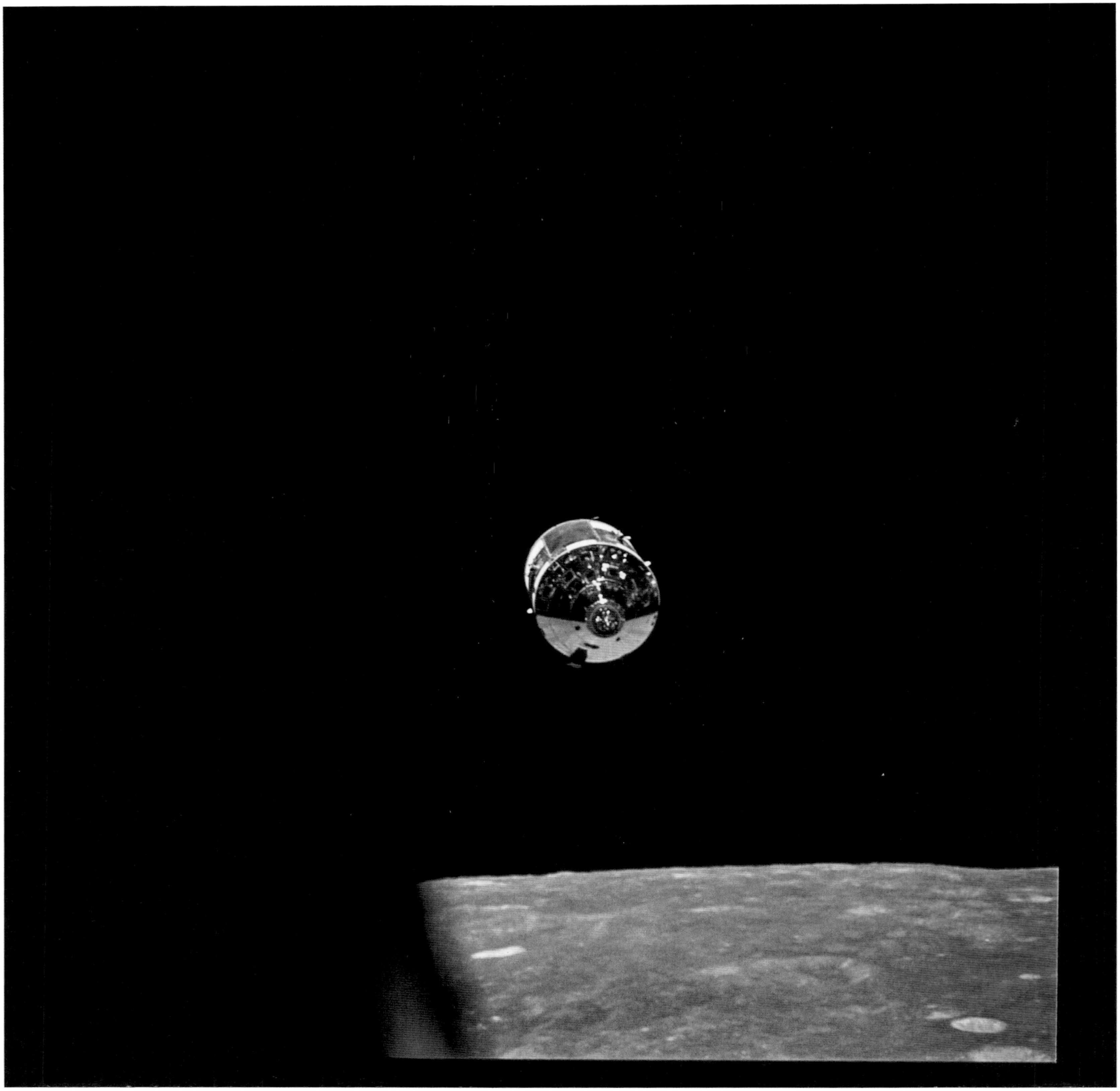

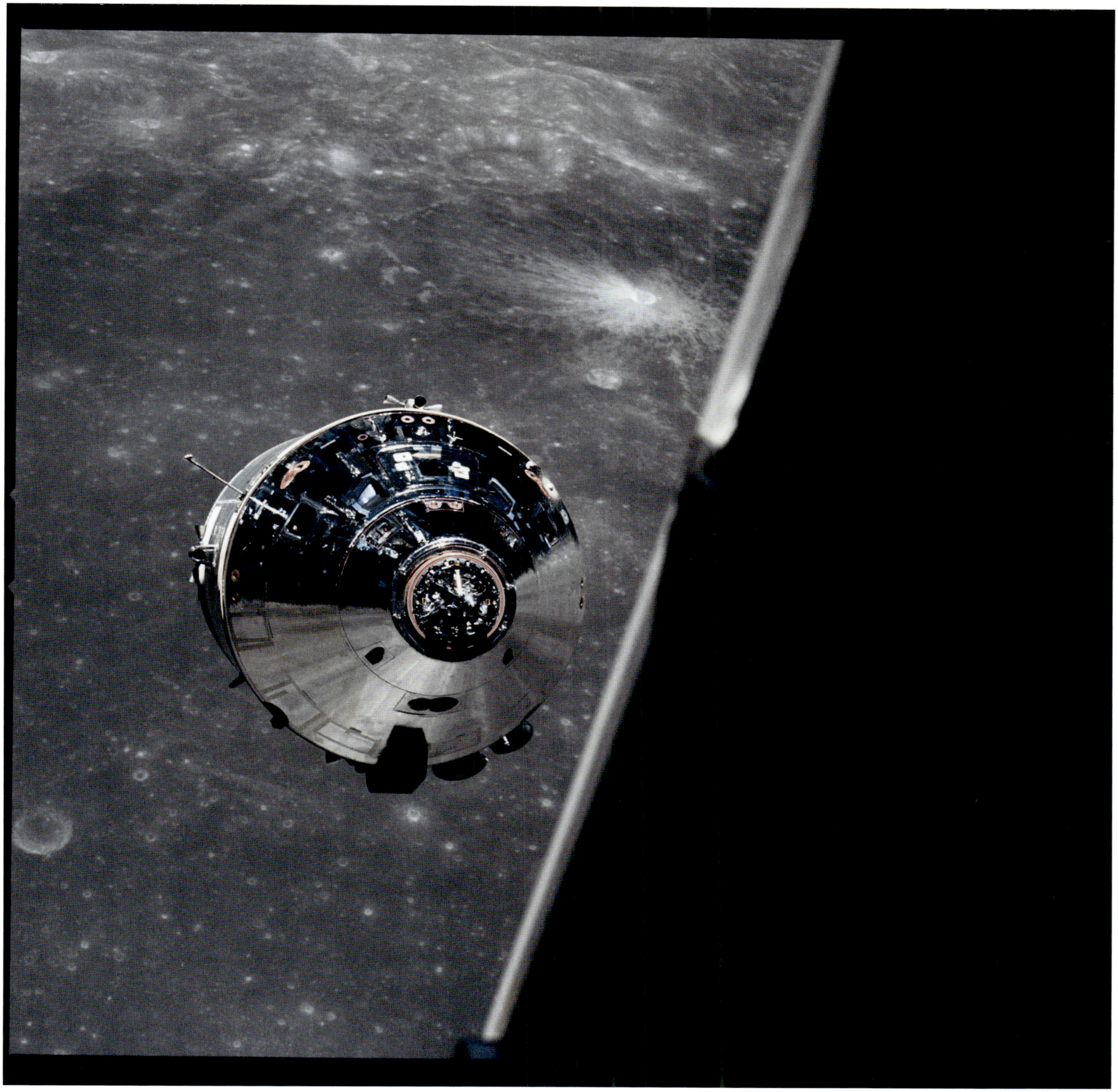

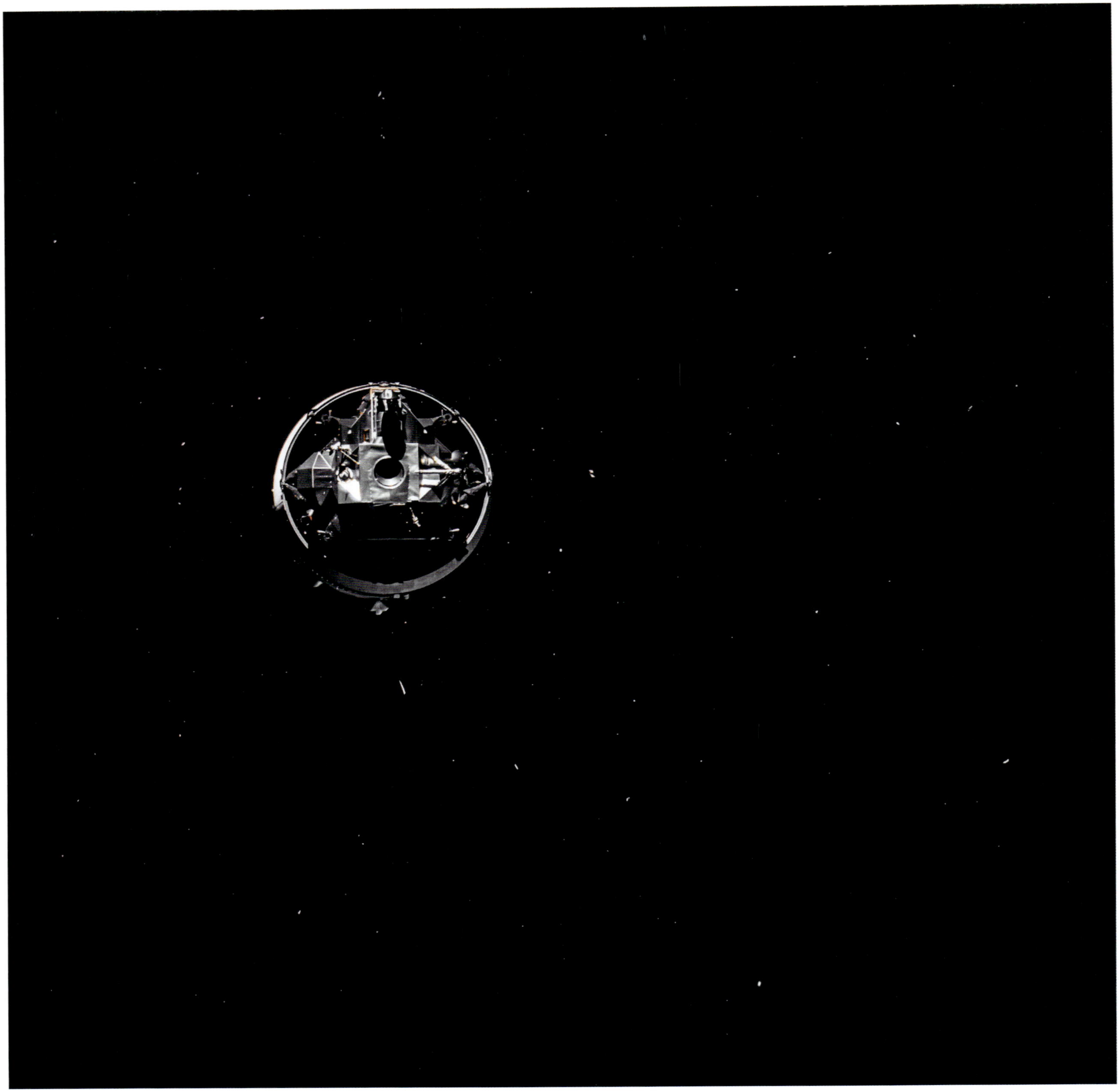

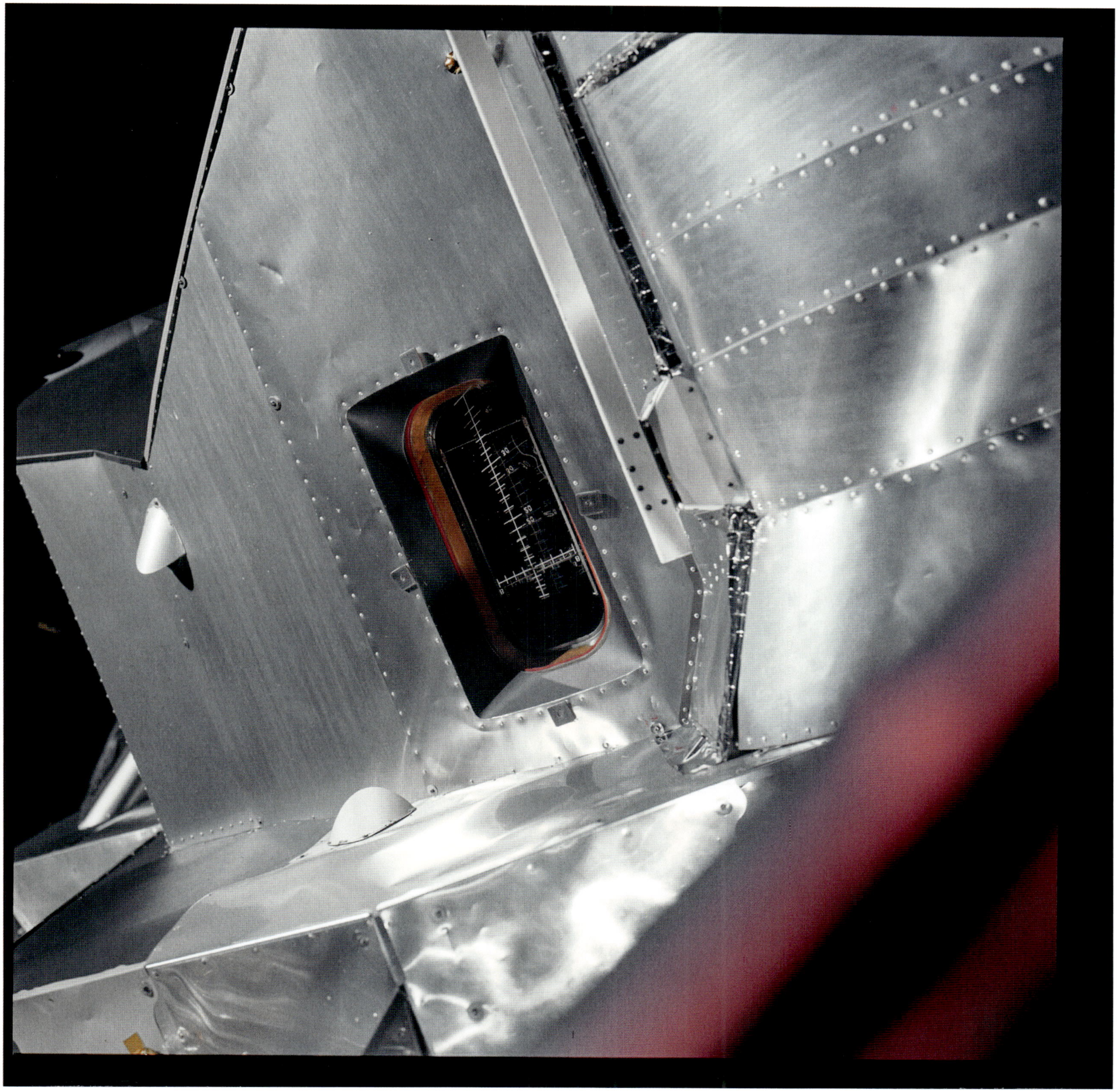

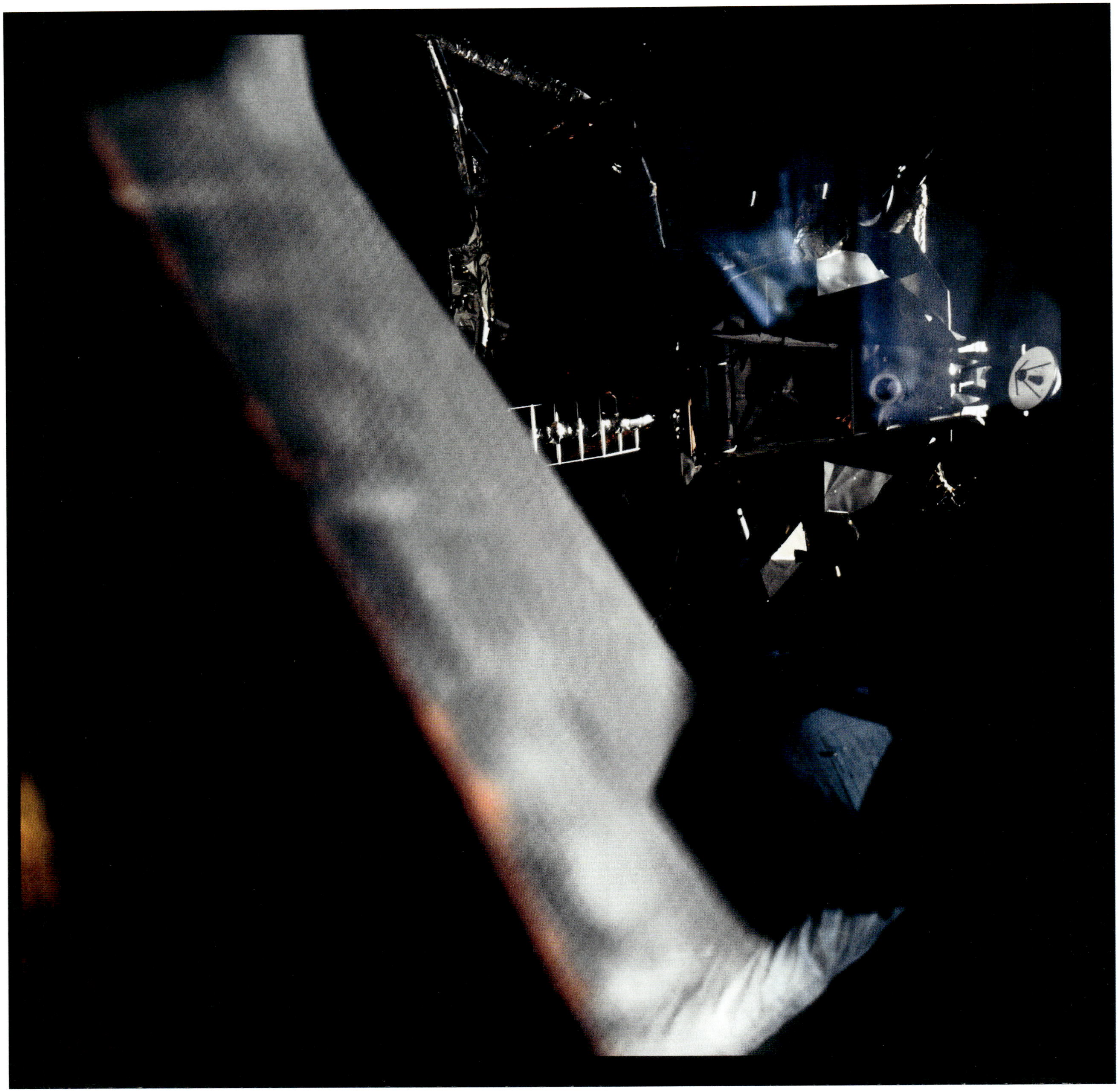

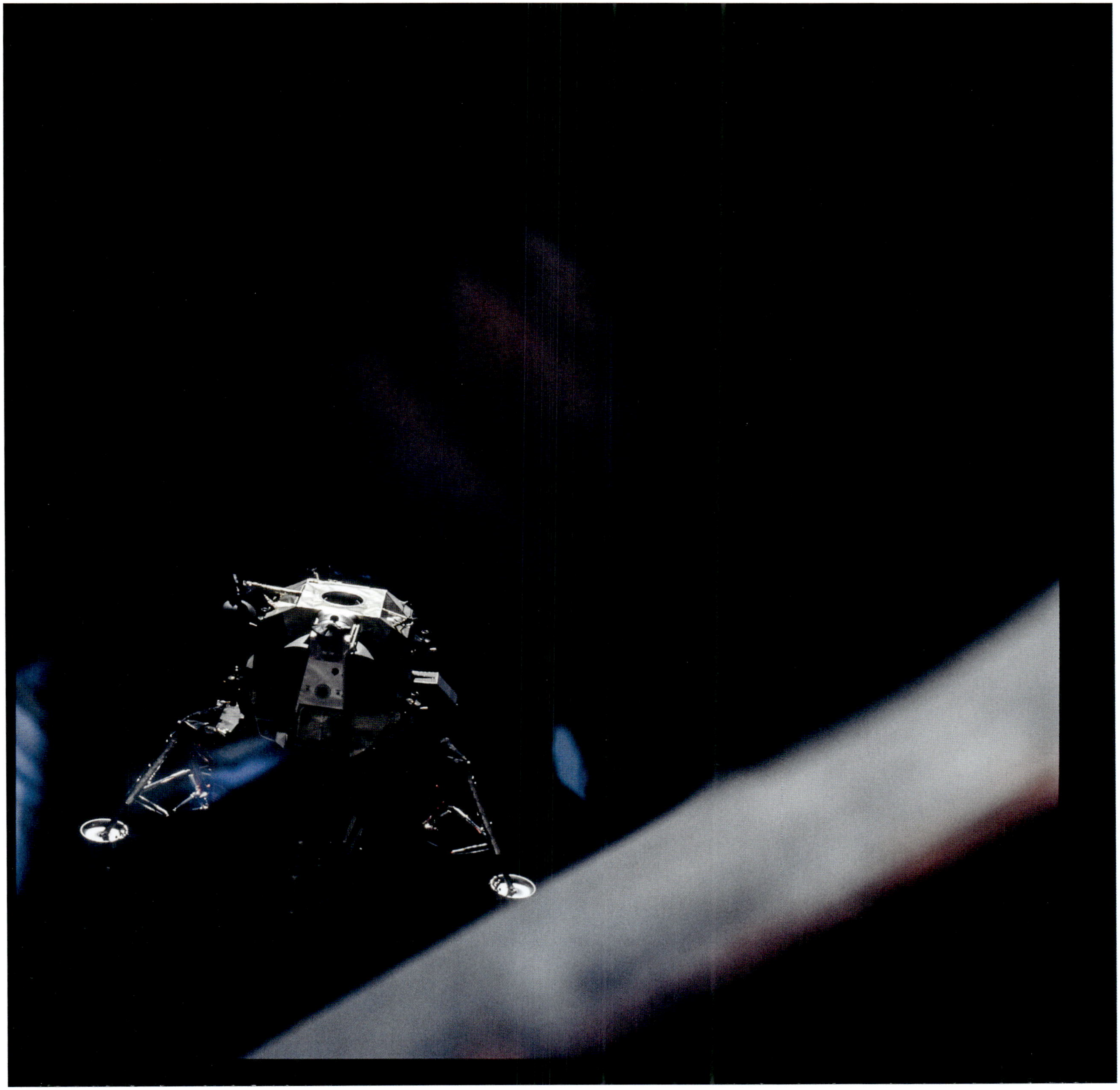

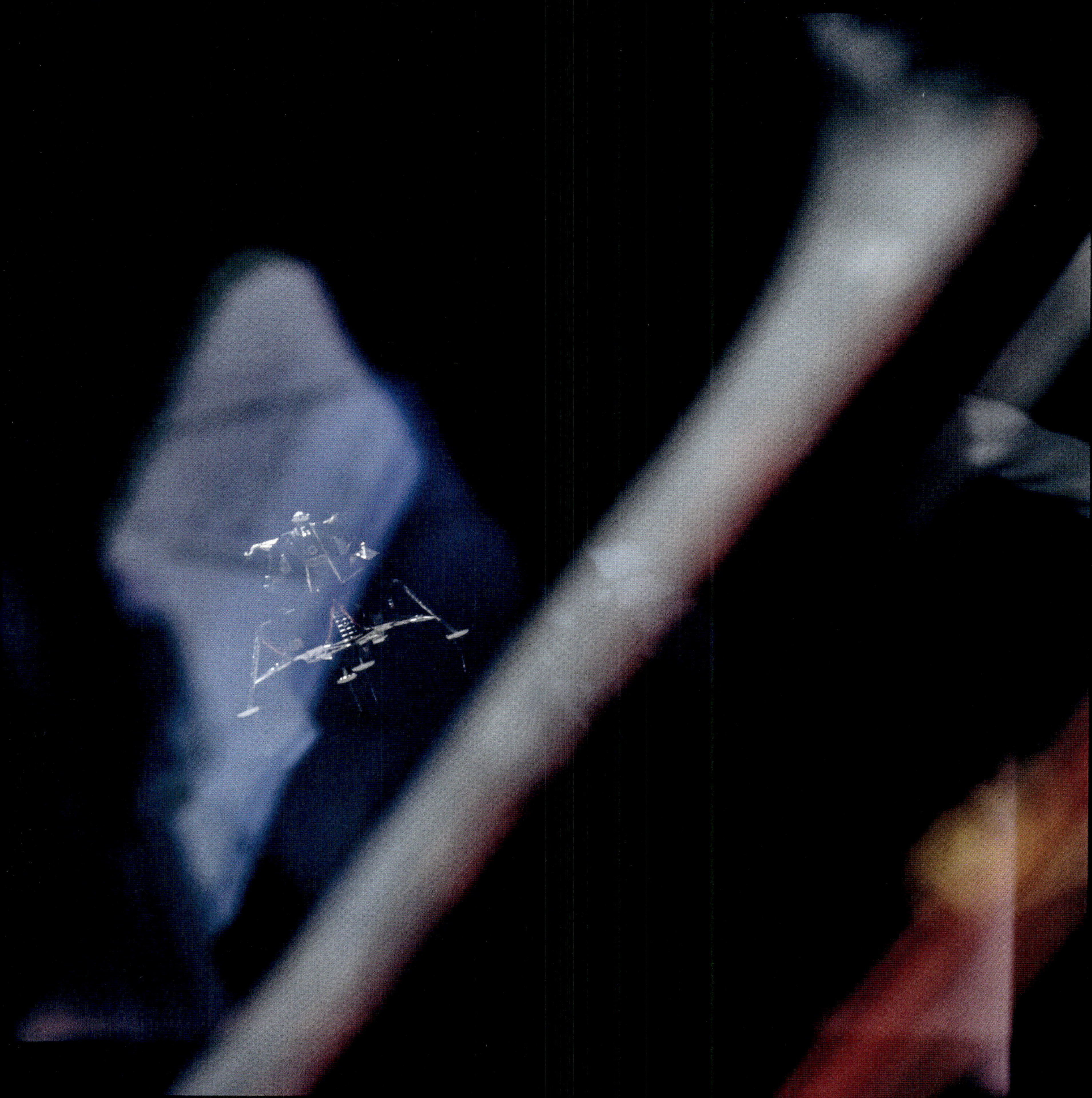

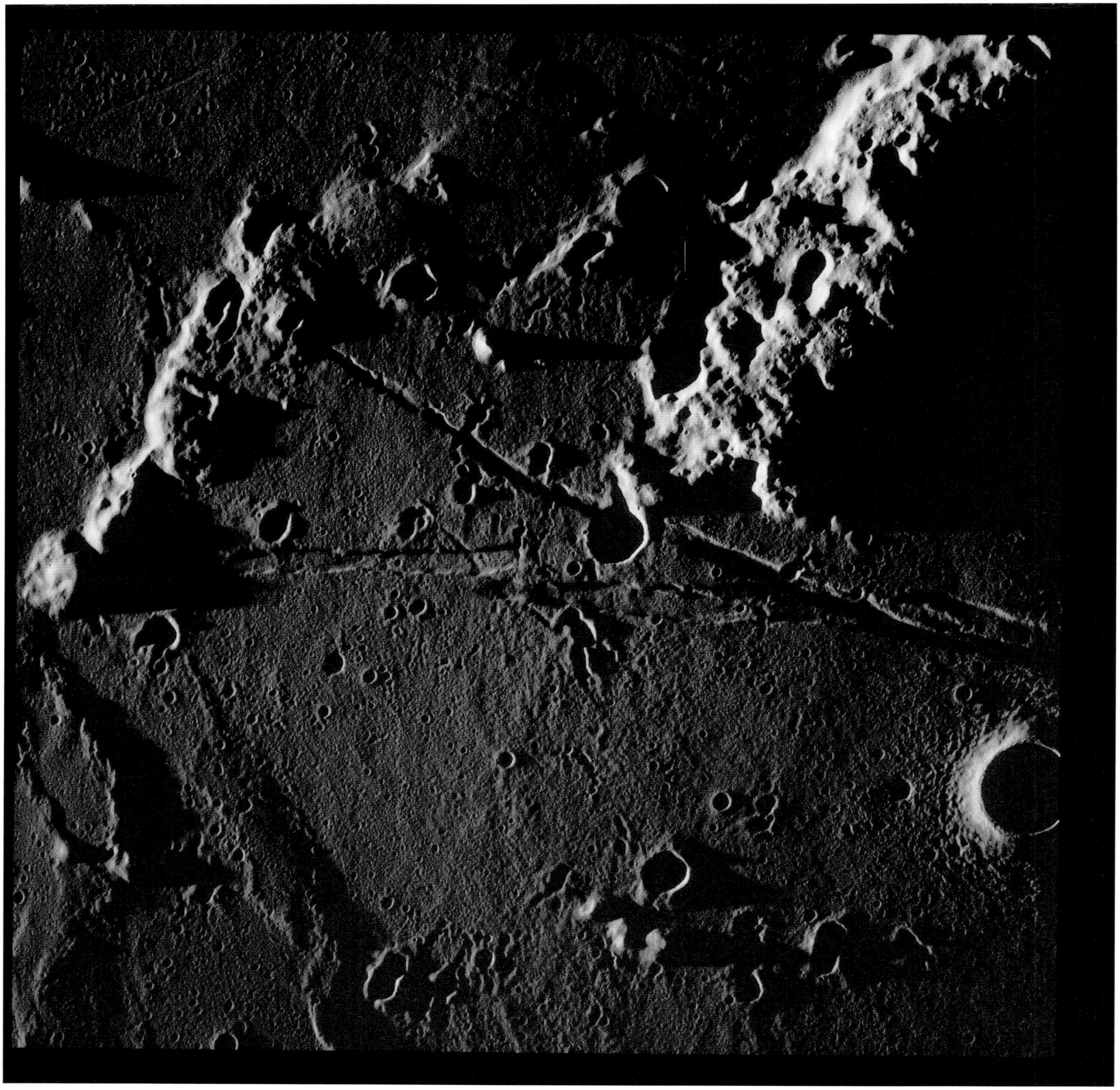

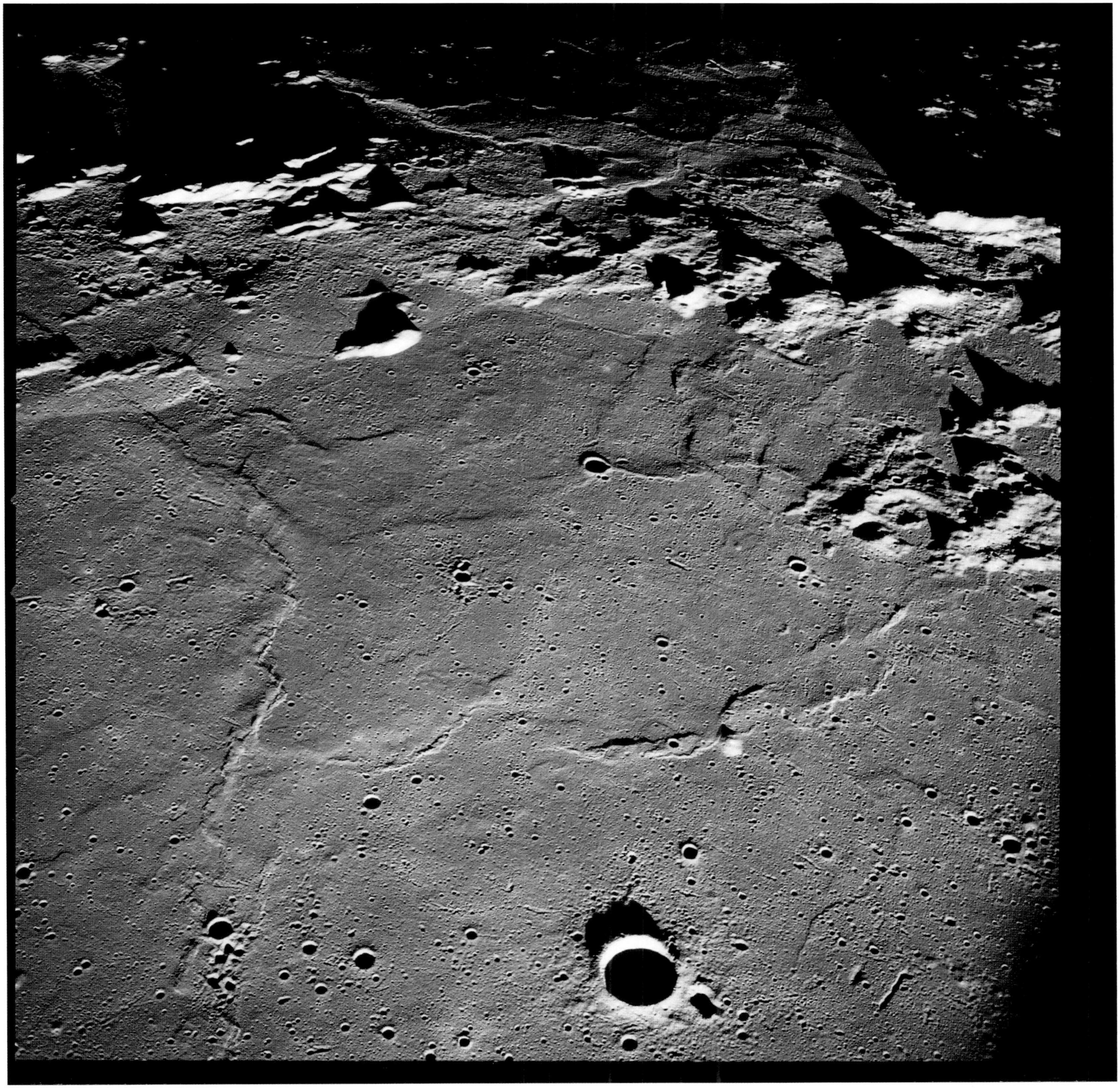

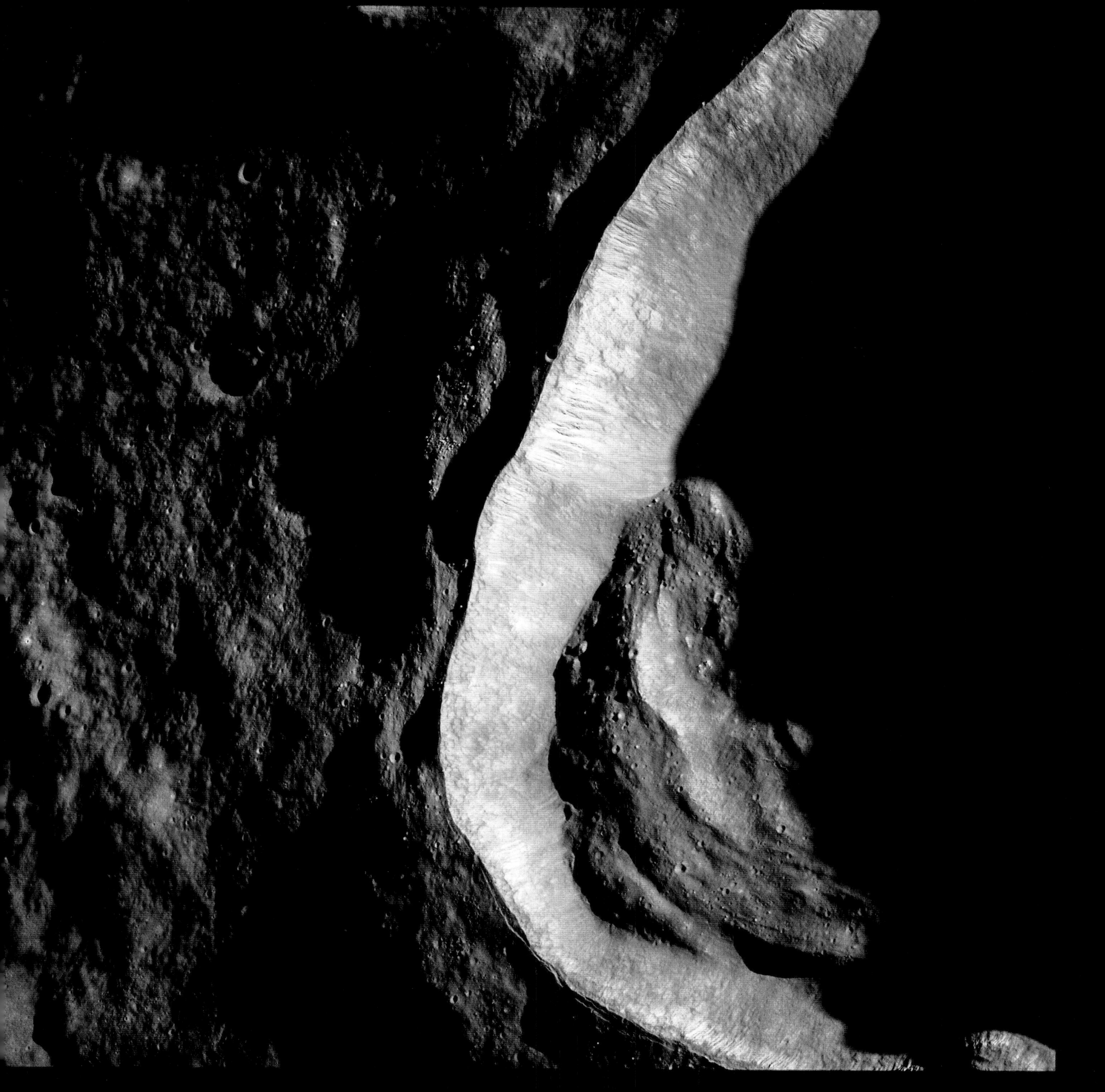

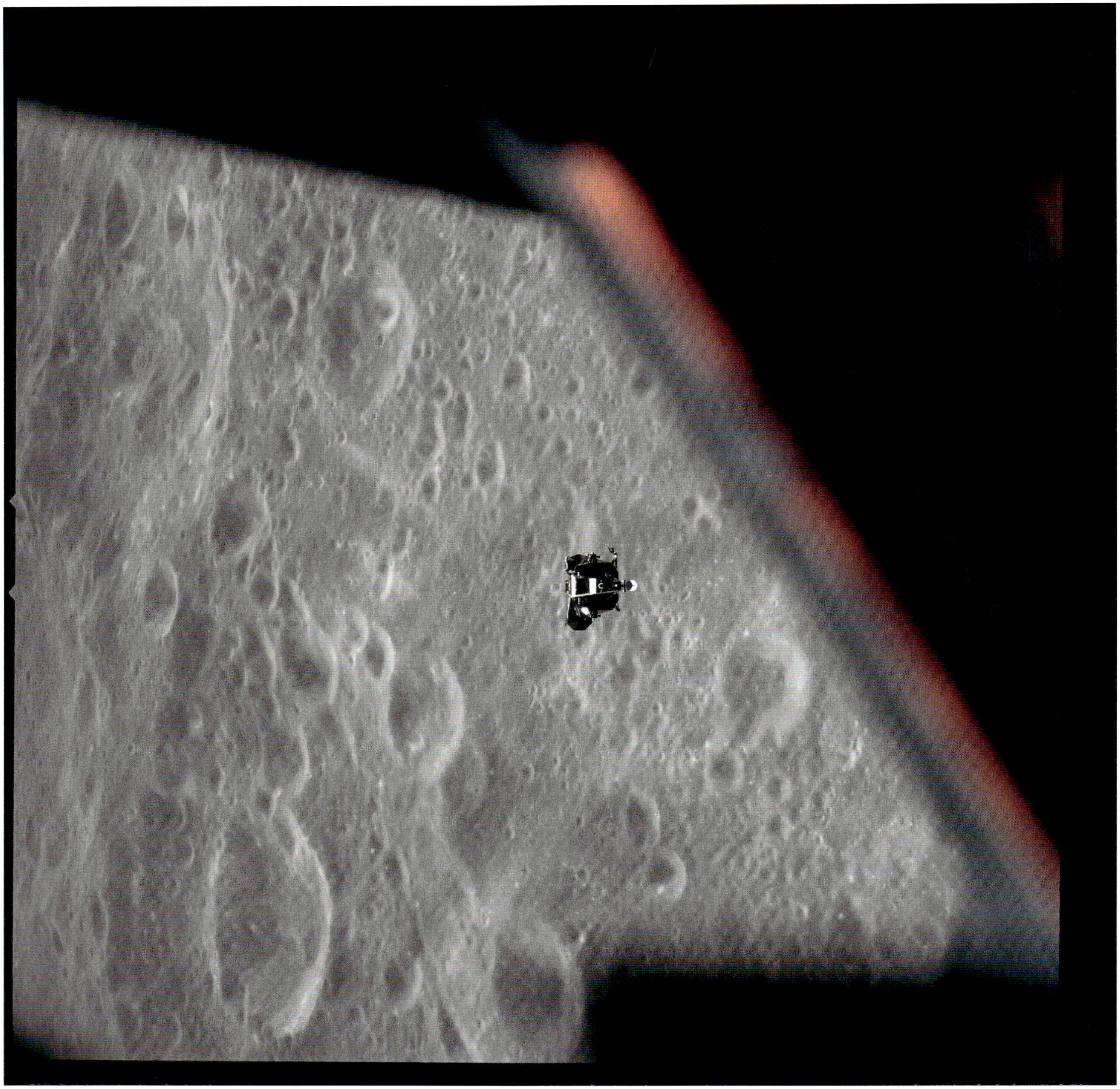

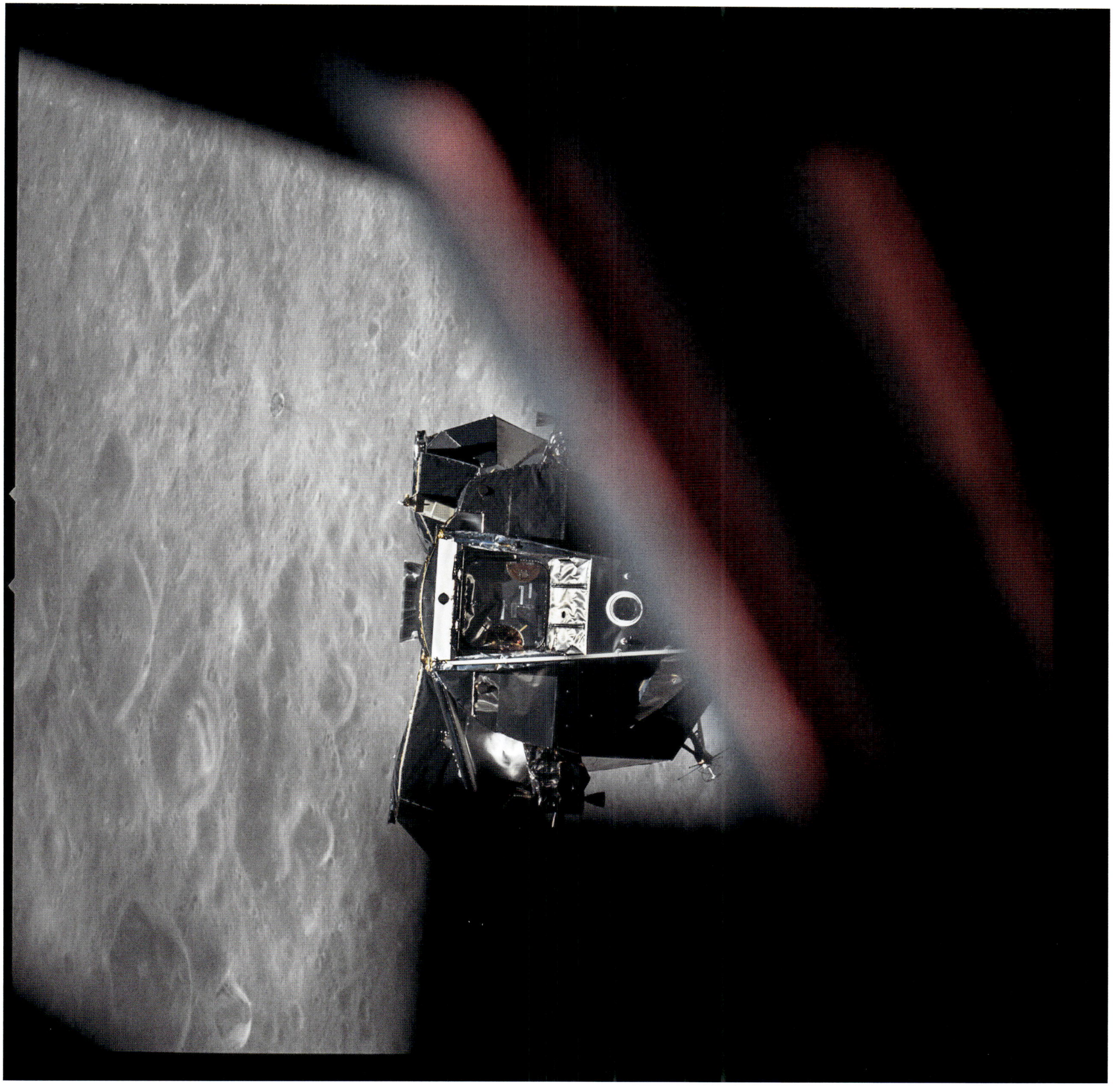

93 (AS10-27-3987)
Suspended in the blackness of space, Earth appears as a delicate, vibrant sphere, with swirling clouds and vast oceans visible. Captured during Apollo 10's return journey from the Moon.

94 (AS10-27-3959)
As Apollo 10 headed home, the crew captured this stunning contrast between the illuminated lunar surface and the vast blackness of space. At this point in the mission, the spacecraft was at one of its slowest speeds, traveling roughly 3,280ft/s (1km/s), before Earth's gravity gradually accelerated it toward reentry.

95 (AS10-27-3916)
An oblique view across the Moon, taken just after Trans-Earth Injection (TEI). The image looks over Goddard, Al-Biruni, Joliot, and Lomonosov Craters, toward Giordano Bruno Crater, one of the youngest large craters on the Moon, estimated to be around four million years old.

96 (AS10-34-5170)
A view of Papaleksi and Spencer Jones Craters on the Moon's far side. Named after renowned astronomers, these impact sites stand against the rugged, ancient terrain of a region never visible from Earth.

97 (AS10-27-3888)
Apollo 10 captured this stunning 'Earthrise' over Mare Smythii, a 232mi (373km)-wide lava-filled basin on the Moon's eastern limb.

98 (AS10-27-3856)
Apollo 10's Command/Service Module (CSM) *Charlie Brown*, captured from the Lunar Module (LM) *Snoopy* during rendezvous in lunar orbit. Below, the bright Glazenap E and F Craters mark the rugged Moon.

99 (AS10-27-3874)
CSM *Charlie Brown*, seen in lunar orbit, captured from LM *Snoopy* during rendezvous operations. This closer view highlights the docking mechanism, including the probe and drogue assembly, which allowed astronauts to transfer between spacecraft.

100 (AS10-34-5011)
LM *Snoopy* is seen inside the S-IVB stage during the Transposition, Docking & Extraction (TD&E) maneuver. This critical step required the CSM to turn, dock with the LM, and pull it free before heading to the Moon.

101 (AS10-34-5066)
A close-up of LM *Snoopy*'s rendezvous window, with reticles on the glass to aid in visual alignment during docking with CSM *Charlie Brown*.

102 (AS10-34-5086)
LM *Snoopy*, captured after undocking from CSM *Charlie Brown* during Apollo 10's lunar landing rehearsal.

103 (AS10-34-5092)
LM *Snoopy* drifts in lunar orbit after undocking, undergoing a visual inspection before descent. Though it never landed, Apollo 10 brought it within 9.7mi (12km) of the surface, performing a final lunar descent test before Apollo 11.

105 (AS10-34-5082)
Bathed in sunlight, LM *Snoopy* glows in space, its golden foil shimmering against the void. This surreal view, captured after undocking, highlights solar flares refracting across the frame, giving the spacecraft an ethereal presence.

106 (AS10-34-5073)
Moltke Crater, Moltke B, and Rima Hypatia, located near Mare Tranquillitatis. Moltke's bright rays indicate it is a relatively young impact crater, while Rima Hypatia's winding trench suggests ancient tectonic or volcanic activity. This site lies just 118mi (190km) from the Apollo 11 landing area.

107 (AS10-27-3907)
Sinus Medii, home to the Surveyor 4 and 6 landing sites. While Surveyor 4 lost contact before touchdown, Surveyor 6 successfully landed and performed a small 'hop,' becoming the first spacecraft to lift off from the Moon, a crucial test for future Apollo landings.

109 (AS10-34-5165)
Apollo 10 captured Goddard Crater, a 55mi (89km)-wide impact site in the central lunar highlands. Its sharp rim and central peak mark it as a well-preserved example of a lunar impact crater. Named after rocketry pioneer Robert H. Goddard, this crater provides insight into the Moon's violent impact history.

110 (AS10-34-5108)
LM *Snoopy*'s ascent stage, seen in the distance, making its way back to CSM *Charlie Brown* after a crucial lunar orbit test.

111 (AS10-34-5115)
LM *Snoopy*'s ascent stage moves closer to docking with CSM *Charlie Brown* after its simulated lunar ascent. The next time astronauts performed this maneuver, it would be Neil A. Armstrong and Buzz Aldrin reuniting with the CSM after walking on the Moon.

Apollo 11

XI

Neil A. Armstrong

Michael Collins

Edwin 'Buzz' Aldrin

Apollo 11

Launch: July 16th, 1969
Lunar landing: July 20th, 1969
Splashdown: July 24th, 1969

Command Module: Columbia
Lunar Module: Eagle

Apollo 11 is surely the most famous mission in the history of all space exploration. It accomplished the 1961 goal proposed by U.S. President John F. Kennedy: "...before this decade is out, of landing a man on the Moon and returning him safely to the Earth."

After a 102-hour journey from Cape Kennedy in Florida, Armstrong and Aldrin landed the Lunar Module *Eagle* in the Sea of Tranquility. Armstrong emerged from the Lunar Module first, becoming the first human to set foot on the Moon. As he descended the ladder he released the Modularized Equipment Stowage Assembly to deploy the camera that recorded his 'giant leap'.

Armstrong and Aldrin spent about 2.5 hours on the Lunar surface collecting samples, performing experiments and taking hundreds of photographs. The main photography objectives of the mission were to capture scientifically interesting sites and potential future Apollo landing sites from orbit and record the Lunar Module as well as the history-making activities on the lunar surface.

Many of the photographs were taken for future training and engineering purposes, among which were pictures to help analyze the Lunar Module in its post-landing condition.

In the short period spent on the surface of the Moon, many historic photographs were taken, such as Armstrong's portrait of Aldrin in which he captured himself in the reflection of Aldrin's visor.

A close-up view of one of Buzz Aldrin's bootprints, taken for an experiment to study the behavior of Moon dust and the effects when pressure was applied to the surface, would turn out to be one of the most iconic photographs in history.

Before taking off, the crew left behind a commemorative stainless steel plaque which read: "Here men from the planet Earth first set foot upon the moon, July 1969 A.D. We came in peace for all mankind." This plaque can be seen on page 139.

Neil A. Armstrong
Commander

Born on August 5th, 1930 in Wapakoneta, Ohio. He was Backup Command Pilot of Gemini 5, Command Pilot of Gemini 8, Backup Command Pilot of Gemini 11 and Backup Commander of Apollo 8. As Commander of Apollo 11 he became the first man to walk on the Moon.

Michael Collins
Command Module Pilot

Born on October 31st, 1930 in Rome, Italy. He was one of the third group of astronauts named in October 1963. He was Backup Pilot of Gemini 7 and Pilot of Gemini 10. He was assigned to the Apollo 8 crew but was removed to undergo surgery. In January 1970, he resigned from NASA and the Air Force.

Edwin 'Buzz' Aldrin
Lunar Module Pilot

Born on January 20th, 1930 in Montclair, New Jersey. He was Backup Pilot of Gemini 9A, Pilot of Gemini 12 and Backup Command Module Pilot of Apollo 8. As the Lunar Module Pilot of Apollo 11 he became the second man to walk on the Moon.

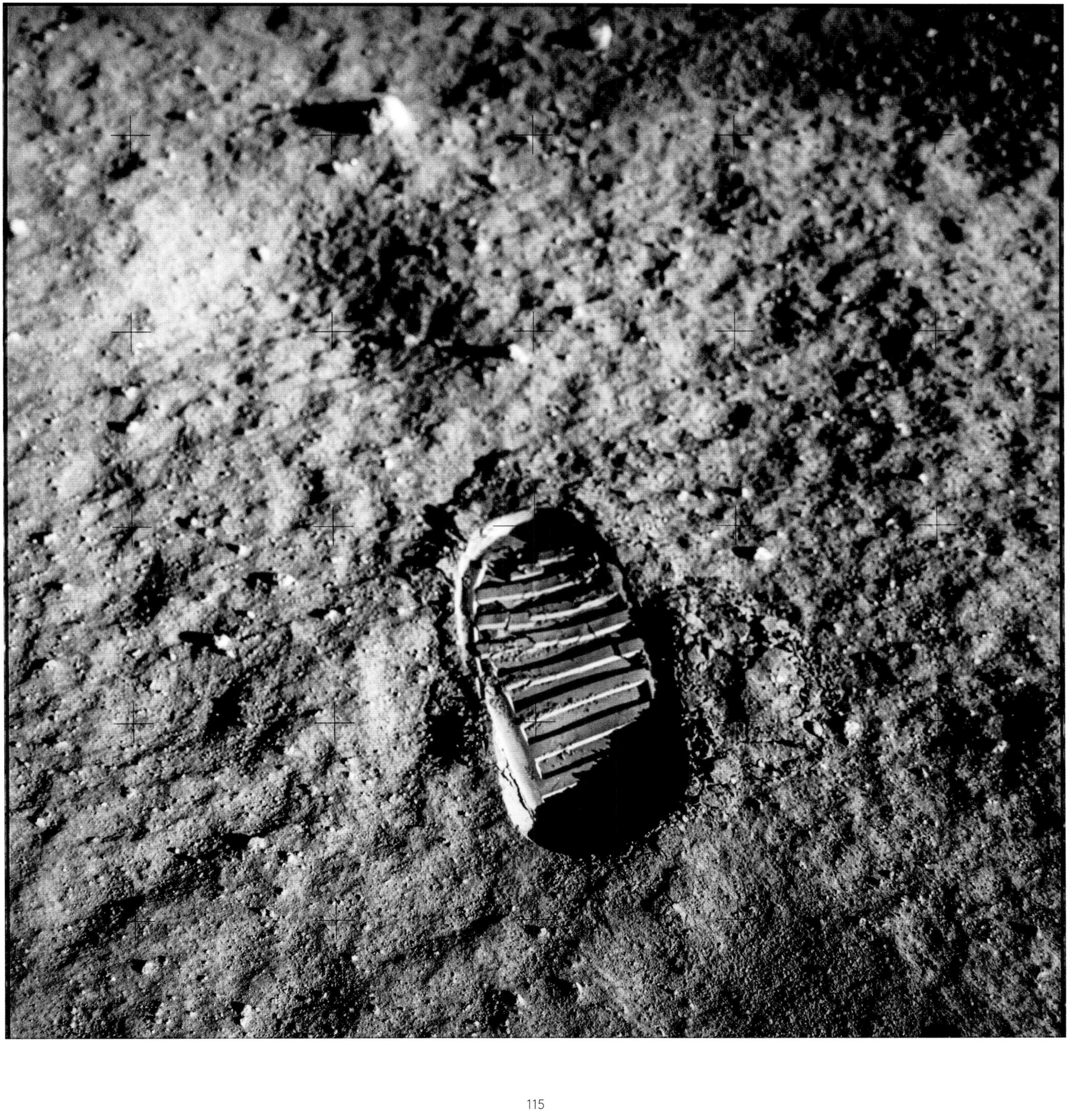

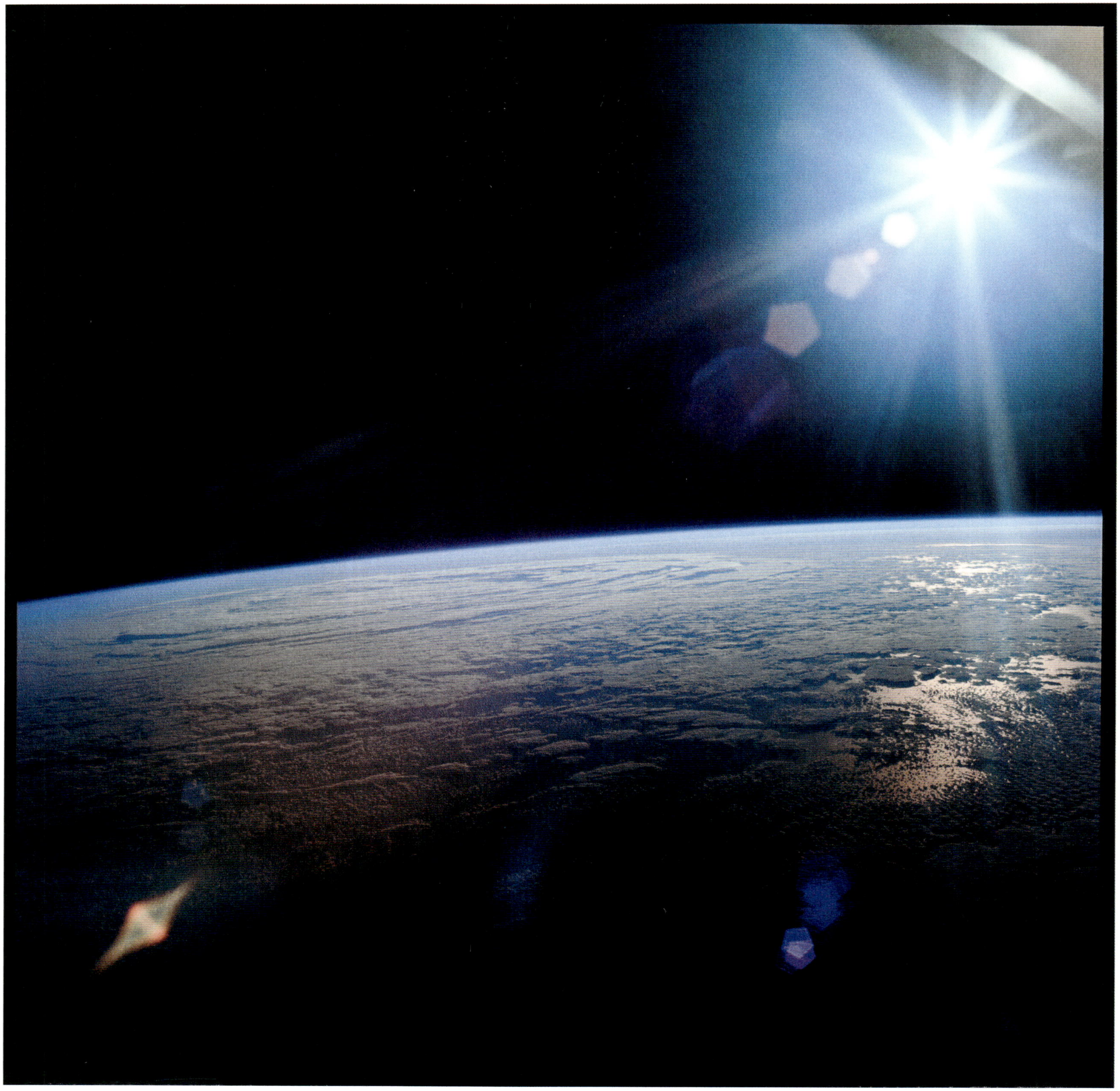

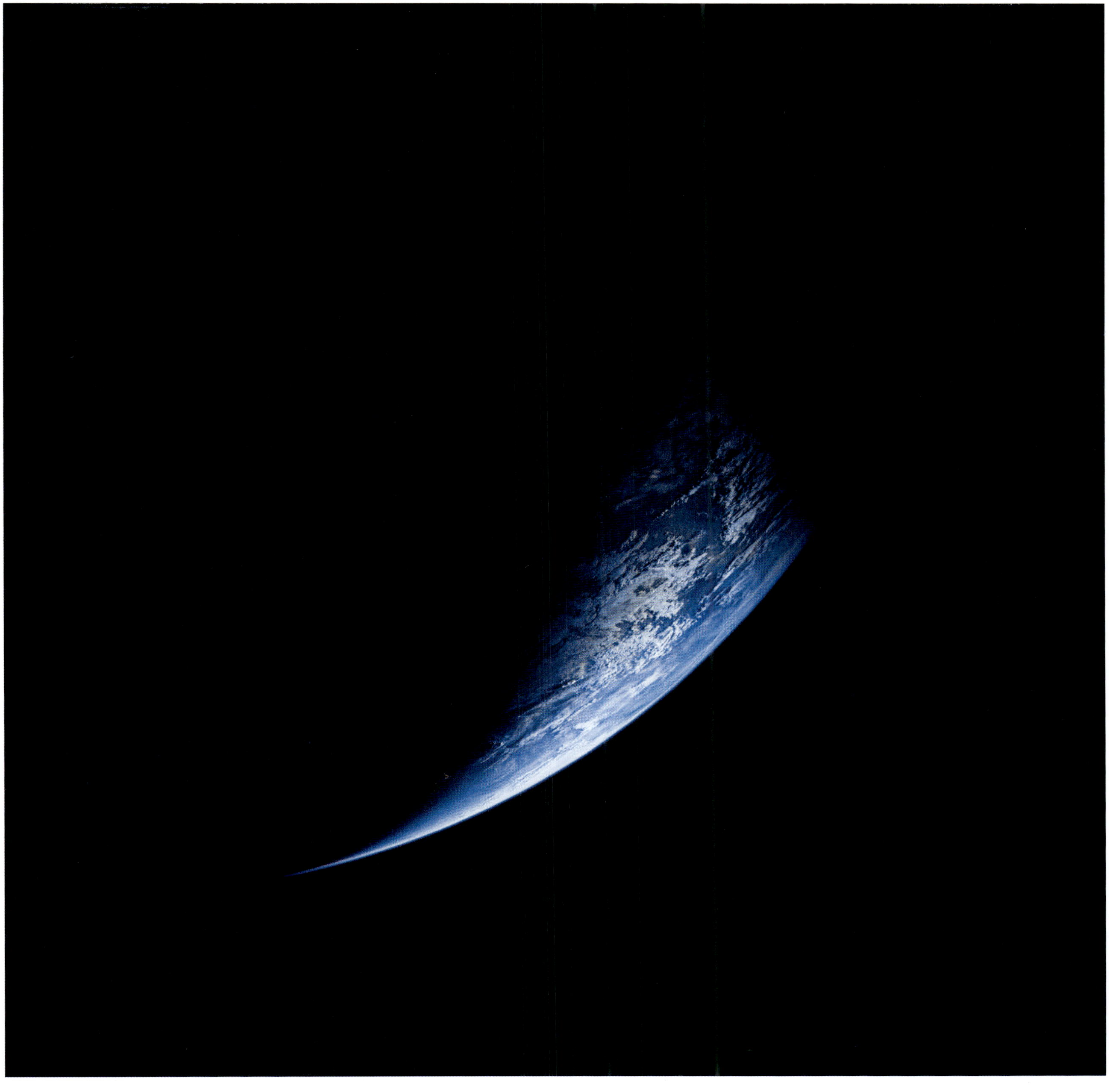

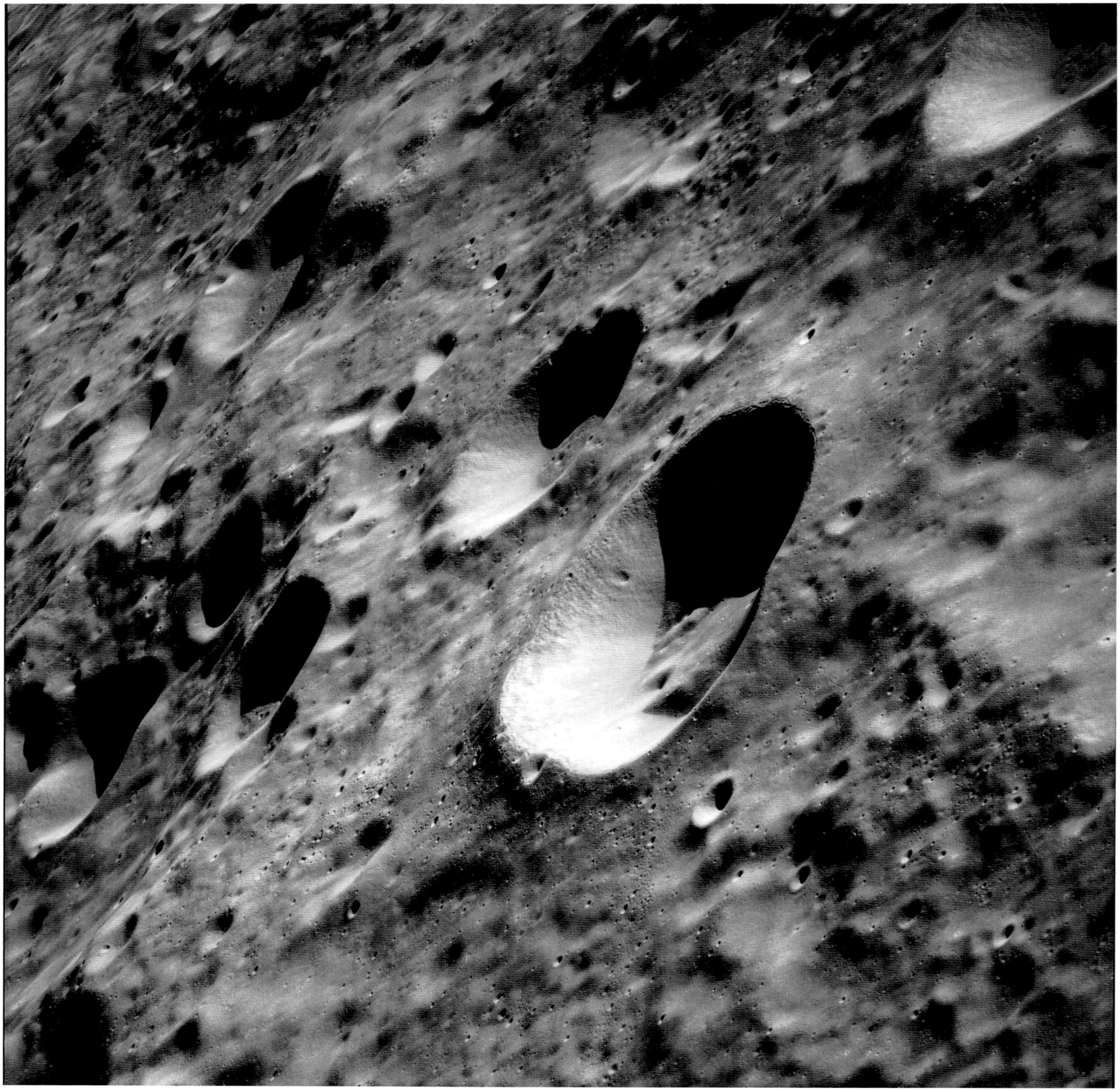

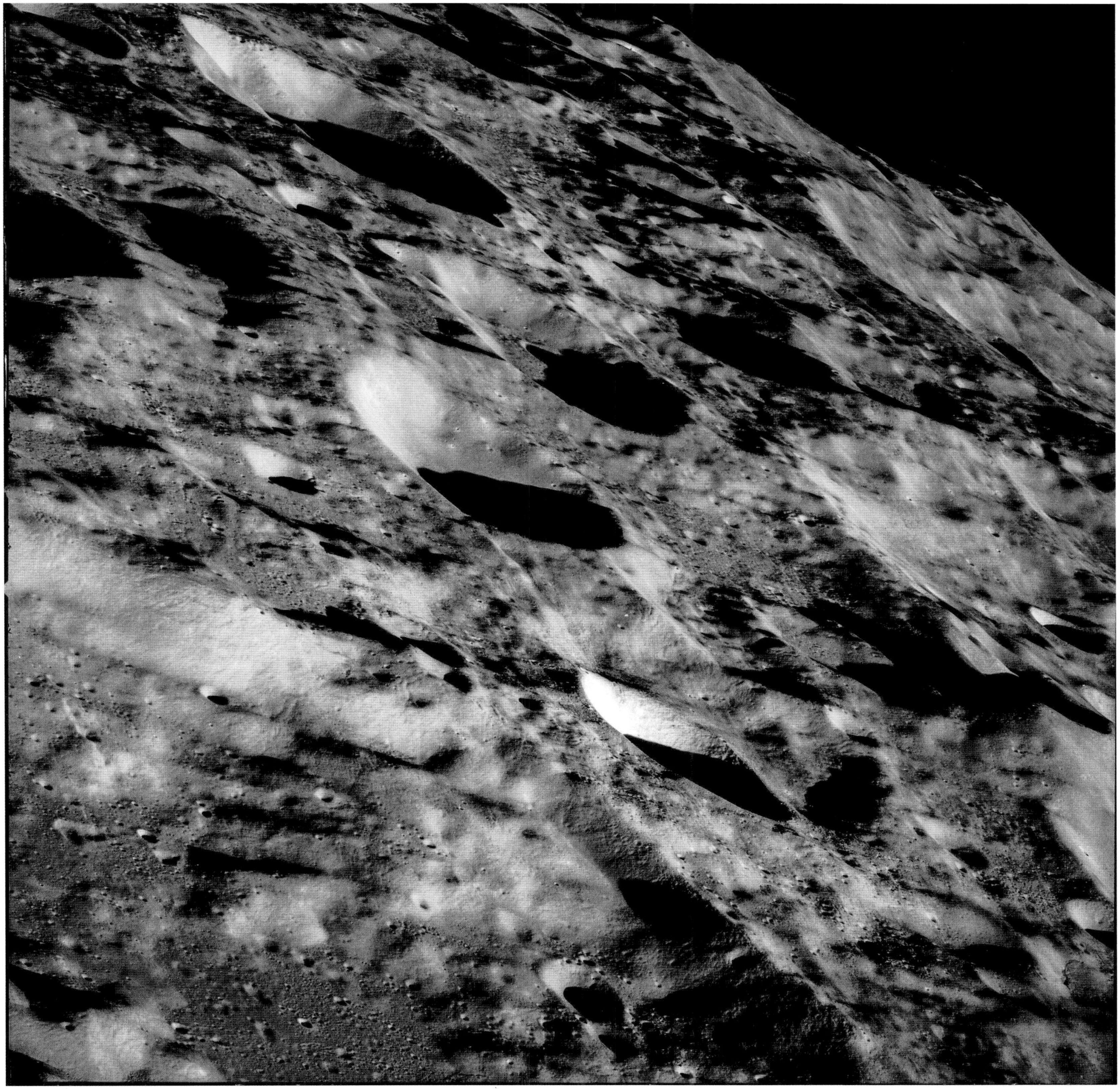

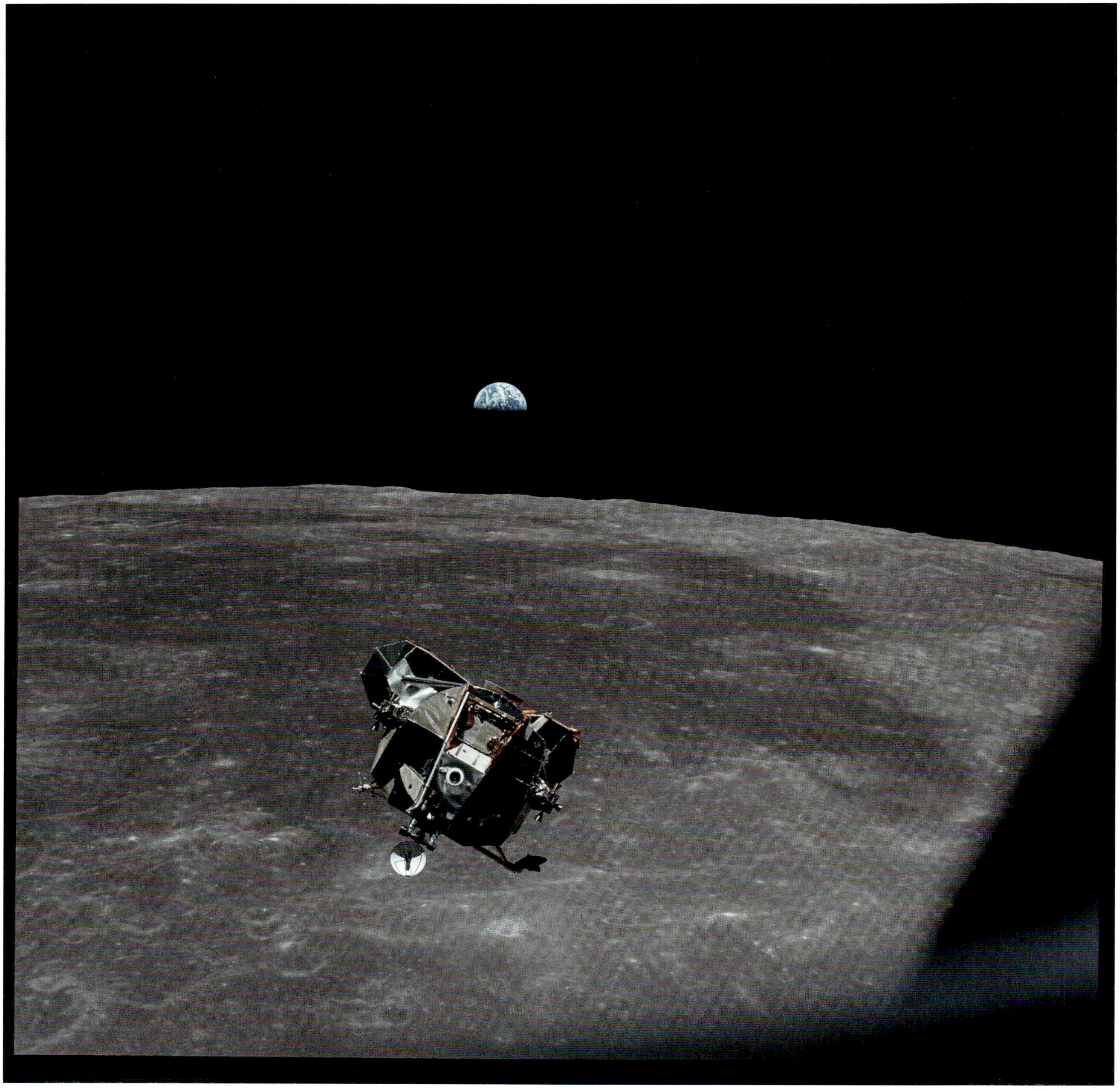

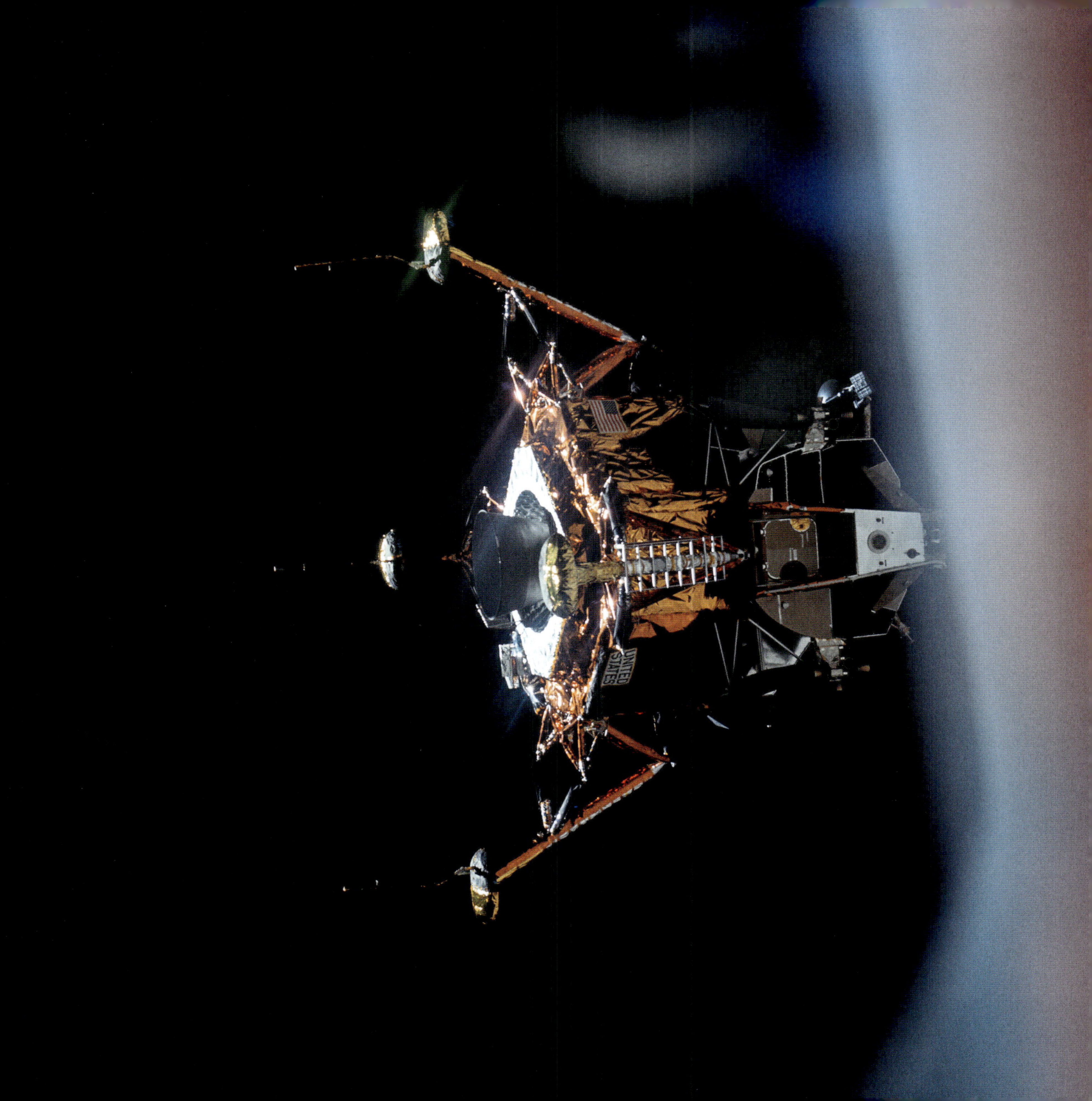
UNITED
STATES

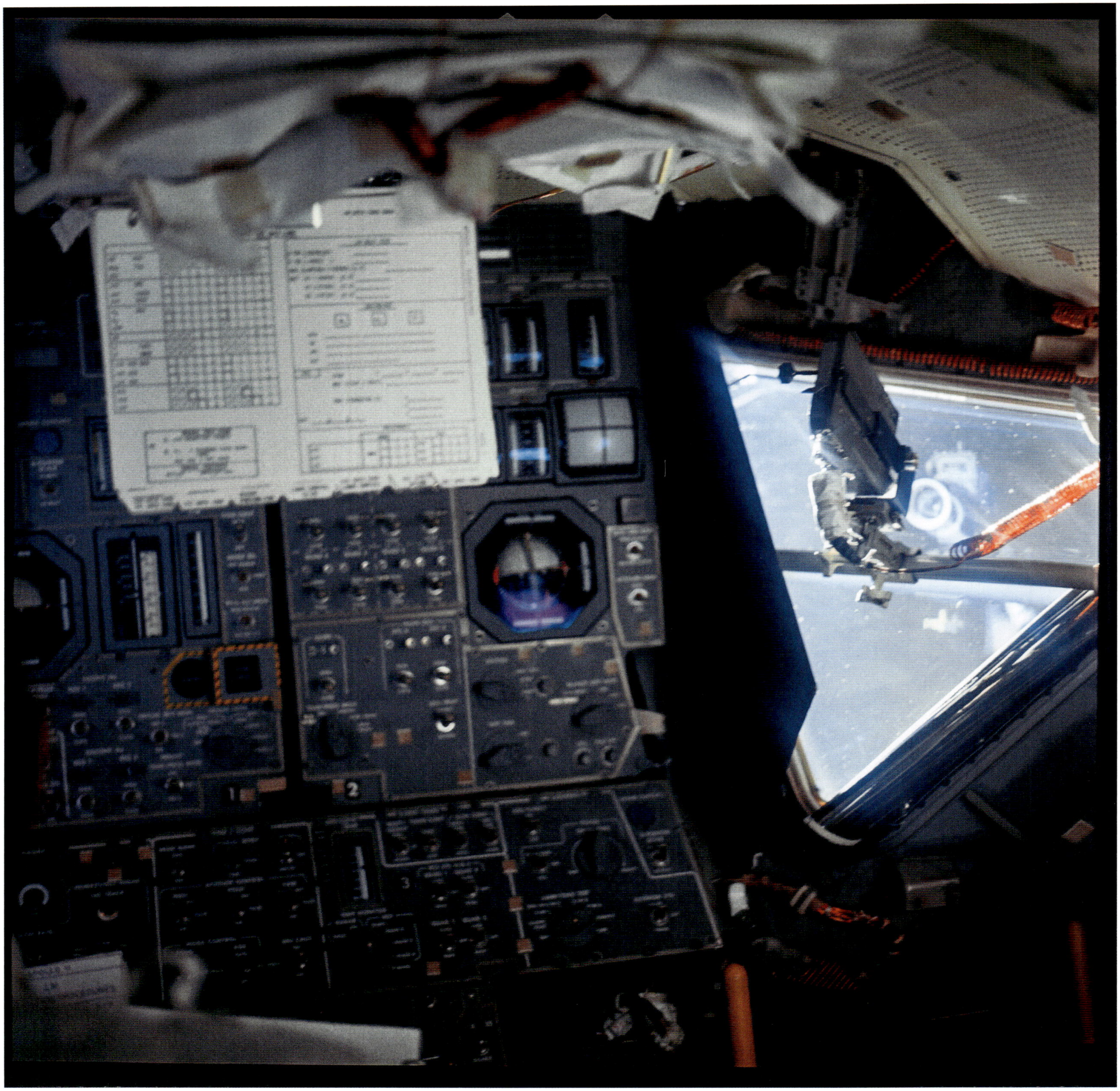

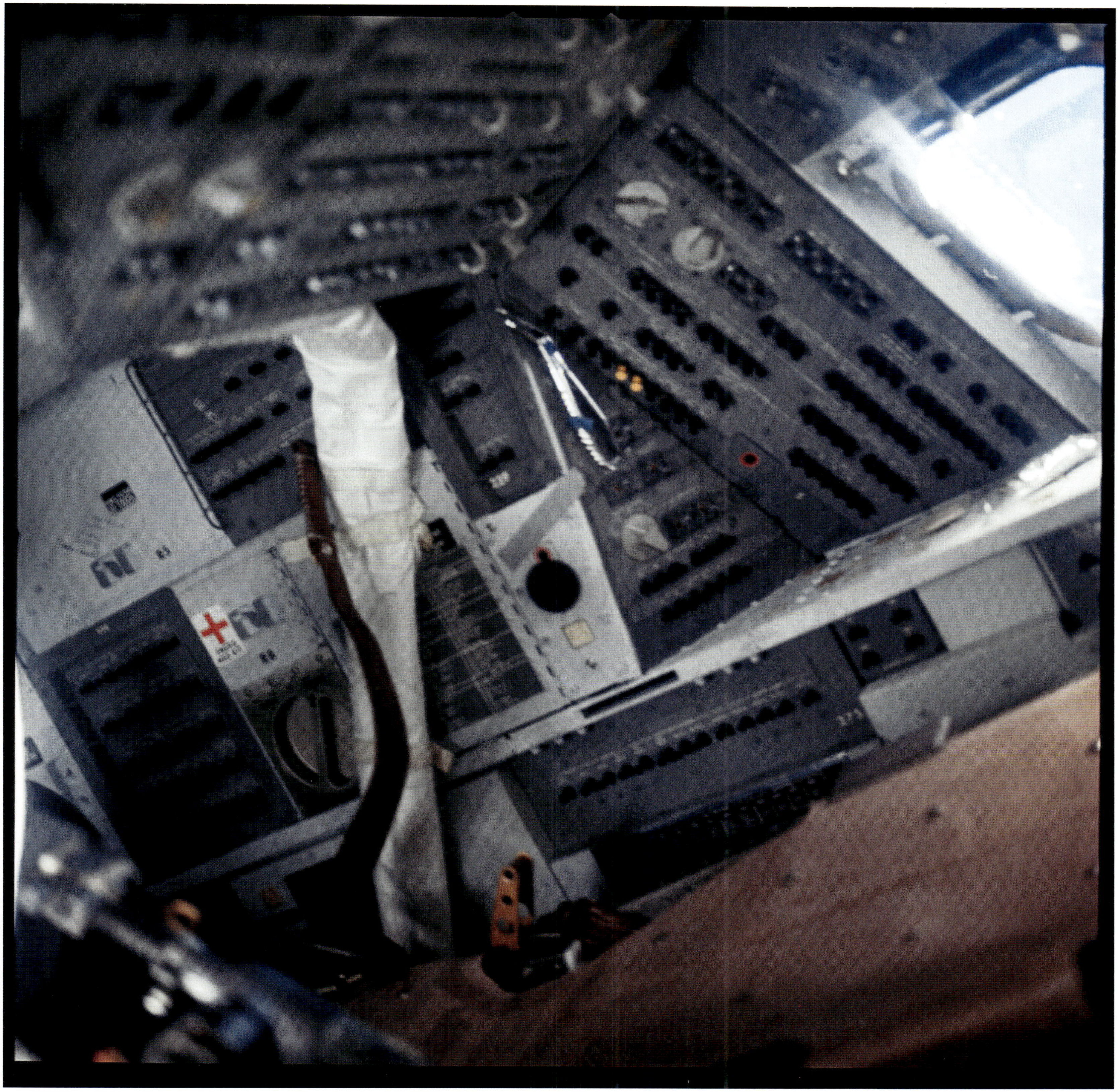

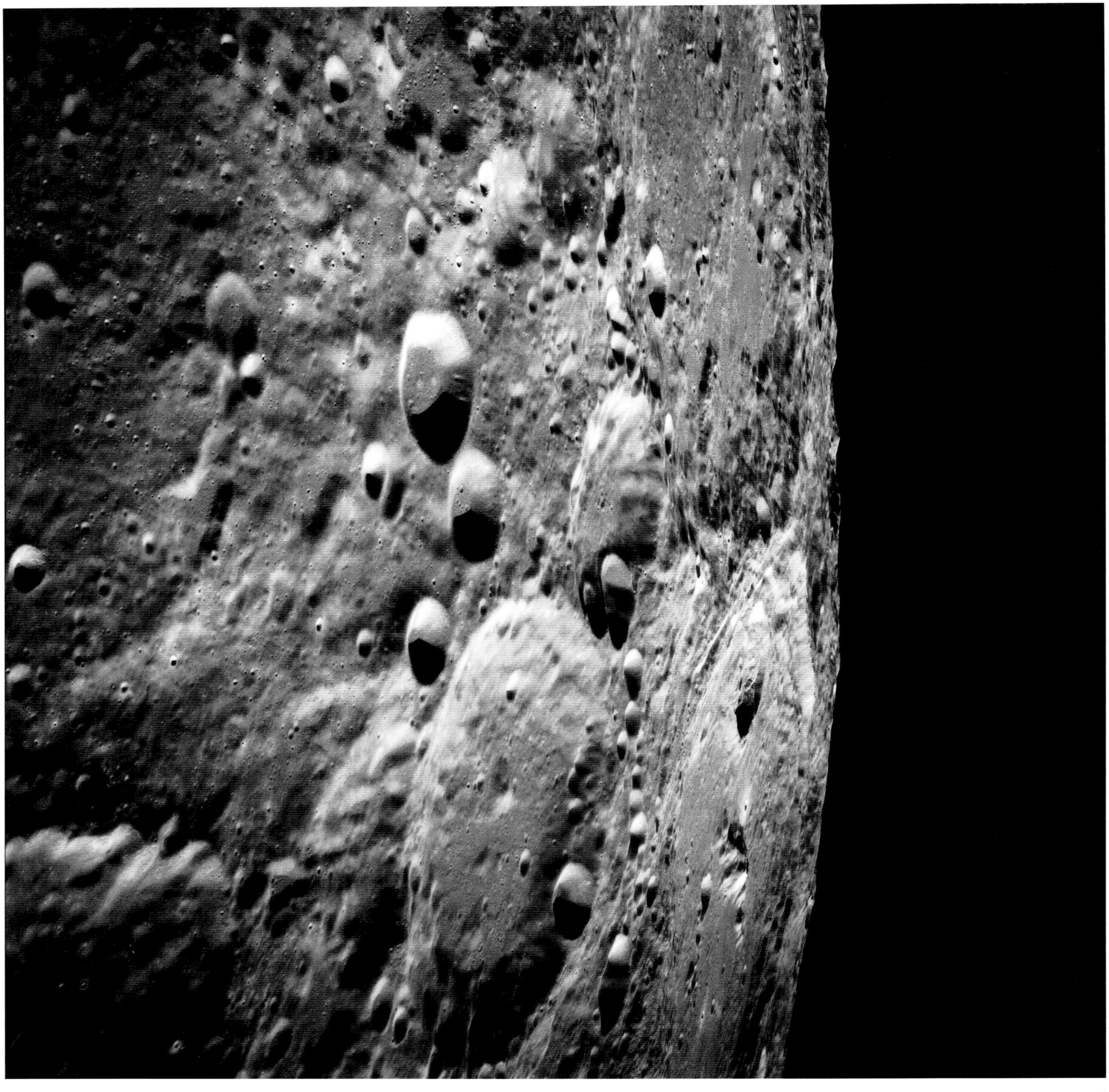

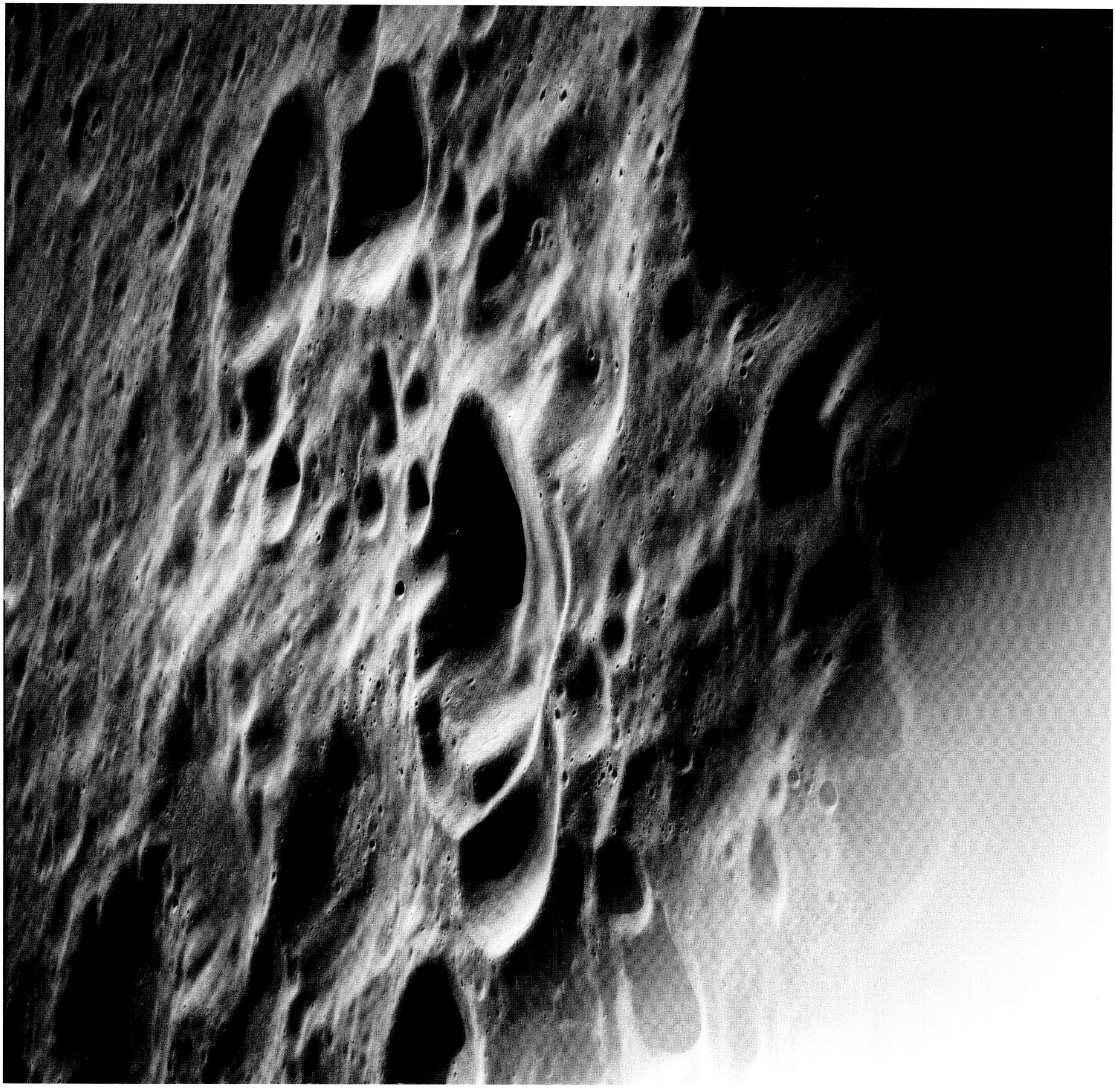

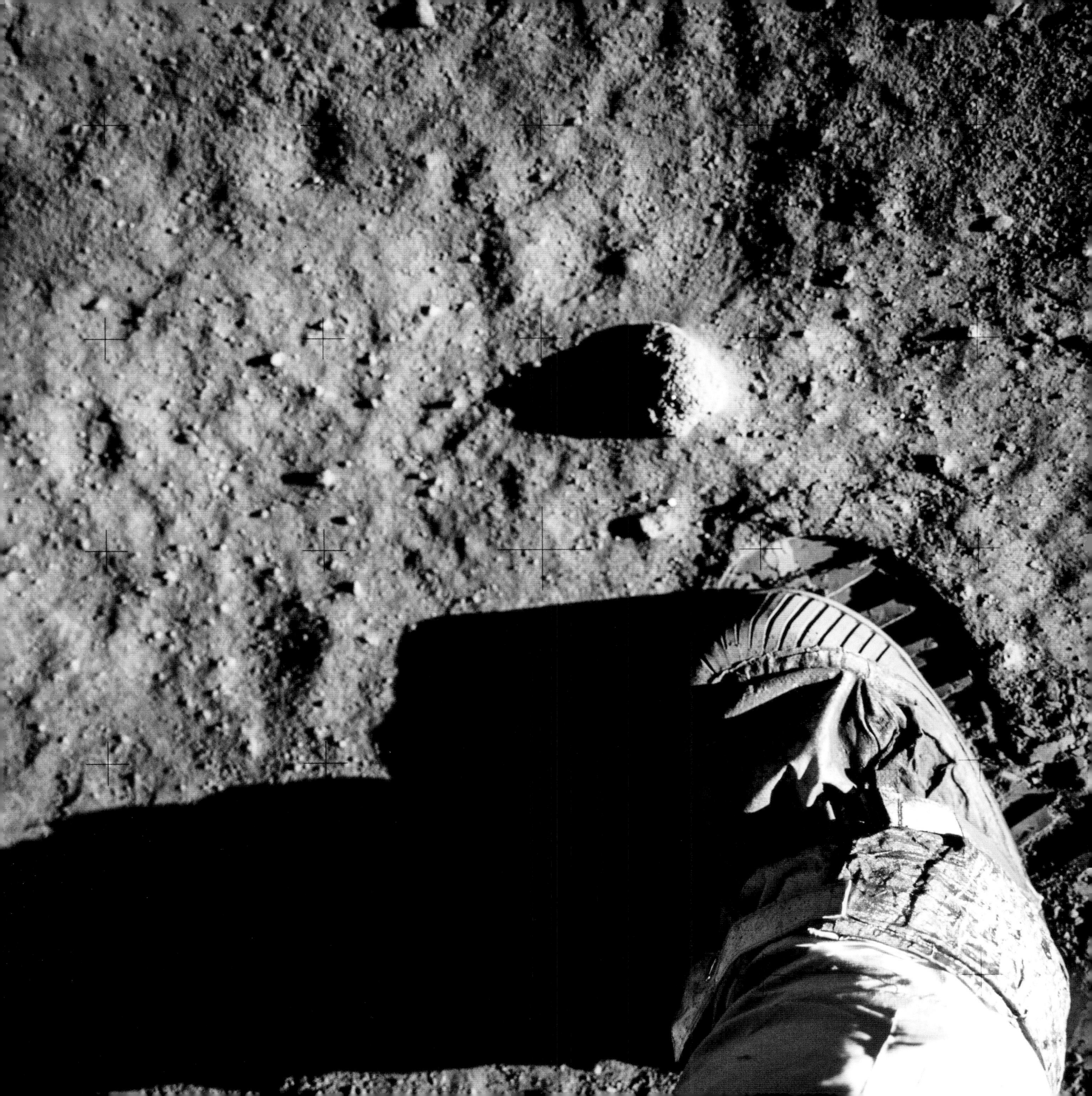

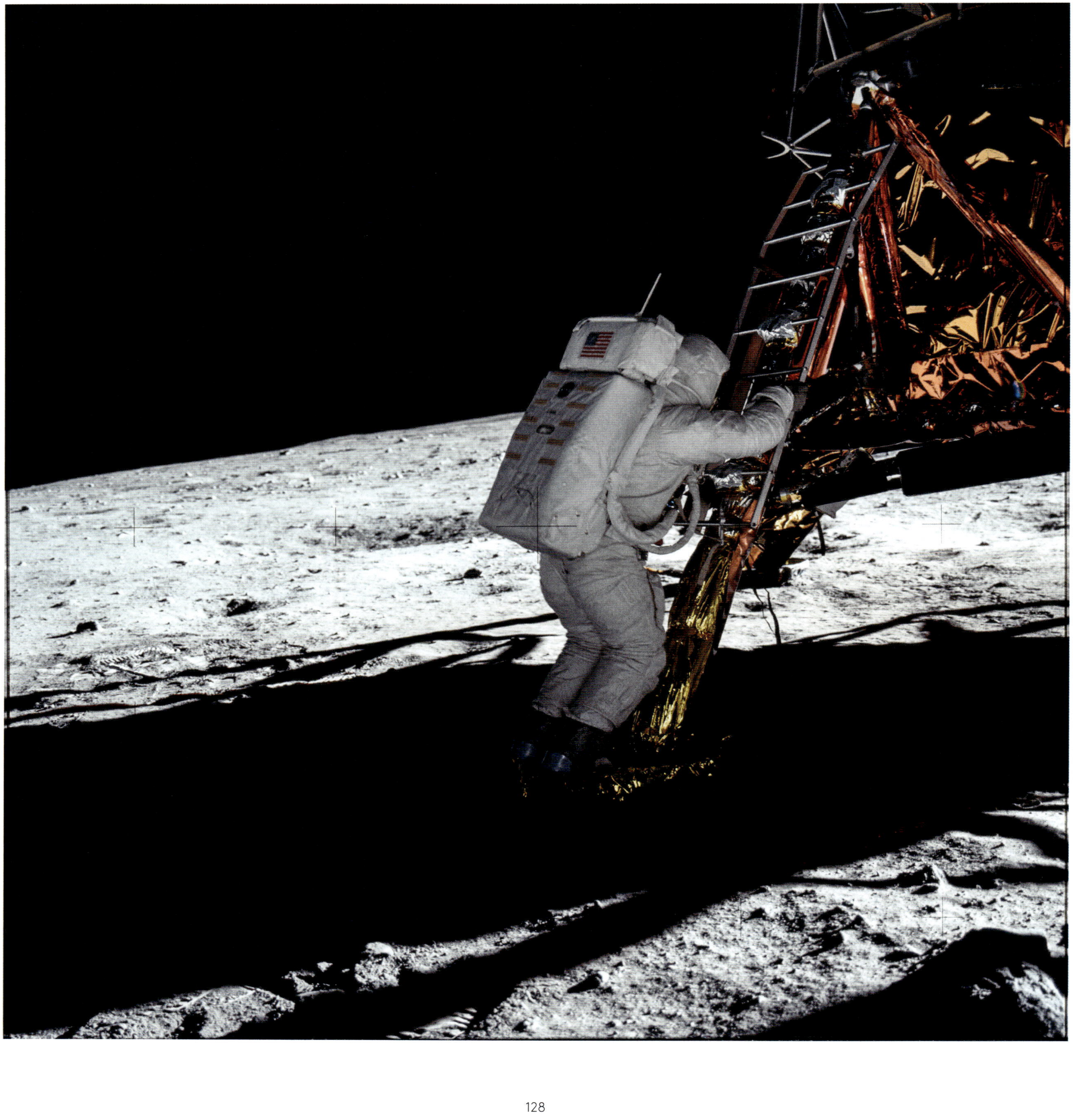

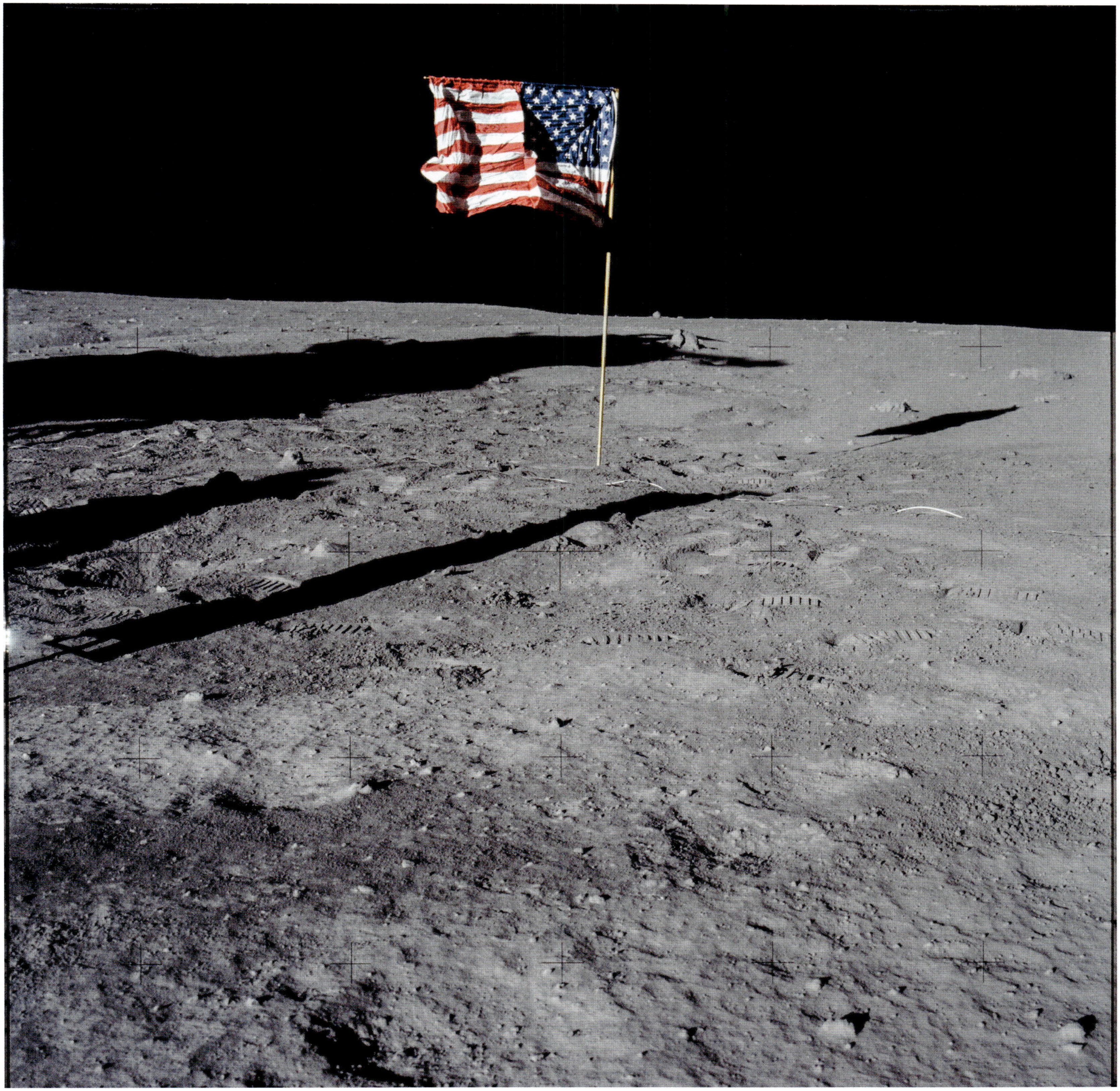

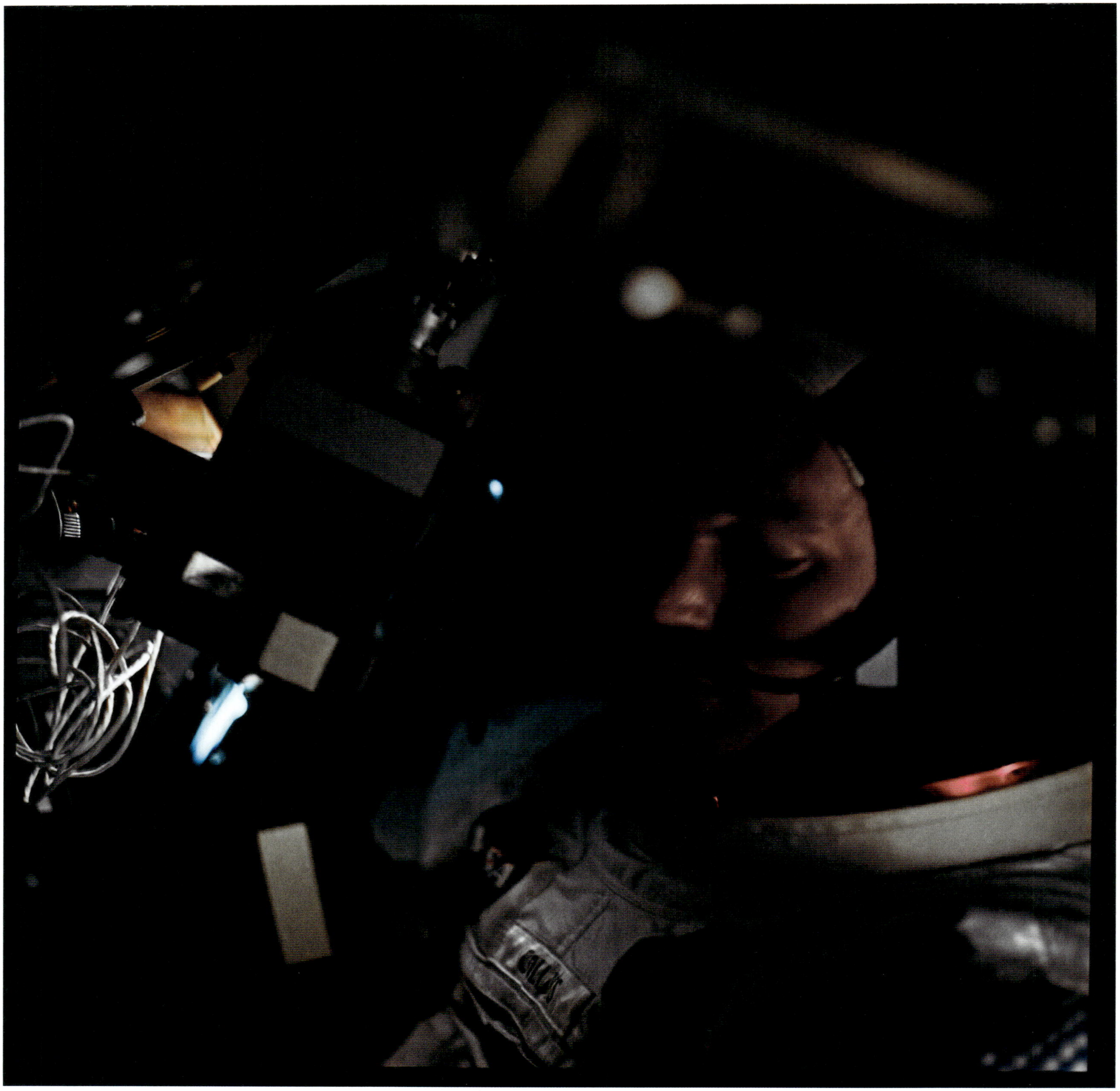

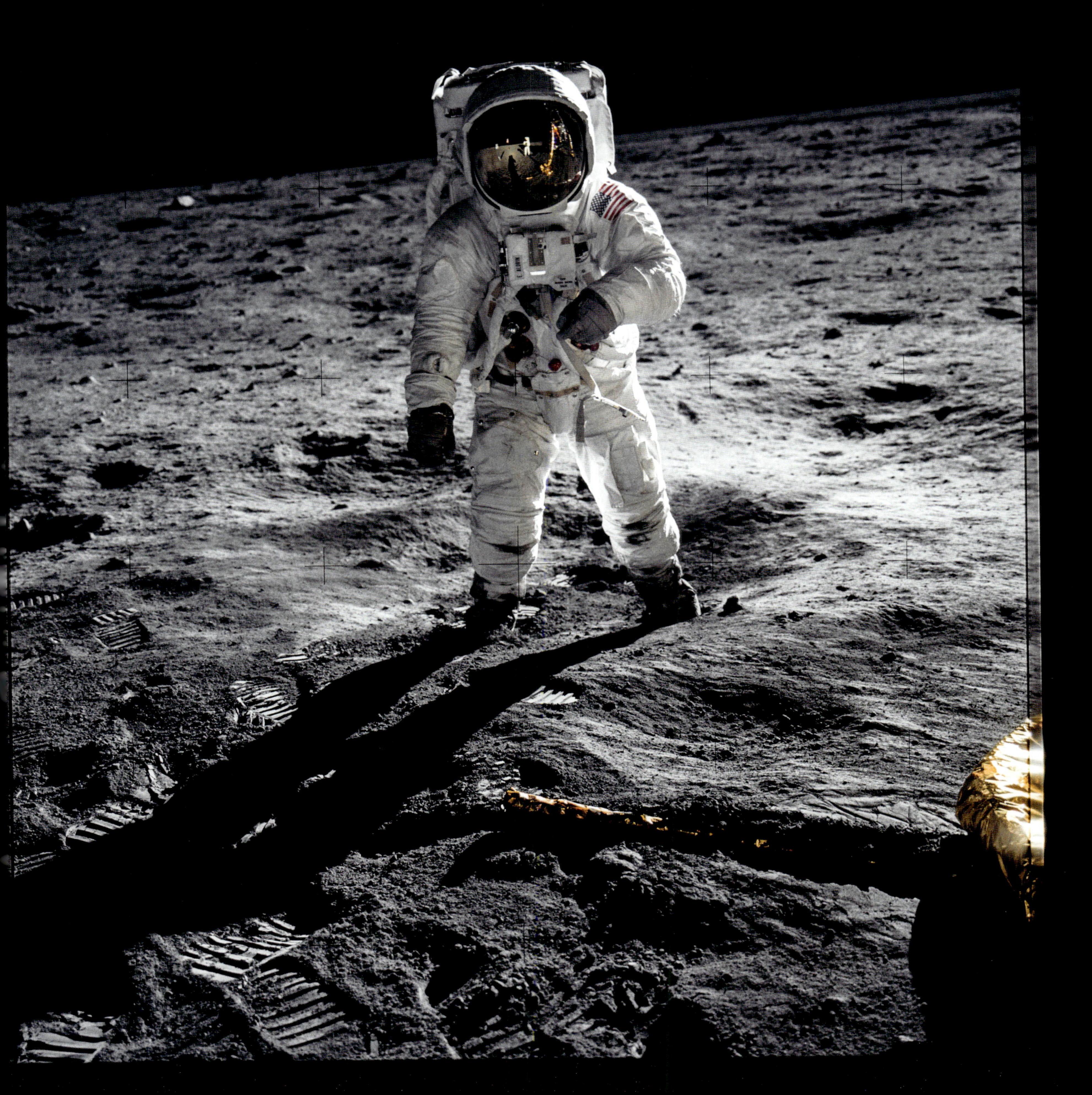

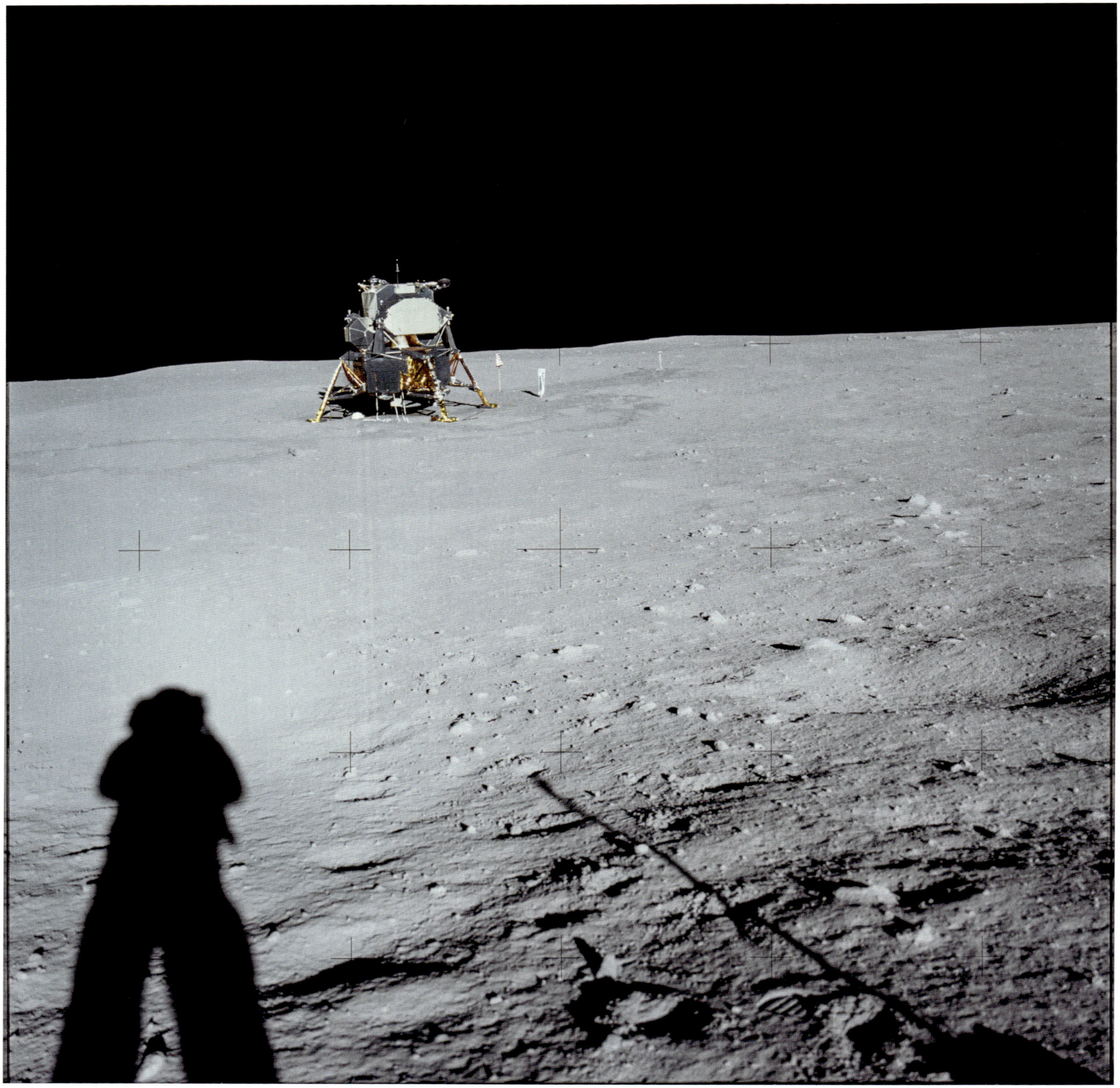

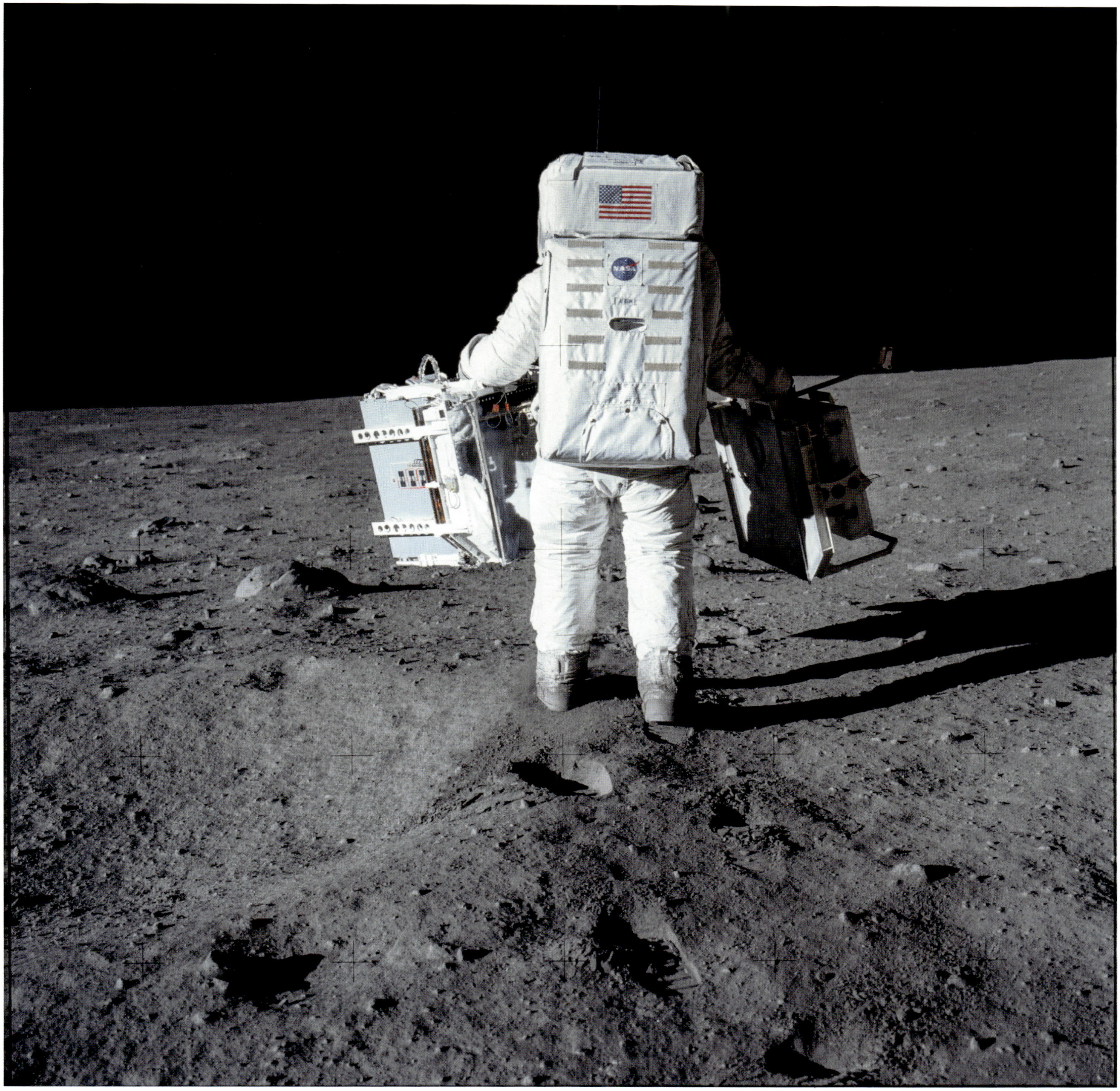

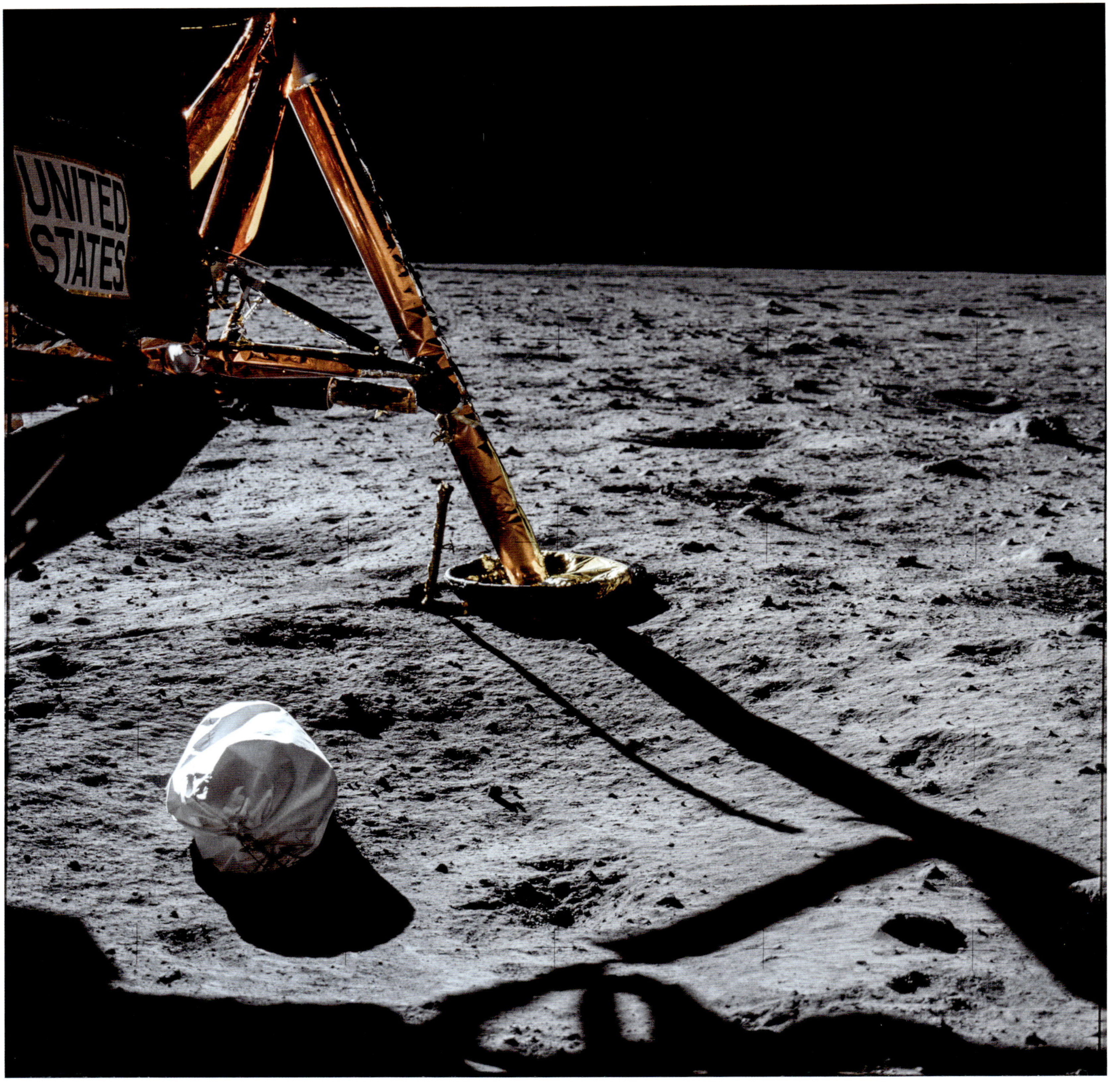
UNITED
STATES

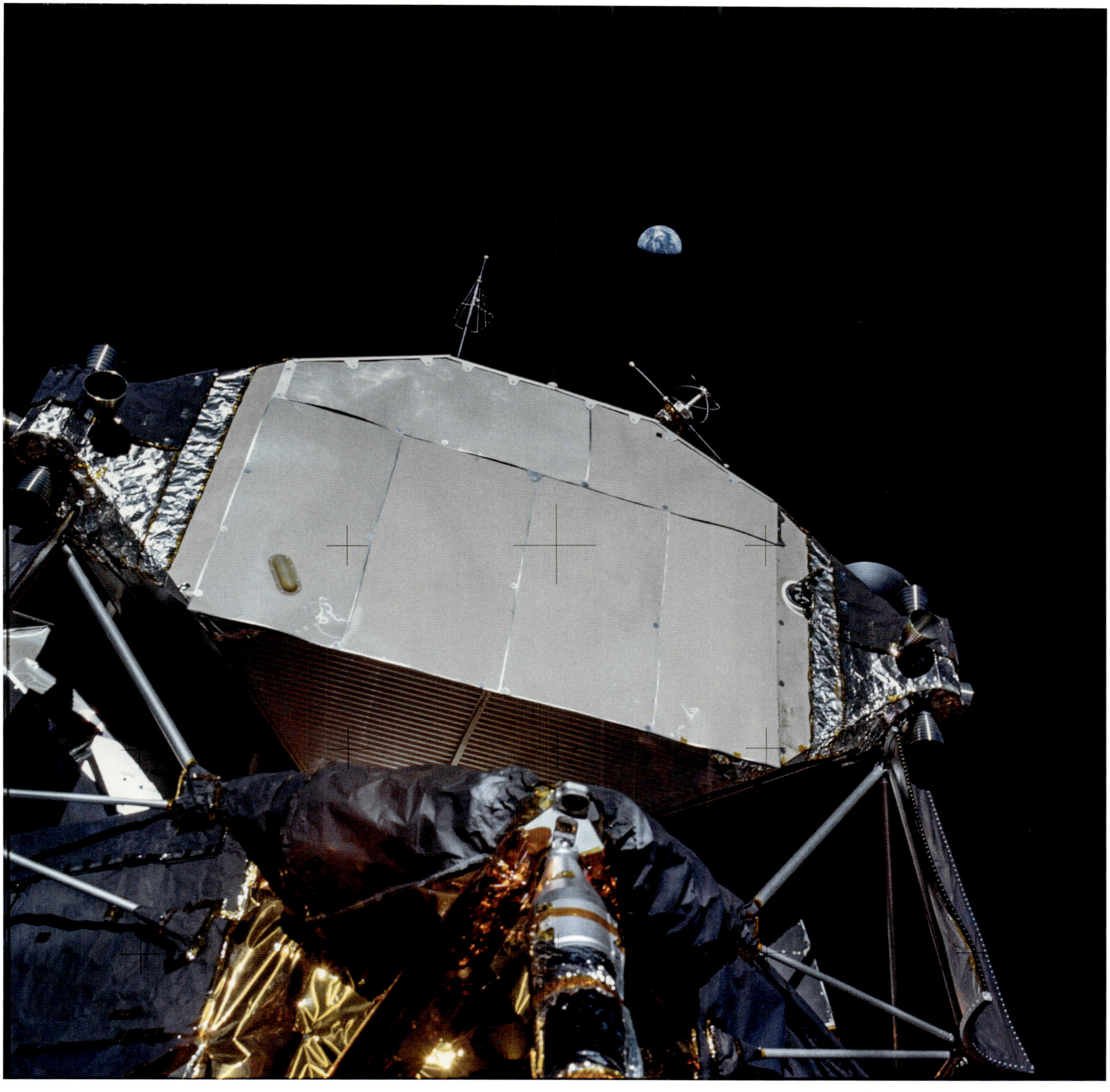

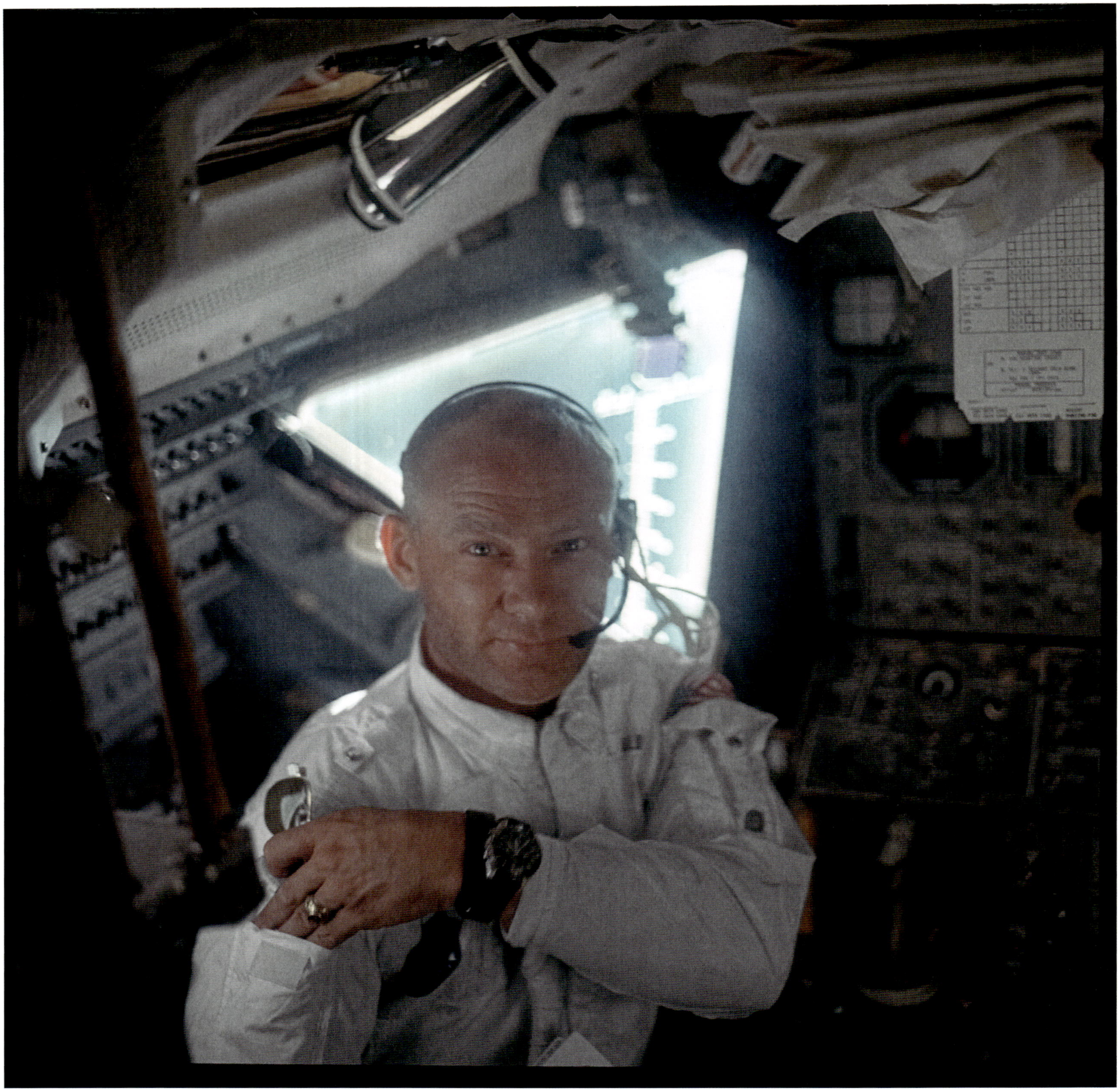

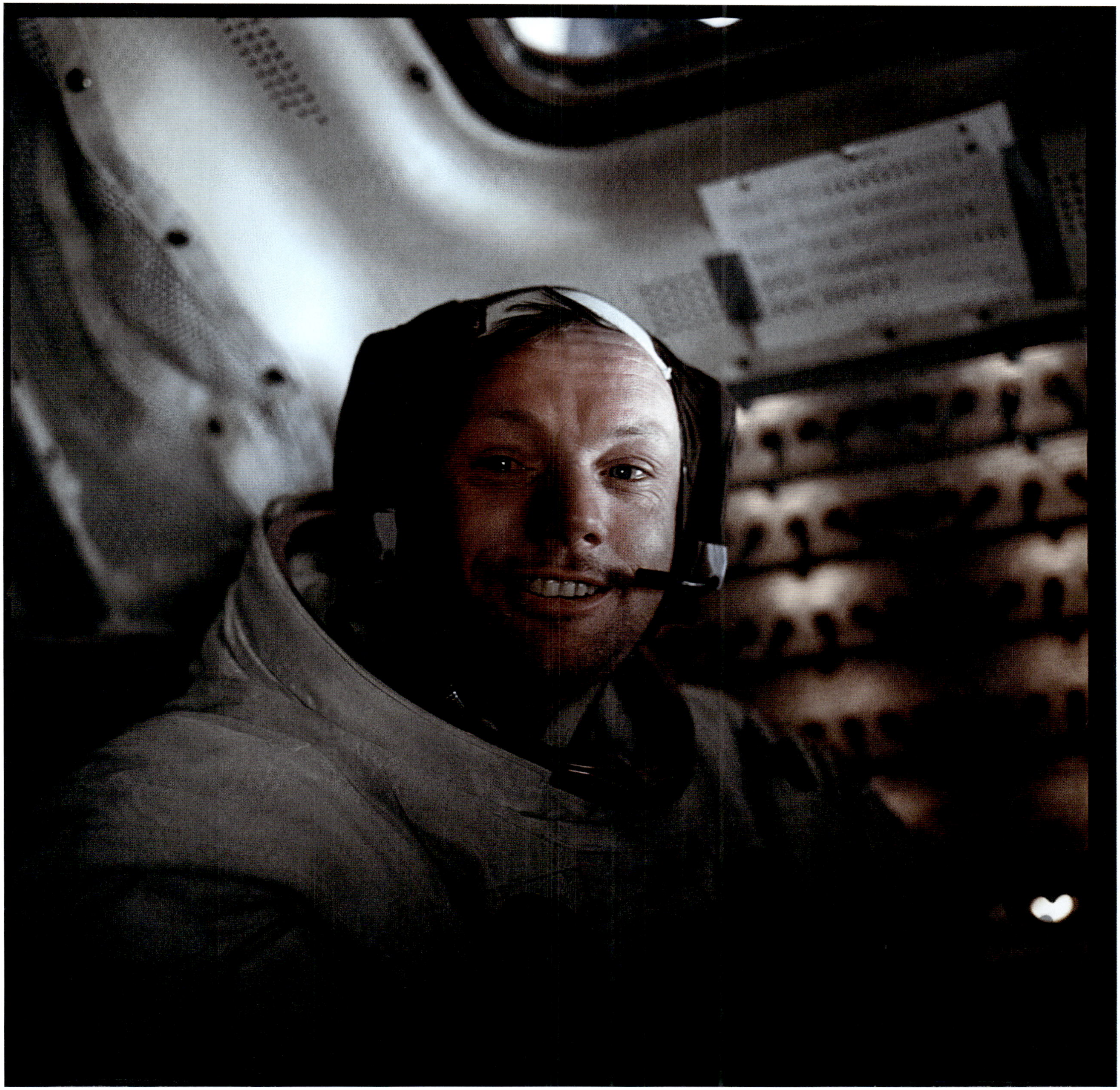

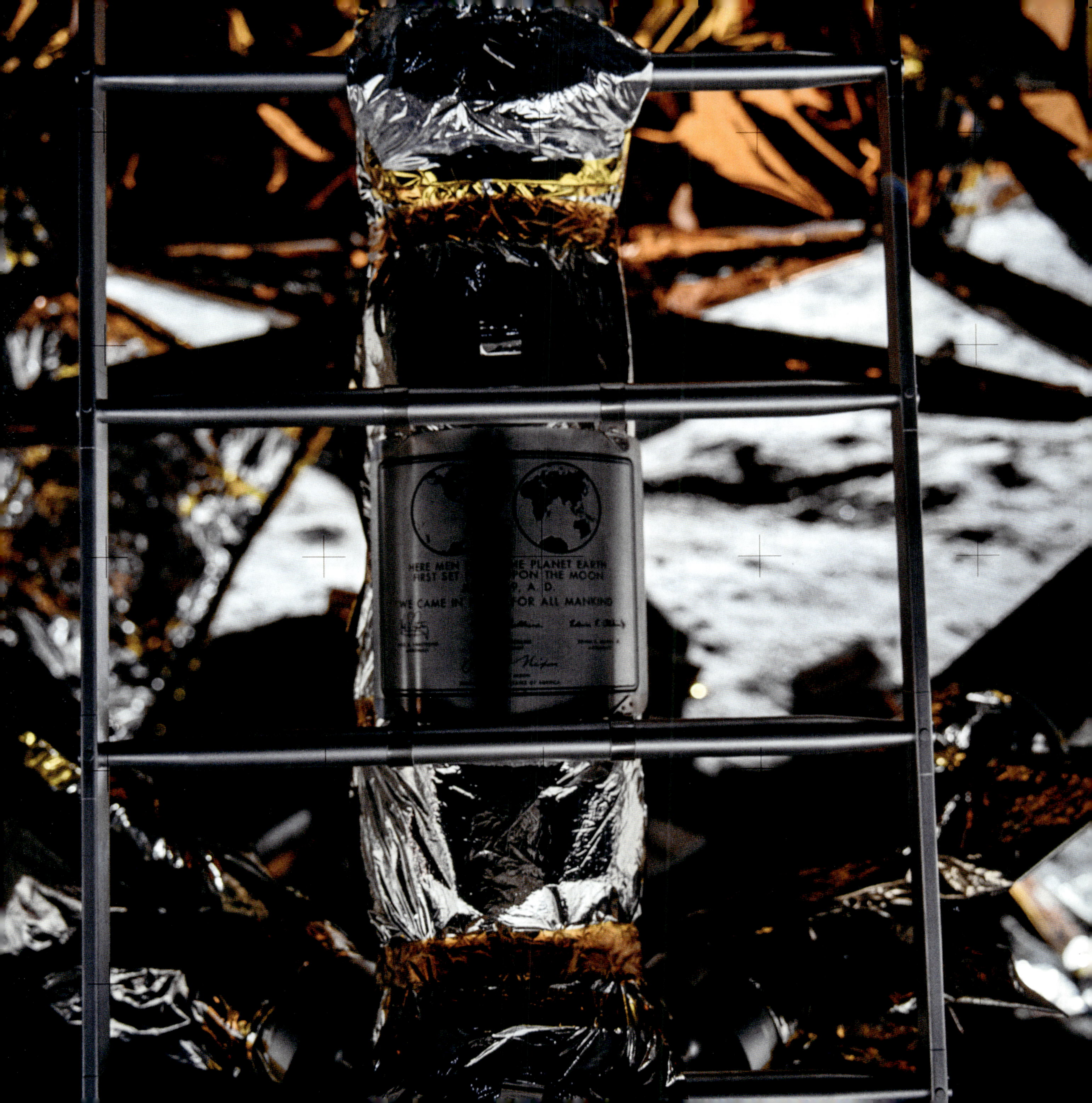
HERE MEN
PLANET EARTH
FIRST SET
THE MOON
FOR ALL MANKIND

115 (AS11-40-5877)
Lunar Module Pilot Buzz Aldrin's bootprint in the fine lunar regolith, captured during Apollo 11. This image was taken as part of a soil mechanics experiment, studying how the Moon's surface responds to pressure.

116 (AS11-36-5293)
A beautiful view of Earth, captured during Apollo 11's initial Earth orbit, offering the crew one final close-up look at their home planet before traveling to another celestial body. This phase allowed for final systems checks before the historic journey to the Moon.

117 (AS11-44-6694)
Earth during Apollo 11's return journey, seen during the Trans-Earth Coast (TEC) phase. The coastline of Somalia and the Indian Ocean are visible in this striking image, taken as the crew made their way back home after the first Moon landing.

118 (AS11-38-5565)
A view of the Moon's far side, captured during Apollo 11. With no direct Earth visibility, this region remained unexplored until spaceflight allowed for detailed imaging, revealing its rugged and heavily cratered terrain.

119 (AS11-41-6152)
A view of the south rim of Daedalus Crater on the Moon's far side, captured from Command Module (CM) *Columbia*. This 58mi (93km)-wide impact crater features steep walls and central peaks, contributing to the study of lunar impact cratering and surface evolution.

120 (AS11-44-6642)
Lunar Module (LM) *Eagle* ascends from the Moon, preparing to dock with Command/Service Module (CSM) *Columbia*. Captured by Command Module Pilot Michael Collins, this image shows the lunar surface below and Earth rising over the horizon.

121 (AS11-44-6574)
The Apollo 11 LM *Eagle*, photographed from CSM *Columbia* after undocking in lunar orbit. This critical maneuver set the stage for the first Moon landing, as Commander Neil A. Armstrong and Aldrin prepared for descent while Collins remained in orbit.

122 (AS11-36-5389)
The interior of LM *Eagle* during Apollo 11's Trans-Lunar Coast (TLC) phase. This image captures the compact workspace where Armstrong and Aldrin prepared for the first Moon landing.

123 (AS11-36-5387)
The right-hand side of CSM *Columbia*, seen during TLC. With over 500 switches, dials, and gauges, the CSM was the central hub for controlling Apollo 11's trajectory, ensuring precise navigation to the Moon.

124 (AS11-38-5583)
This image showcases Stratton Crater, named after British astronomer Dr. Frederick Stratton. Located on the Moon's far side, its sharp rim and bright ejecta highlight ancient impact forces, revealing the rugged, unseen lunar terrain.

125 (AS11-38-5559)
Captured in lunar orbit during Apollo 11, this image shows the rugged highlands south of Coriolis Crater, named after Gaspard-Gustave de Coriolis, who described the Coriolis effect — the force influencing winds and currents on Earth. The Moon, lacking an atmosphere, experiences no such effect.

127 (AS11-40-5880)
This close-up image from Apollo 11 captures Aldrin's boot pressing into the lunar soil during a soil mechanics test. The imprint shows the Moon's fine, powdery regolith and low gravity. With no wind or water to erode it, this bootprint could last for millions of years.

128 (AS11-40-5869)
A historic moment — Aldrin climbing down the LM ladder, moments before stepping onto the Moon. Taken by Armstrong, this image highlights the LM's structure and the stark lunar landscape beyond.

129 (AS11-40-5905)
Part of Aldrin's panorama, this image shows the first of six U.S. flags planted on the Moon. With no atmosphere, the flag stands motionless, held up by a horizontal rod. The vast lunar landscape stretches beyond.

130 (AS11-36-5292)
Collins, calm and focused, floats inside the CSM *Columbia* just over an hour after liftoff. Captured by Aldrin, the image shows Collins with the onboard TV camera.

131 (AS11-40-5903)
One of the most famous photos ever taken, this image captures Aldrin on the Moon, with Armstrong, the LM, and the lunar surface reflected in his visor.

132 (AS11-40-5961)
Armstrong captured this view of LM *Eagle* from Little West Crater, a 108ft (33m)-wide impact site and one of his farthest excursions from the LM. Near the lander, an array of scientific experiments is visible — including the retroreflector, which still helps measure the Earth-Moon distance using lasers.

133 (AS11-40-5942)
Aldrin carries the Early Apollo Surface Experiments Package (EASEP), which included a seismometer that operated for 21 days, detecting ground movements and slight tremors. Unlike Earth, the Moon has no plate tectonics, yet moonquakes occur due to gravitational forces, temperature shifts, meteoroid impacts, and deep seismic activity.

134 (AS11-40-5850)
A close-up of LM *Eagle*'s south landing gear, its Mylar and Kapton foil insulation and wide footpad resting on lunar dust. Engineers designed the crushable footpads to prevent excessive sinking, ensuring the LM could land safely on an unknown and potentially soft surface.

135 (AS11-40-5924)
The rear of LM *Eagle*, standing against the stark lunar surface, with Earth visible in the sky above. From the Moon, Earth appears 3.7 times larger than the Moon does from Earth and remains nearly fixed in place due to the Moon's tidal lock.

136 (AS11-36-5390)
A moving portrait of Aldrin inside the LM *Eagle*, taken by Armstrong during Apollo 11's translunar phase. You can sense the anticipation as the crew prepared for their historic descent to the Moon.

137 (AS11-37-5528)
After completing the first moonwalk, Armstrong sits inside the LM *Eagle* in a rare moment of reflection. His face is calm, yet behind his eyes is the weight of history—he had just taken humanity's first steps on another world.

139 (AS11-40-5899)
A close-up of the LM's ladder and commemorative plaque, which reads: "We came in peace for all mankind." Signed by the Apollo 11 crew and President Richard Nixon, it remains on the Moon as a message to future explorers.

Apollo 12

XII

Charles Conrad, Jr.

Richard F. Gordon, Jr.

Alan L. Bean

Apollo 12

Launch: November 14th, 1969
Lunar landing: November 19th, 1969
Splashdown: November 24th, 1969

Command Module: Yankee Clipper
Lunar Module: Intrepid

Apollo 12 was the first rocket launch attended by an incumbent United States President. Richard Nixon witnessed a successful but dramatic launch. Thirty-six seconds after liftoff, the Saturn V launch vehicle became electrically charged, triggering a number of lightning bolts that traveled down the length of the rocket, through its ionized plume and down to the Earth. Luckily, the strikes did not damage the Saturn V and it continued on its journey to the Moon.

The objective of Apollo 12 was to land the Lunar Module in a precise location in the Ocean of Storms, near the previously landed Surveyor 3 spacecraft. The astronauts were tasked with traversing to the Surveyor 3 probe to collect experiments and parts of the spacecraft to analyze how it was affected by the exposure to the lunar environment. Conrad is seen standing next to the Surveyor 3 probe on the opposite page. This is the only occasion to date of astronauts 'catching-up' with an object that has been sent to another celestial body.

Instead of the monochrome video camera that was used on Apollo 11, the Apollo 12 crew brought the first color camera to send live footage of the Moon and the astronauts back to Earth. Unfortunately, color TV coverage of their mission ended after just 42 minutes when Bean accidentally pointed the camera directly at the sun while setting it up near the Lunar Module, burning out the video pickup tube and rendering the camera useless.

Conrad and Bean clearly enjoyed photographing each other on the surface of the Moon. Many pictures feature their enormous Moon shadows, while lens flares and overexposures almost give these images a feeling of casual vacation photography. One of the mission's most iconic images features a stunning self-portrait by Conrad in the reflection of Bean's visor, with a perfectly composed lunar horizon in the background. This image can be found on page 151. Just over half of the photographic targets were documented with Bean unfortunately forgetting to take several rolls of film he exposed back with him to the Lunar Module, leaving them on the surface of the Moon.

Charles 'Pete' Conrad, Jr.
Commander

Born on June 2nd, 1930 in Philadelphia, Pennsylvania. He was the Pilot of Gemini 5, Backup Command Pilot of Gemini 8, Command Pilot of Gemini 11, and Backup Commander of Apollo 9. As Commander of Apollo 12, he became the third man to walk on the Moon. He later became Commander of Skylab 2.

Richard F. Gordon, Jr.
Command Module Pilot

Born on October 5th, 1929 in Seattle, Washington. He was the Backup Pilot of Gemini 8, the Pilot of Gemini 11, and Backup Command Module pilot of Apollo 9. He was also the Backup Commander of Apollo 15 before retiring from NASA and the U.S.Navy in 1972.

Alan L. Bean
Lunar Module Pilot

Born on March 15th, 1932 in Wheeler, Texas. He was selected for the third group of astronauts in 1963, and was Backup Command Pilot of Gemini 10 and Backup Lunar Module Pilot of Apollo 9. He was the fourth man to walk on the Moon. He later became Commander of Skylab 3 and Backup Commander for the Apollo-Soyuz Test Project.

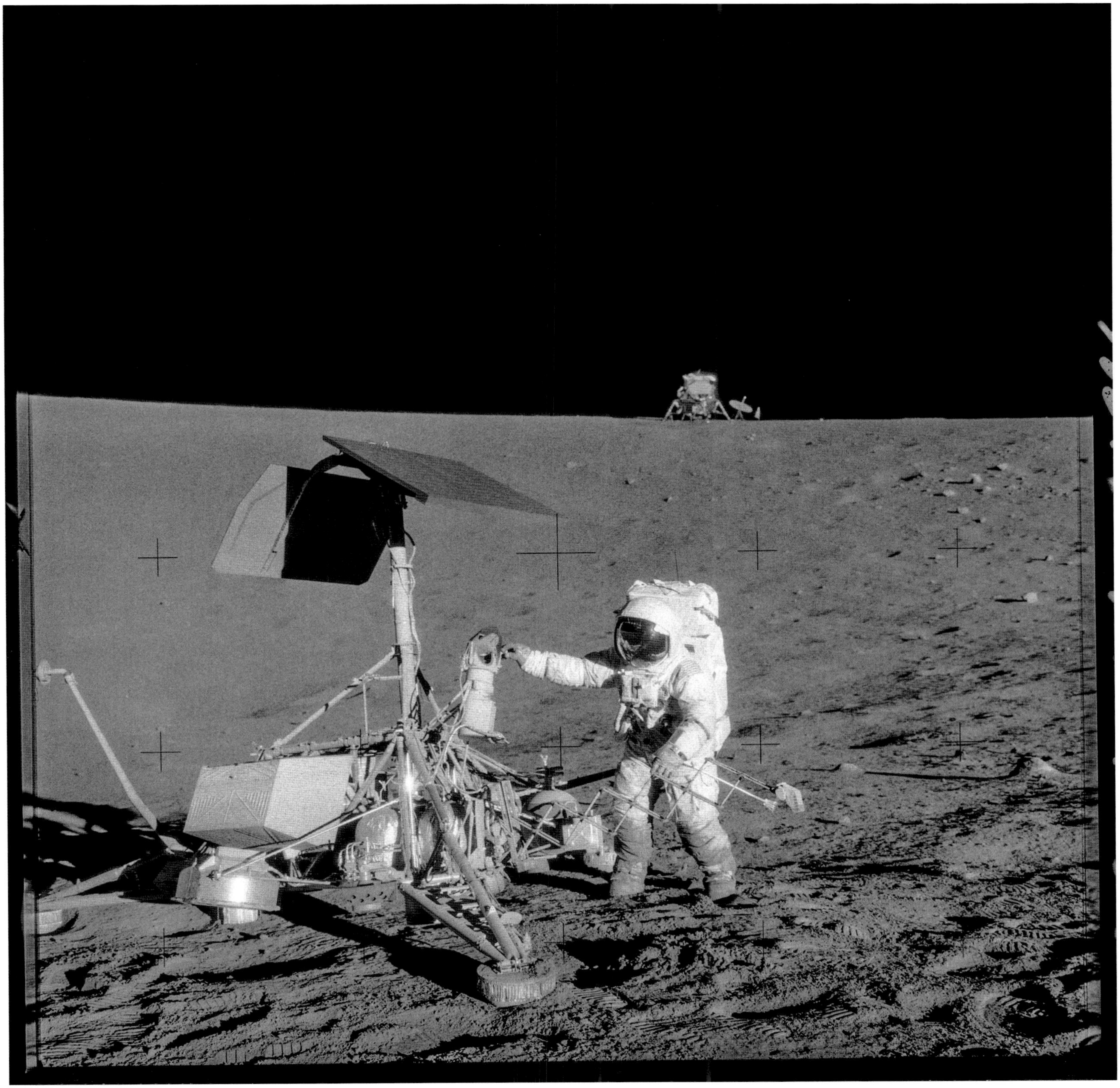

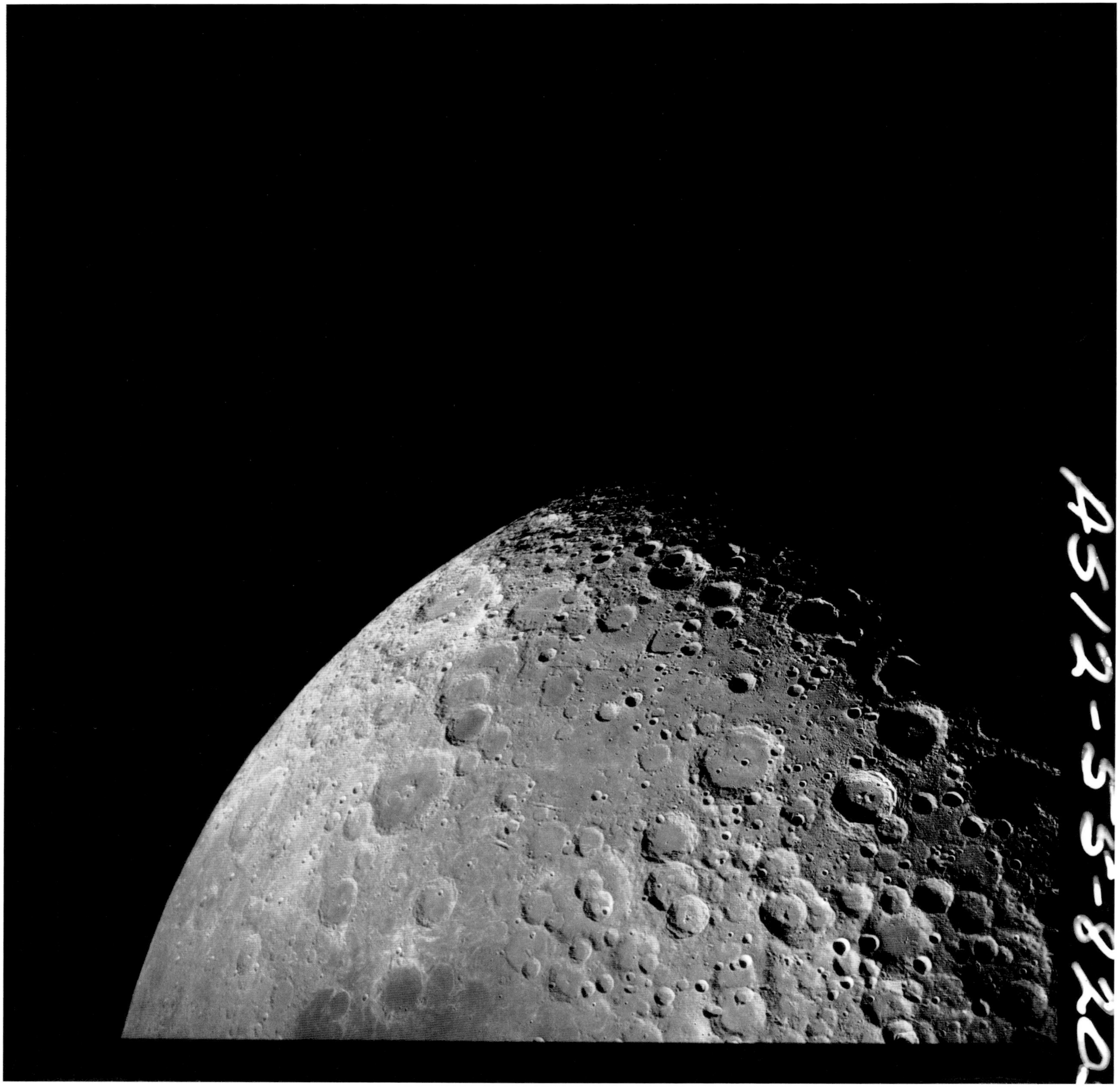

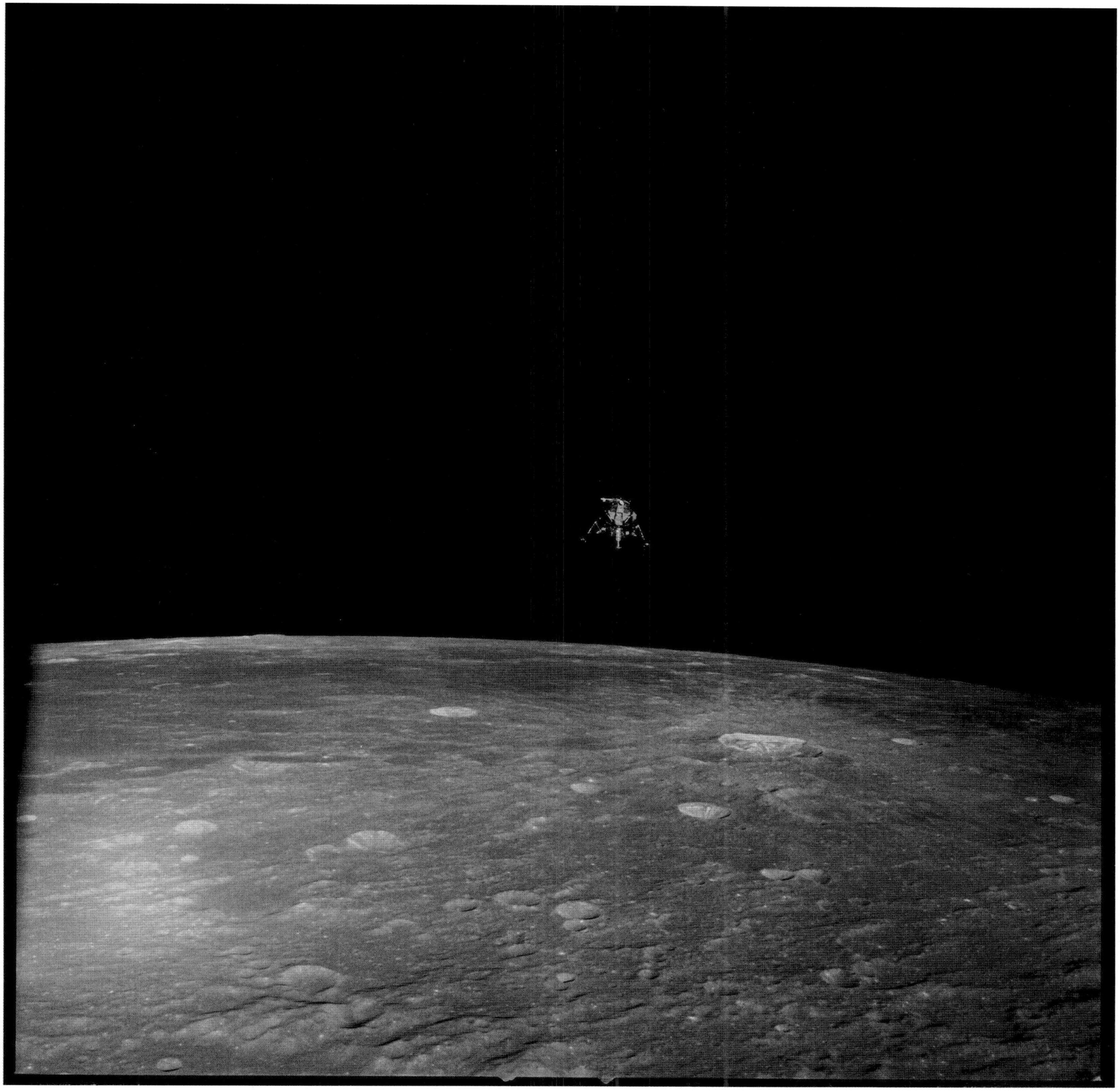

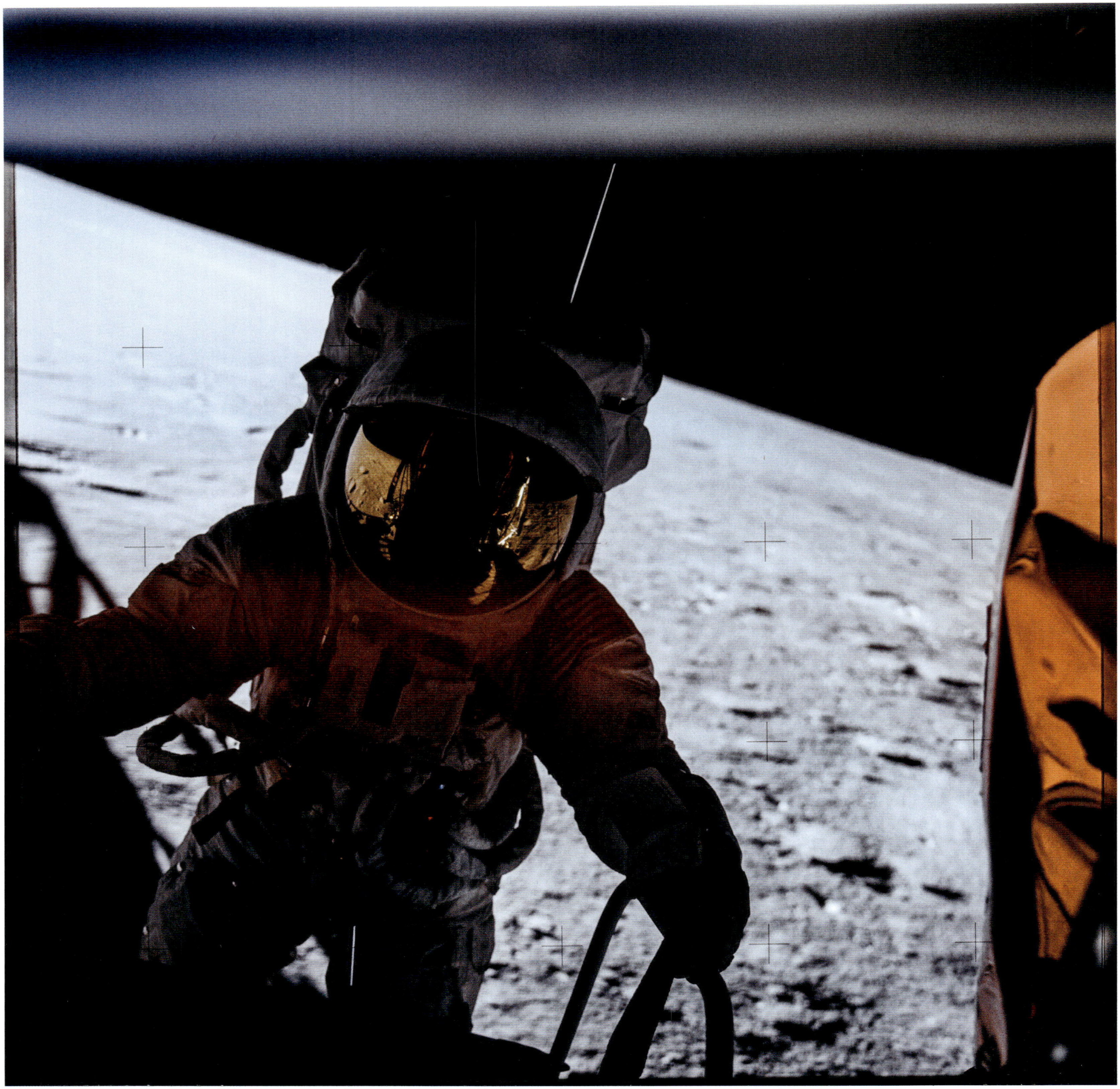

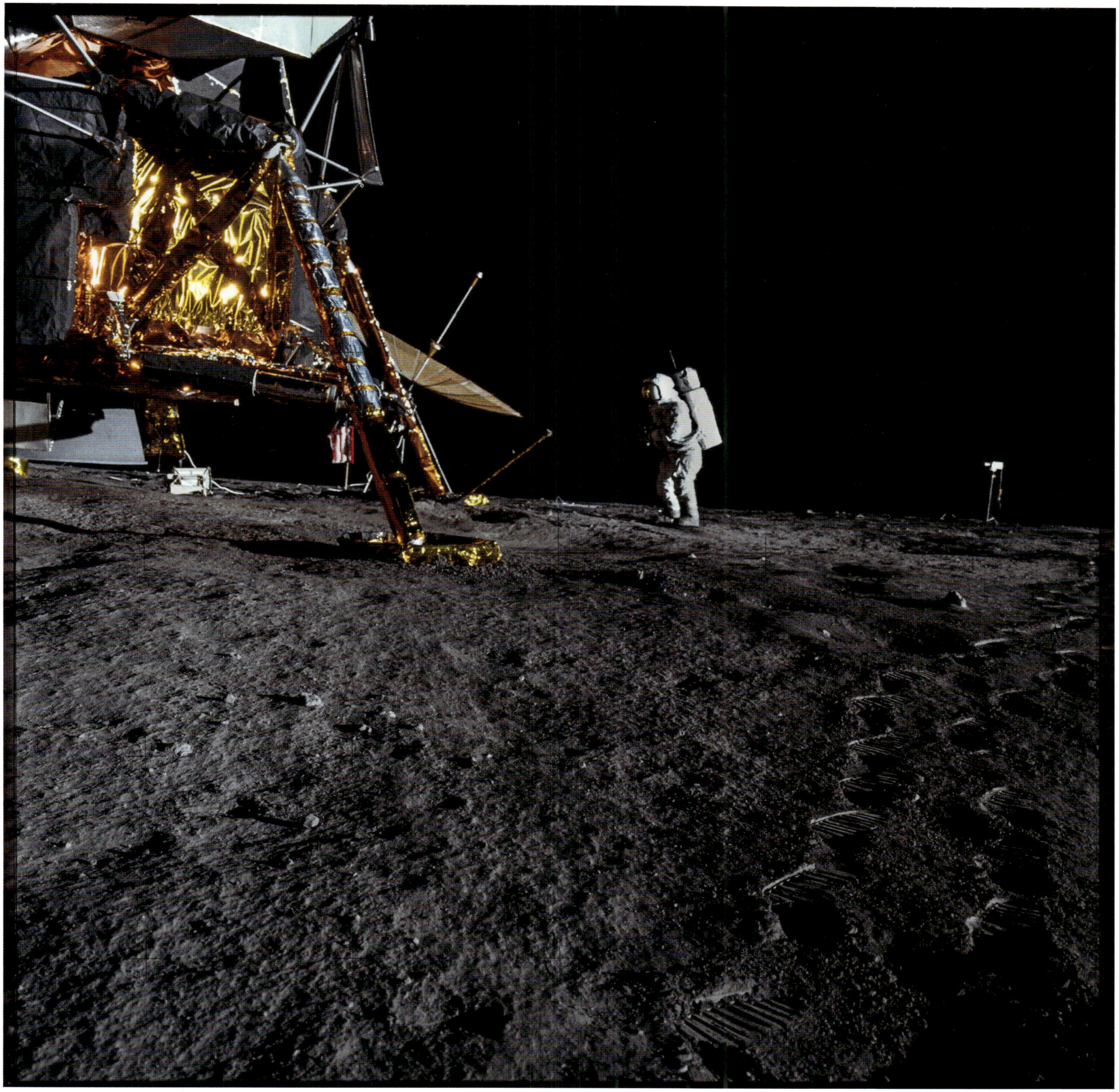

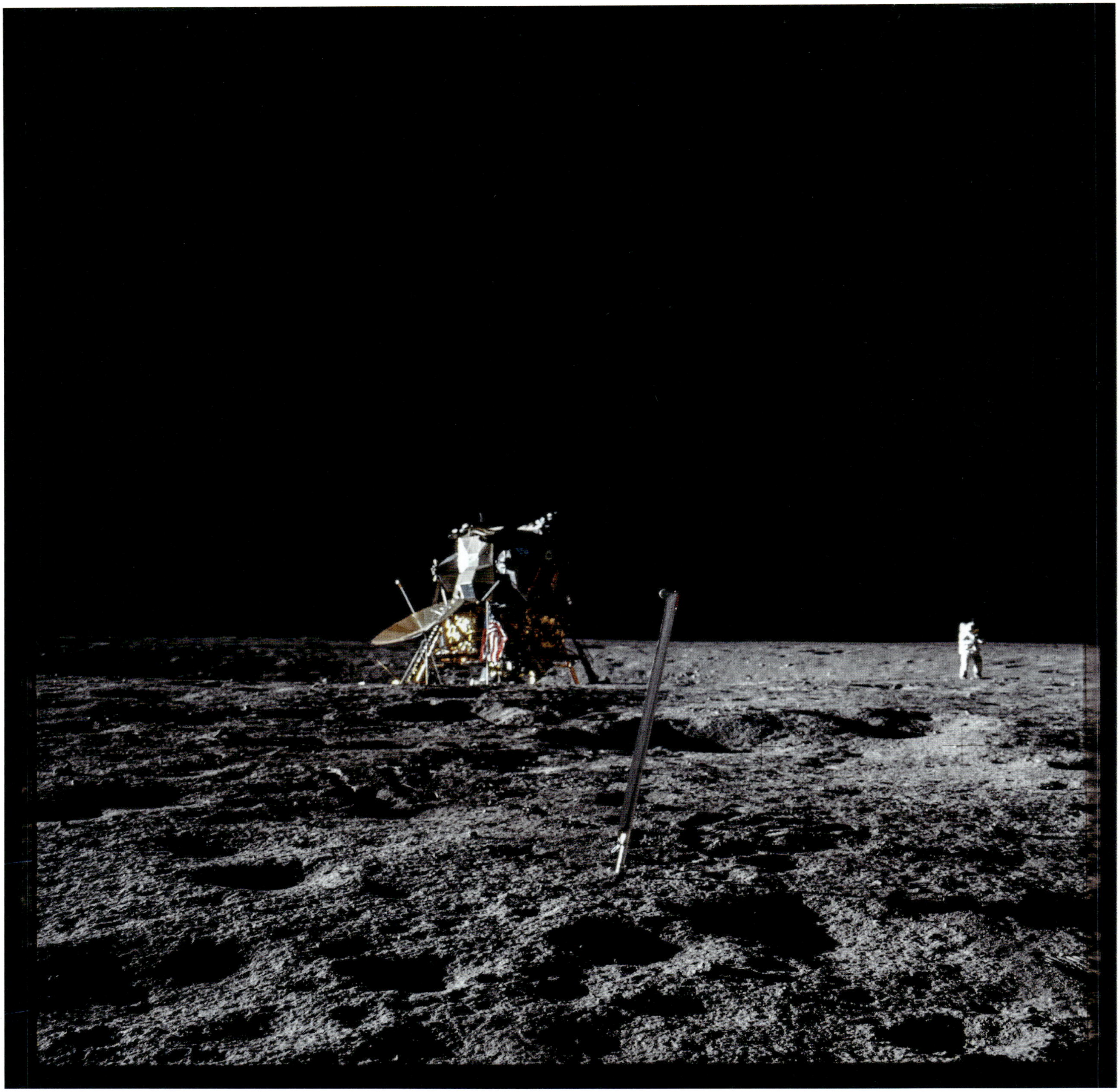

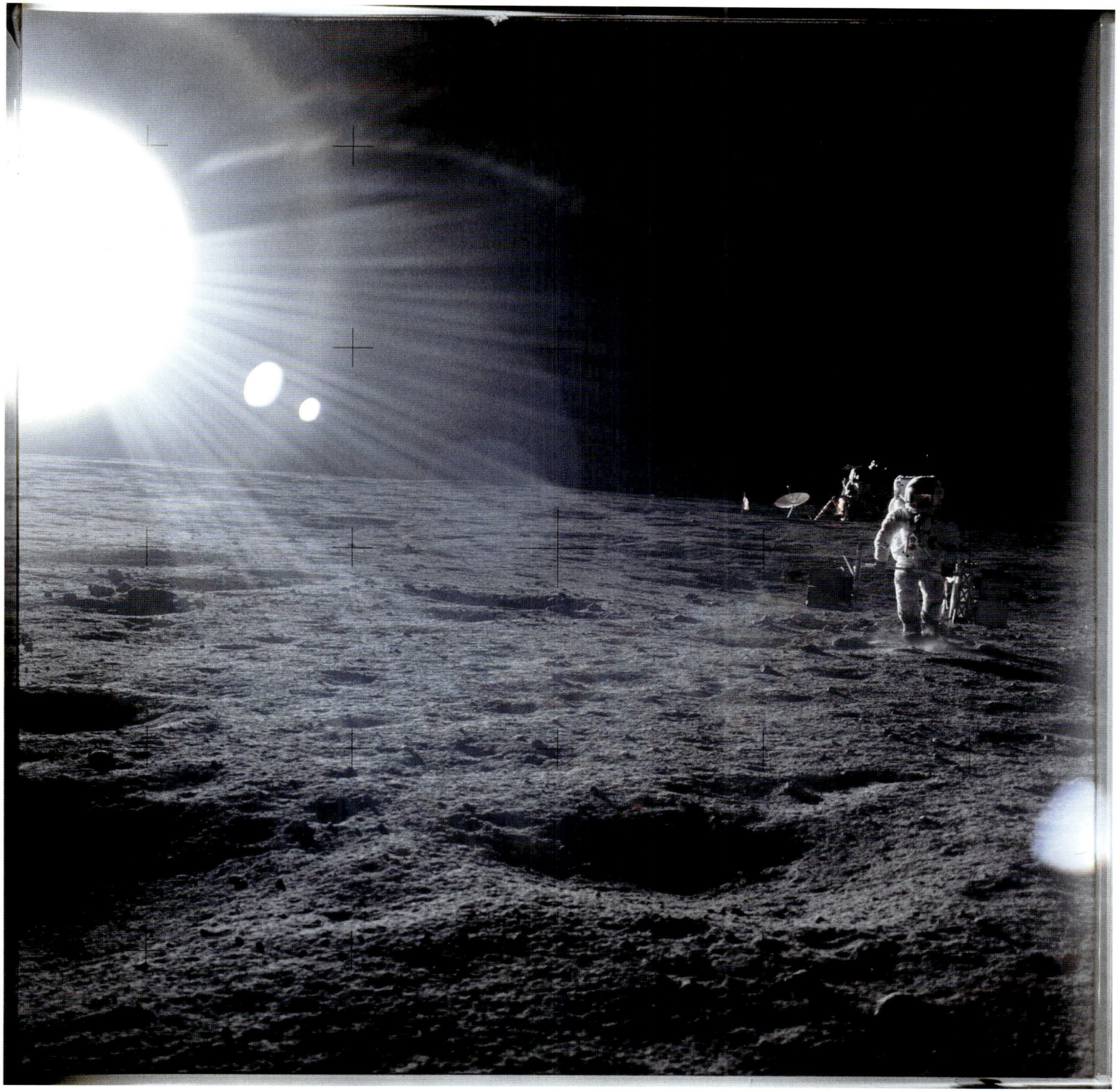

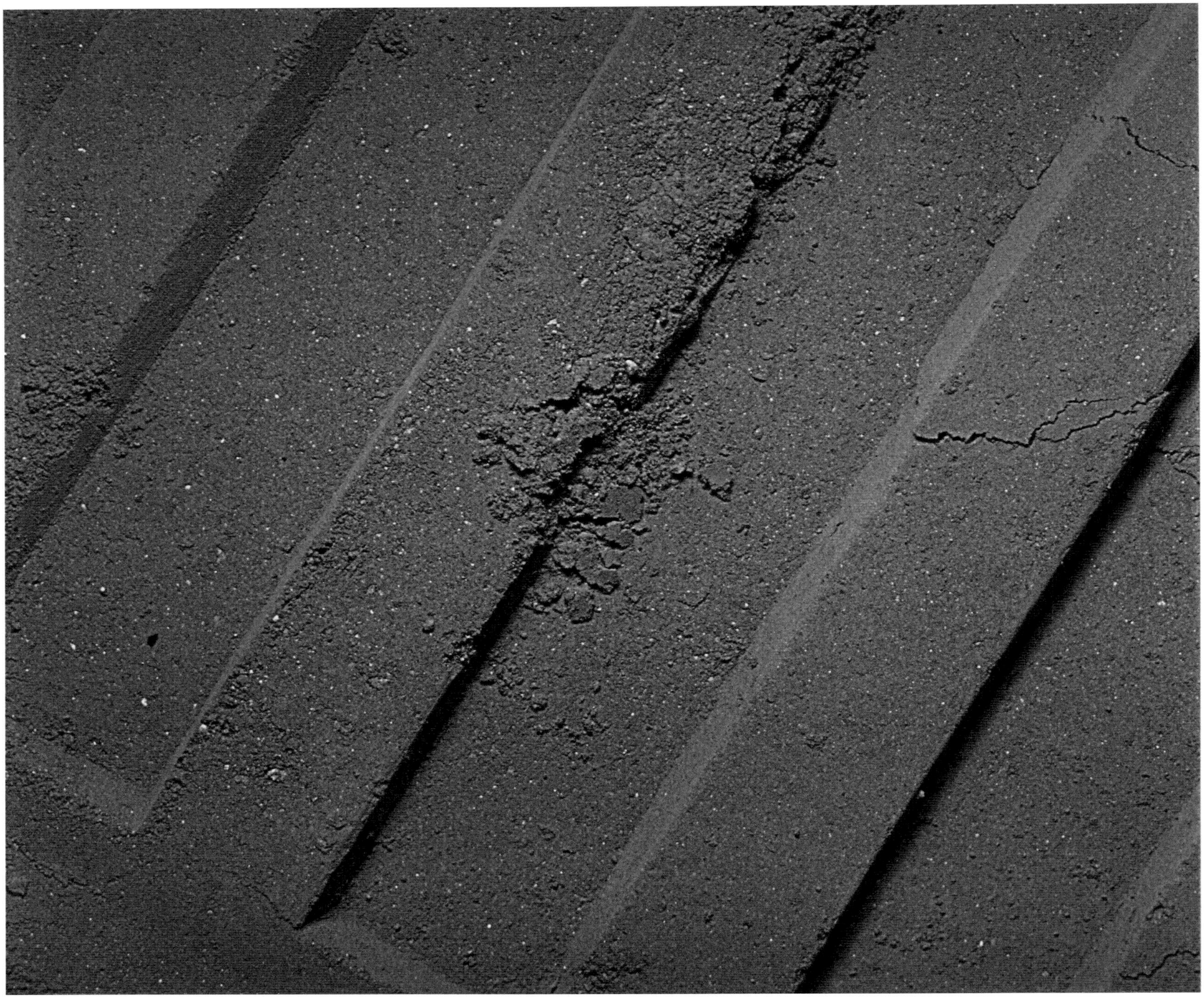

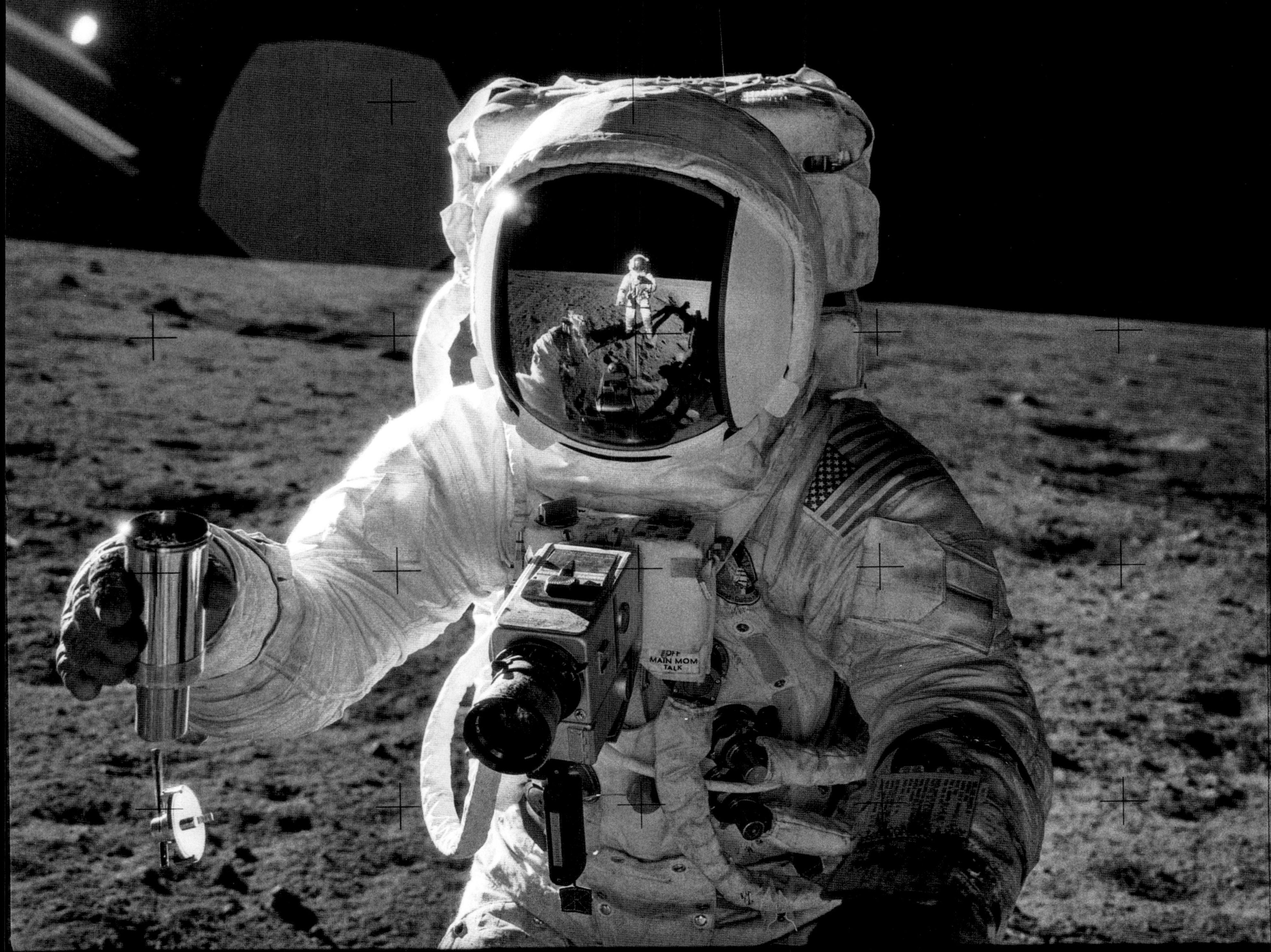
MAIN MOM
TALK

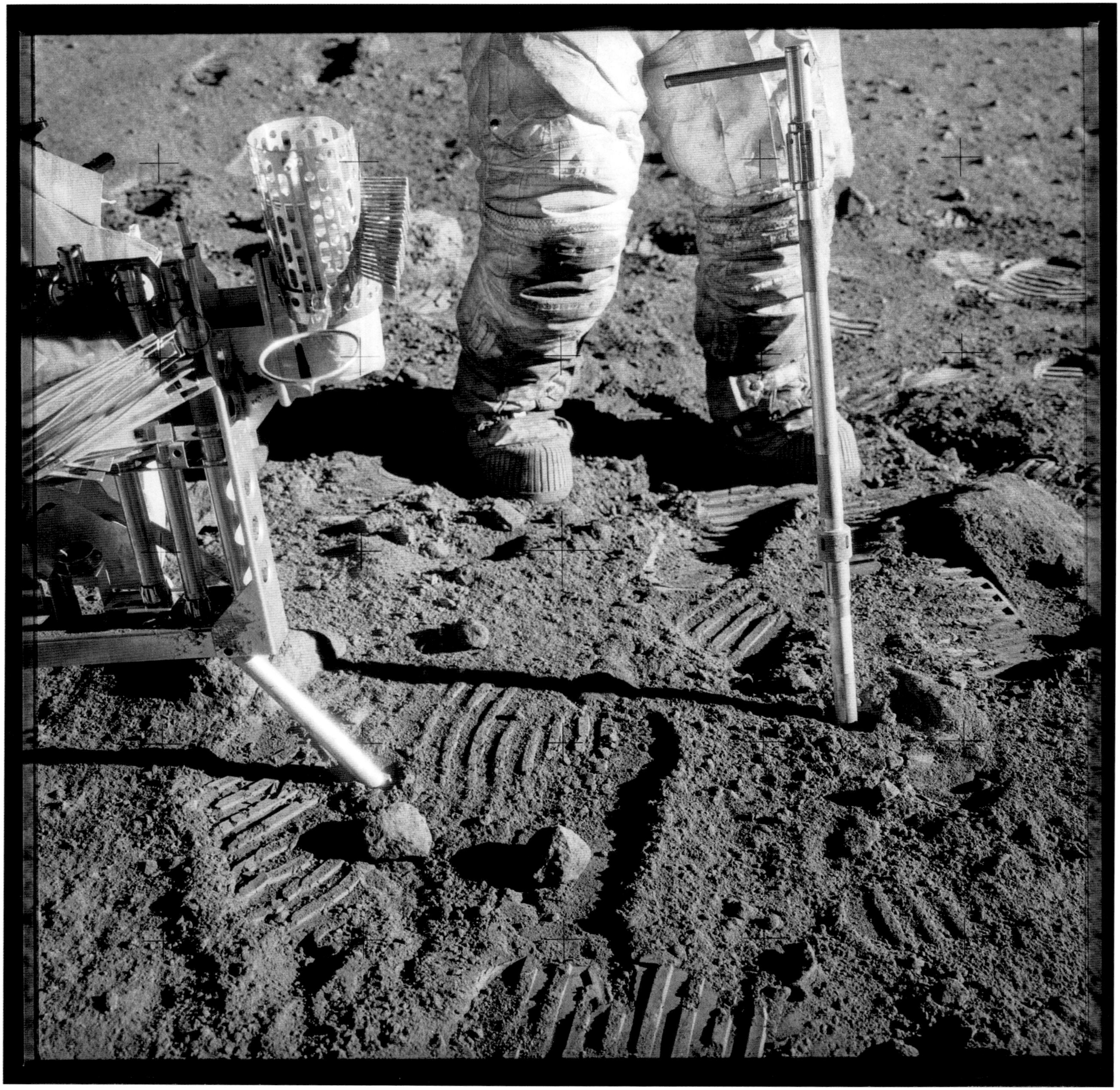

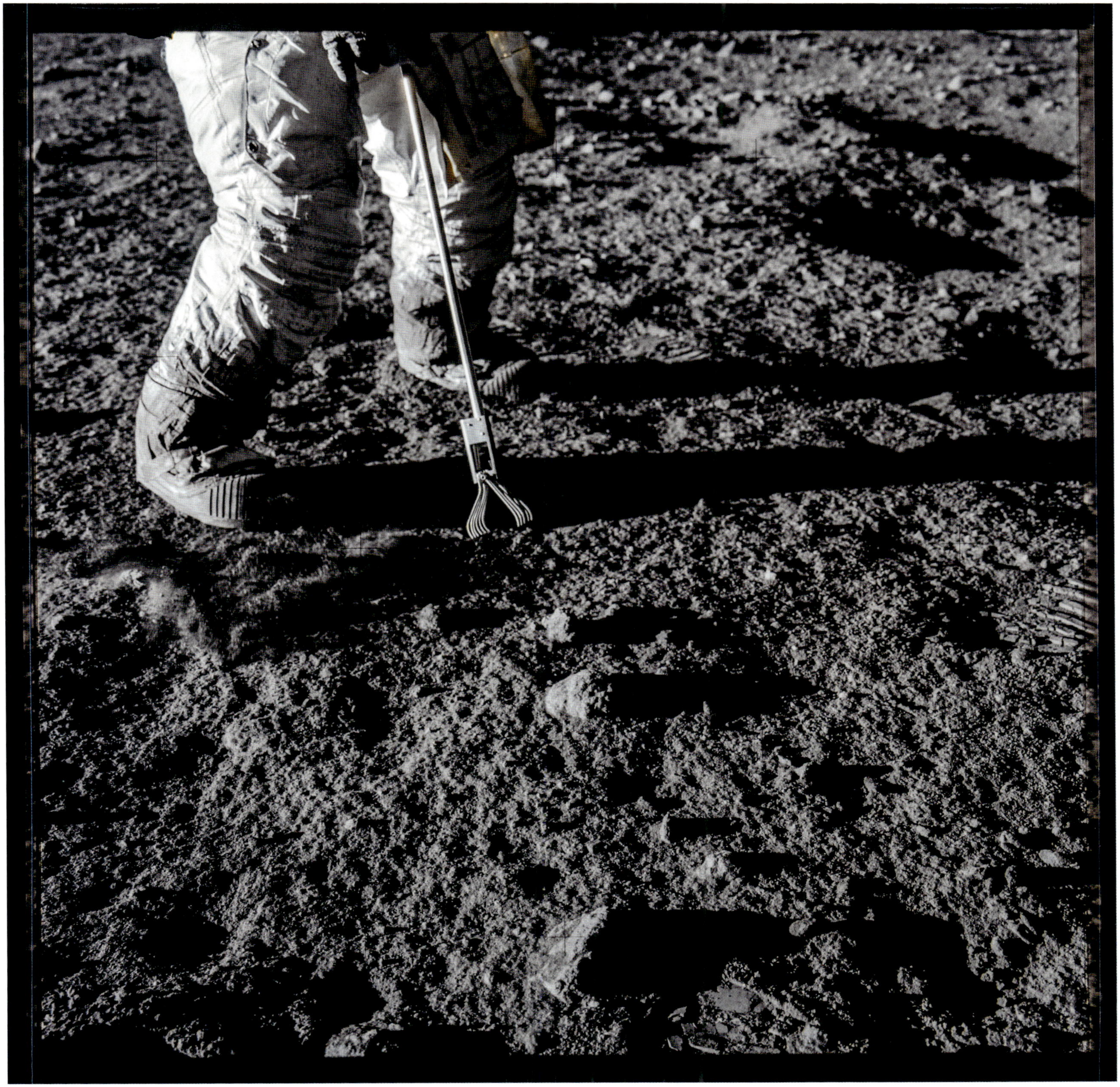

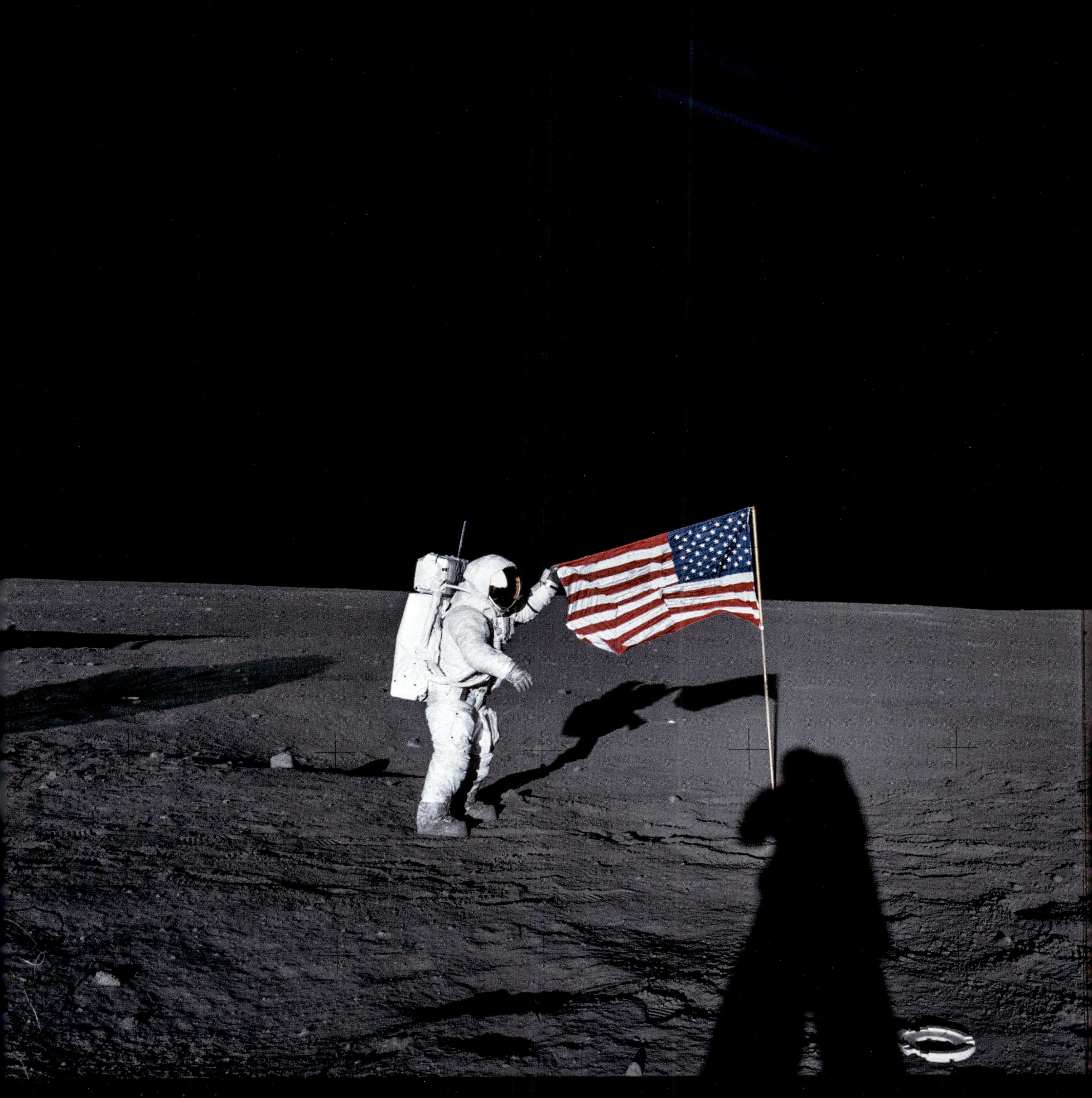

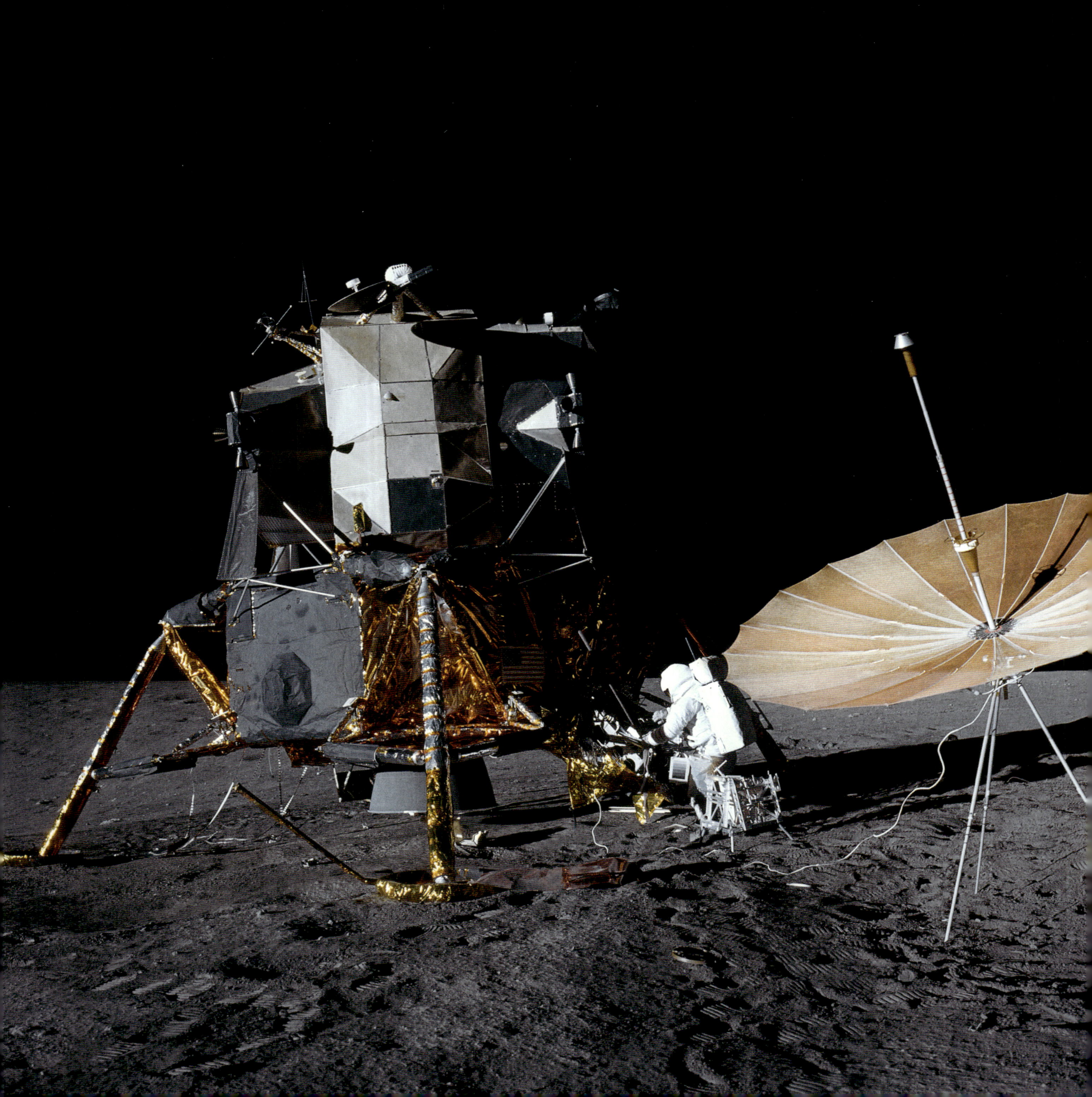

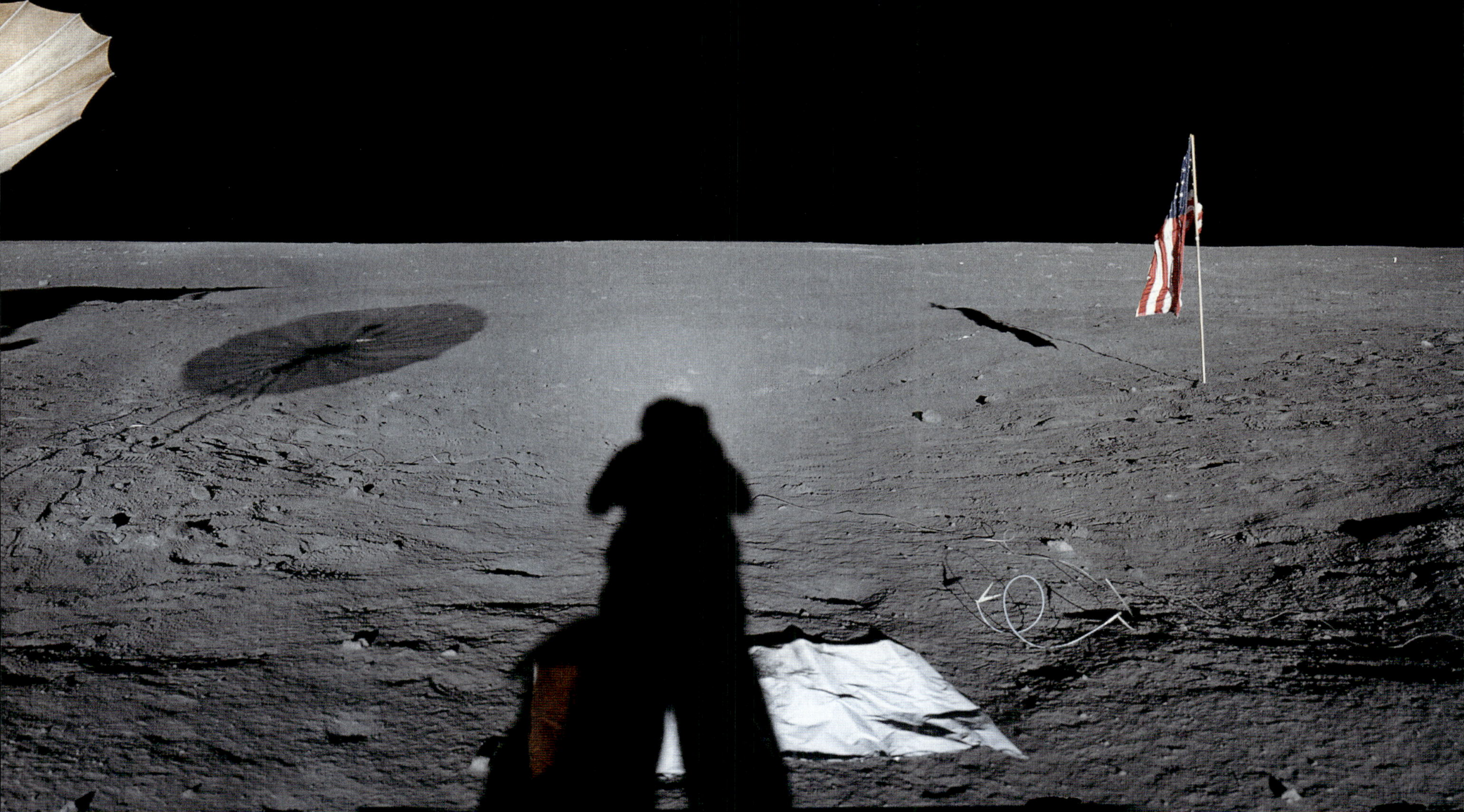

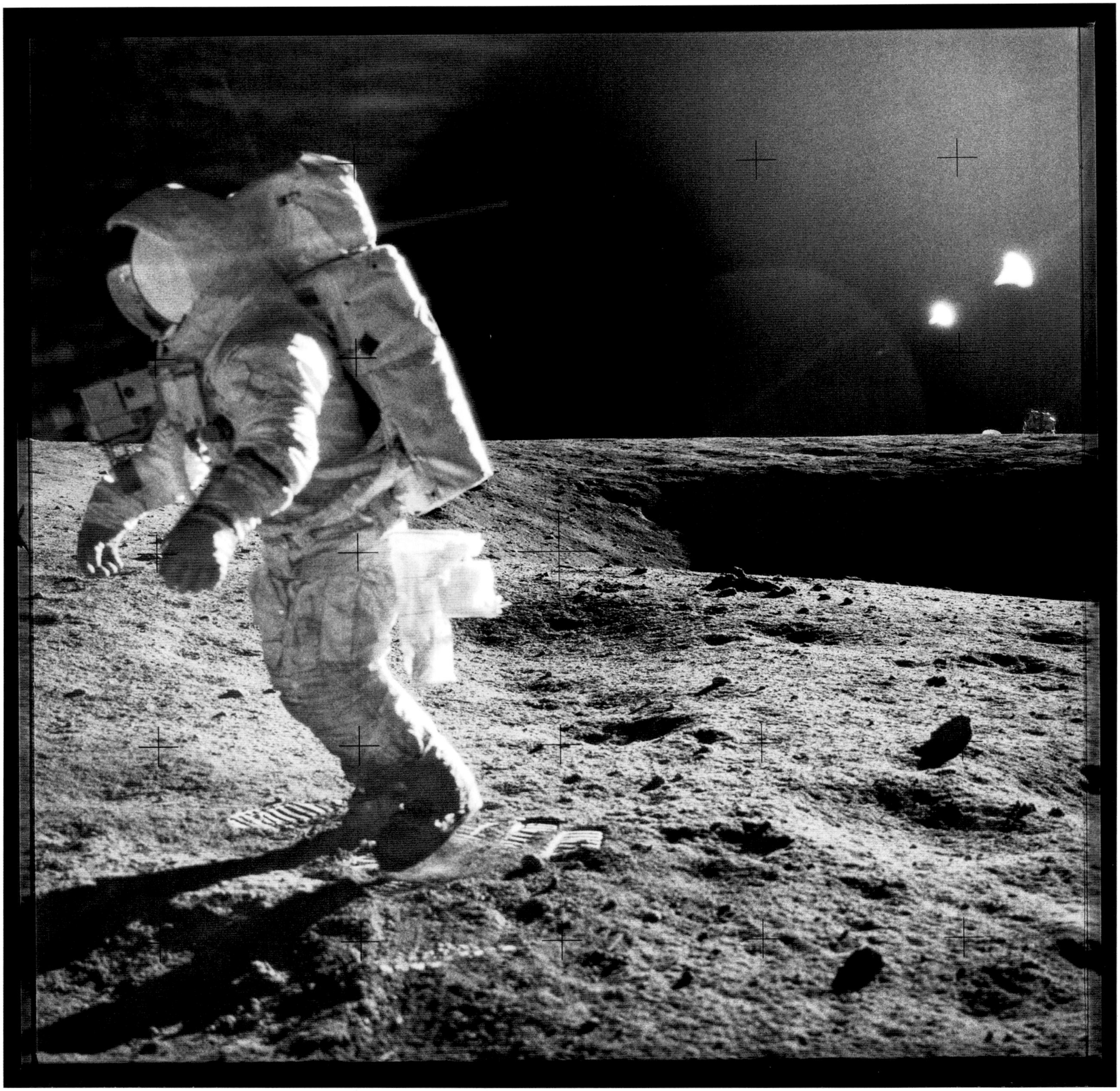

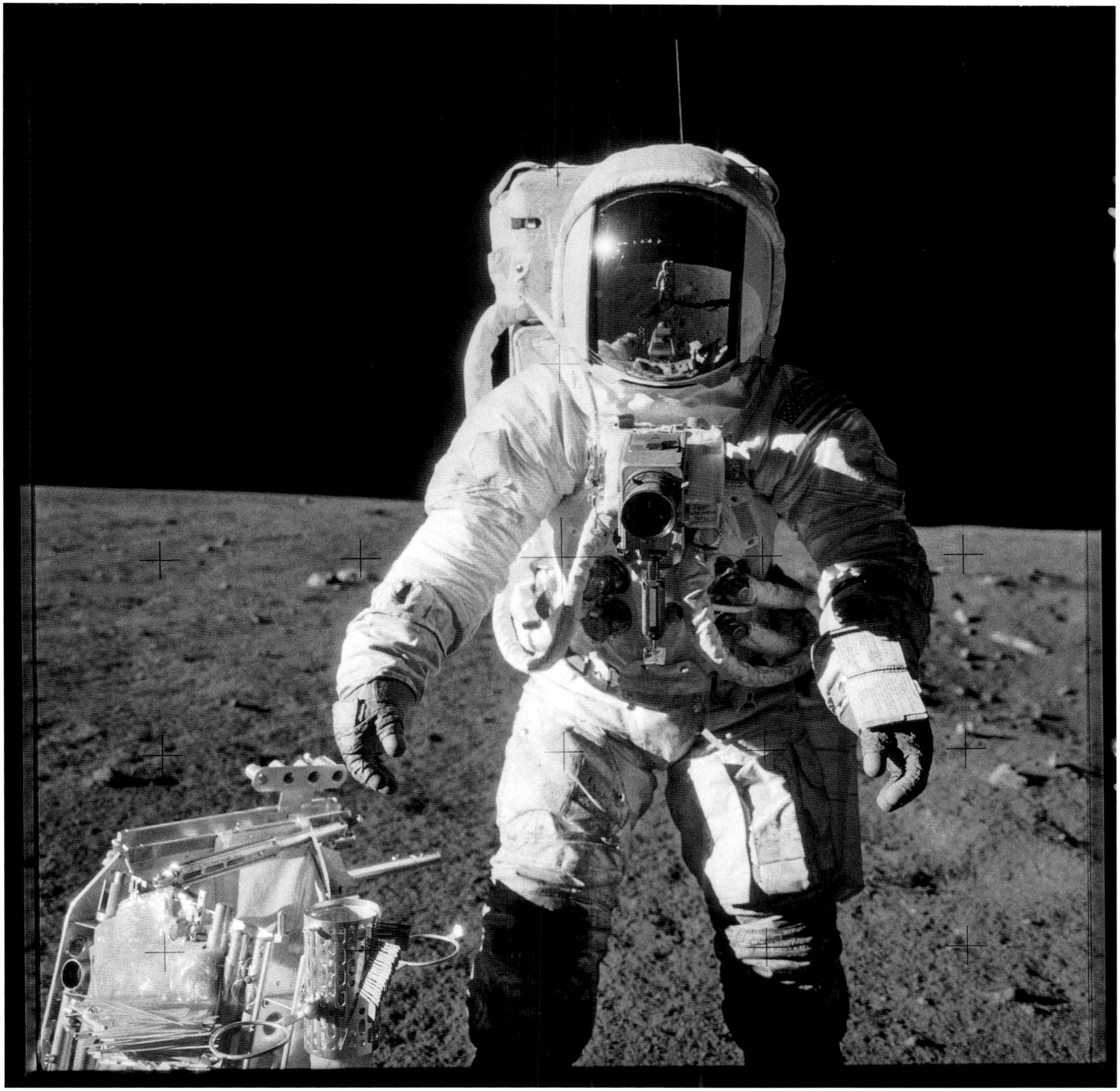

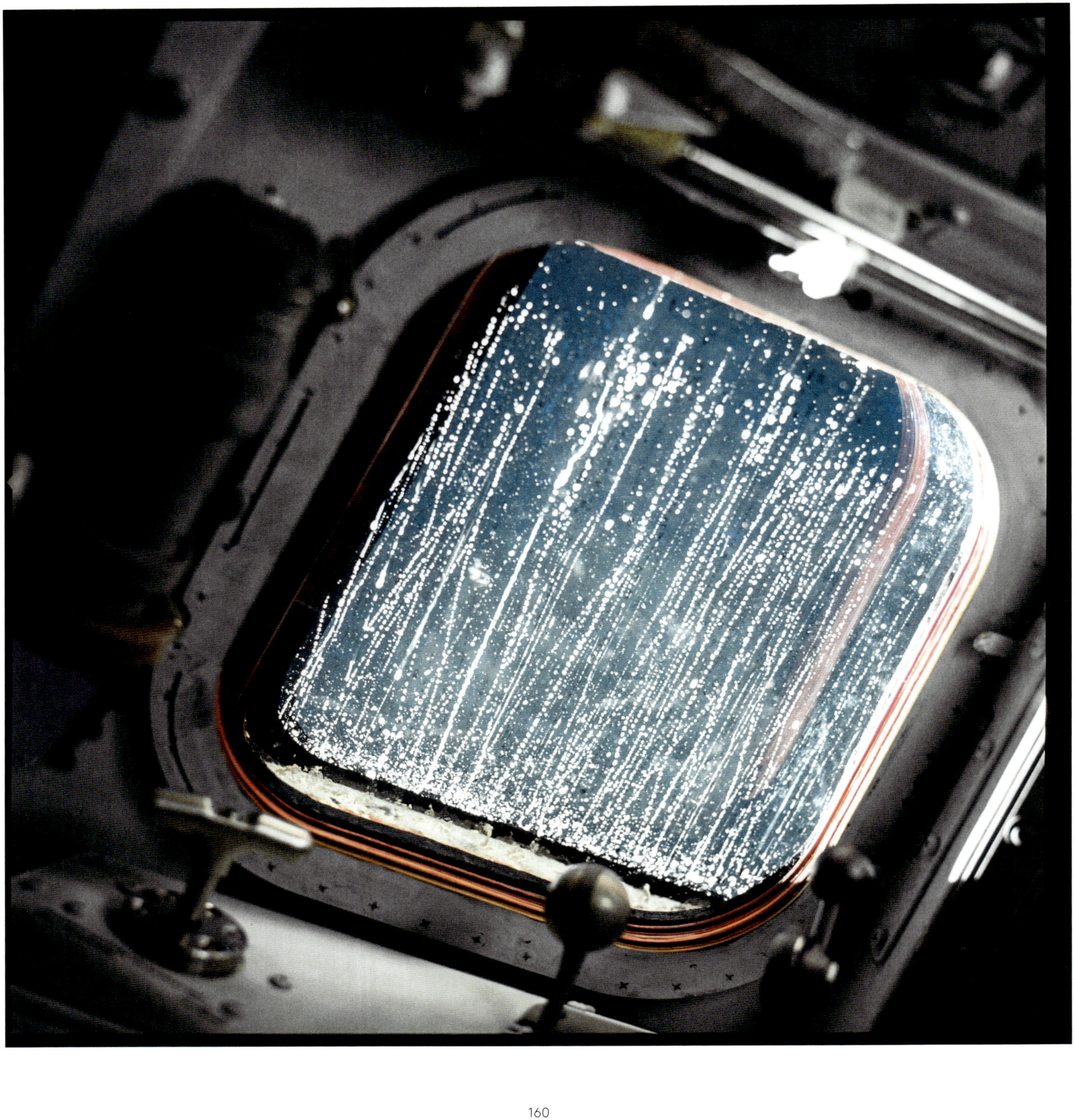

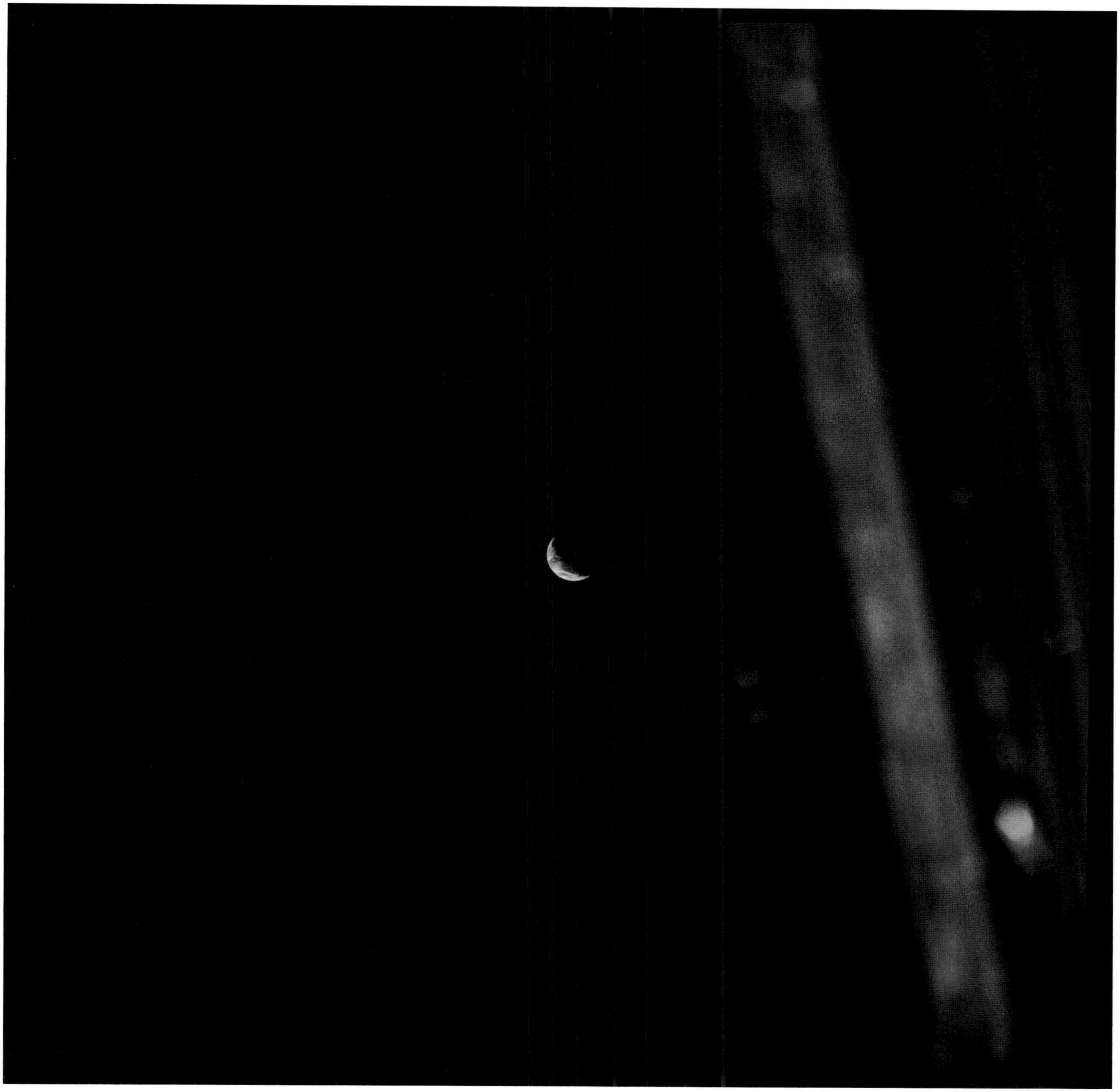

143 (AS12-48-7133)
Apollo 12 Commander Charles 'Pete' Conrad Jr. stands beside the Surveyor 3 probe — marking the only time to date that astronauts have 'caught up' with a robotic spacecraft previously sent to another celestial body. In the background, the Lunar Module *Intrepid* is visible, evidence of Apollo 12's remarkably precise landing near the unmanned lander.

144 (AS12-55-8205)
Apollo 12's Trans-Earth Injection (TEI) phase captured the eastern half of the Moon in stunning detail. The image reveals rugged highlands and deep impact basins, illuminated by long shadows as the spacecraft departed for Earth.

145 (AS12-51-7511)
Intrepid descends toward the Ocean of Storms, its descent engine firing as it prepares for a precision landing near Surveyor 3.

146 (AS12-46-6715)
Conrad descends *Intrepid*'s ladder, becoming the third human to walk on the Moon. Joking about his height, he quipped: "Whoopee! Man, that may have been a small one for Neil, but that's a long one for me."

147 (AS12-46-6779)
Lunar Module Pilot Alan L. Bean stands near Surveyor 3 on the Moon's Ocean of Storms, capturing images of the LM footpad. In the background, the TV camera — accidentally pointed at the Sun by Conrad — was damaged beyond use.

148 (AS12-47-6899)
Apollo 12's Solar Wind Collector (SWC) stands in the foreground, with *Intrepid* and the American flag behind it. The S-band antenna, used for communications, points toward Earth, while Conrad works on the right side of the frame.

149 (AS12-46-6806)
Bean struggles to carry the Apollo Lunar Surface Experiments Package (ALSEP), its mass nearly equal to his own suited weight in lunar gravity. The American flag, S-band antenna, and the LM's gold foil insulation shine under the harsh sunlight.

150 (AS12-57-8448)
A crisp bootprint impression in the fine, powdery lunar soil, captured during Apollo 12. These images were taken to study how the Moon's regolith compacts under pressure, contributing to soil mechanics research that helped scientists understand lunar surface properties better.

151 (AS12-49-7278)
Bean holds the Special Environmental Sample Container (SESC), nearly full with lunar soil. Reflected in his visor, Conrad stands nearby after helping to fill the vacuum-sealed container. The Hasselblad camera, checklist, and Omega watch on Bean's sleeve offer a detailed look at the astronauts' equipment during lunar surface operations.

152 (AS12-49-7243)
Bean's suit is covered in lunar dust as he collects a core sample at Bench Crater. The stiff spacesuit made bending down difficult, while the fine lunar soil clung to everything.

153 (AS12-47-6932)
Conrad collects a sample using an Apollo Lunar Hand Tool near the ALSEP site. The spray of dust near his foot shows how lunar soil reacts differently in the Moon's airless environment.

155 (AS12-47-6896)
Conrad holds up the American flag on the lunar surface. His chest-mounted camera is missing — he had placed it on the Modular Equipment Stowage Assembly (MESA) while setting up the S-band antenna, whose shadow appears in the background.

156-157
(AS12-47-6988–AS12-47-6988)
Conrad stands beside the MESA at *Intrepid*'s side, preparing tools during the first EVA. The S-band antenna, prominently centered in the panorama, beautifully glows in the sunlight. To the right, the long shadow of Bean — who took the image — stretches across the lunar surface, capturing a moment of teamwork on the Ocean of Storms.

158 (AS12-49-7213)
Bean walks near Head Crater, carrying sample bags attached to his saddlebag, while the trigger assembly of his Hasselblad camera is clearly visible. In the background, *Intrepid* stands on the lunar surface.

159 (AS12-49-7281)
Bean stands on the Moon with the Hand Tool Carrier (HTC) at his right hand, loaded with sample bags, a large collection bag, and a core-tube-cap dispenser.

160 (AS12-50-7372)
A view through the left window of Apollo 12's Command Module (CM) *Yankee Clipper*, showing how fluids behave in microgravity.

161 (AS12-50-7396)
A distant, partially illuminated Earth (quarter phase) seen from the LM window. The approximate photo scale is 1:15,000,000, showing Earth as a fragile sphere against the vastness of space.

XIII

Apollo 13

Launch: April 11th, 1970
Splashdown: April 17th, 1970

Command Module: Odyssey
Lunar Module: Aquarius

The Apollo program's seventh manned mission was the third intended to land on the lunar surface. That did not happen, however, and the story of Apollo 13 remains one of the most well-known in space exploration.

The dramatic events began around 56 hours after launch, when an oxygen tank exploded, damaging a large part of the Service Module and putting the crew and spacecraft in serious danger. By aborting their plan to reach the Moon and instead powering up the Lunar Module as a 'lifeboat' to conserve resources, the crew made it back to Earth by the slimmest of margins. The mission became one of history's best-known survival stories, later popularized by the 1995 Hollywood film Apollo 13.

Not only did the crew and Mission Control on Earth pull off one of the greatest rescue missions ever, but the astronauts also managed to document some of their achievements in the few photos they were able to take.

The crew of Apollo 13 still holds the record for traveling the farthest away from Earth. From nearly 250,000mi (400,000km) out, they used the gravitational force of the Moon to slingshot back to Earth and achieve a safe landing in the South Pacific Ocean.

A photograph of the improvised carbon dioxide removal system the crew constructed can be seen on page 173. The damage to the exterior of the Service Module, which exposed its fuel cells, can be found on page 178.

James A. 'Jim' Lovell, Jr.
Commander

Born on March 25th, 1928 in Cleveland, Ohio. He was Backup Pilot of Gemini 4, Pilot of Gemini 7, Backup Command Pilot of Gemini 9A, Command Pilot of Gemini 12, Command Module Pilot of Apollo 8 and Backup Commander of Apollo 11.

John L. Swigert, Jr.
Command Module Pilot

Born August 30th, 1931 in Denver, Colorado. He was scheduled to be the Backup Command Module Pilot of Apollo 13 but was moved to the prime crew 72 hours before the launch in place of Ken Mattingly, who had been exposed to measles.

Fred W. Haise, Jr.
Lunar Module Pilot

Born November 14th, 1933 in Biloxi, Mississippi. He was Backup Lunar Module Pilot of Apollo 8 and Apollo 11 and Backup Commander of Apollo 16. He also commanded one of the crews that flew Space Shuttle Approach and Landing Tests.

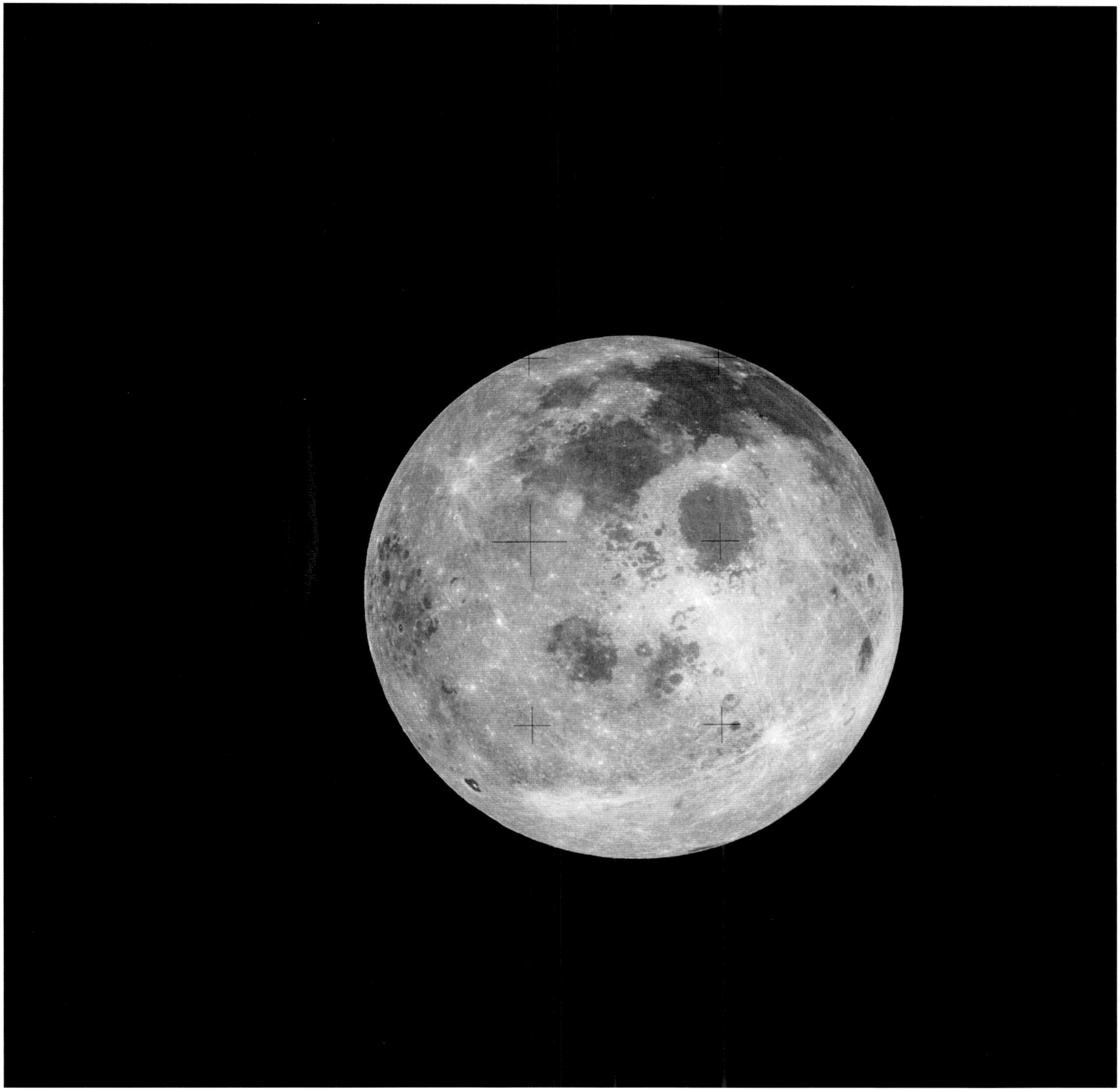

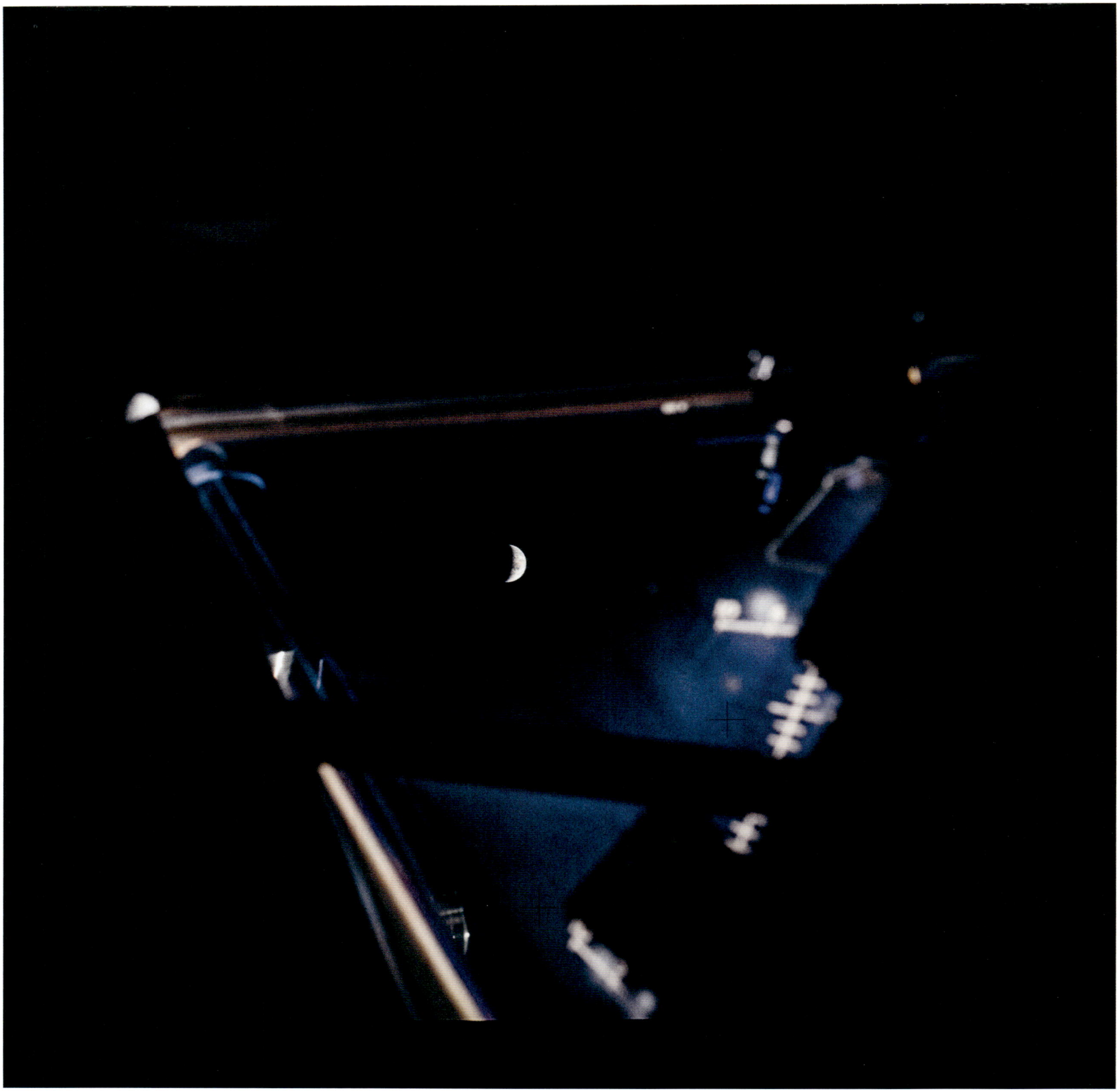

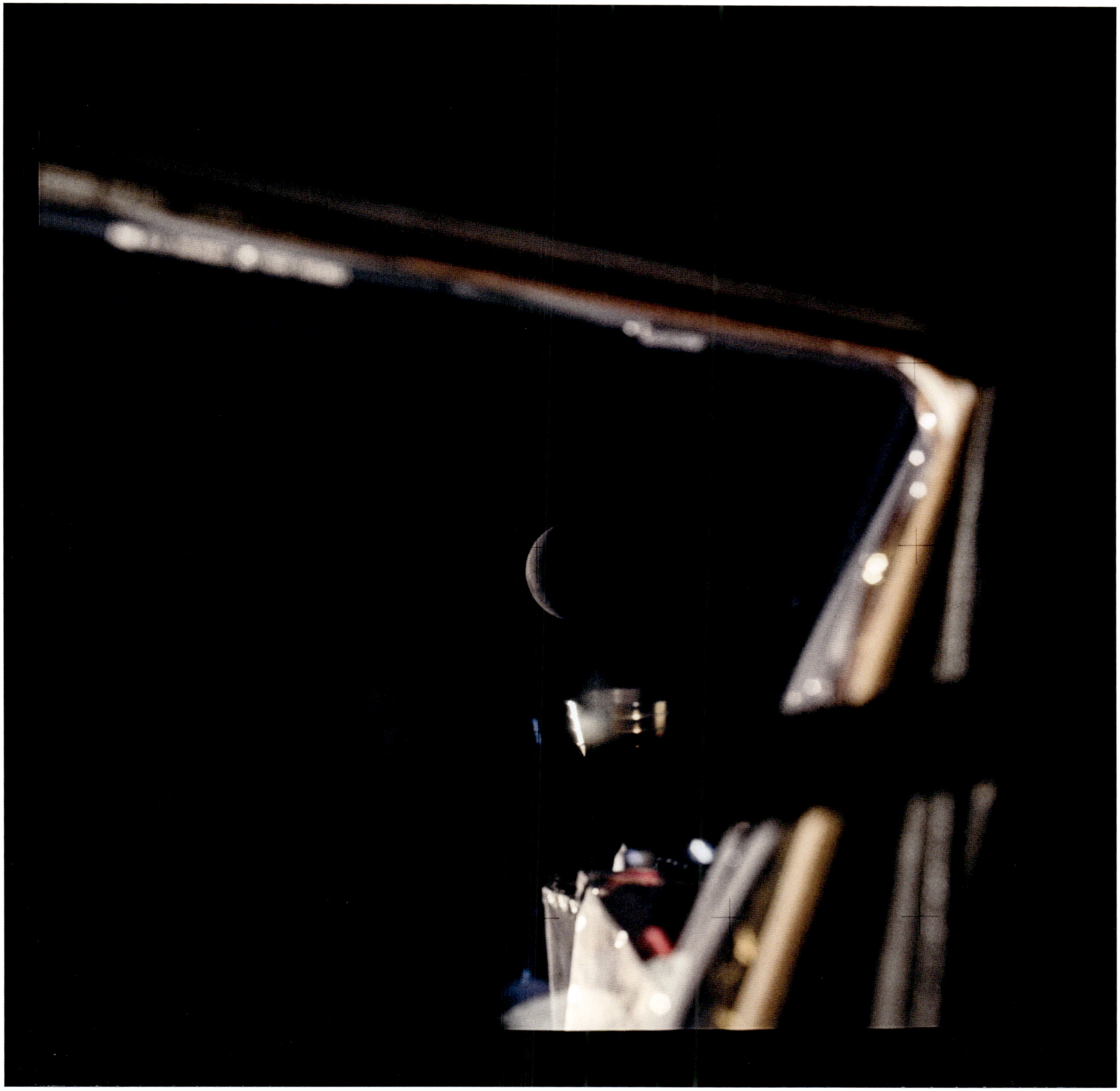

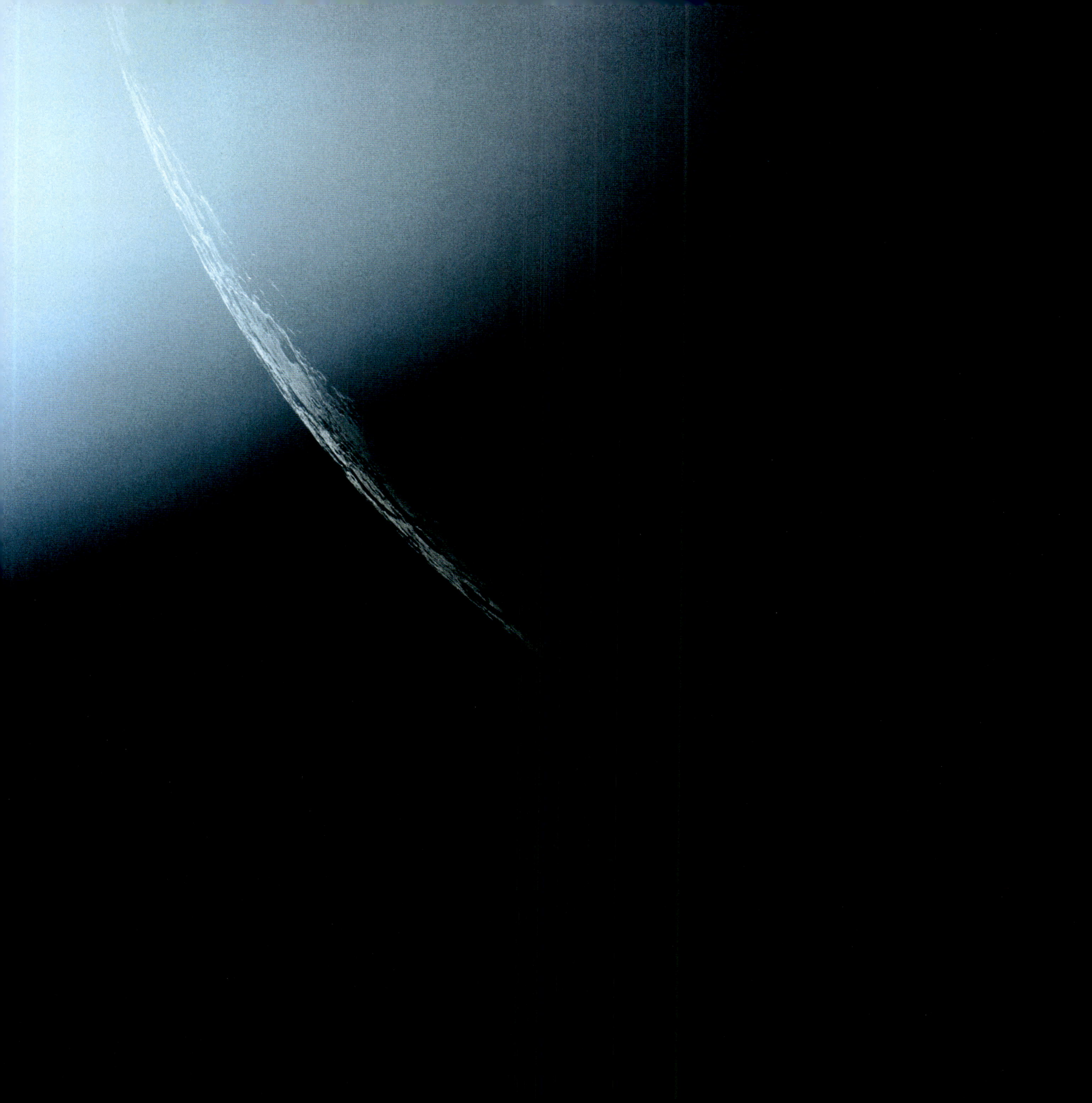

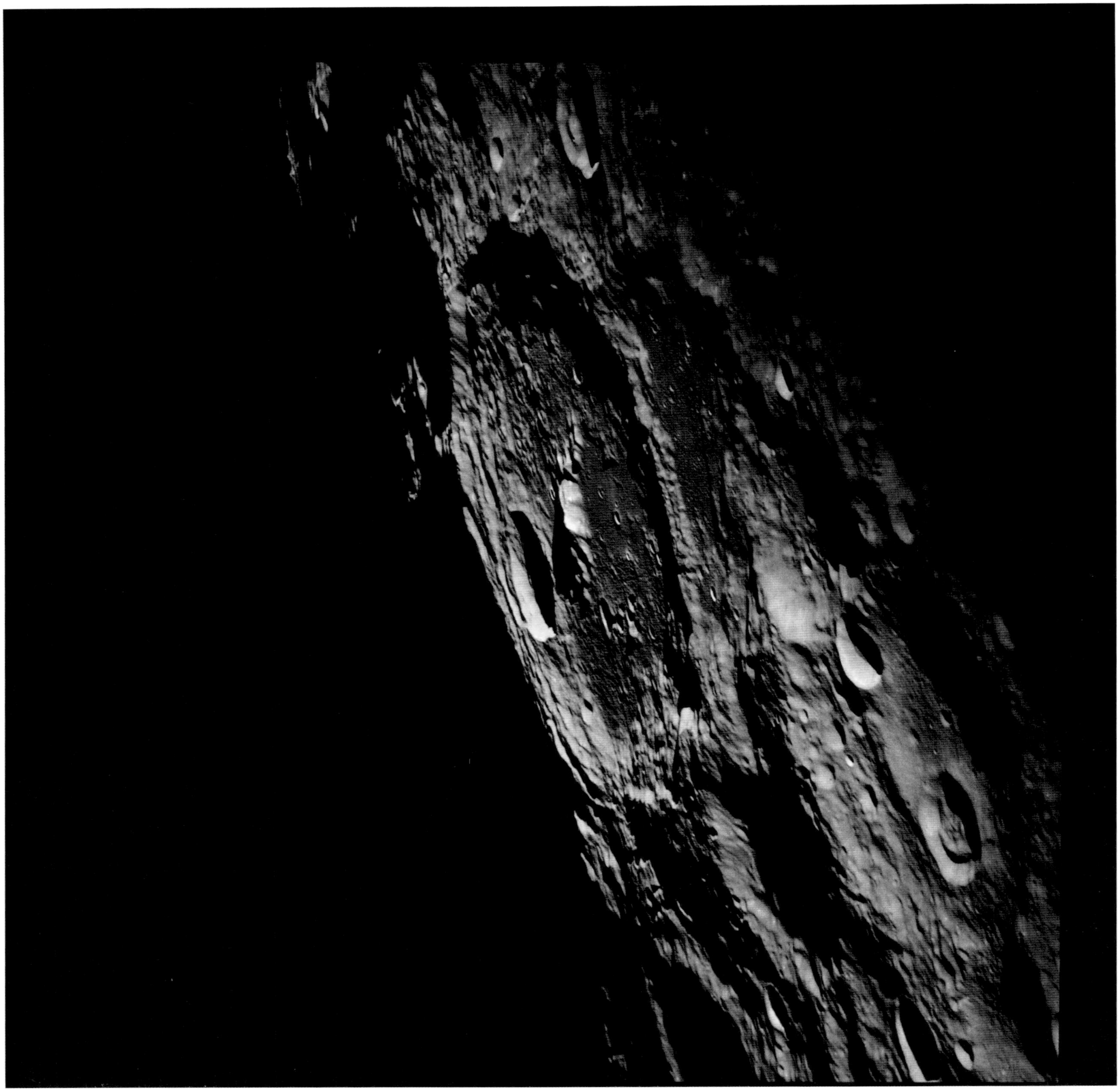

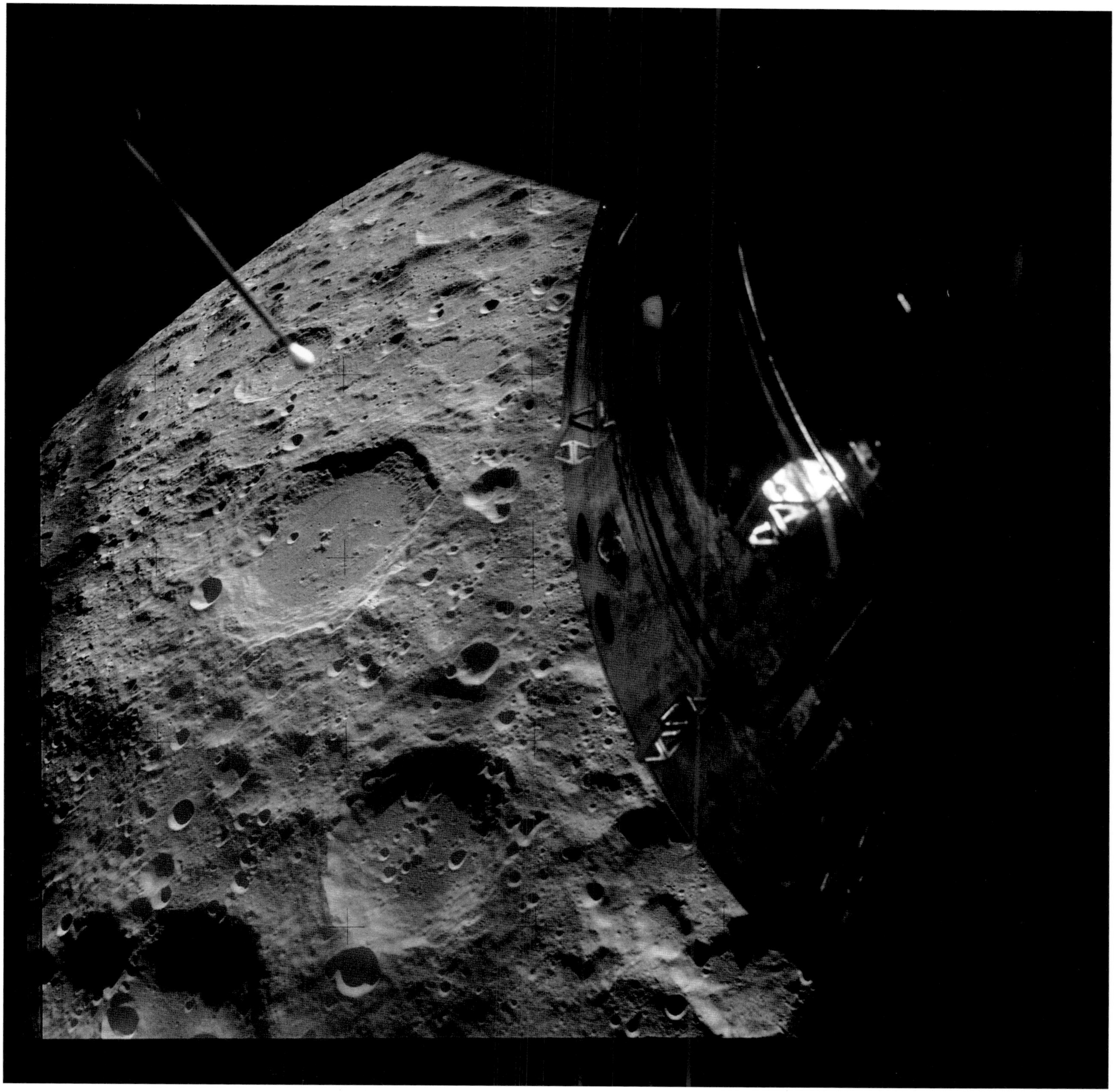

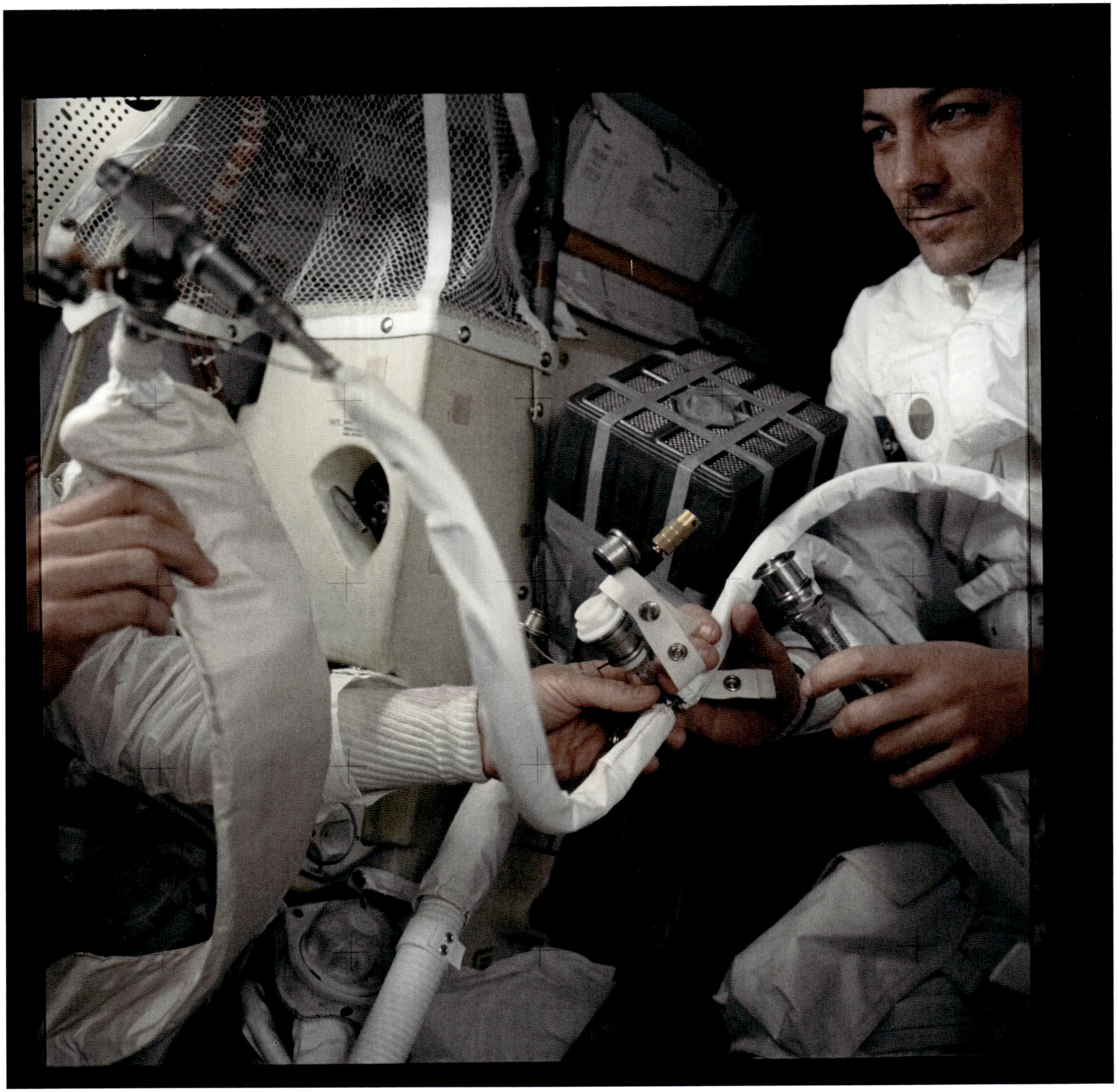

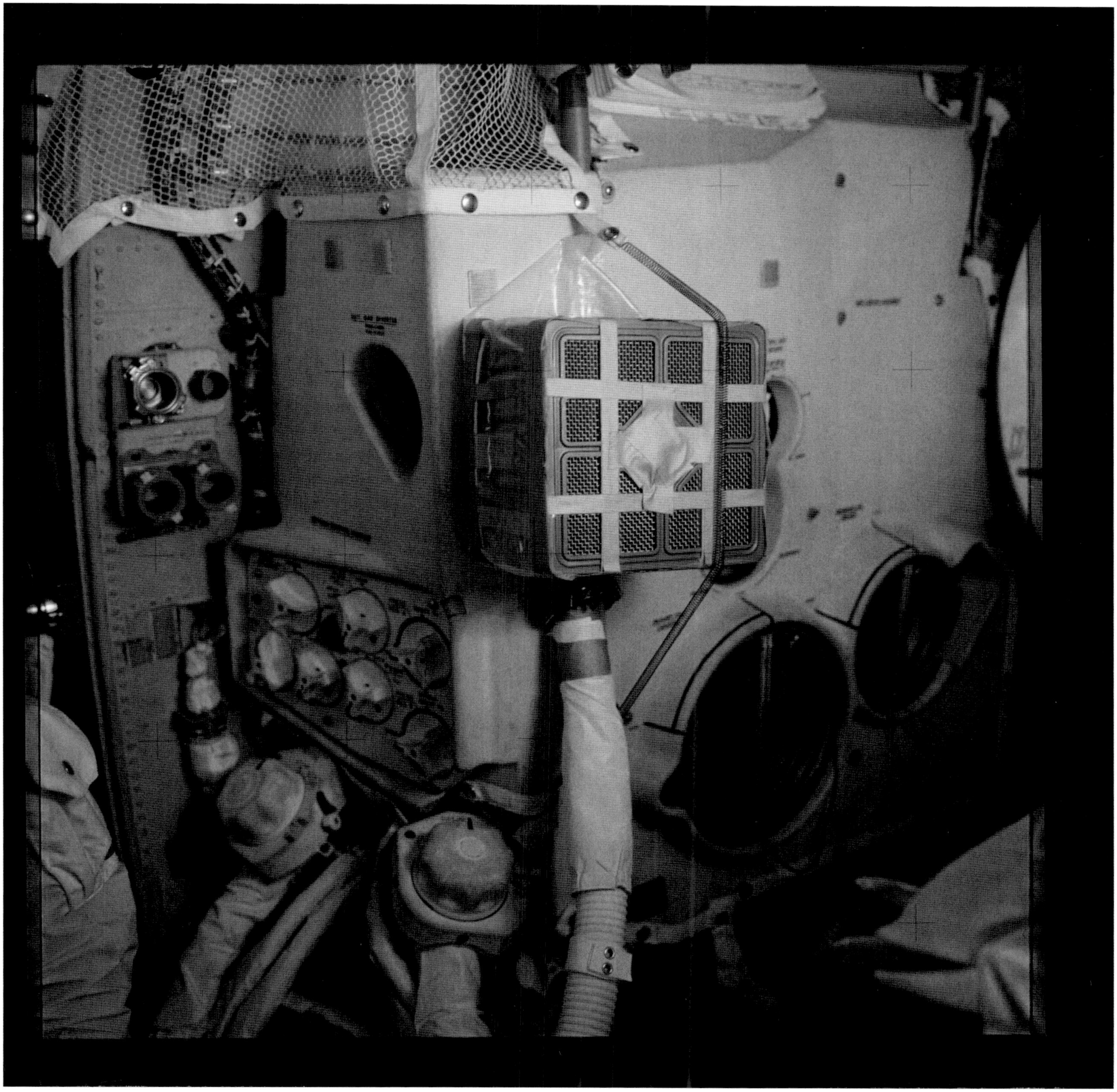

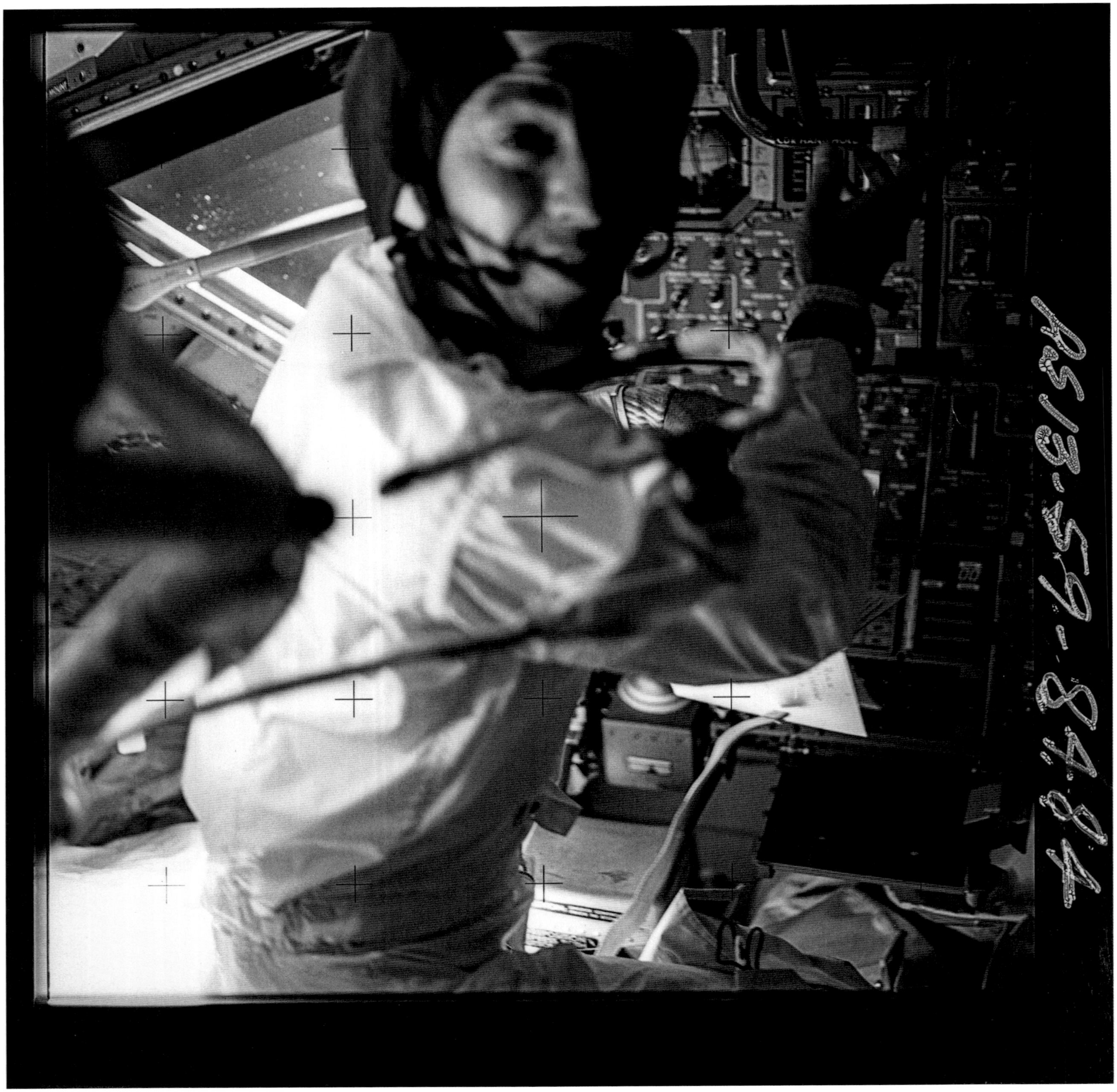
AS13-59-8484

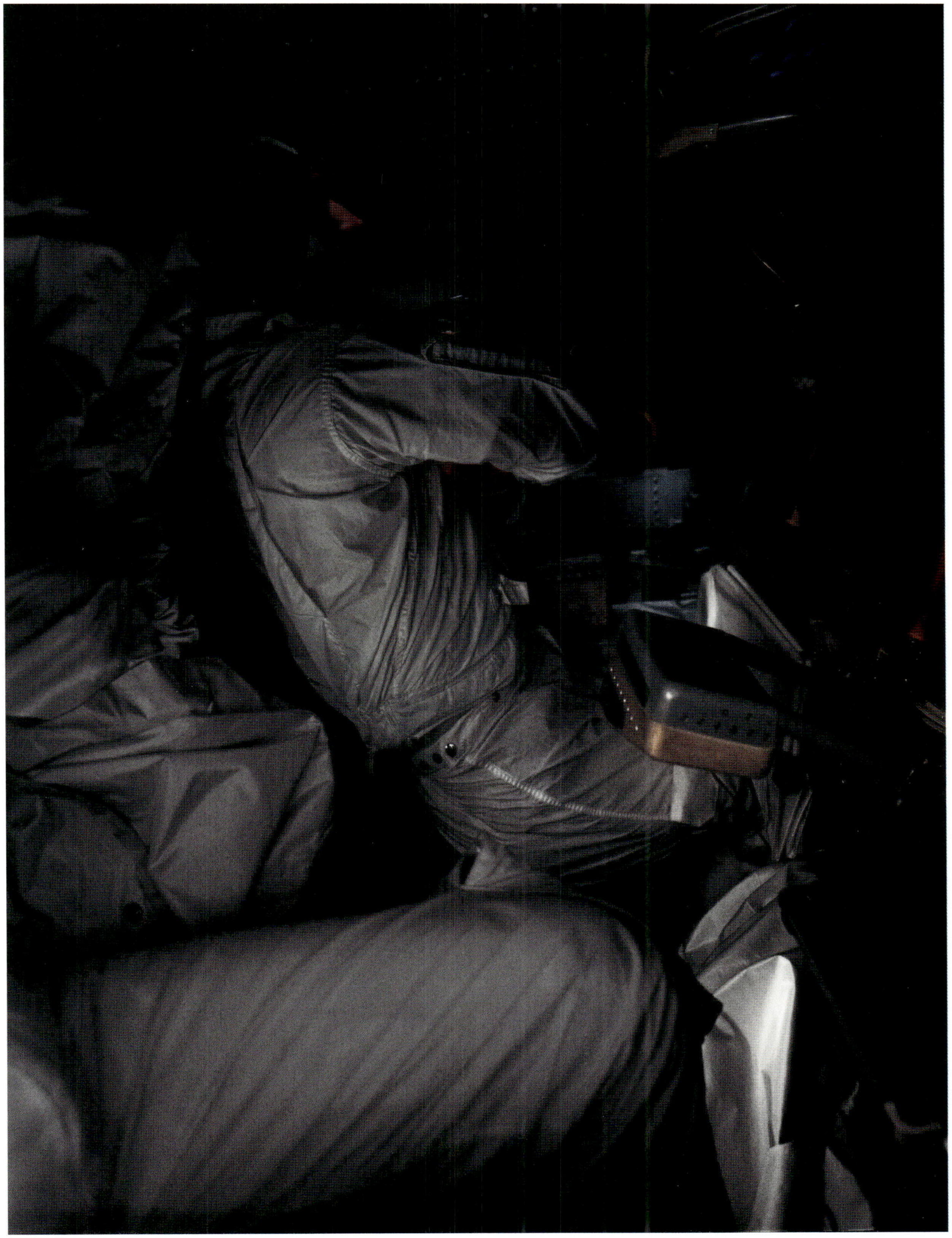

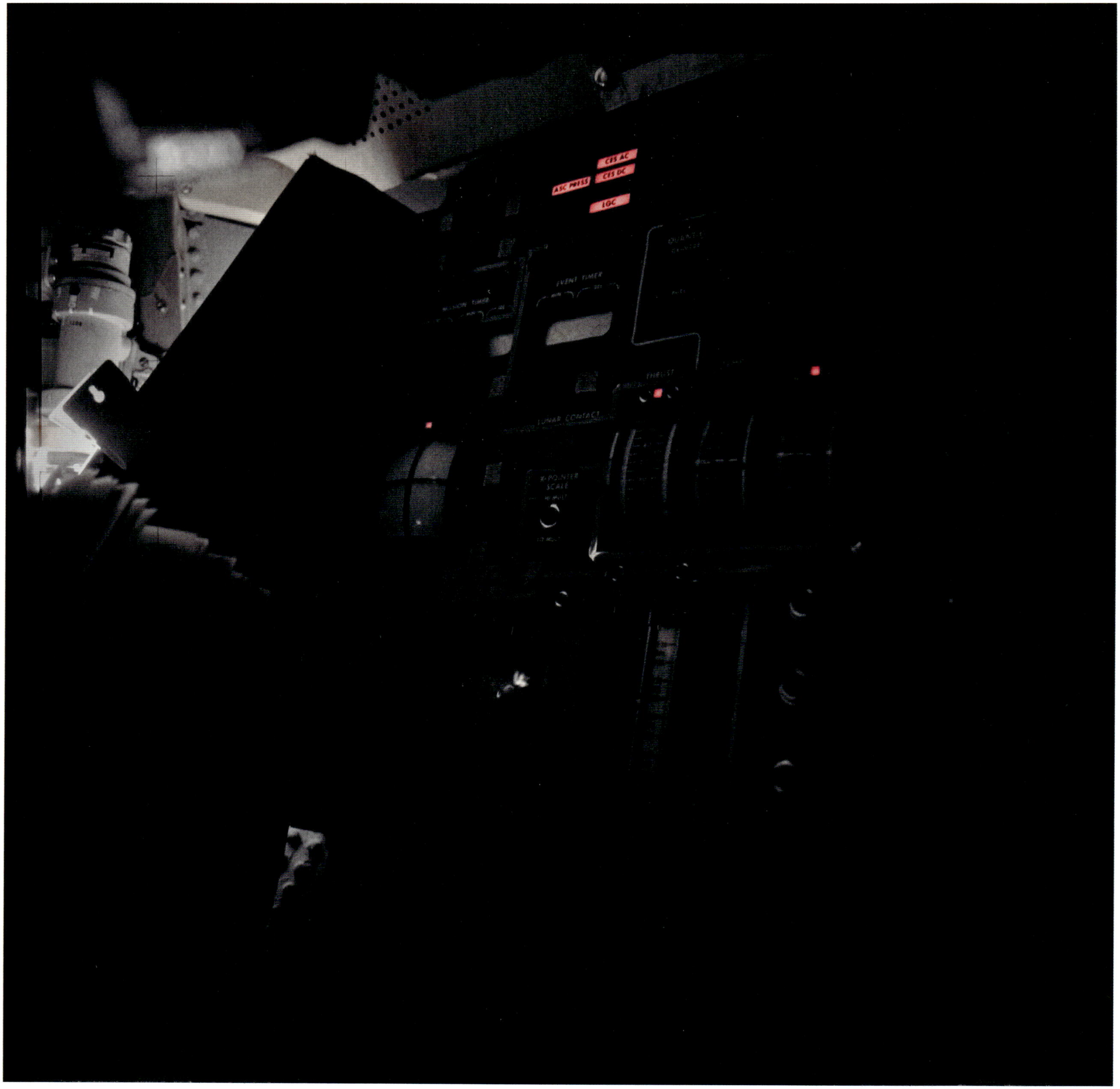
CES AC
ASC PRESS
CES DC
LGC
EVENT TIMER

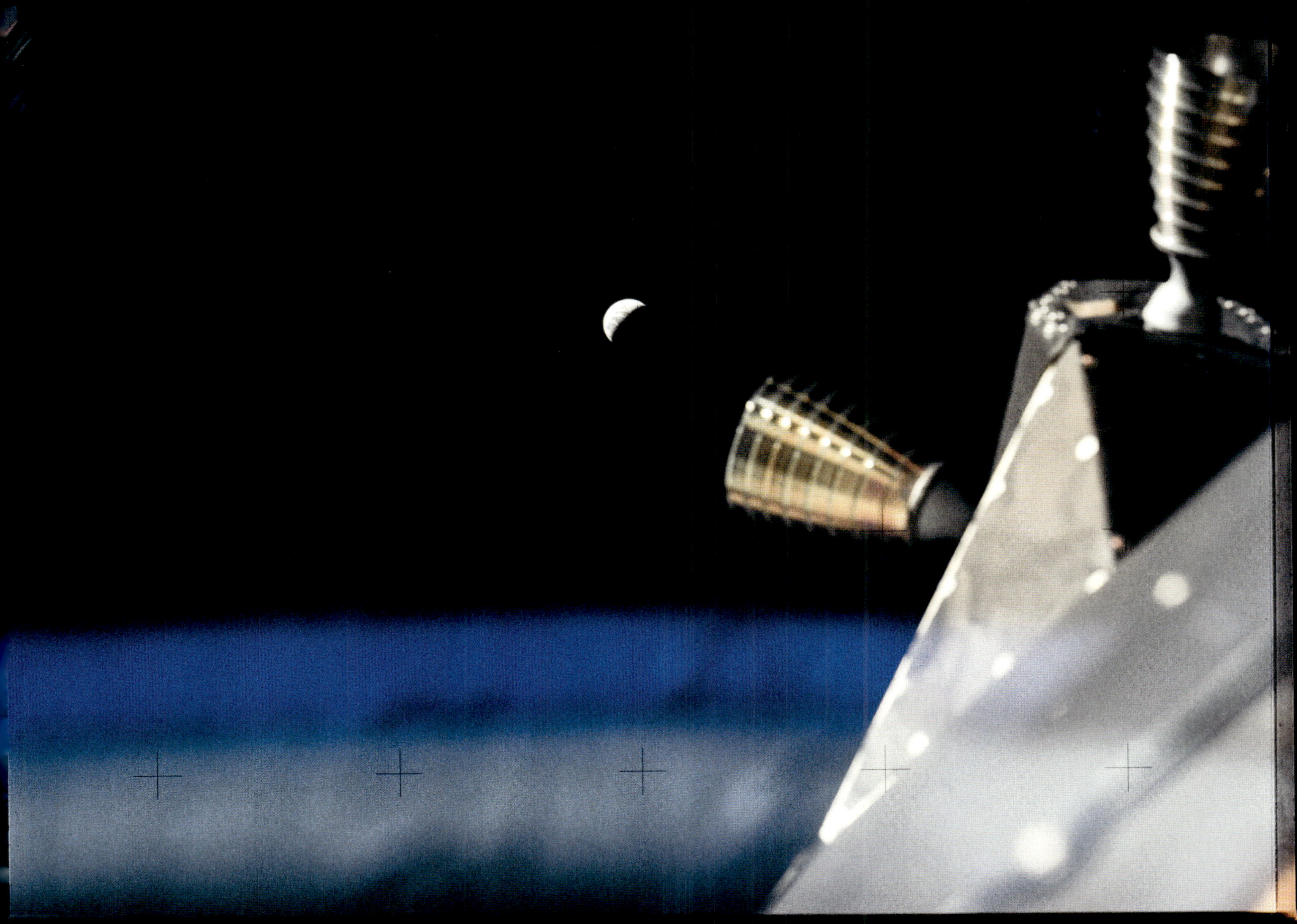

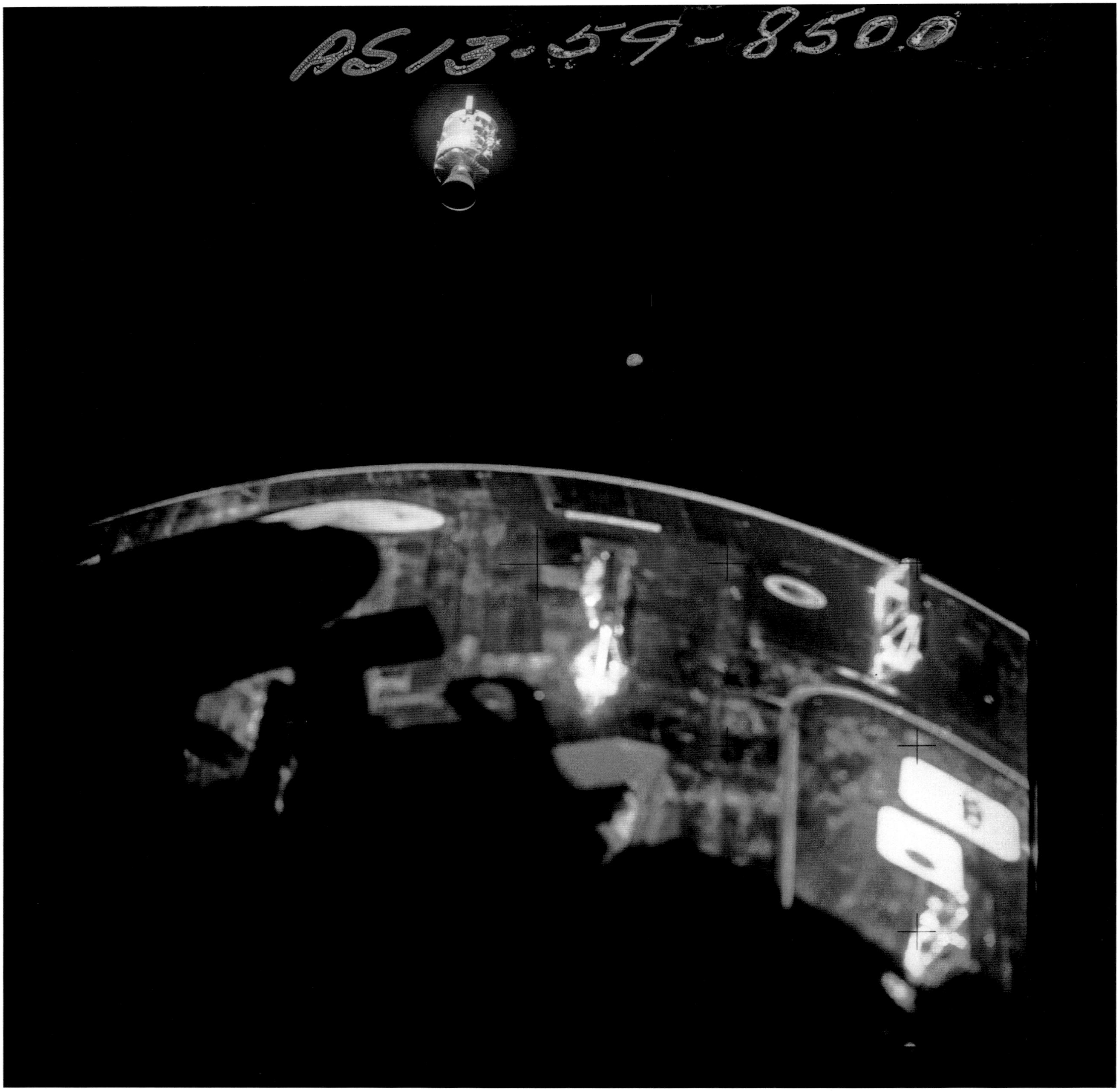
AS13-59-8500

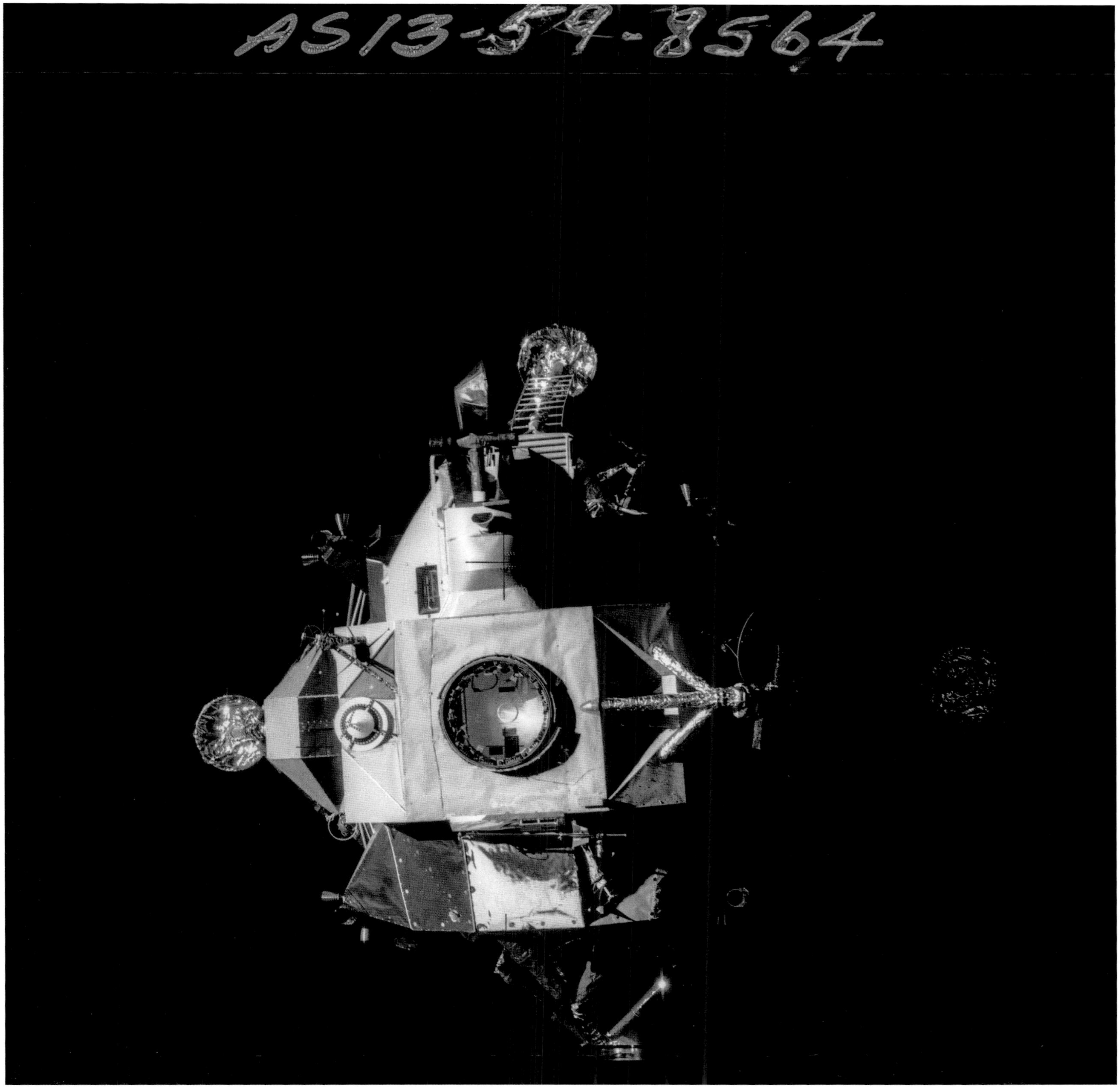
AS13-59-8564

165 (AS13-61-8795)
A bright, fully illuminated lunar disc, seen during Apollo 13's return journey to Earth. Captured as the crew navigated their crippled spacecraft home, this striking image highlights the Moon's rugged surface against the vast darkness of space.

166 (AS13-62-8888)
A delicate crescent Earth, seen from Lunar Module (LM) *Aquarius* during Apollo 13's return journey.

167 (AS13-62-8881)
A crescent Moon, seen from *Aquarius* as Apollo 13 looped around the Moon. This striking view was captured during the crew's tense return journey.

169 (AS13-60-8606)
A striking crescent Moon, captured from Apollo 13 as the crew rounded the lunar far side. The image's subtle blue hue results from lighting conditions and film exposure, adding an ethereal quality to the view.

170 (AS13-60-8635)
Keeler Crater, a 59mi (95km)-wide impact site on the Moon's far side, captured during Apollo 13. The image showcases the crater's rugged rim and central peak.

171 (AS13-62-8912)
Chaplygin Crater, an 84mi (135km)-wide impact feature on the Moon's far side, as seen by Apollo 13. Its eroded walls and smooth floor suggest its ancient origins.

172 (AS13-62-9004)
Command Module Pilot John L. Swigert Jr. (right) and Commander James A. Lovell Jr. repurpose equipment for survival inside LM *Aquarius*. The jury-rigged "mailbox" in the background allowed them to scrub carbon dioxide from the air, a life-saving innovation during their emergency return to Earth.

173 (AS13-62-8929)
A close-up of Apollo 13's improvised "mailbox," a life-saving carbon dioxide scrubber made from plastic bags, cardboard, and duct tape. Designed by engineers at Mission Control and built by the crew in space, this fix adapted CM lithium hydroxide canisters for use in LM *Aquarius*, preventing carbon dioxide buildup and ensuring survival.

174 (AS13-59-8484)
A candid, slightly out-of-focus portrait of Lovell at his station inside *Aquarius*. This poignant image captures one of his final moments in the LM before the crew moved back to CM *Odyssey*, leaving behind the spacecraft that had sustained them through their harrowing return journey.

175
(AS13-62-8988–AS13-62-8990)
An unusual stitched series of photos, revealing the cramped interior of LM *Aquarius*, with Lovell seen resting during Apollo 13's return to Earth. As the LM served as a lifeboat, the crew endured freezing temperatures and limited power.

176 (AS13-62-8932)
A detailed view of the LM control panel, showing how Apollo 13's crew conserved power by running systems at minimal levels. With most displays dark, only essential instruments remained active, helping to stretch the LM's limited resources and ensure survival on the return to Earth.

177 (AS13-61-8867)
A striking view of a crescent Earth, with *Aquarius* in the foreground. Captured during Apollo 13's return journey, this image highlights the fragile beauty of Earth from deep space.

178 (AS13-59-8500)
The damage to Apollo 13's Service Module (SM) is clearly visible in this photo, taken after it was jettisoned before reentry. The explosion of an oxygen tank tore through the module, exposing internal structures and crippling key systems.

179 (AS13-59-8564)
An iconic moment as the Apollo 13 crew jettisons their lifeboat, *Aquarius*, before reentry. Originally built for a Moon landing, *Aquarius* instead sustained the astronauts after the oxygen tank explosion, making its final separation a poignant farewell before their safe return to Earth.

Apollo 14

XIV

Alan B. Shepard, Jr.

Stuart A. Roosa

Edgar D. Mitchell

Apollo 14

Launch:	January 31st, 1971
Lunar landing:	February 5th, 1971
Splashdown:	February 9th, 1971

Command Module:	Kitty Hawk
Lunar Module:	Antares

Apollo 14 was the third Apollo mission to land on the Moon, with the crew spending two days on the lunar surface. Getting there proved challenging when two significant problems occurred during descent.

The first issue arose when the Lunar Module's computer received a false 'ABORT' signal caused by a faulty switch. If not resolved immediately, this would have triggered an automatic sequence separating the ascent stage from the descent stage — effectively aborting the landing. With real-time guidance from NASA and MIT, Lunar Module Pilot Edgar Mitchell manually entered software modifications to override the signal just in time.

A second issue emerged when the landing radar failed to lock onto the Moon's surface, depriving the guidance computer of altitude and descent rate data. With only moments to spare, the radar finally acquired a signal at 18,000ft (5.48km), enabling Commander Alan Shepard to manually pilot the Lunar Module to a near-perfect landing — within 100ft (30m) of the target — making it the most precise touchdown of all the Apollo lunar missions.

Shepard's A7-L spacesuit was the first to feature red stripes on the arms, legs, and helmet, making it easier to identify him in photographs during extravehicular activities. Many of the scientific instruments and experimental setups used during the mission were documented in photographs, several of which appear in this chapter.

Alan B. Shepard, Jr.
Commander

Born on November 18th, 1923 in East Derry, New Hampshire. He became America's first man in space as the pilot of the Mercury-Redstone 3 spacecraft named Freedom 7. He was Backup Pilot of Mercury 9 and became the fifth man to walk on the Moon where he famously hit two golf balls.

Stuart A. Roosa
Command Module Pilot

Born on August 16th, 1933 in Durango, Colorado. He was chosen with the fifth group of astronauts in 1966 and was the Backup Command Module Pilot of Apollo 16 and 17. He resigned from NASA in 1976.

Edgar D. Mitchell
Lunar Module Pilot

Born on September 17th, 1930 in Hereford, Texas. He was chosen with the fifth group of astronauts in 1966, and was Backup Lunar Module Pilot of Apollo 10 and Apollo 16. He was the sixth man to walk on the Moon. He retired from the Navy and NASA in October 1972.

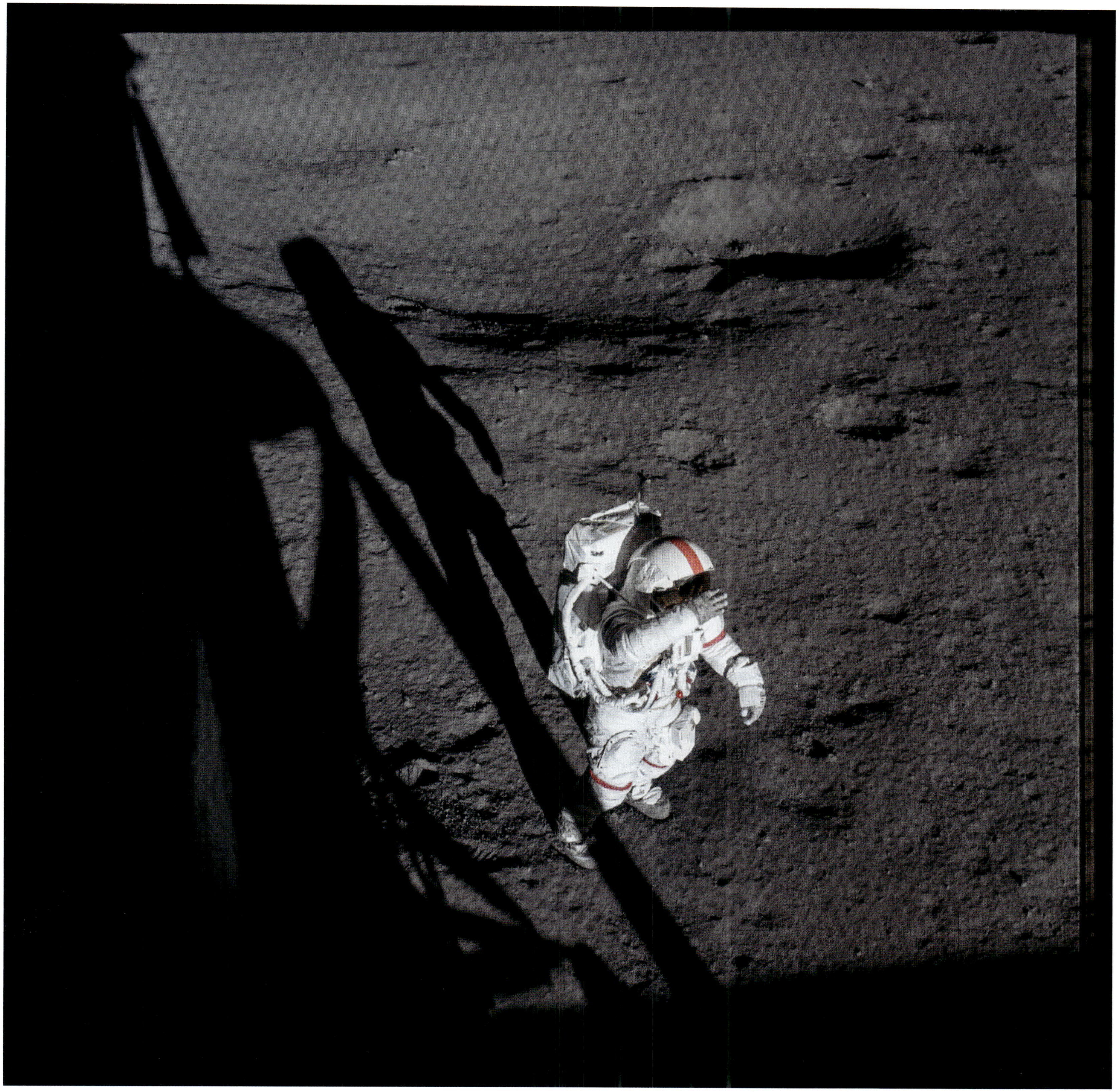

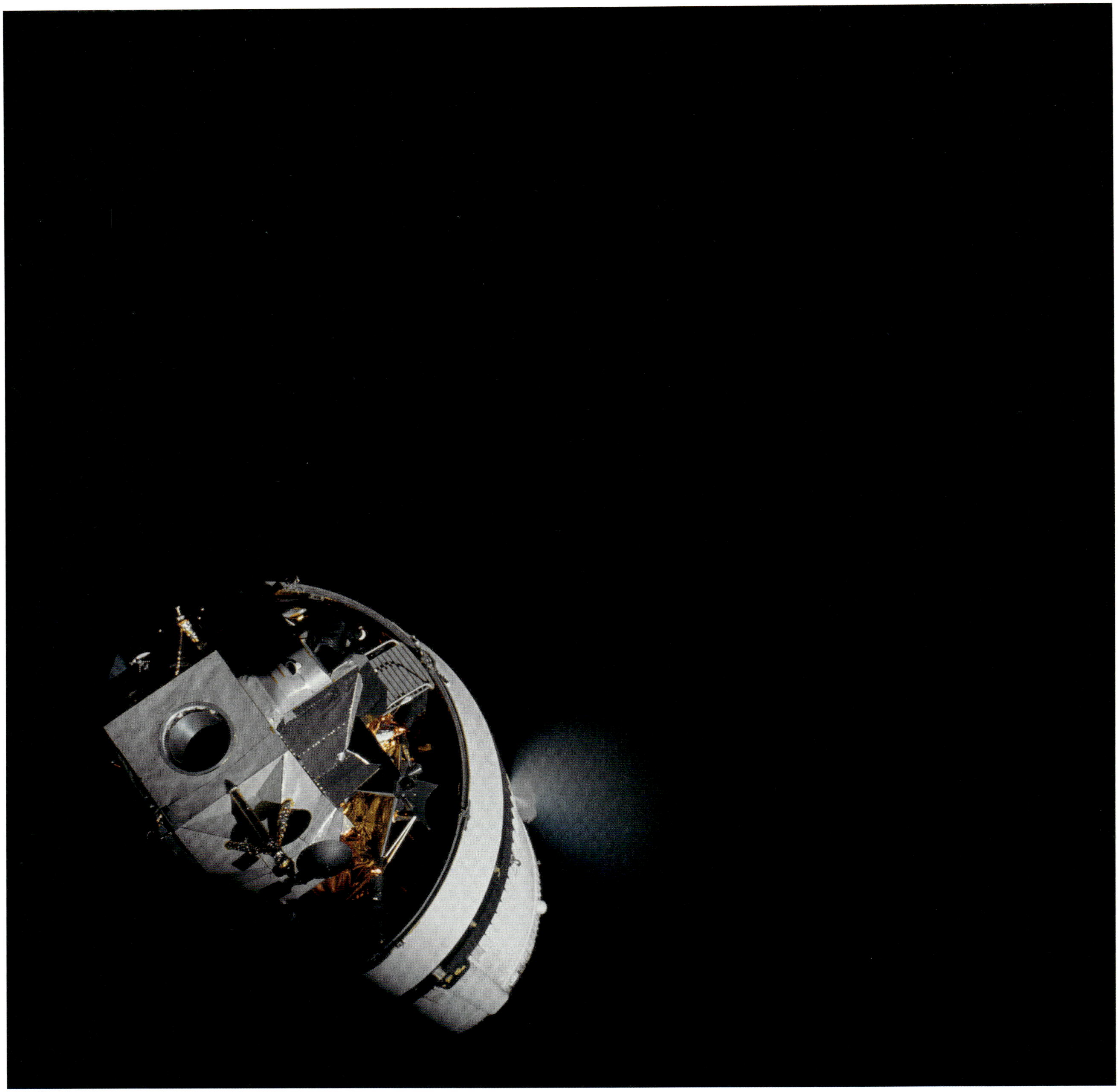

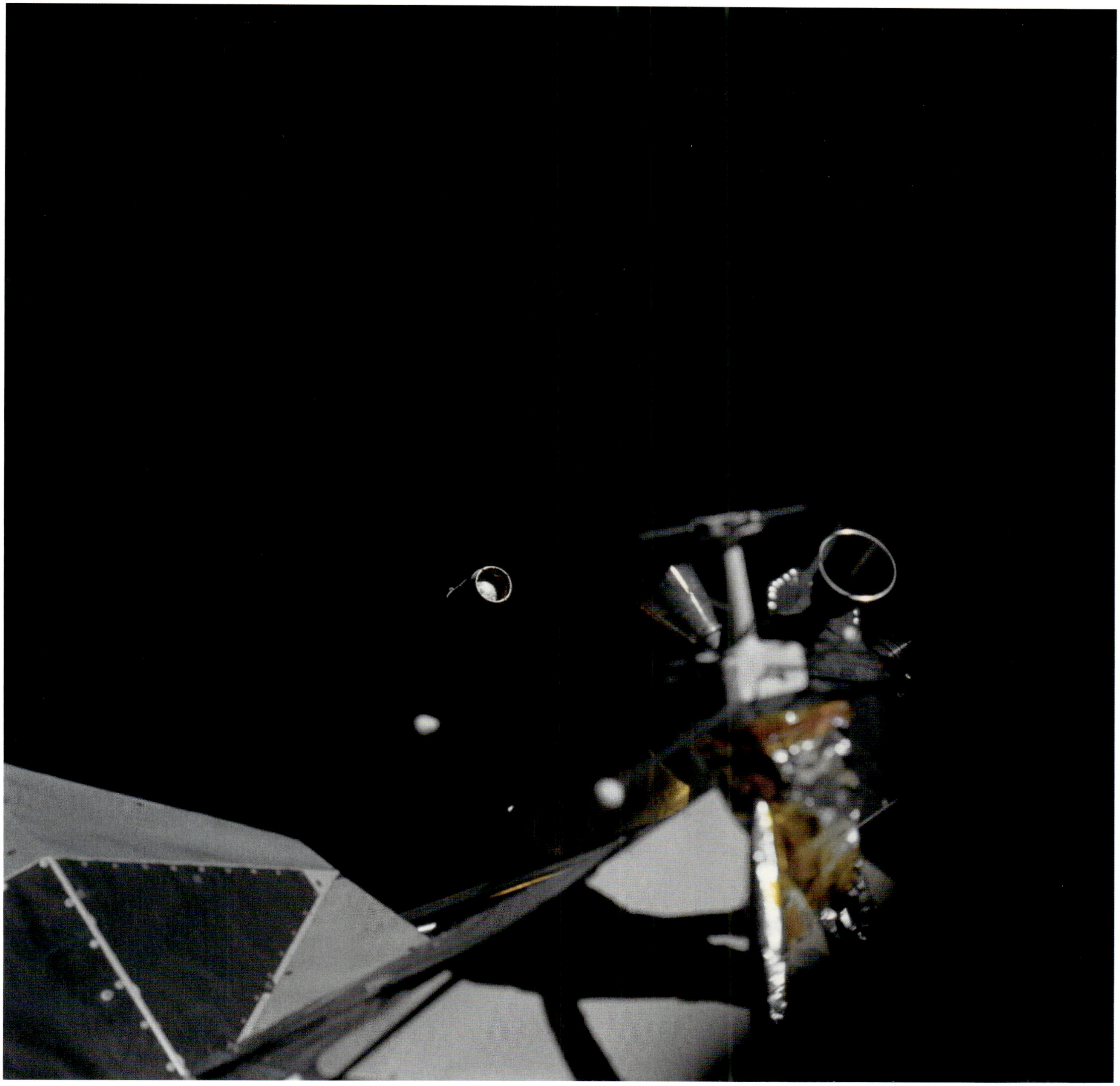

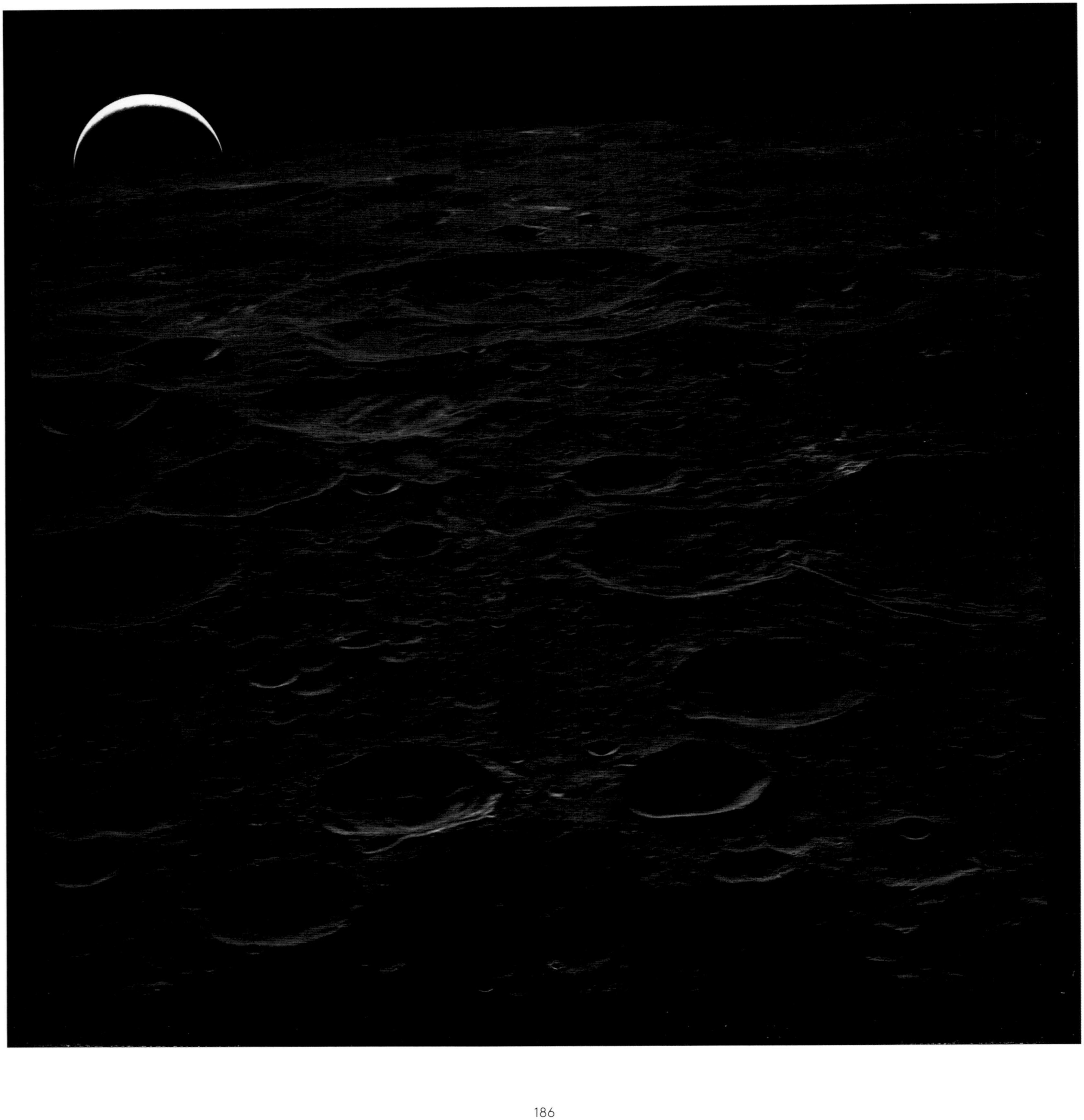

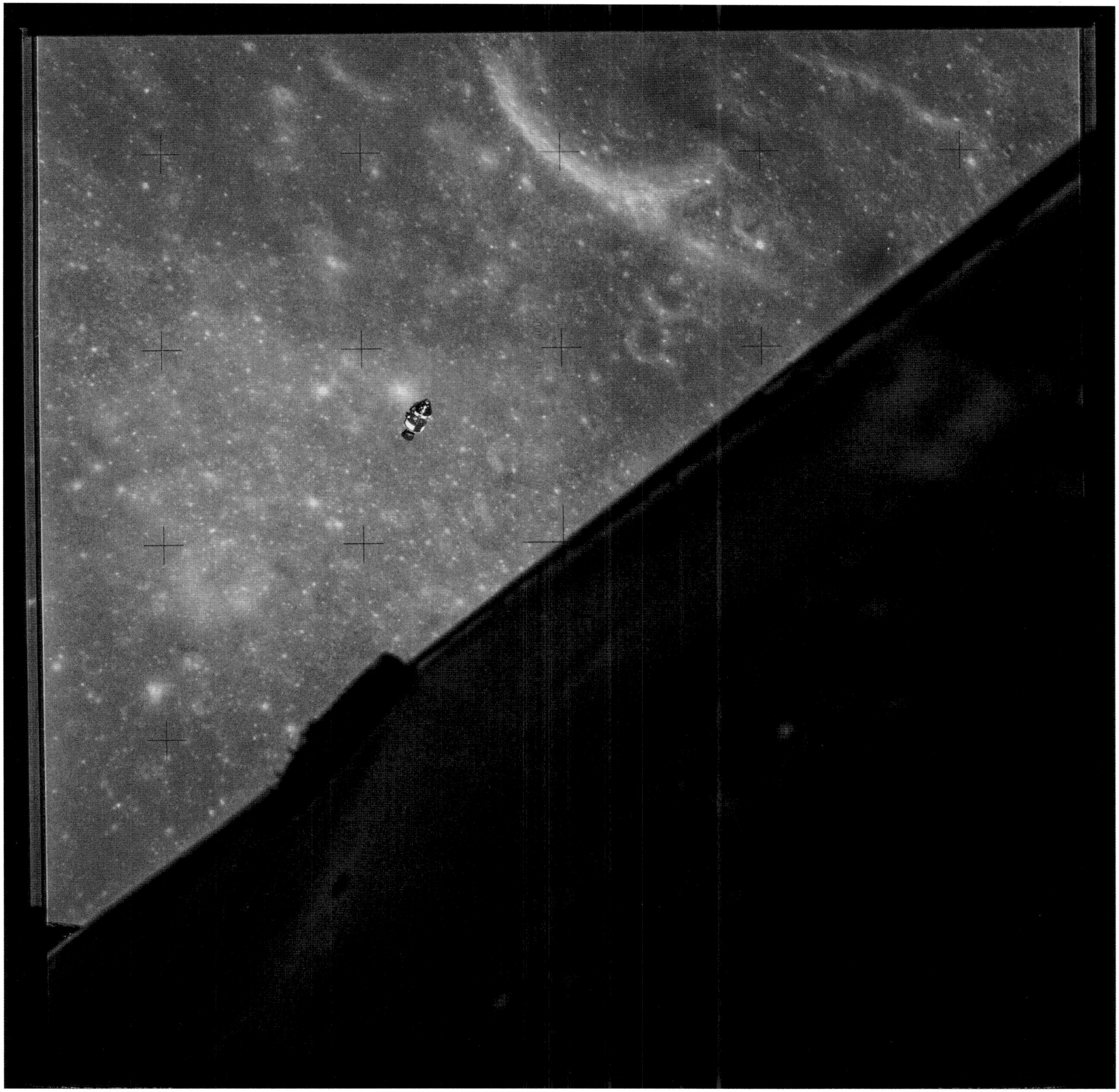

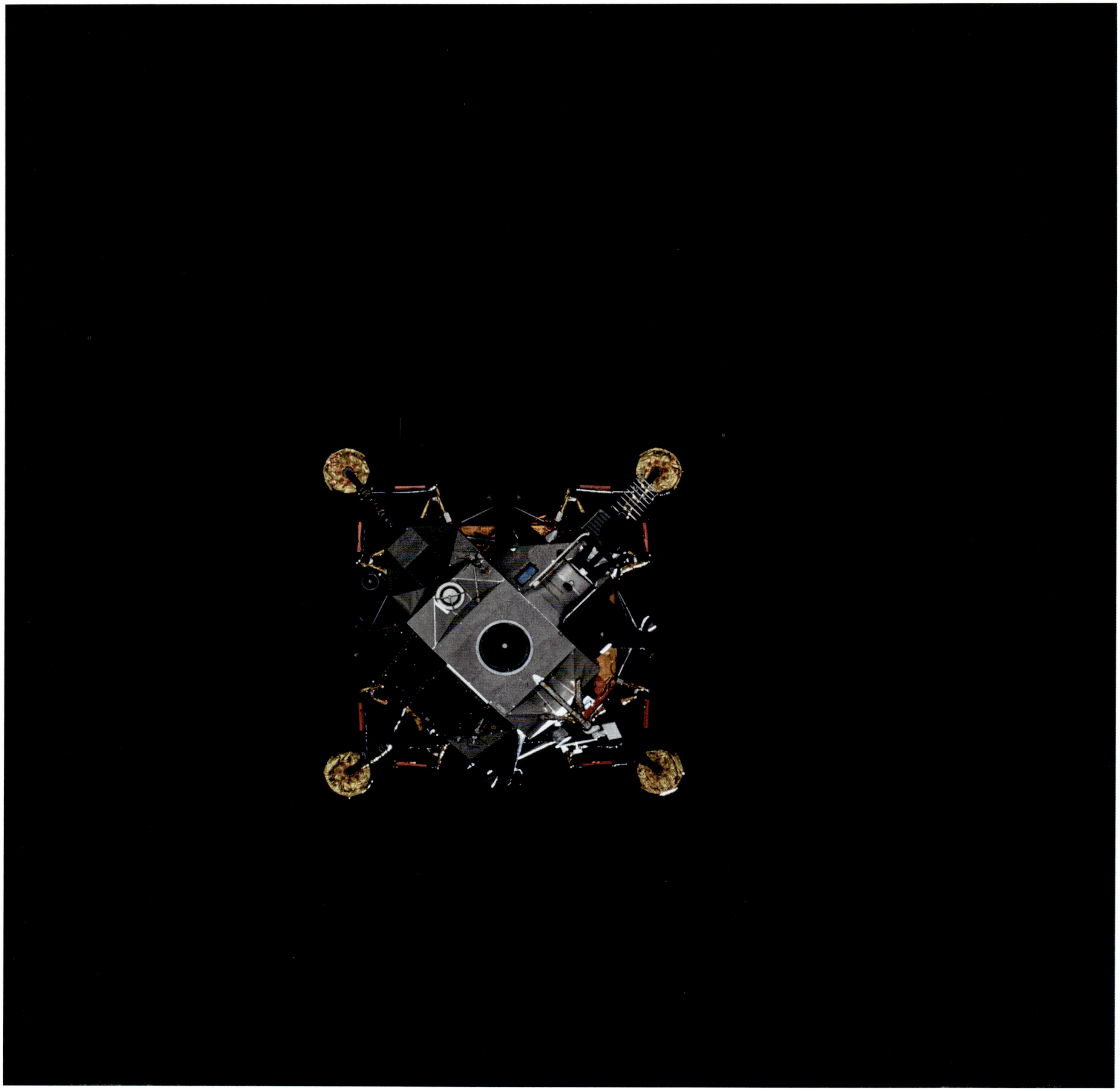

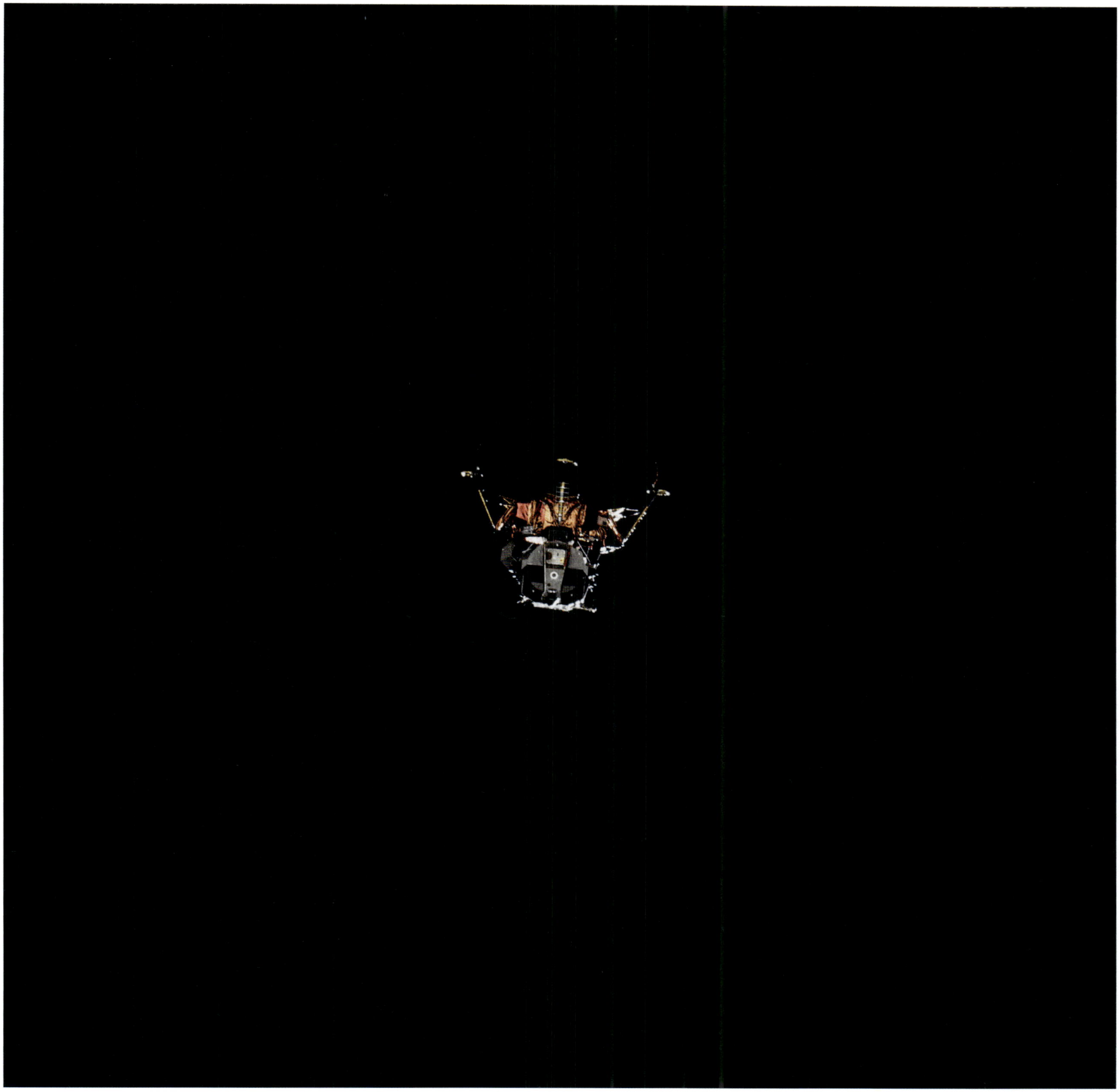

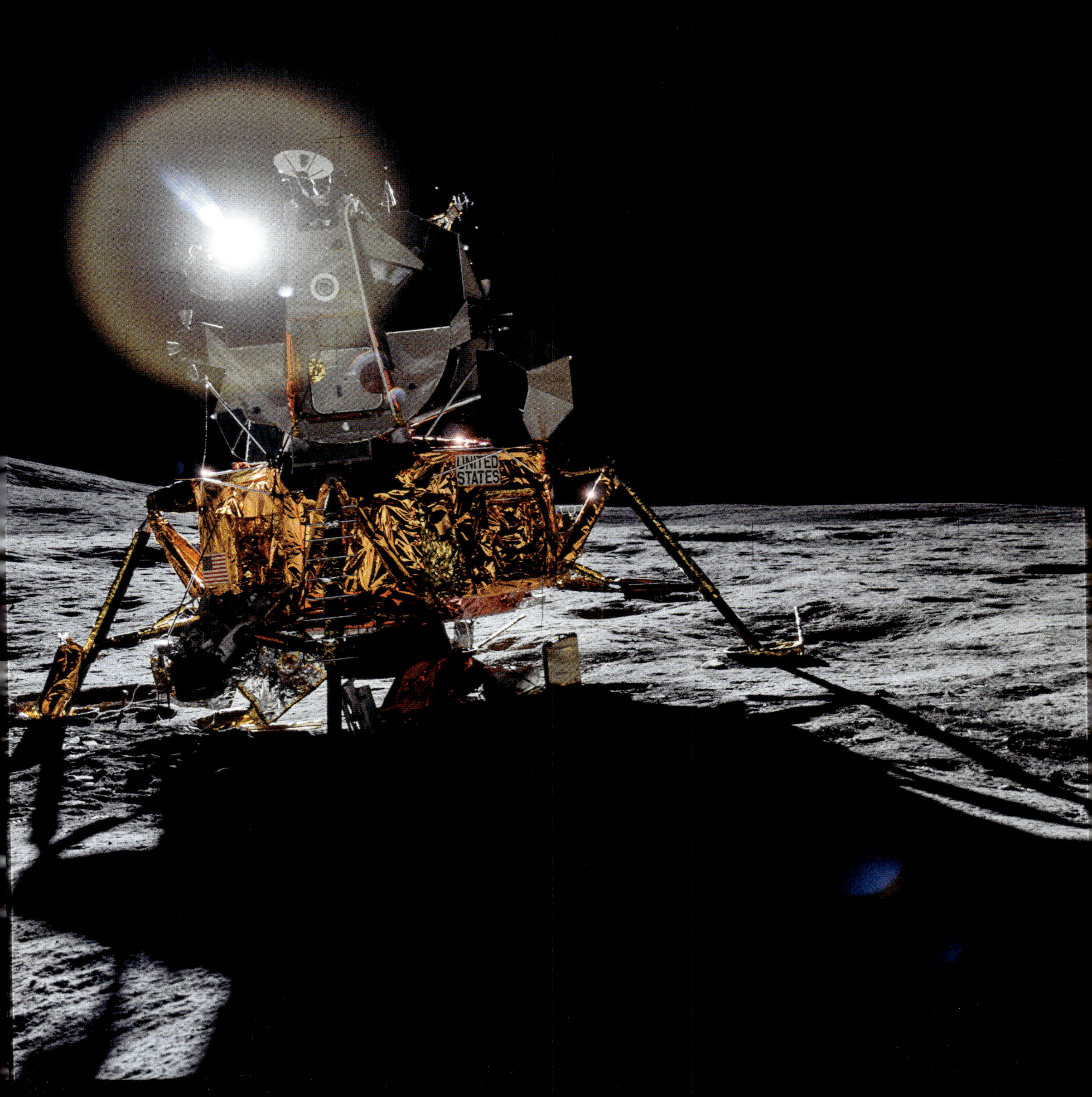
UNITED
STATES

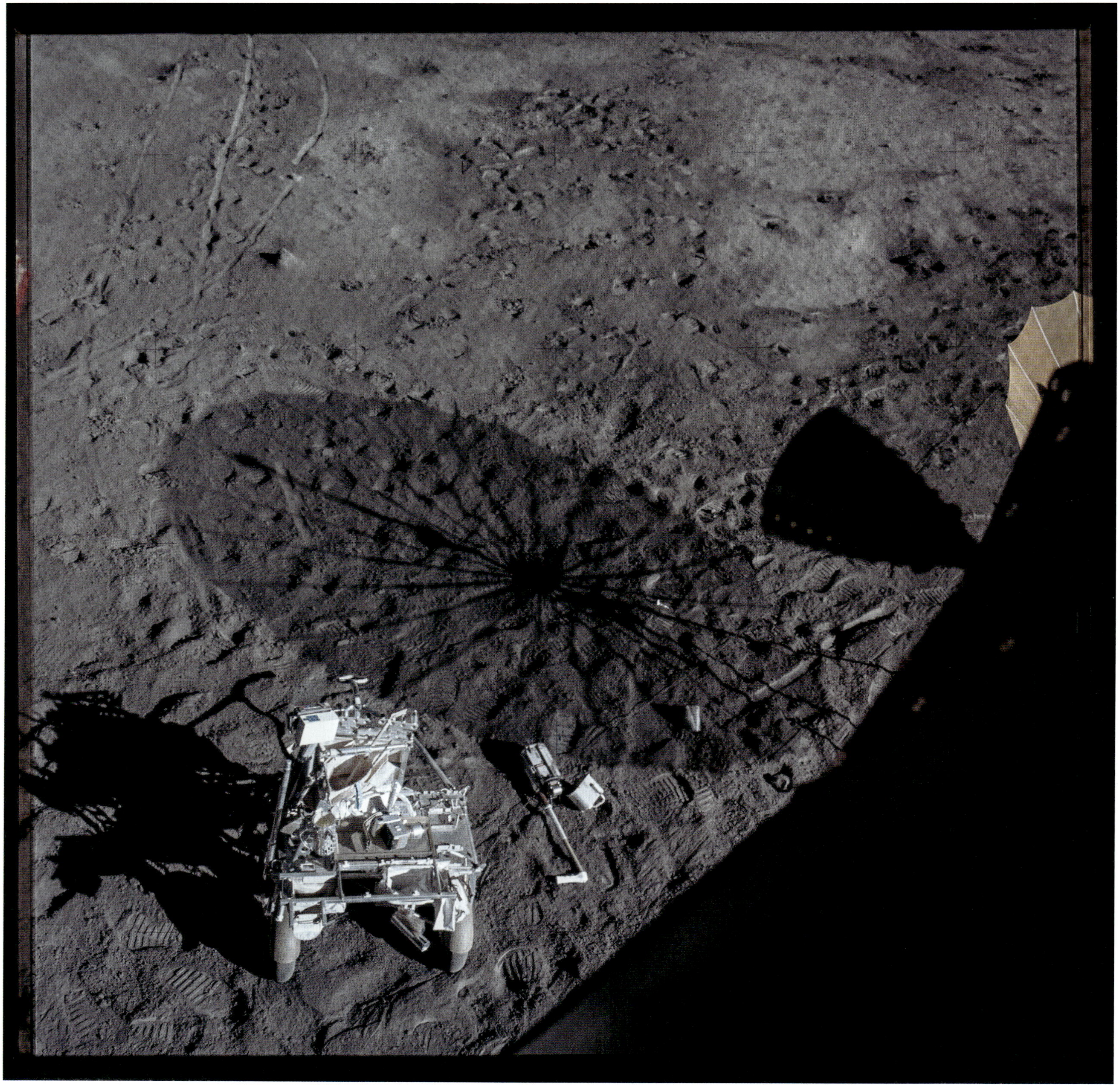

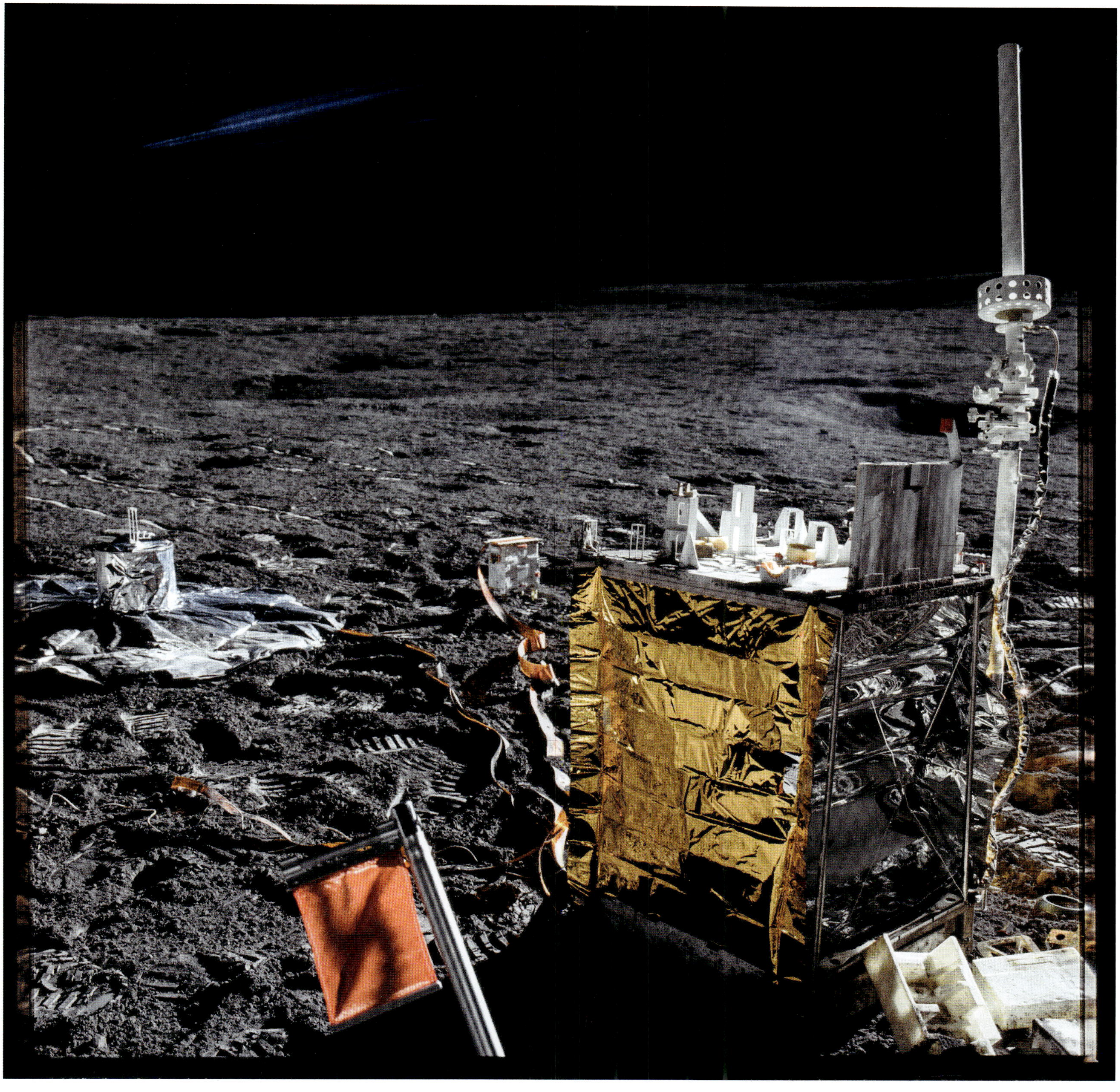

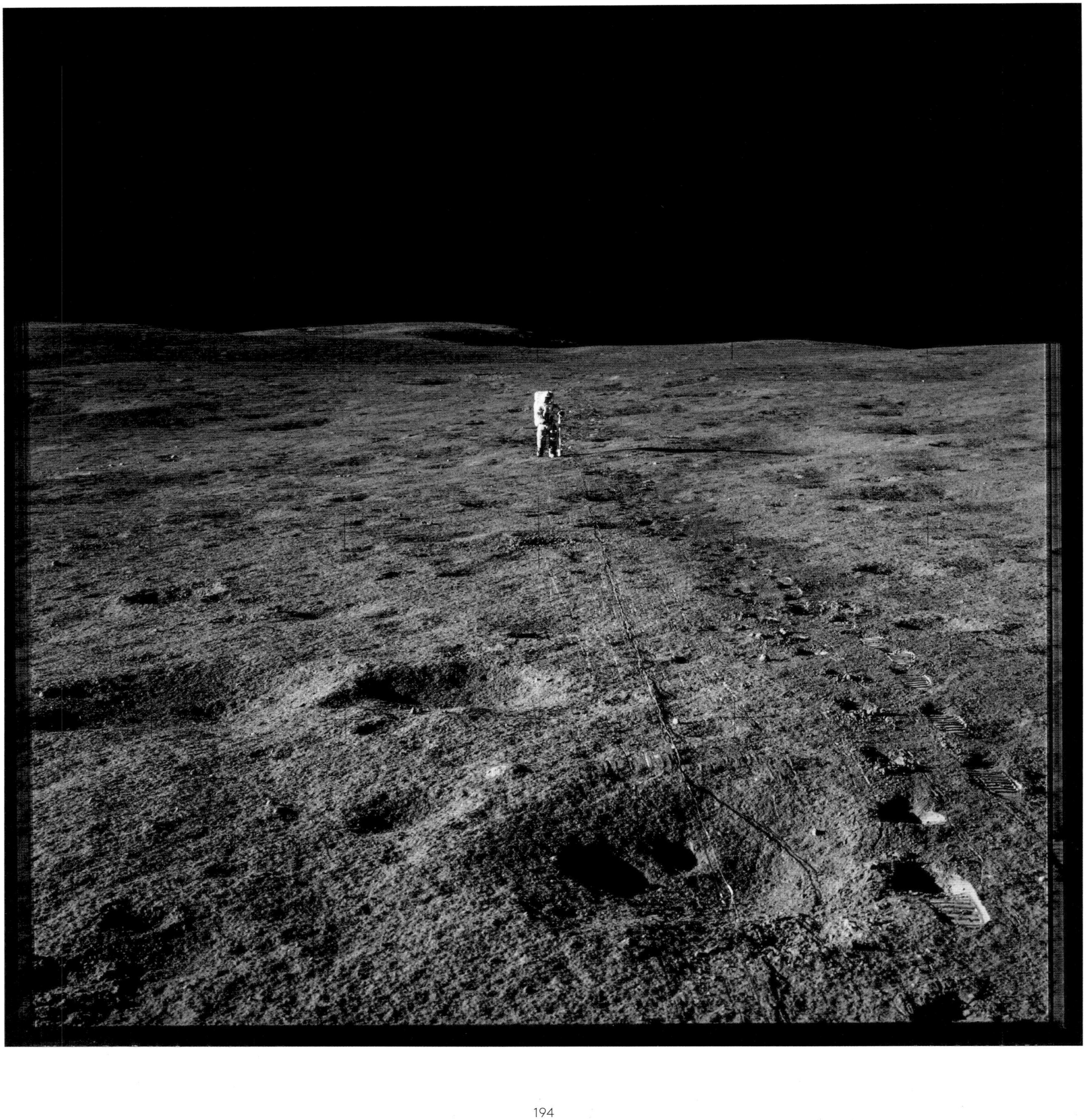

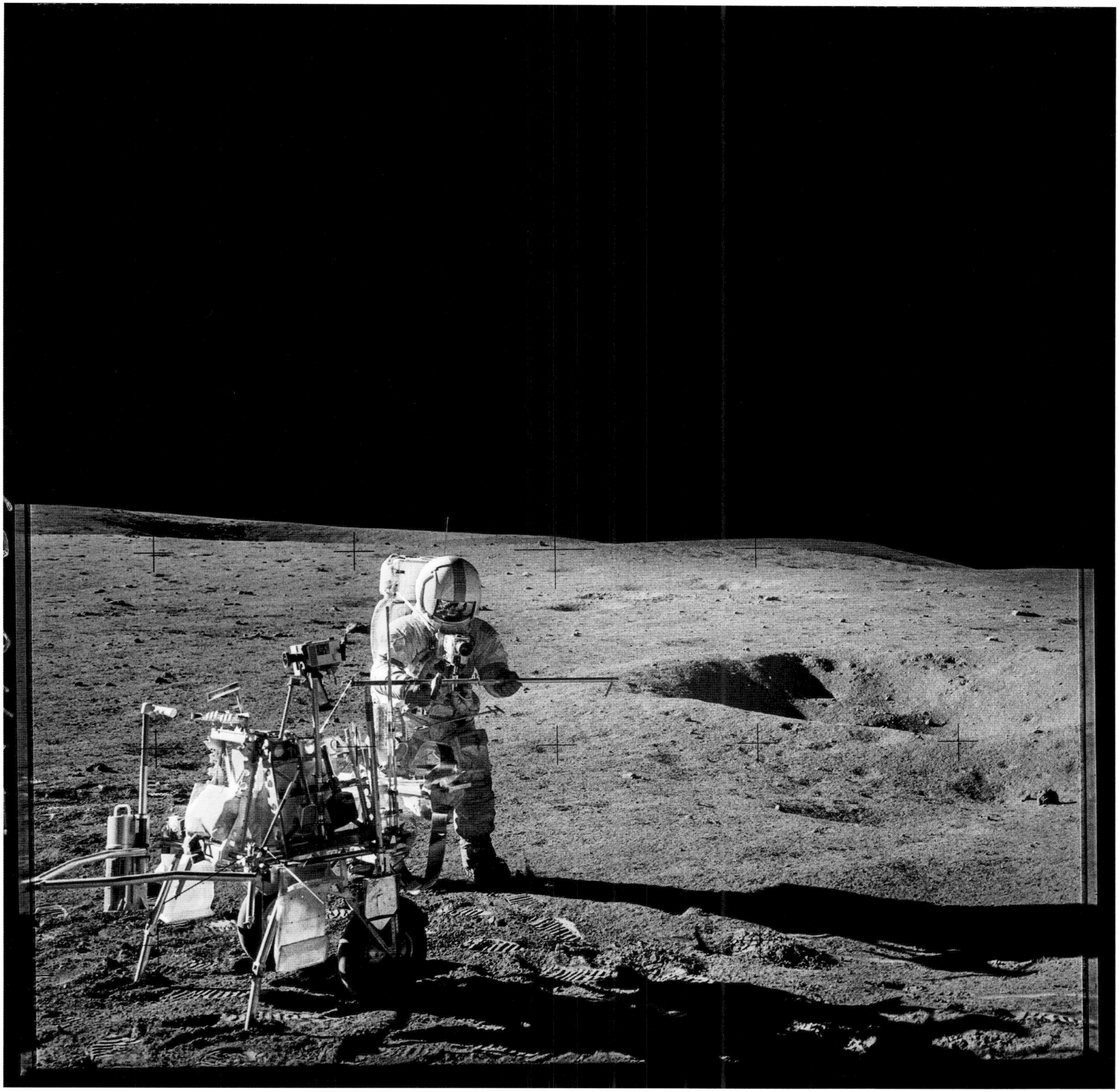

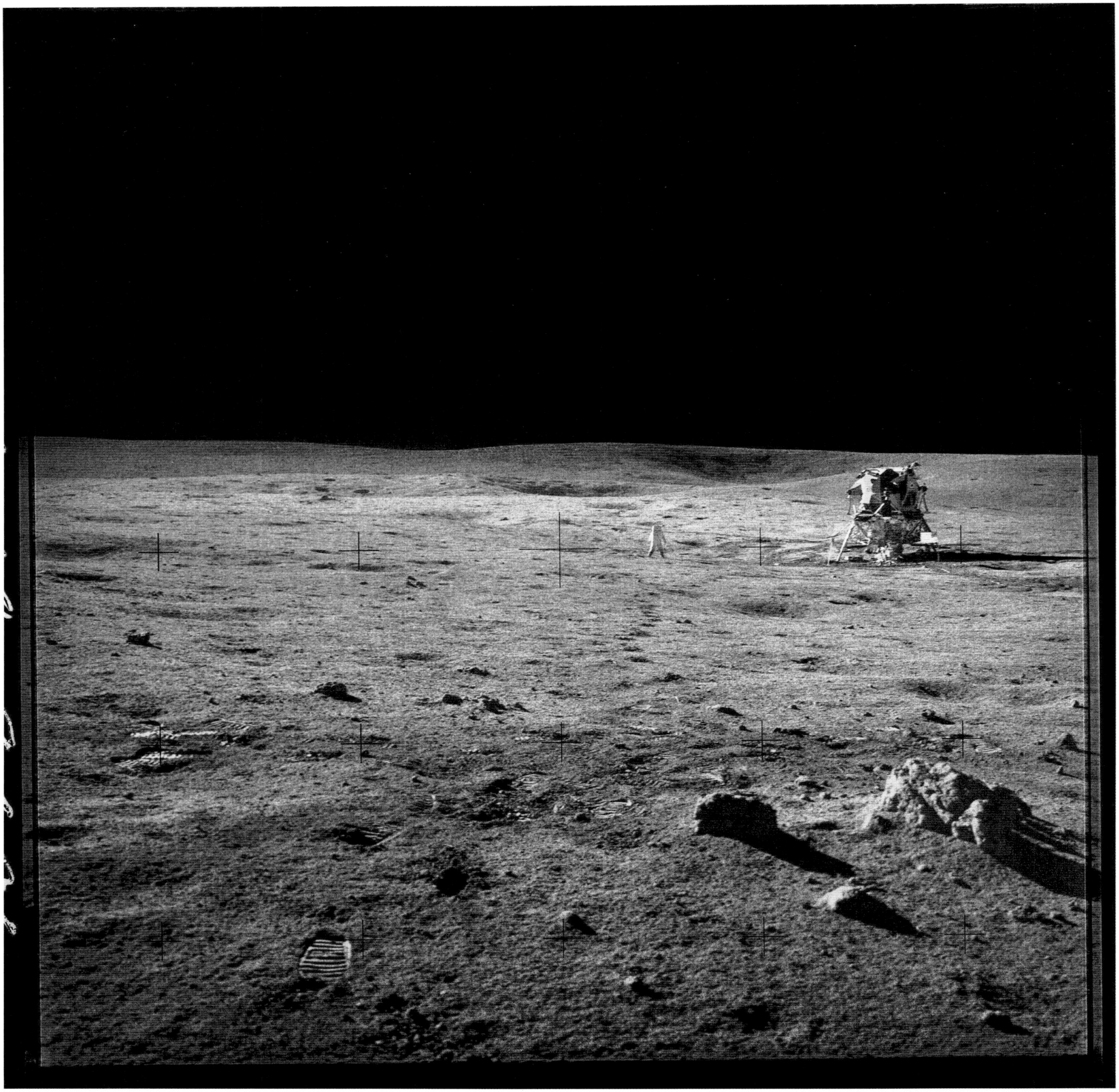

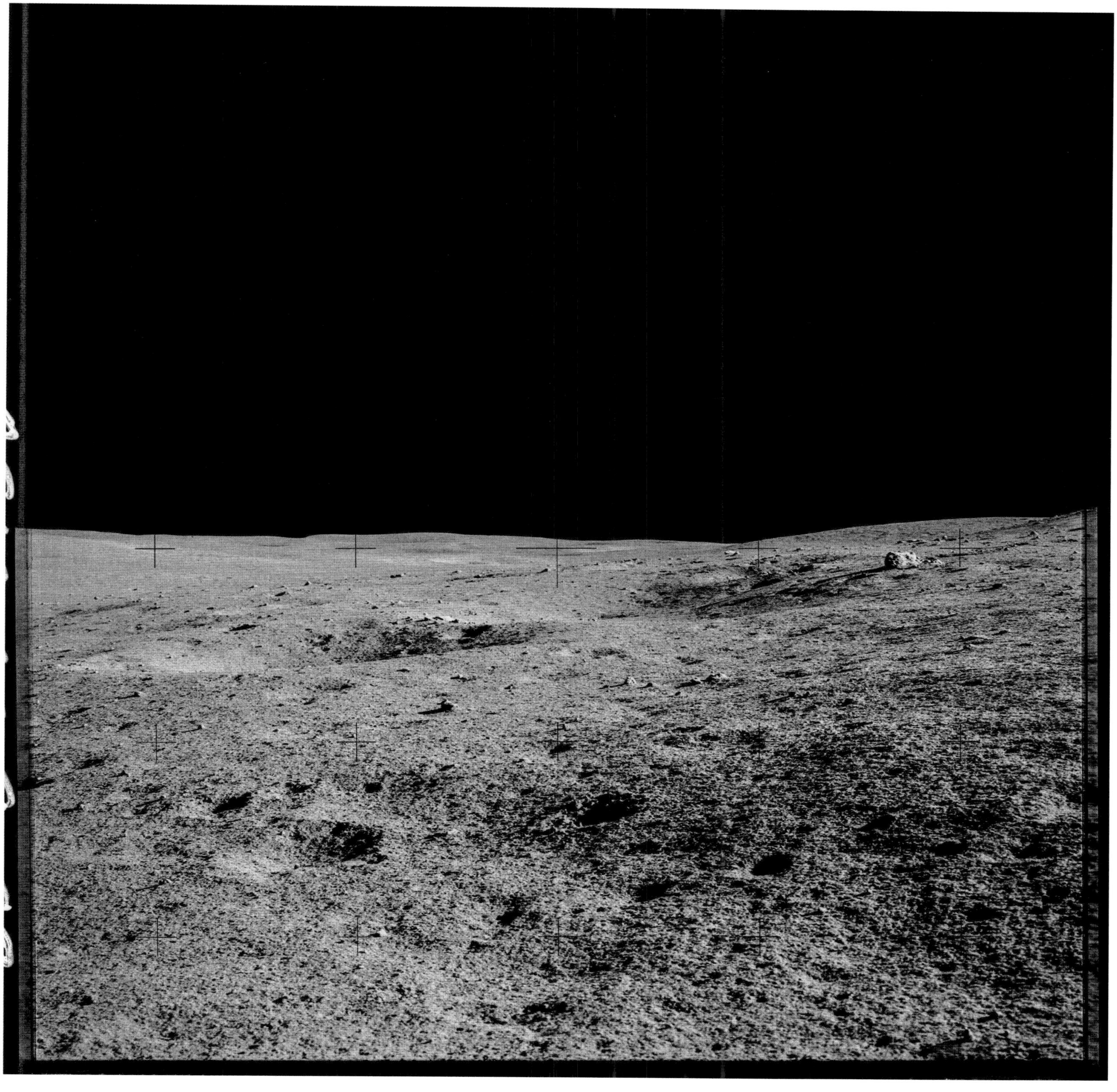

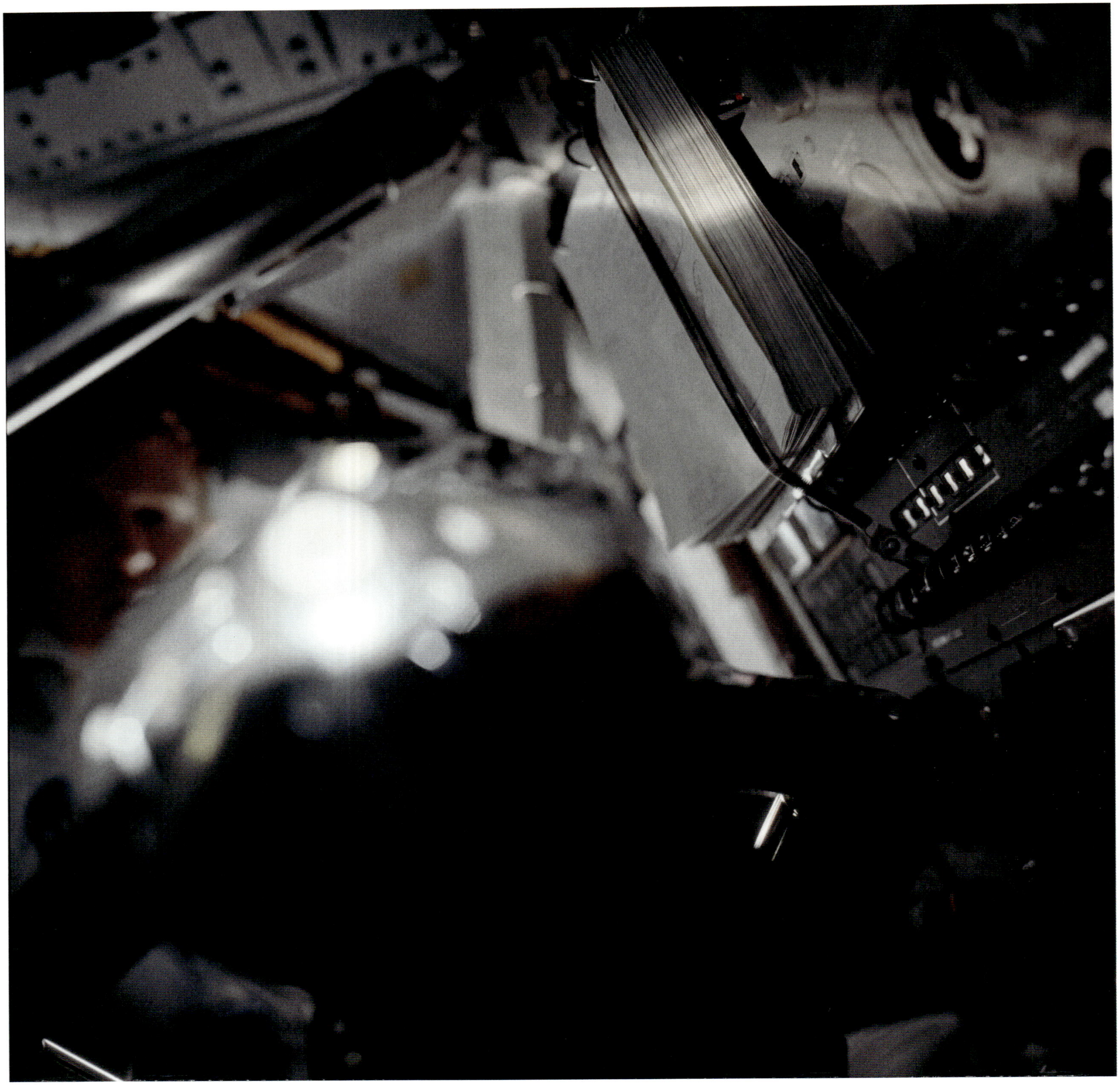

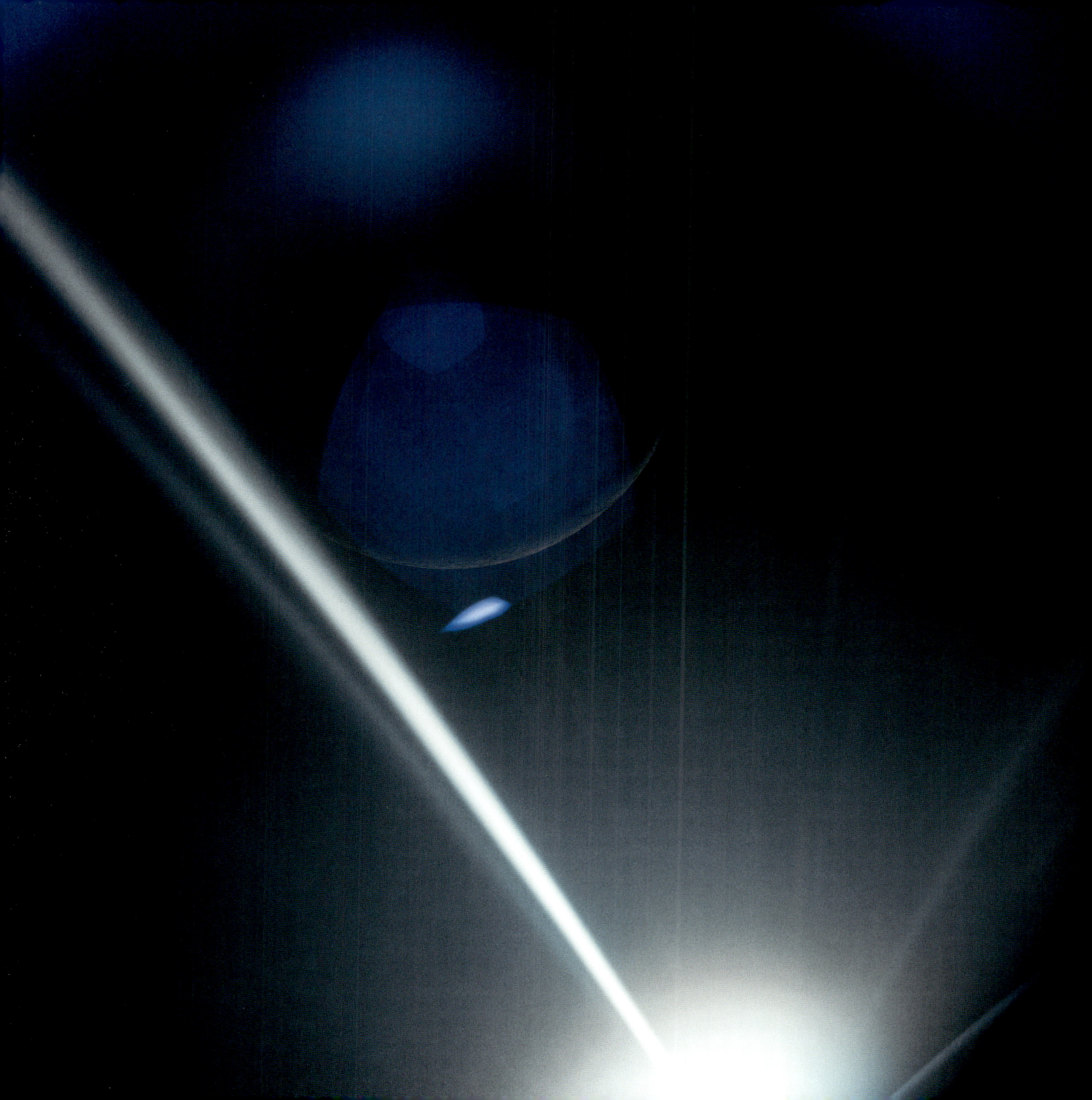

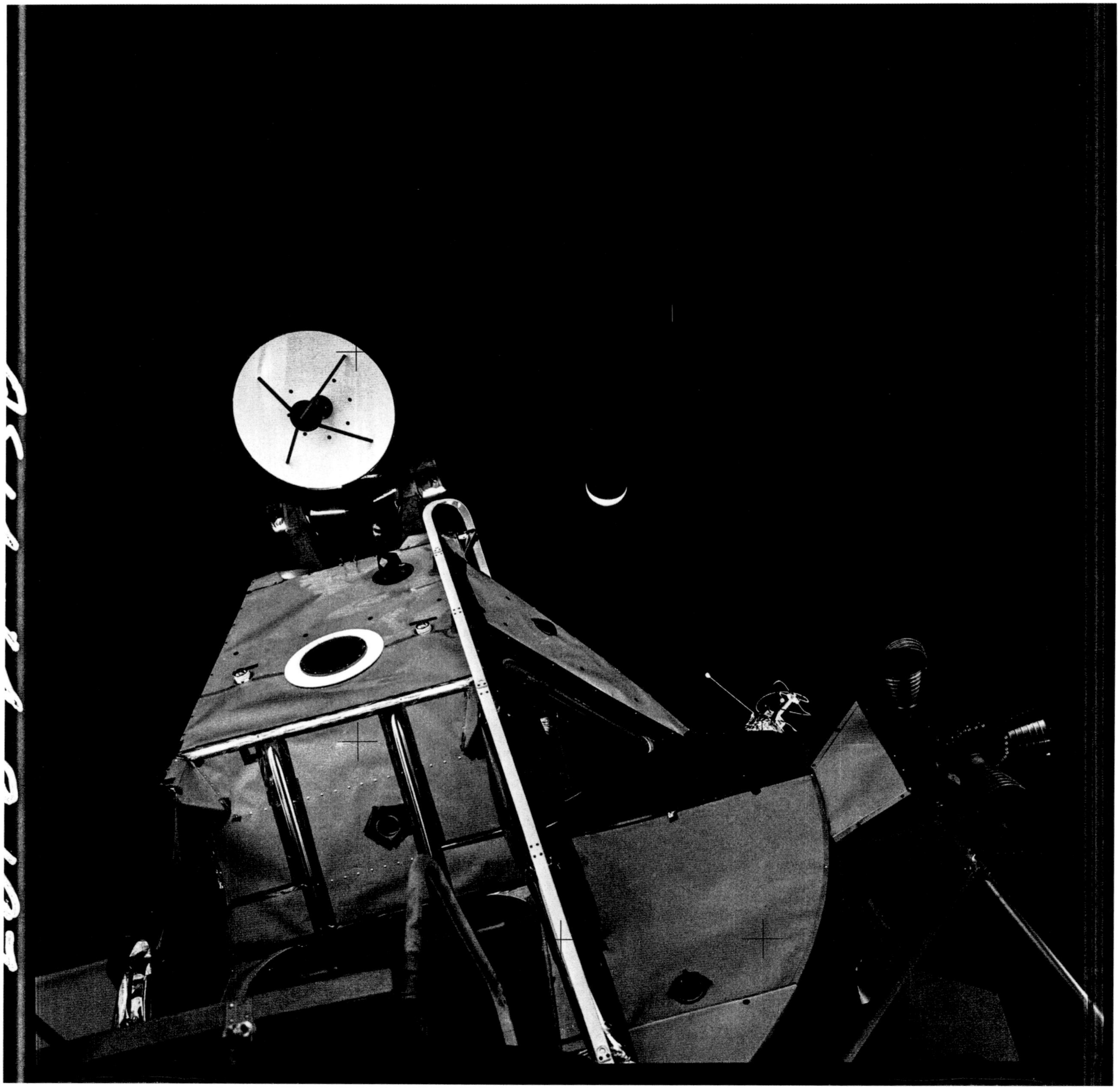

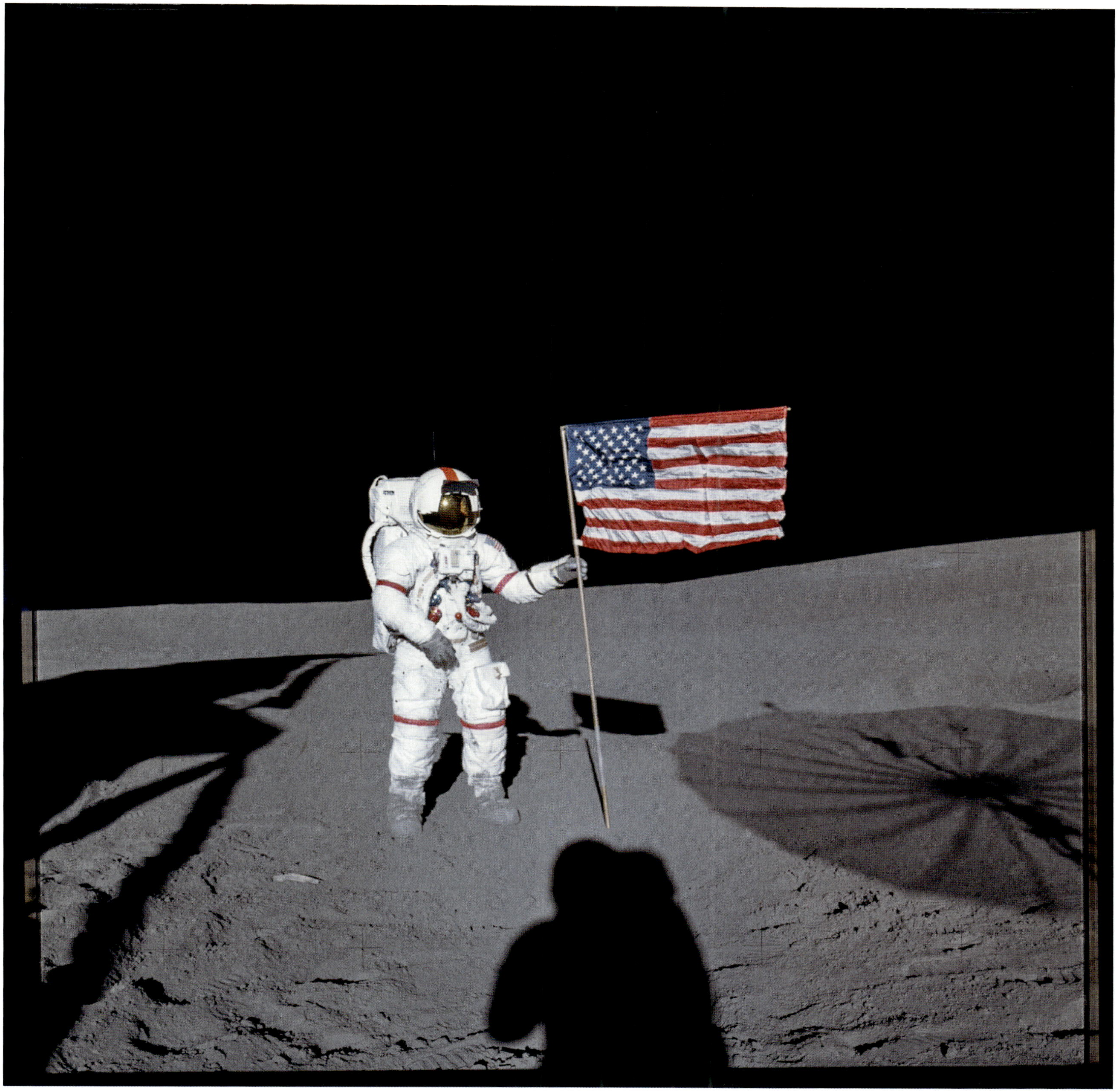

183 (AS14-66-9229)
Apollo 14 Commander Alan B. Shepard Jr., seen from the Lunar Module (LM) *Antares*, stands on the Moon during his exploration of the Fra Mauro Highlands. His red-striped spacesuit, introduced on this mission, made it easier to identify the commander during Extravehicular Activities (EVAs).

184 (AS14-72-9919)
LM *Antares*, seen inside the S-IVB stage during the Trans-Lunar Injection (TLI) phase of Apollo 14. The crew encountered docking difficulties when trying to extract the LM from the S-IVB. After two failed attempts, Shepard and Command Module Pilot Stuart A. Roosa successfully docked on the third try, demonstrating both their skill and the spacecraft's robust design.

185 (AS14-72-9924)
A view from the Command Module (CM) *Kitty Hawk*, showing LM *Antares* after emerging from the S-IVB stage during Apollo 14's Transposition, Docking & Extraction (TD&E) maneuver.

186 (AS14-71-9845)
A dark, moody black-and-white photograph of 'Earthrise' over the Moon, with Firsov, Lobachevsky, and 201 Craters visible below. Captured during Apollo 14's Trans-Earth Coast (TEC).

187 (AS14-66-9222)
A view from *Antares* of CM *Kitty Hawk* in lunar orbit.

188 (AS14-74-10206)
A view from *Kitty Hawk*, showing LM *Antares* after separation, preparing for its descent to the Fra Mauro Highlands.

189 (AS14-74-10209)
A view from *Kitty Hawk*, showing LM *Antares* drifting toward its descent to the Fra Mauro Highlands.

191 (AS14-66-9306)
A front view of *Antares*, its gold and black thermal foil gleaming under the harsh lunar sunlight. A dramatic circular lens flare dominates the frame, reflecting off the spacecraft's surface and contrasting against the blackness of space.

192 (AS14-66-9340)
A view from *Antares*' window, showing the Modular Equipment Transporter (MET) on the Fra Mauro surface. Used by Shepard and Lunar Module Pilot Edgar D. Mitchell, the MET allowed for greater mobility and sample collection during Apollo 14's lunar exploration.

193 (AS14-67-9384)
Shepard and Mitchell deploy the Passive Seismic Experiment (PSE), designed to detect moonquakes and meteorite impacts. This experiment provided key insights into the Moon's interior structure, confirming its solid crust and partially molten mantle.

194 (AS14-67-9374)
Mitchell deploys the geophone line for the Active Seismic Experiment (ASE). As he fires the hand-held thumper, seismic waves travel through the lunar surface, helping scientists study its subsurface structure.

195 (AS14-68-9405)
Shepard assembles a double core tube next to the MET on the Fra Mauro Highlands. This tool helped collect deep lunar soil samples.

196 (AS14-68-9486)
A stunning landscape composition featuring Shepard and LM *Antares*, with Triplet Crater in the background. Captured during Mitchell's Station H panorama, this image highlights the dramatic terrain of the Fra Mauro Highlands.

197 (AS14-64-9082)
A panoramic frame from Station B1, taken during Apollo 14's EVA-2. This site provided key geological samples of Imbrium impact ejecta. Shepard and Mitchell collected core tube samples, revealing layered lunar regolith that helped scientists study the Moon's impact history and surface evolution.

198 (AS14-76-10331)
A close-up, out-of-focus view of the docking probe, with the CM instrument panel in the background. Roosa is visible in the frame.

199 (AS14-72-9938)
A lunar crescent illuminated by the Sun, with striking lens flares creating a dramatic effect. Captured by the Apollo 14 crew during their return journey to Earth.

200 (AS14-64-9197)
A dark, grainy black-and-white photograph creates a dramatic composition, showing LM *Antares* silhouetted against a crescent Earth.

201 (AS14-66-9232)
A proud moment as Shepard holds the American flag on the Fra Mauro lunar surface during Apollo 14. Captured by Mitchell, the image includes Mitchell's shadow and the shadow of *Antares*.

Apollo 15

XV

David R. Scott

Alfred M. Worden

James B. Irwin

Apollo 15

Launch: July 26th, 1971
Lunar landing: July 30th, 1971
Splashdown: August 7th, 1971

Command Module: Endeavour
Lunar Module: Falcon

Apollo 15 was the first of the 'J' missions — longer-duration lunar stays with a heightened focus on science. The Boeing/GM-built Lunar Roving Vehicle (LRV), in development since May 1969, debuted on this ninth crewed Apollo mission. Fully loaded with two astronauts and their gear, the rover weighed over 1,500lb (680kg) and reached speeds up to 8mph (12.8kmh). This expanded mobility allowed the crew to venture farther from the landing site, conduct more scientific experiments, and capture a wider variety of photographic scenes.

The mission aimed to return more varied, high-quality photographs than any previous Apollo flight. The crew was the first to receive extensive field geology training, enabling them to perform in-depth analysis of the Moon's surface. Apollo 15 landed in the Apennine Mountains — rising nearly 3.1mi (5km) — near Hadley Rille, an 80mi (130km) long, 1,200ft (370m) deep volcanic canyon. The dramatic slopes and chasms of this area provided some of the most photogenic backdrops of any lunar landing site.

Hadley Rille appears prominently in orbital imagery captured from the Command Module, along with other stunning photographs that highlight low-sun illumination on the Moon's surface.

Kept secret until after the mission, the 'Fallen Astronaut' was a small aluminum sculpture depicting a spacesuited figure, created by Belgian artist Paul van Hoeydonck. It was placed on the lunar surface beside a plaque honoring astronauts and cosmonauts who had died advancing space exploration. A photograph of this memorial appears later in this chapter on page 214.

David R. Scott
Commander

Born on June 6th, 1932 in San Antonio, Texas. Scott flew with Neil Armstrong on the Gemini 8 mission and later served as Backup Senior Pilot in Apollo 1's first backup crew, Command Module Pilot of Apollo 9 and Backup Commander of Apollo 12. He retired from the Air Force in March 1975 with the rank of Colonel and over 5,600 hours of flying time.

Alfred M. Worden
Command Module Pilot

Born on February 7th, 1932 in Jackson, Michigan. He was Backup Command Module Pilot of Apollo 12. In September 1972 he was assigned to the NASA Ames Research Center as Director of Advanced Research and Technology. He resigned from NASA and the Air Force on September 1st, 1975.

James B. Irwin
Lunar Module Pilot

Born on March 17th, 1930 in Pittsburgh, Pennsylvania. He was chosen in the fifth group of astronauts in 1966. He was Backup Lunar Module Pilot of Apollo 12, and was the eighth man to walk on the Moon. He resigned from NASA and the Air Force in July 1972.

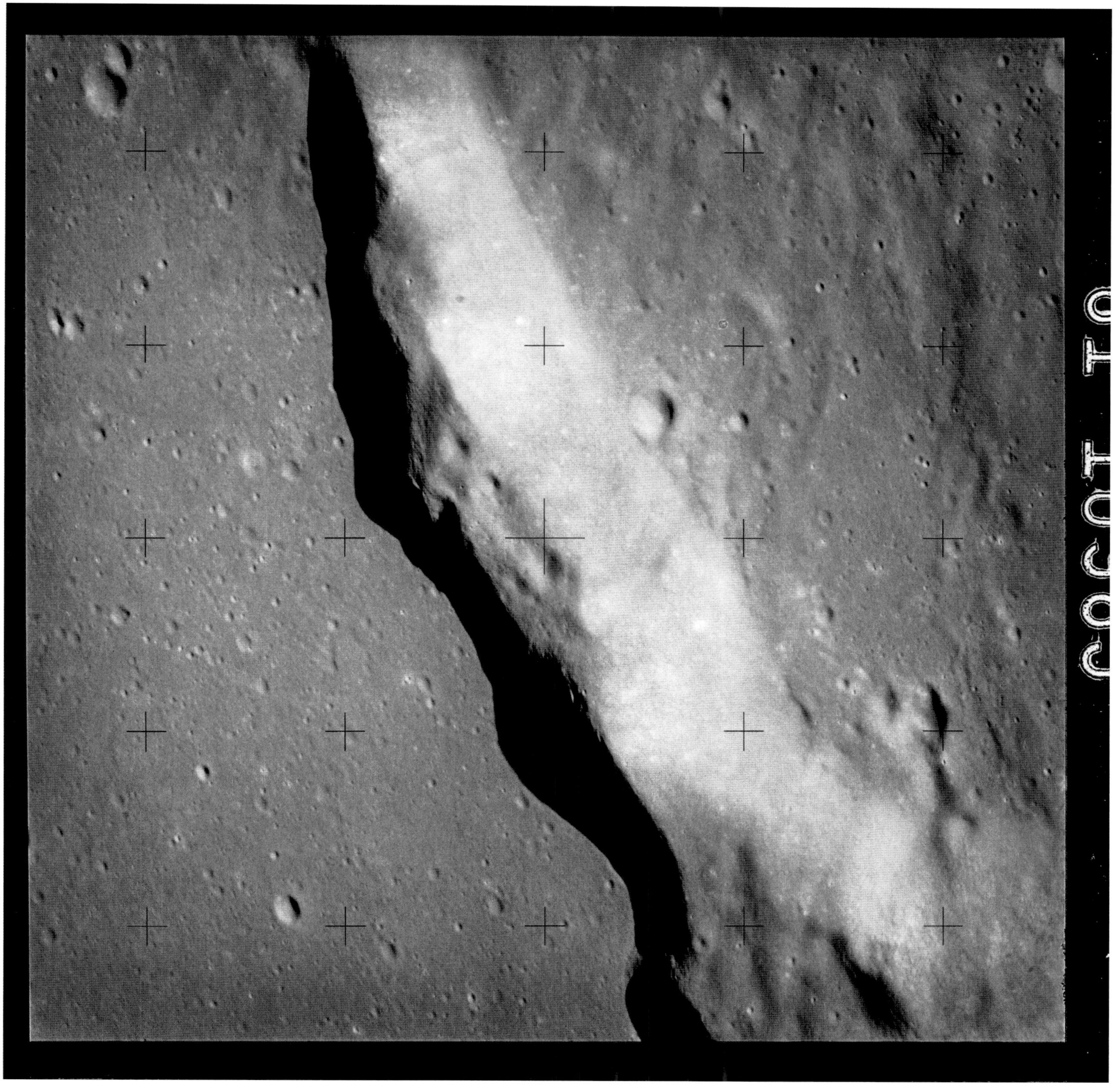

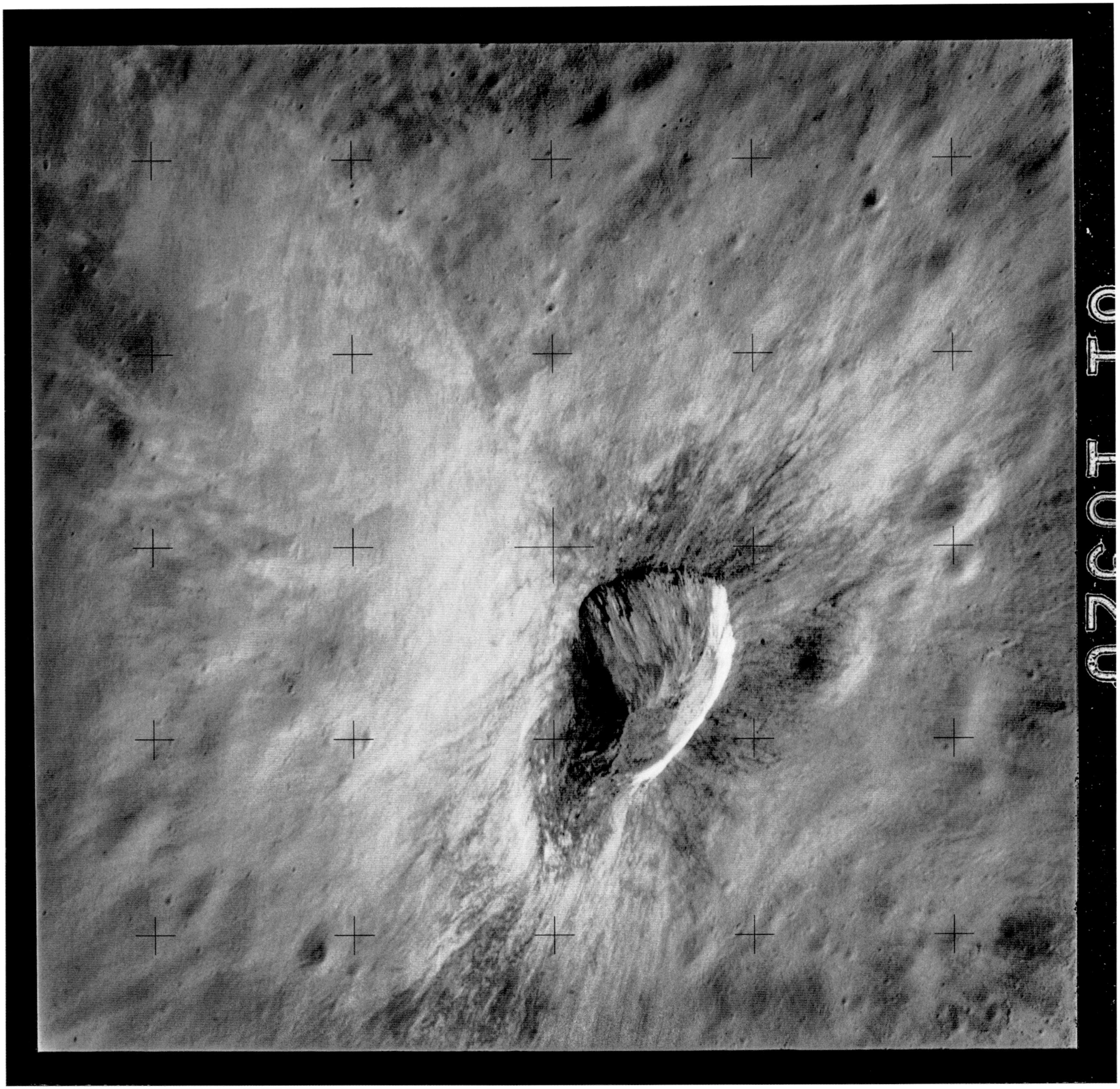

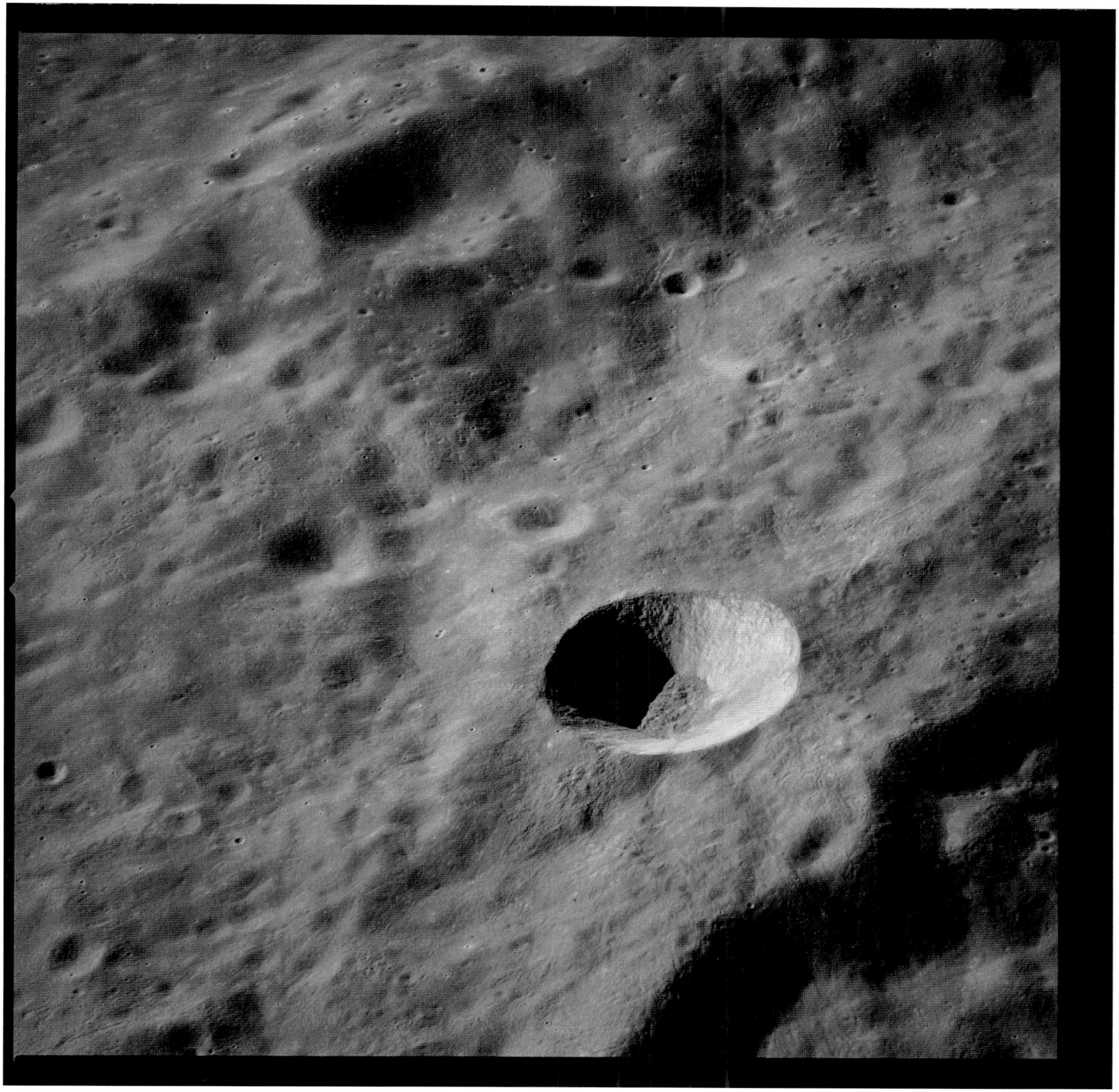

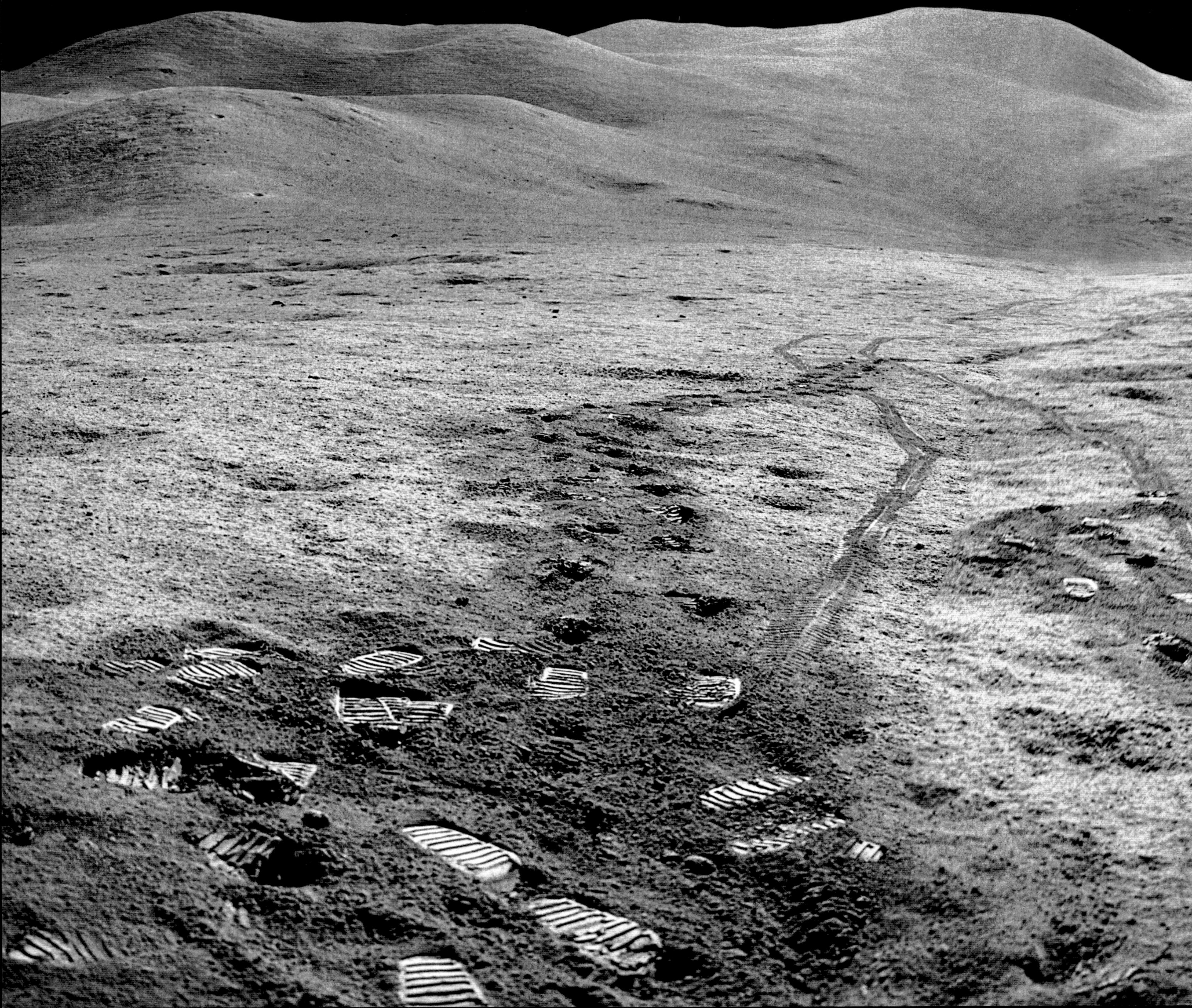

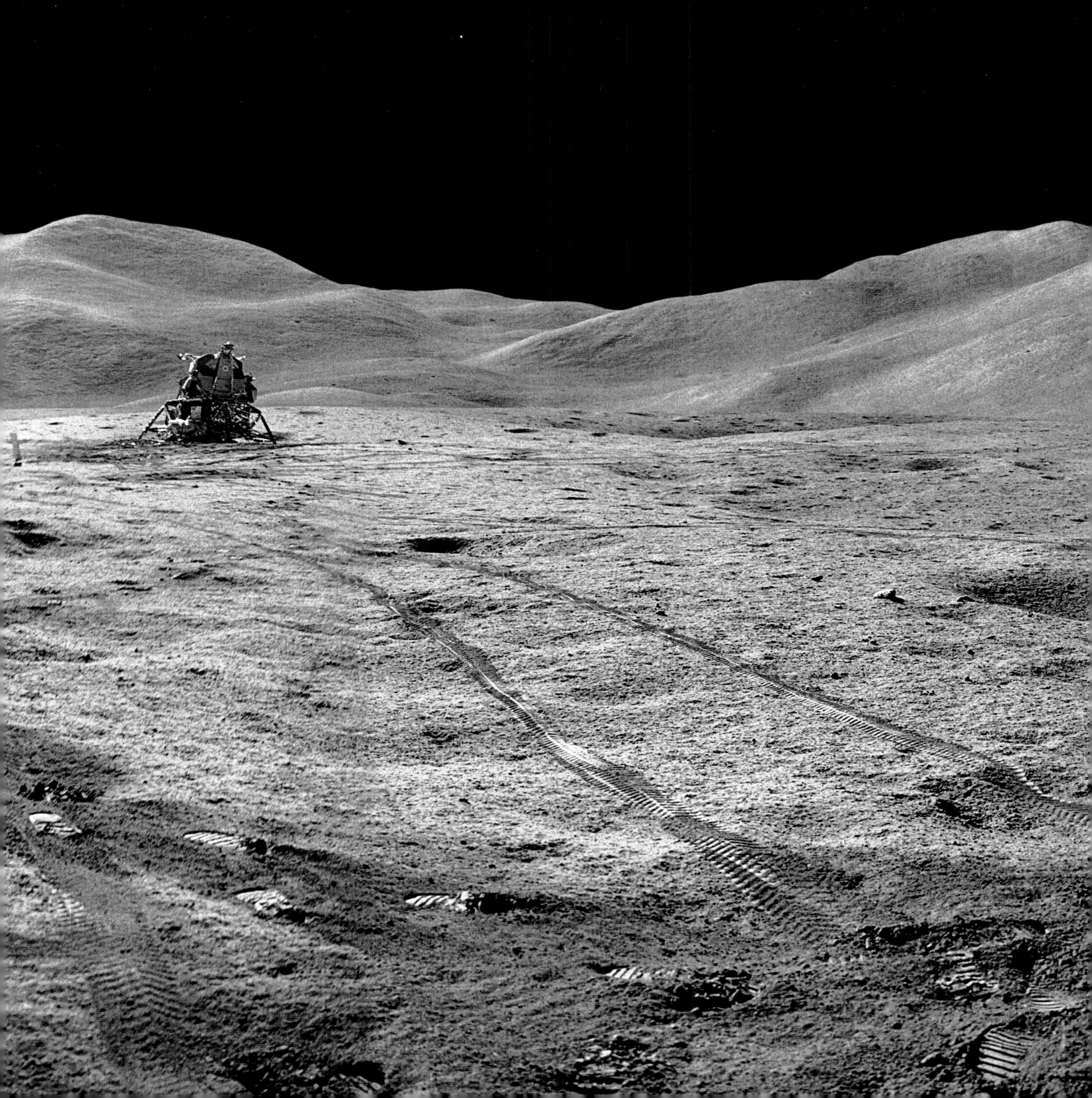

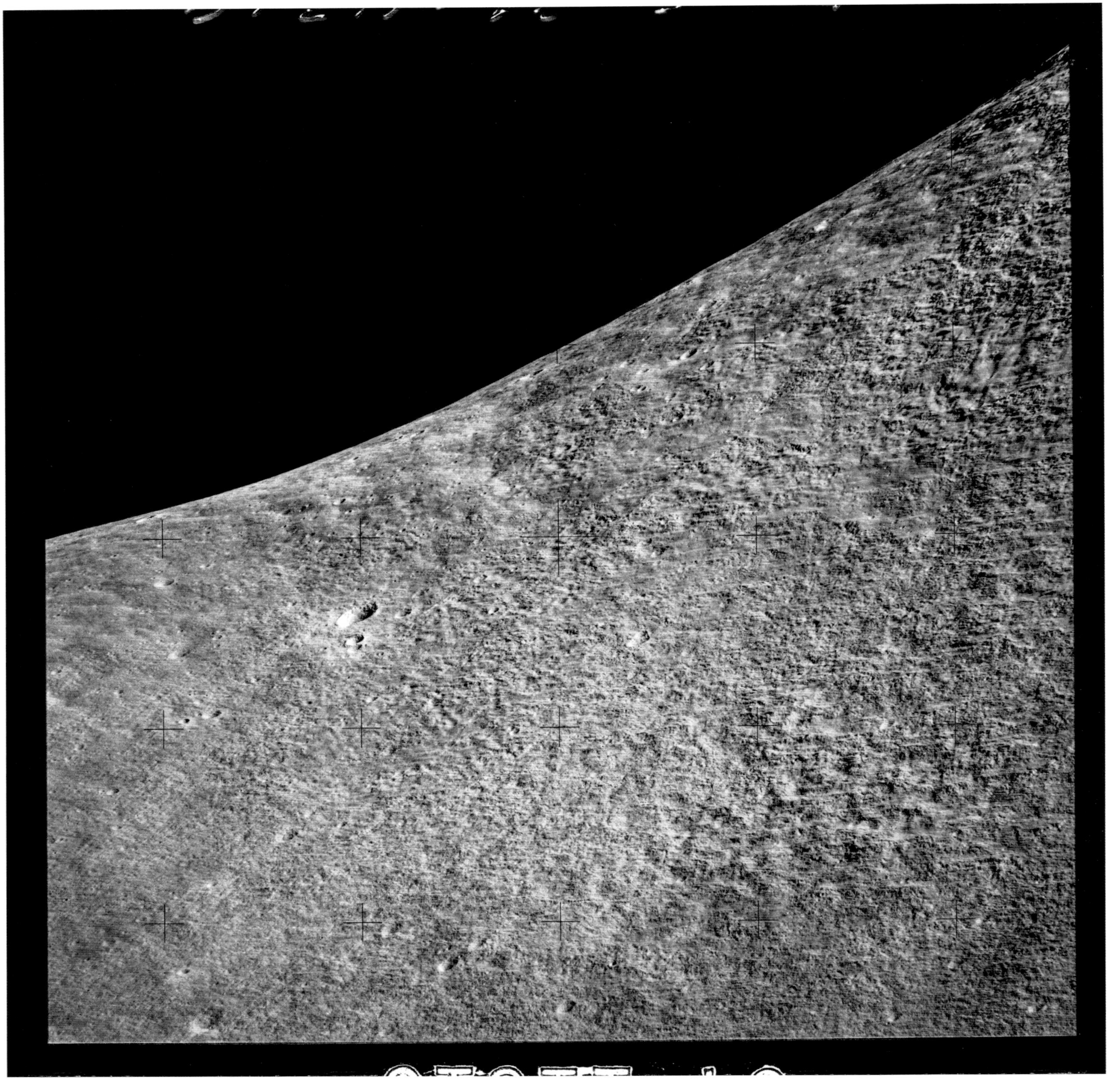

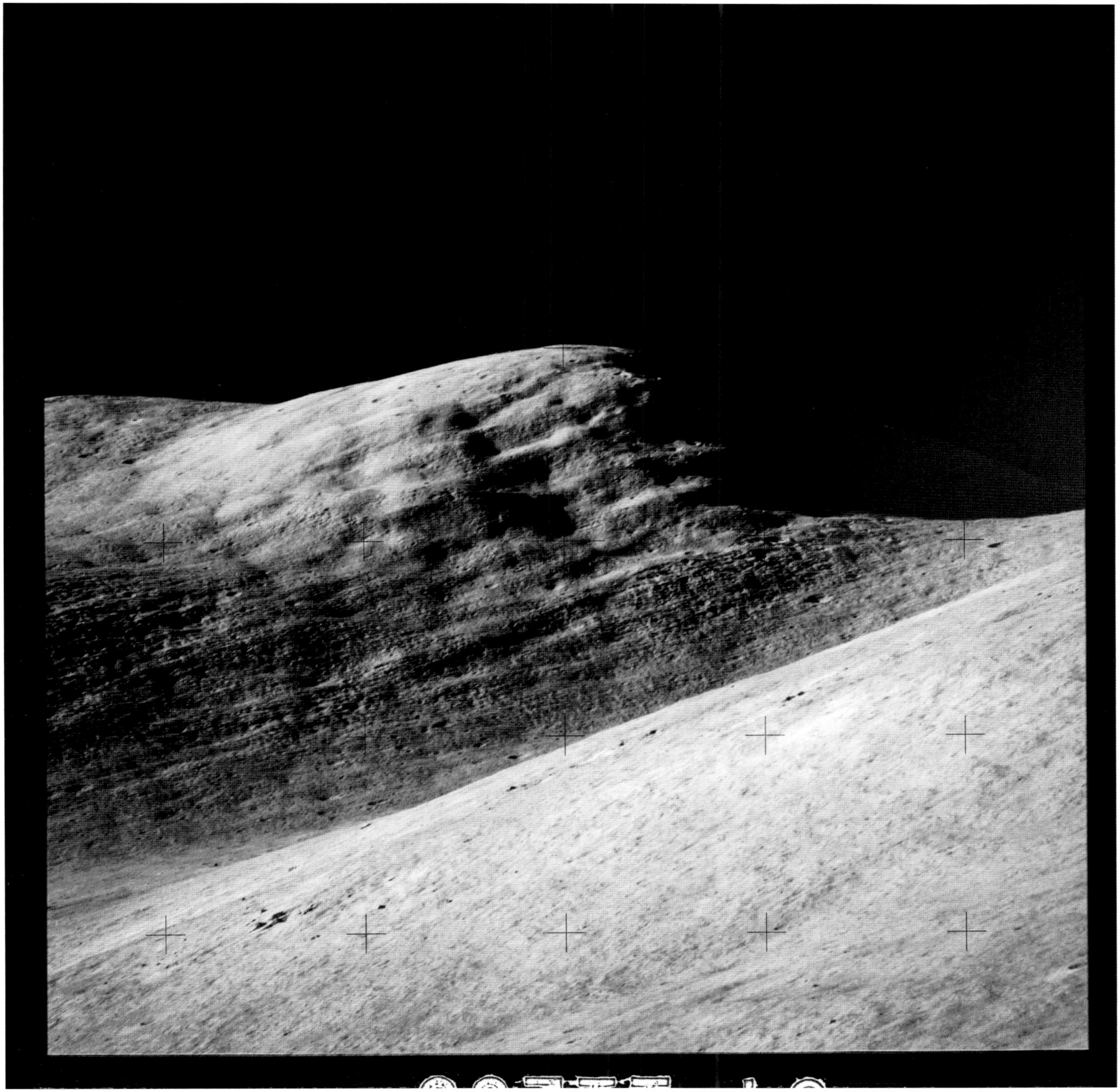

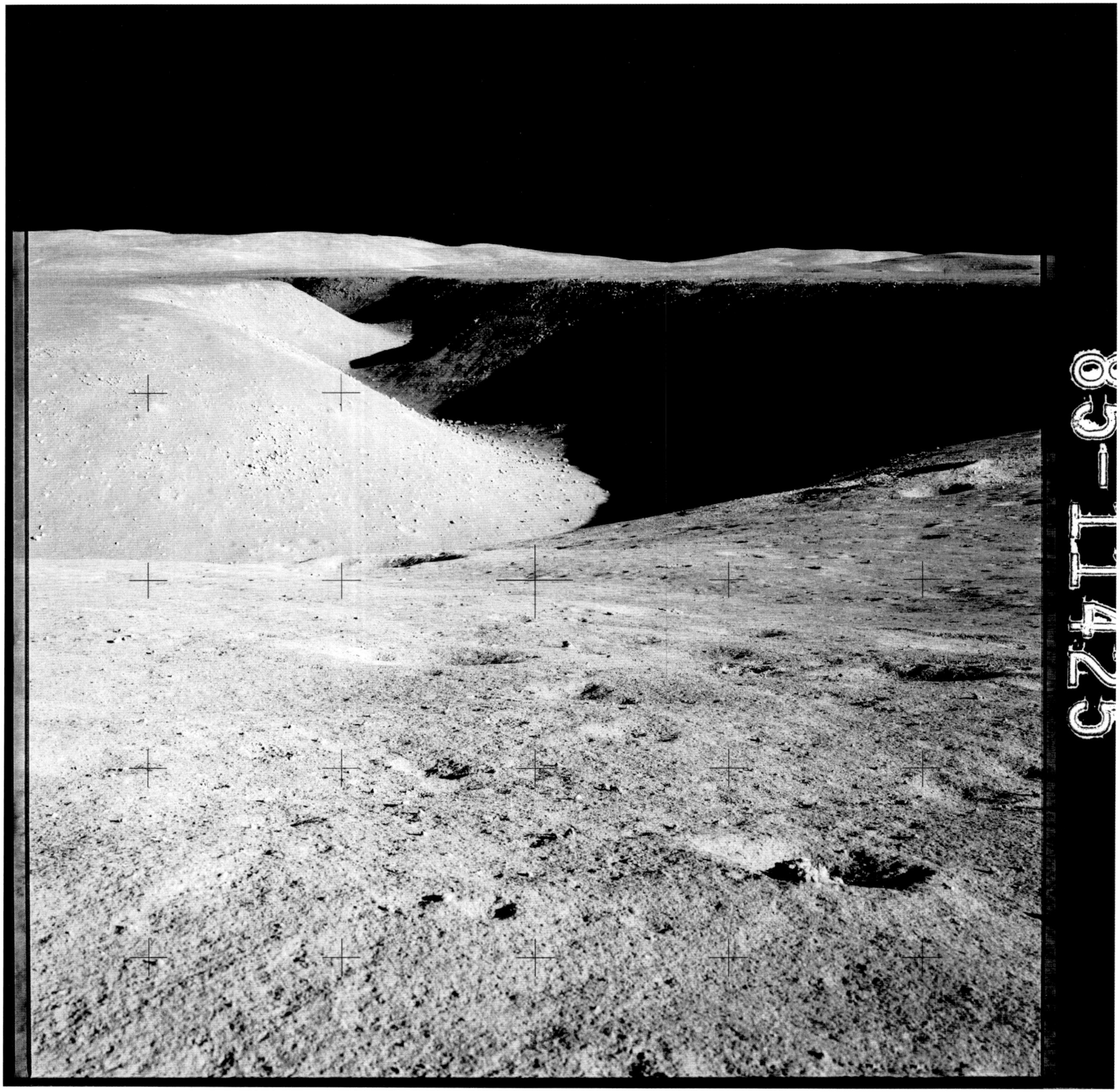
85-11425

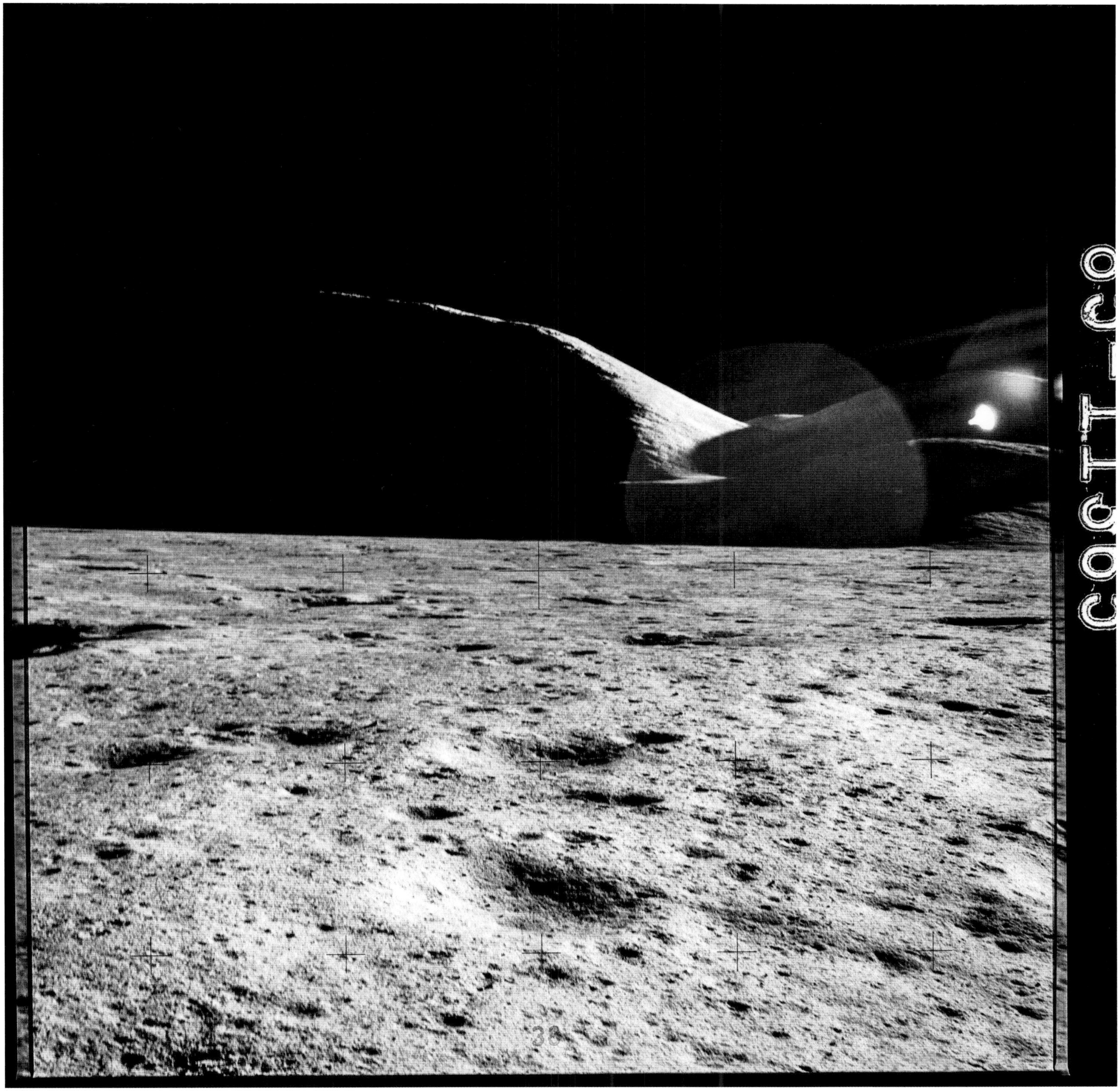

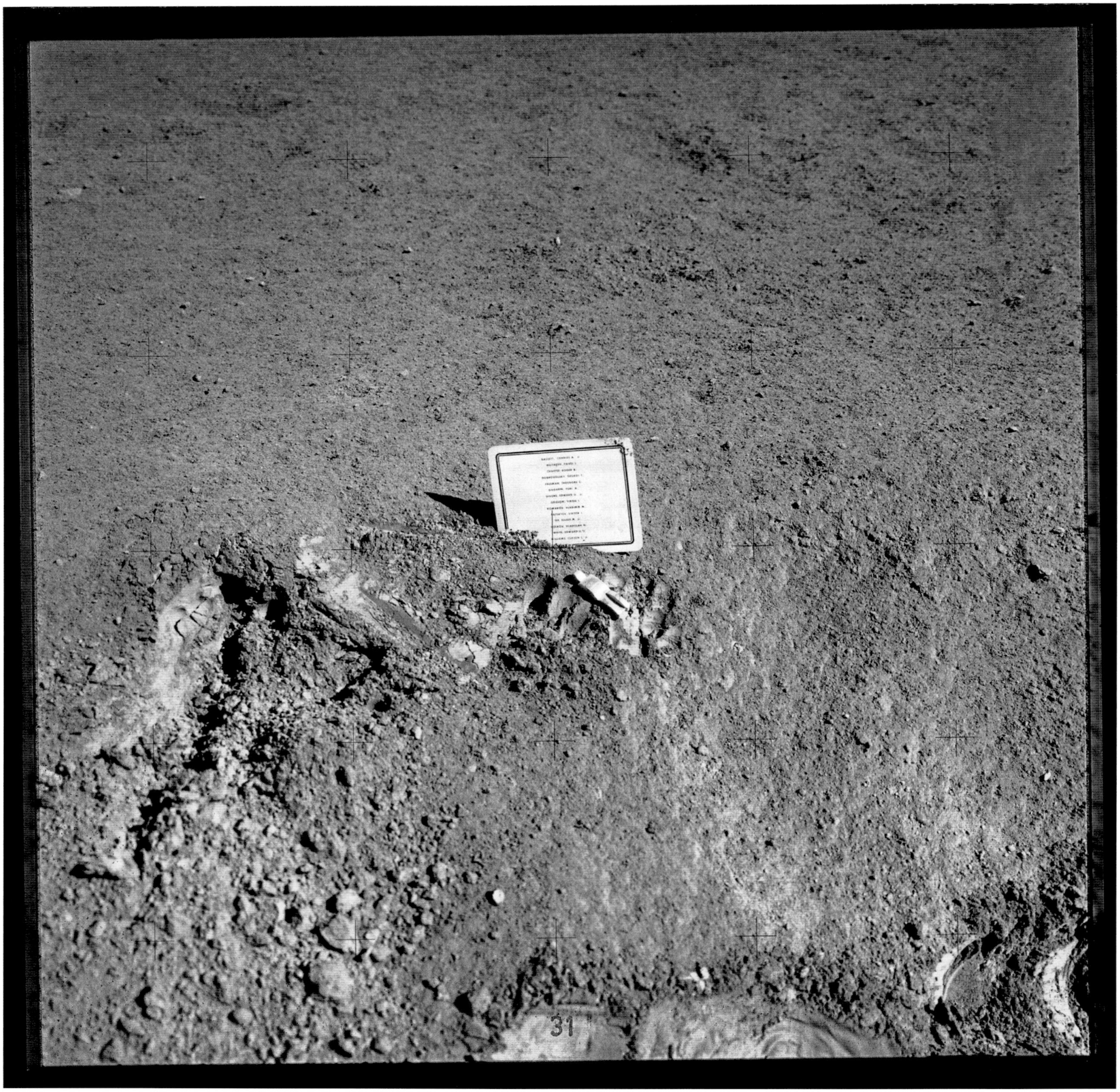

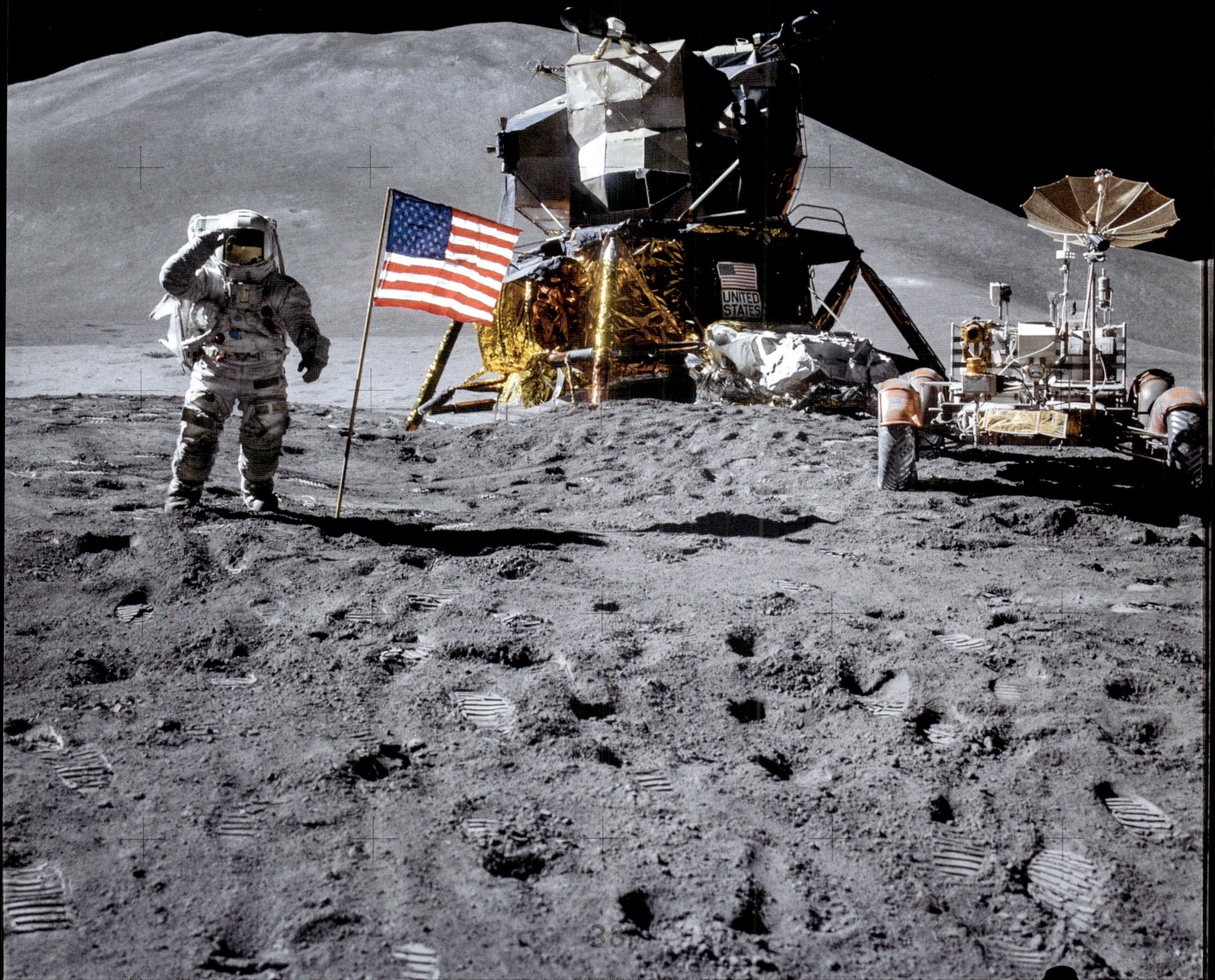
UNITED
STATES

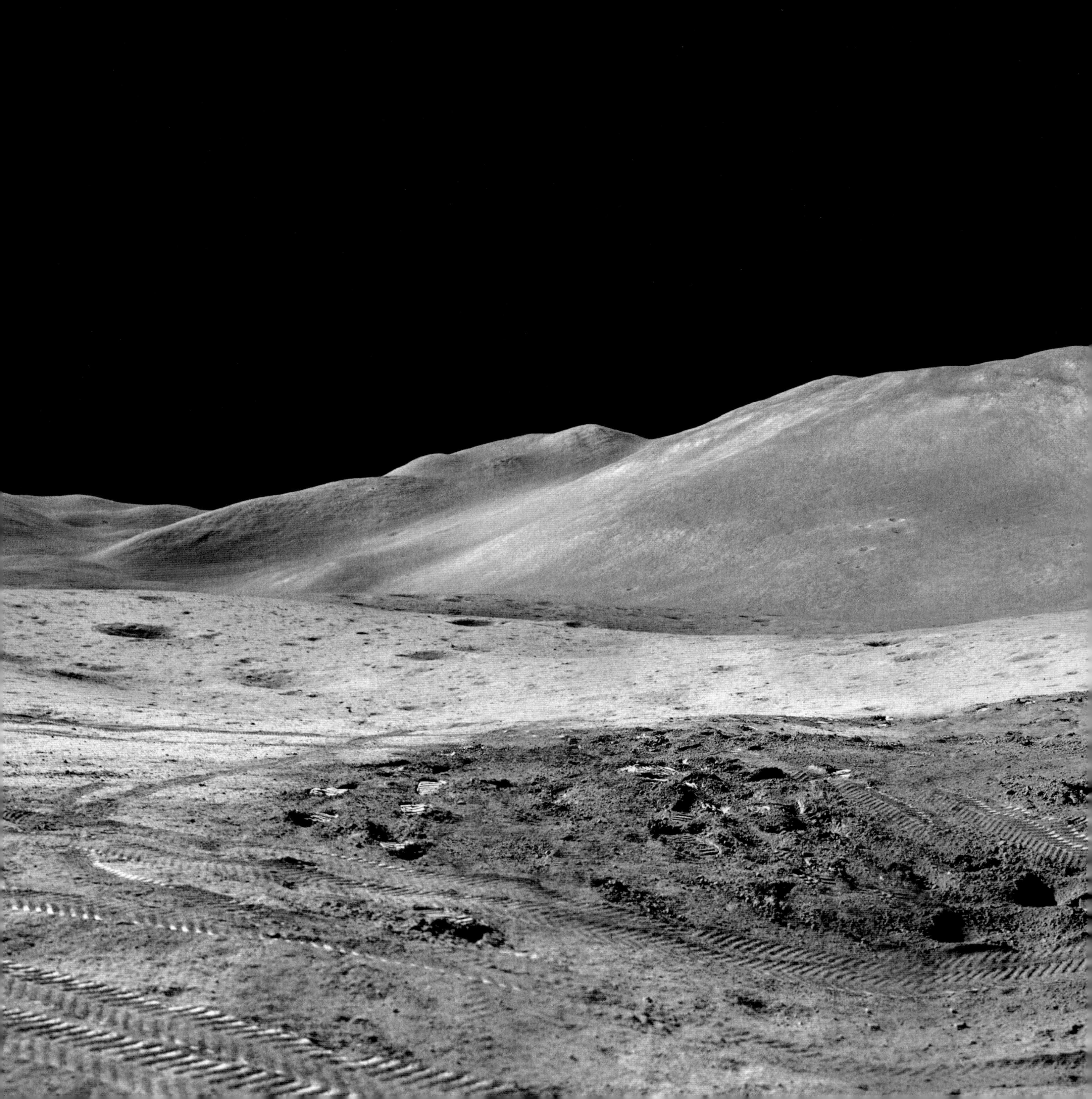

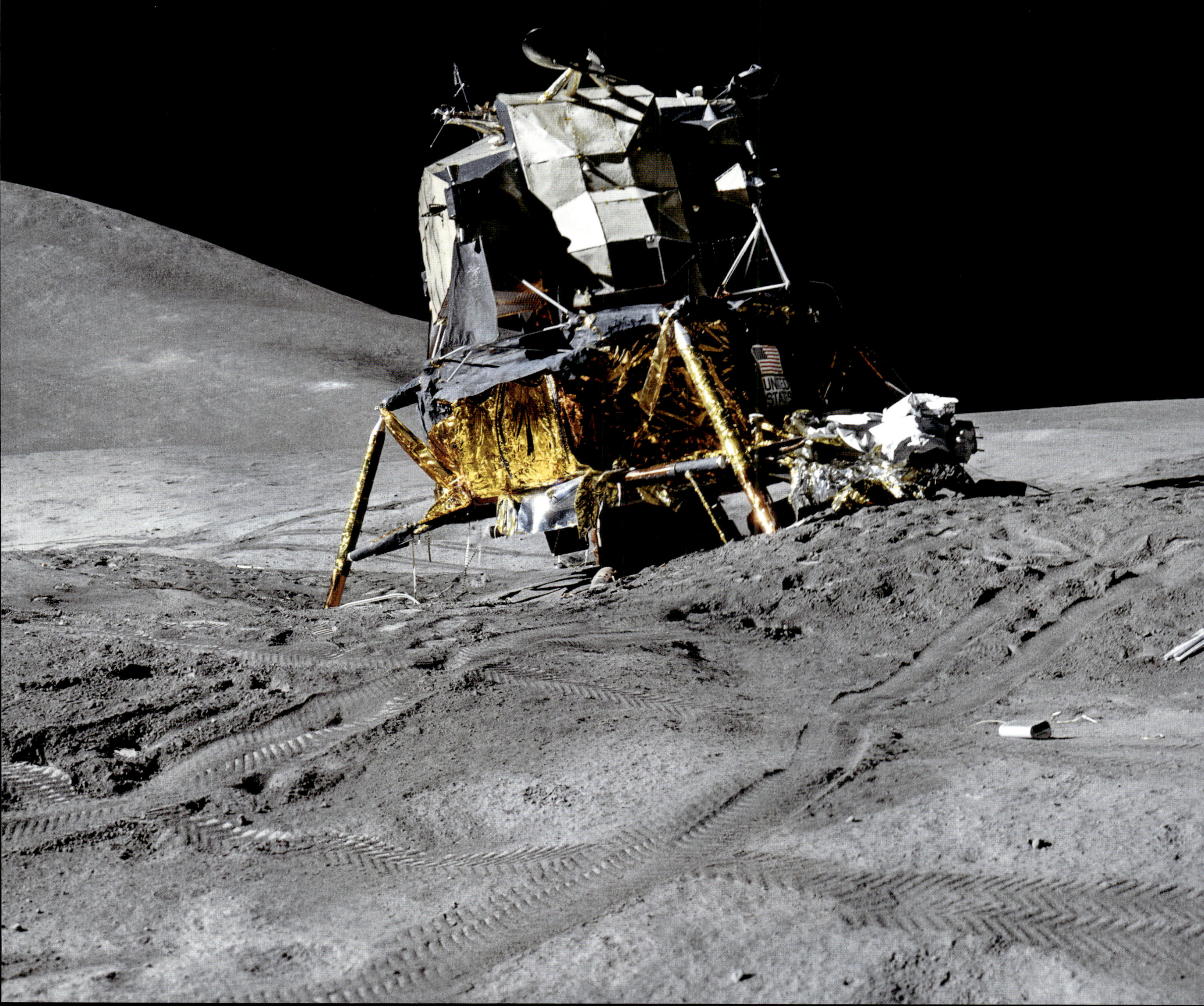

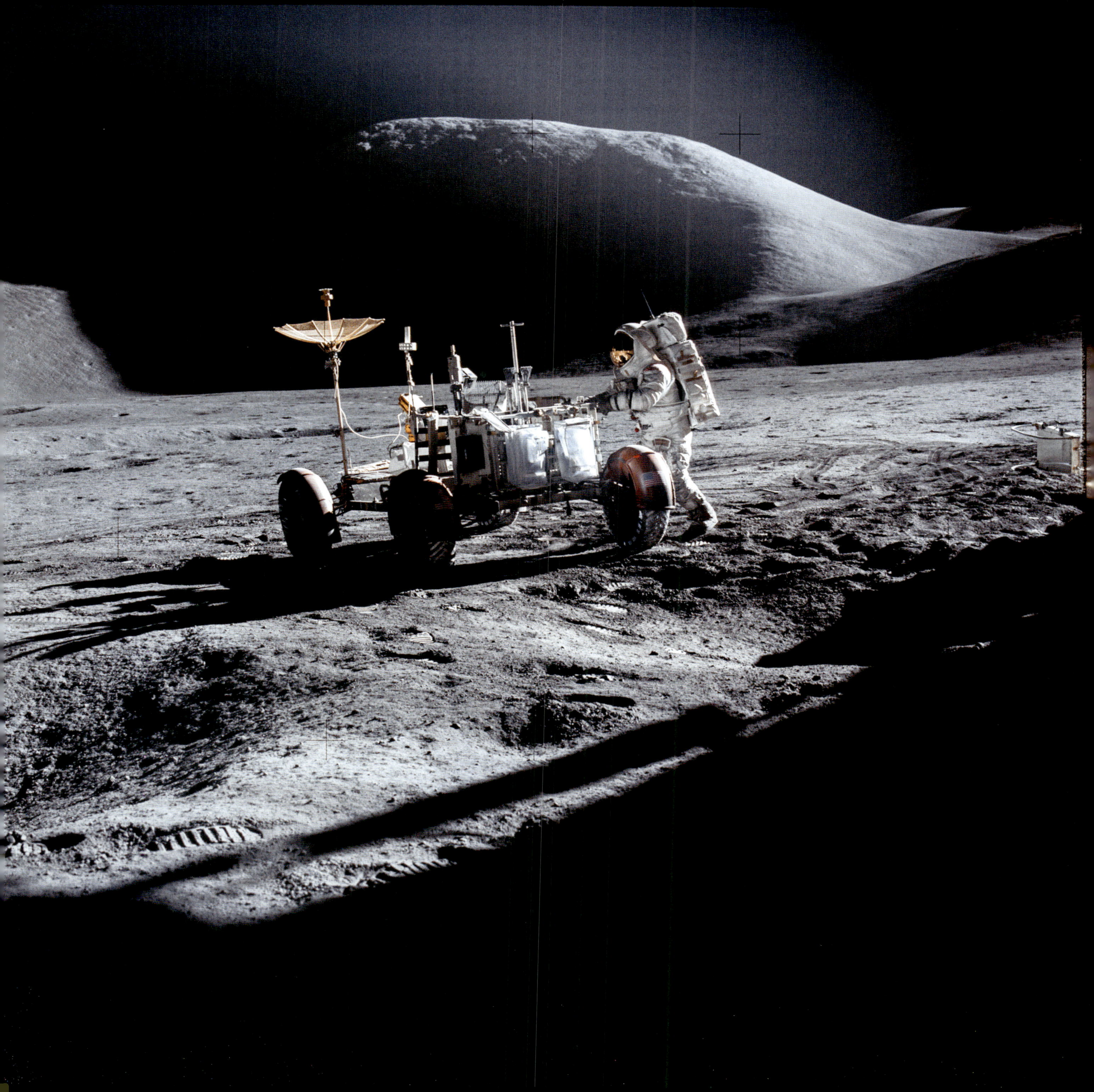

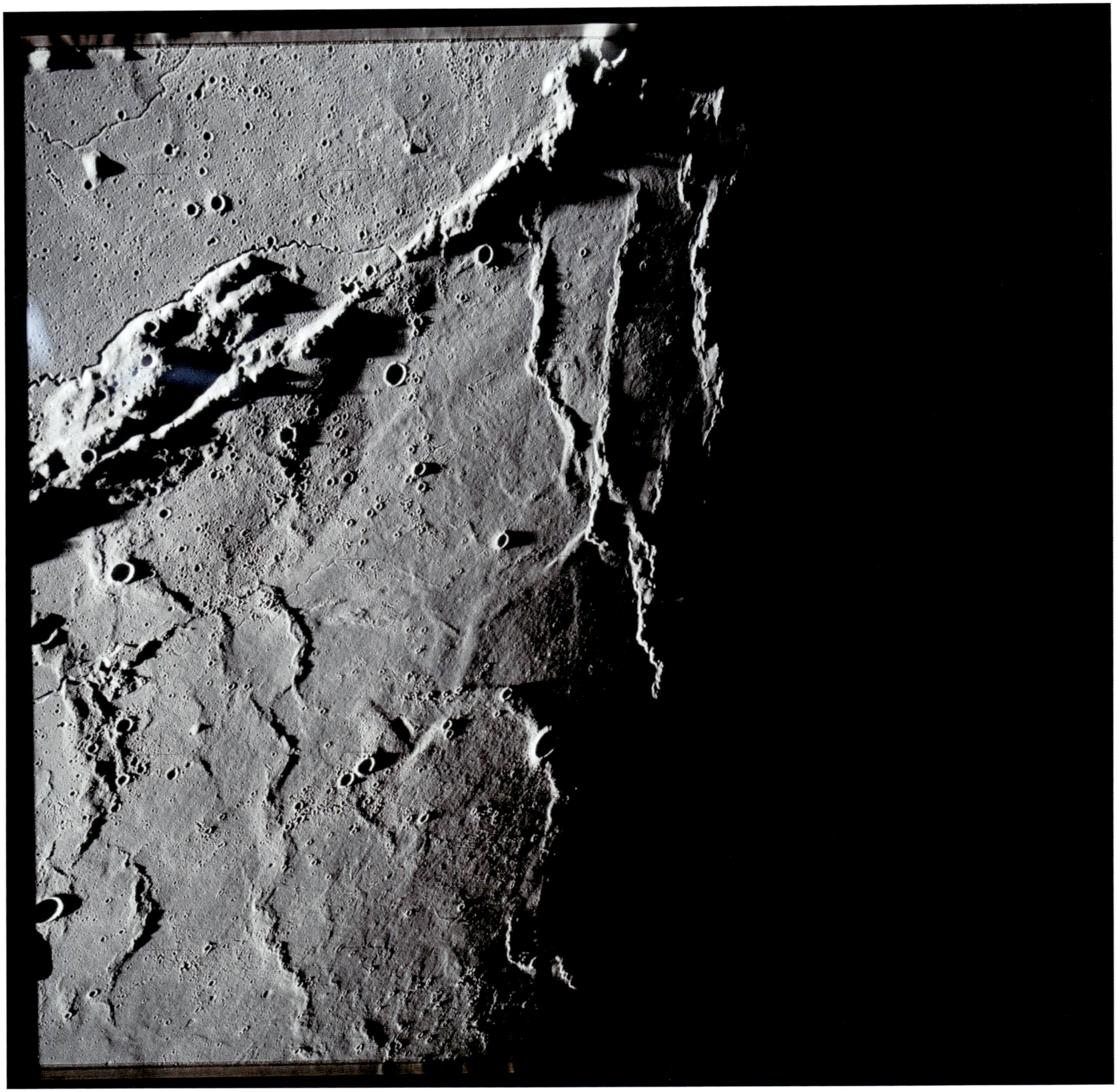

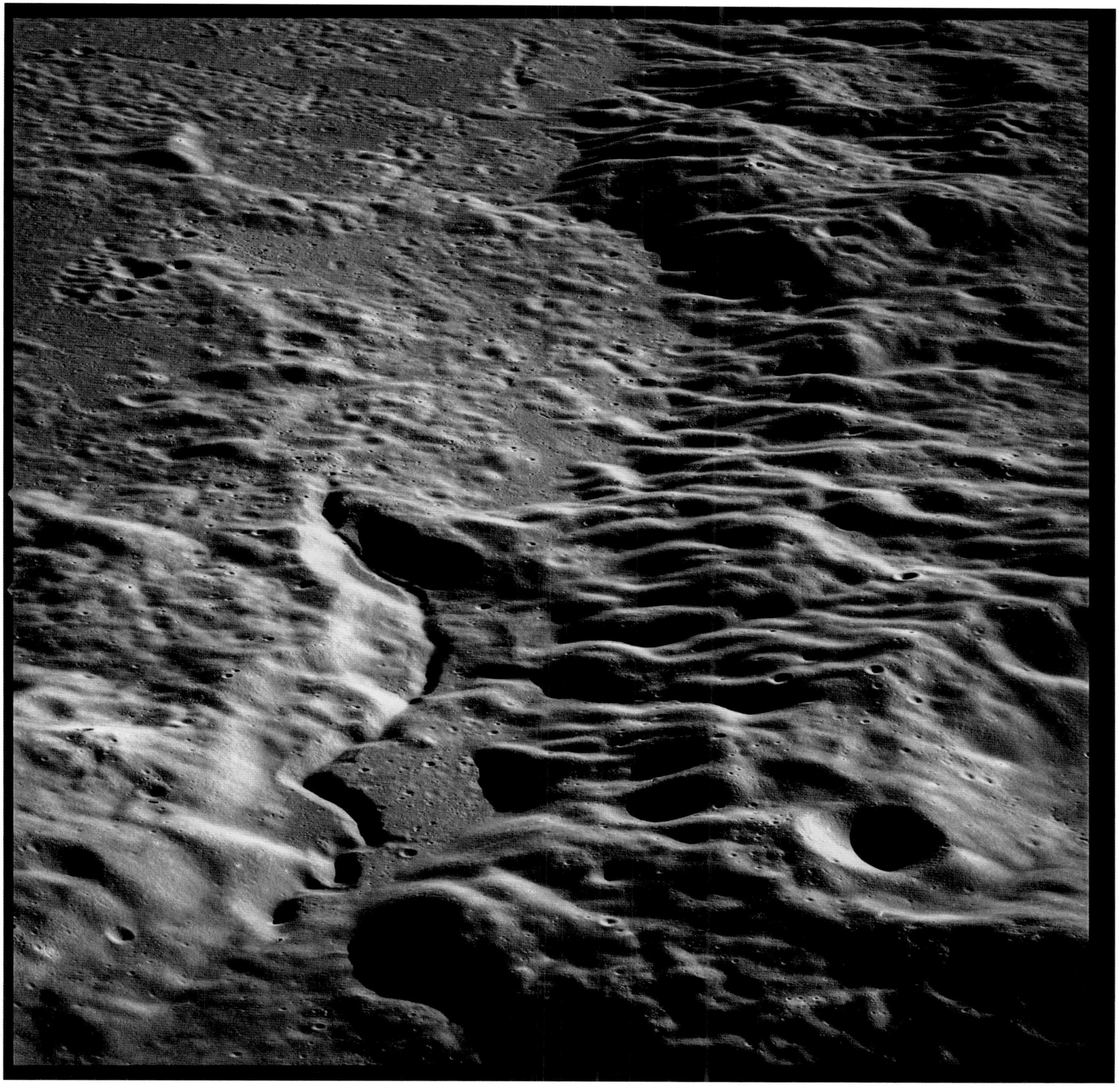

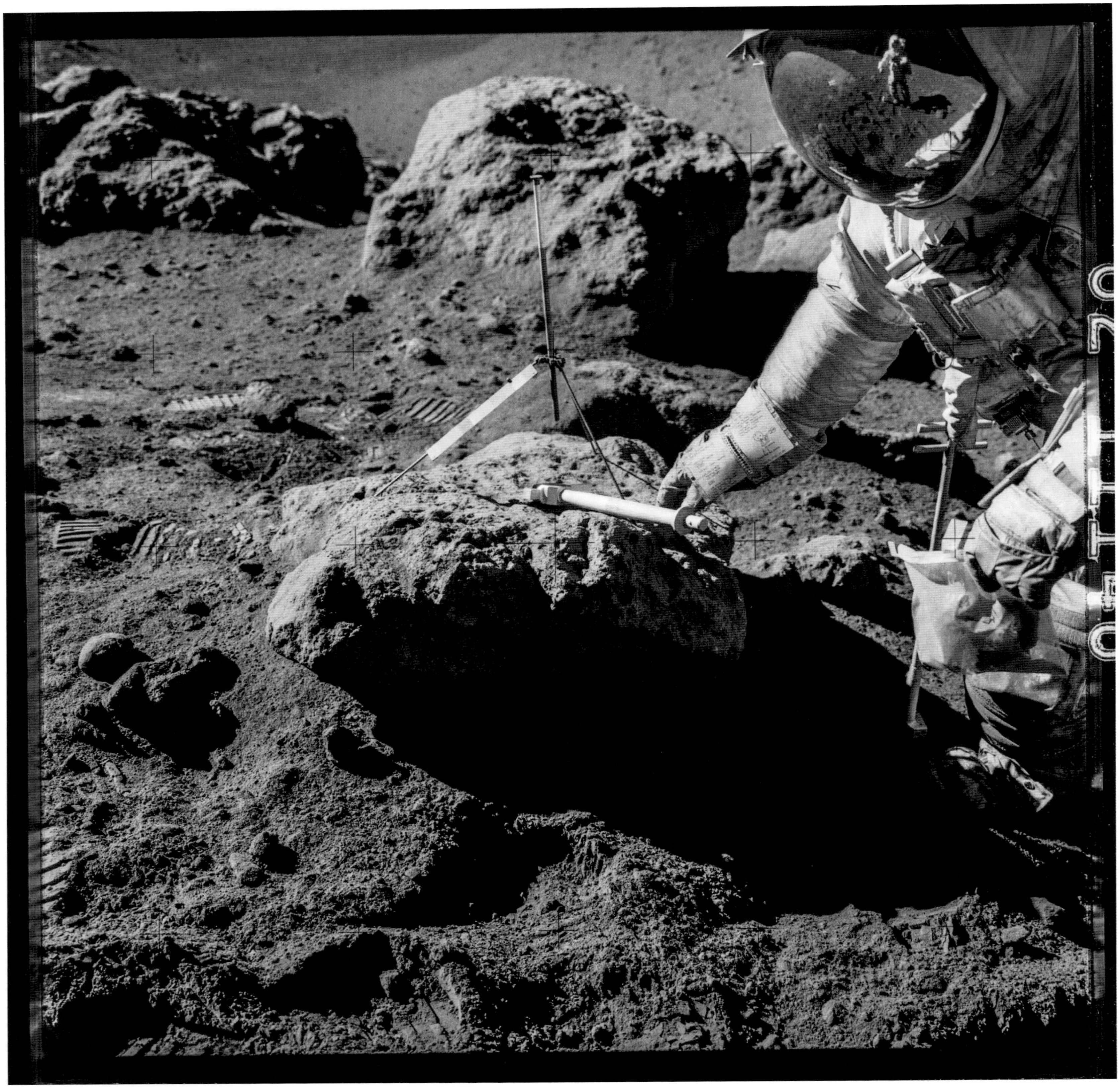

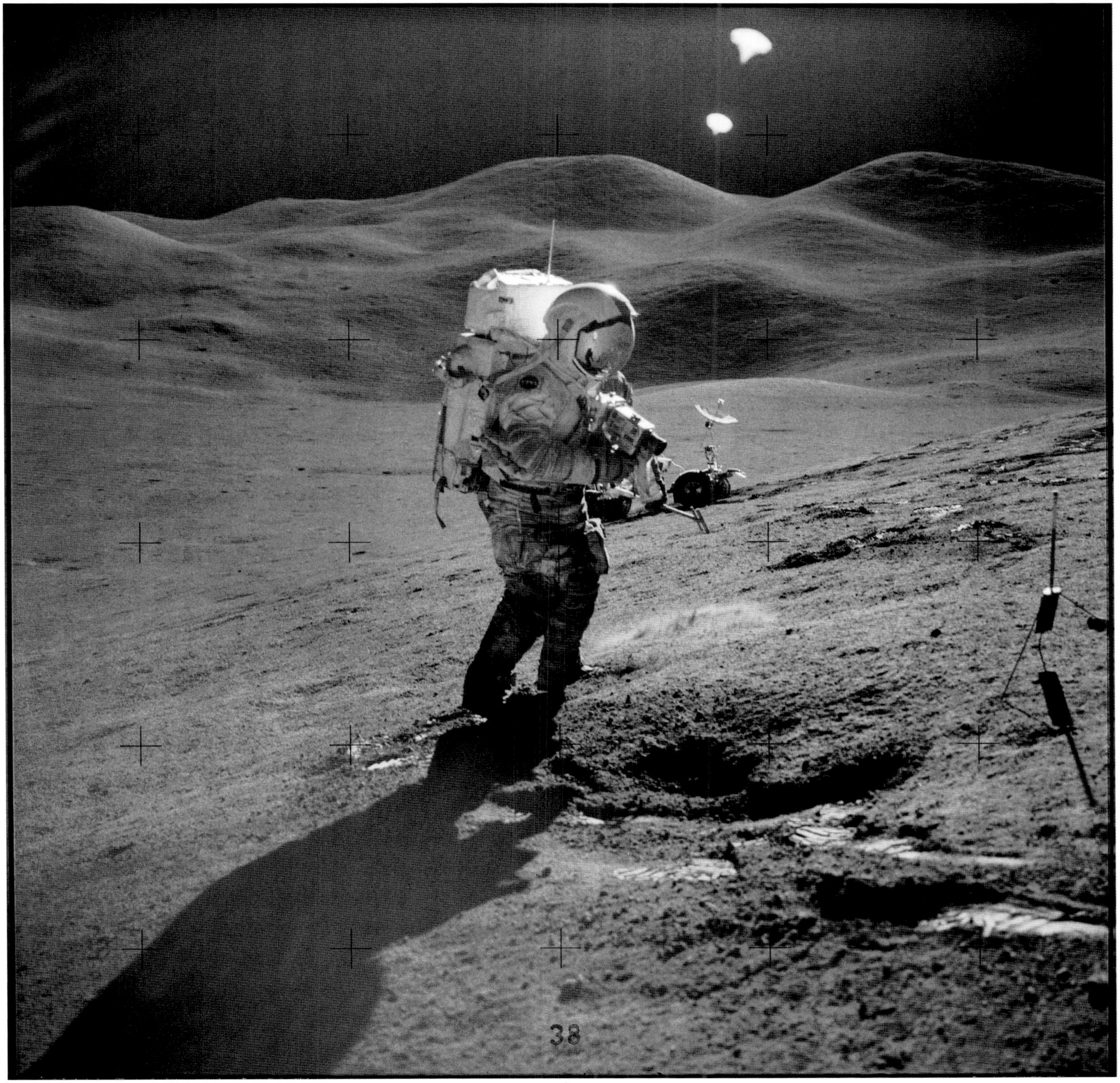
38

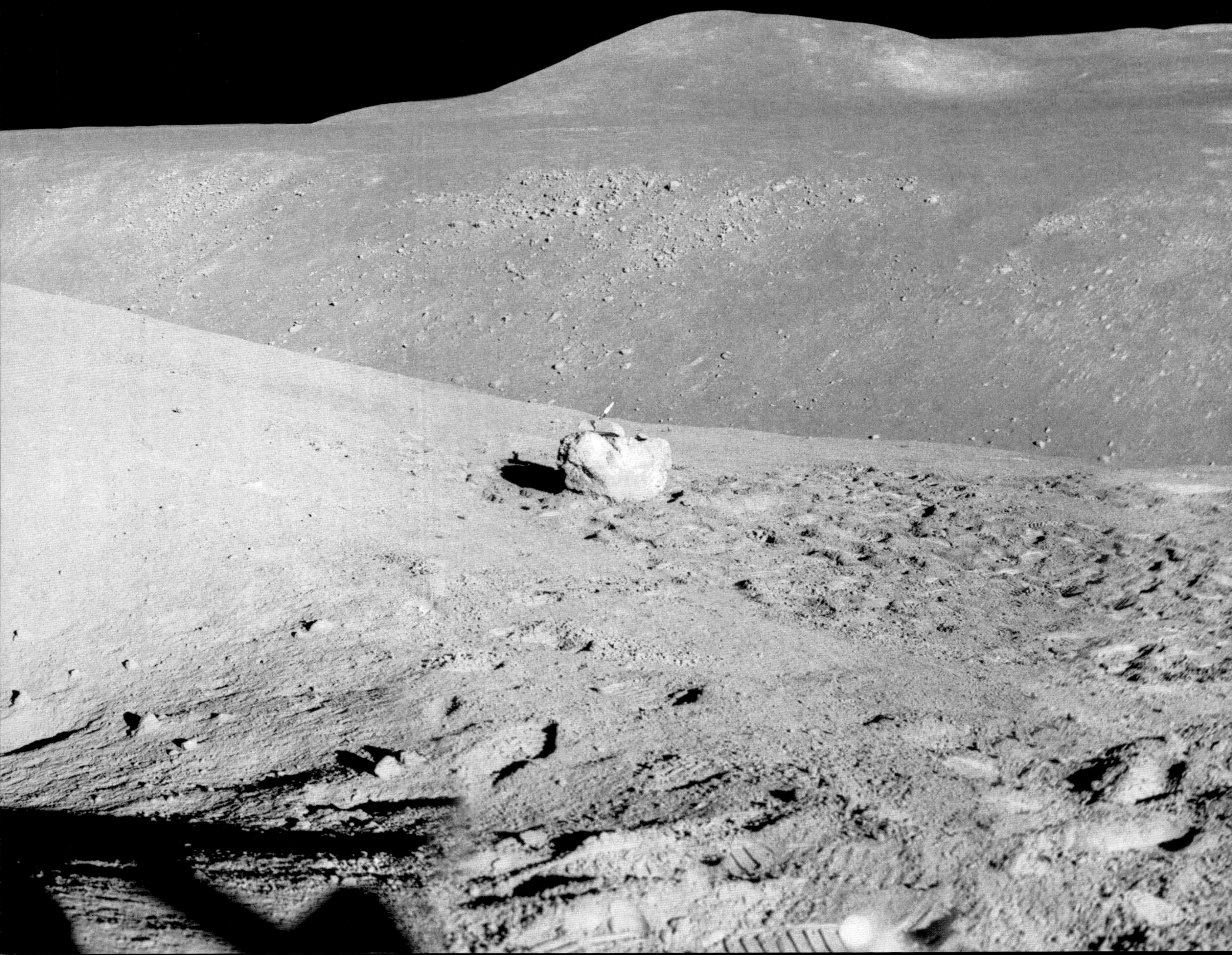

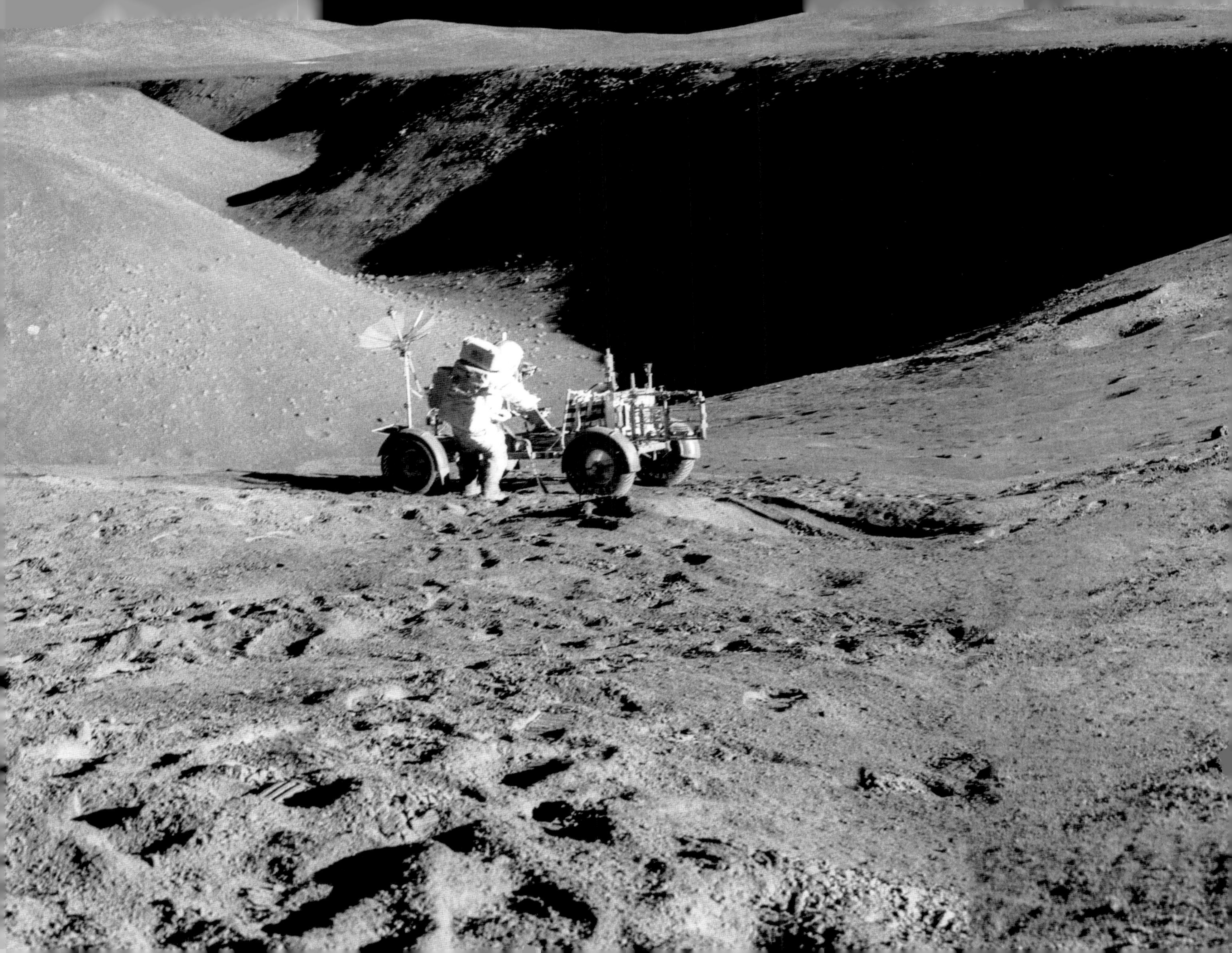

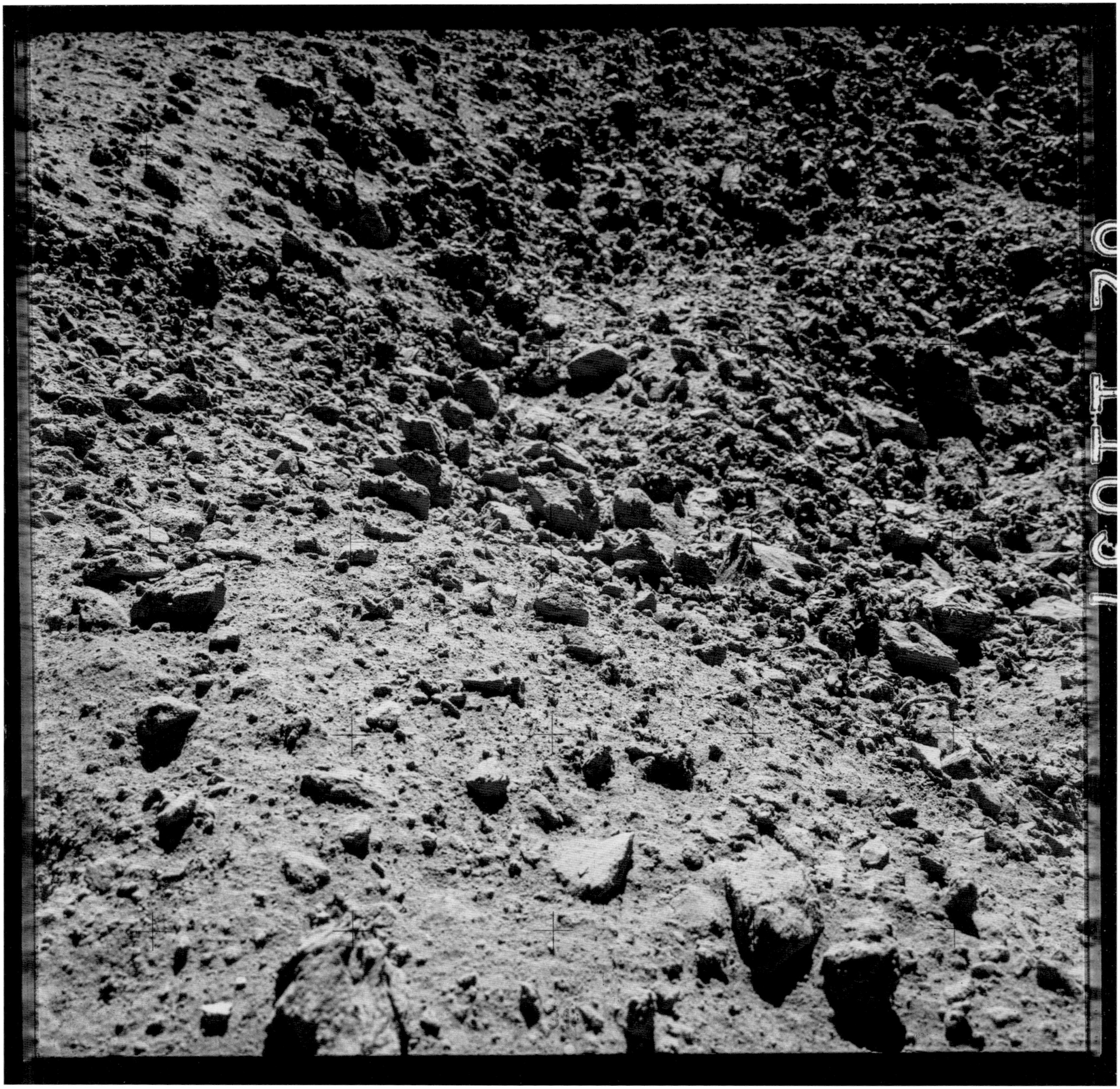

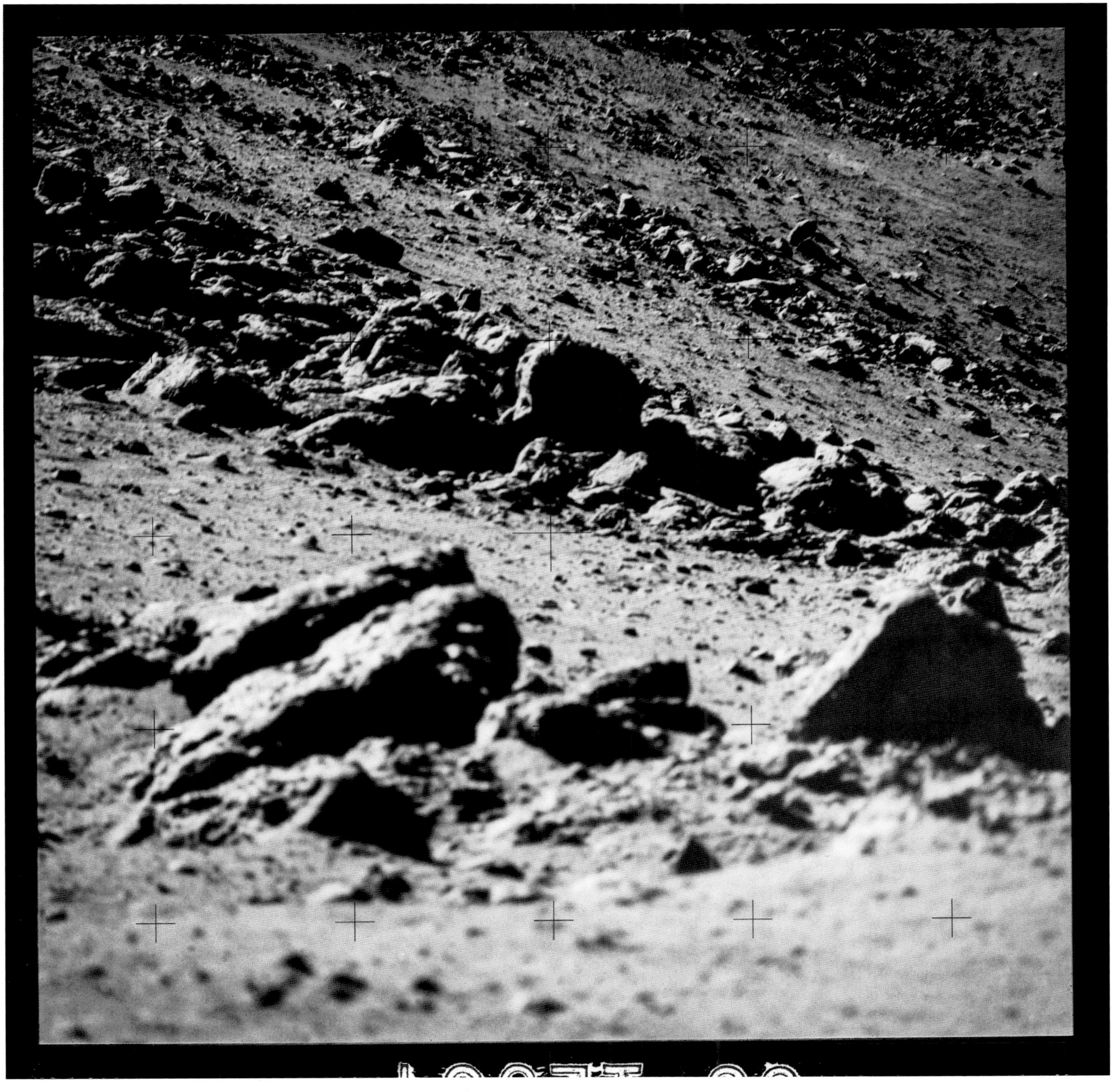

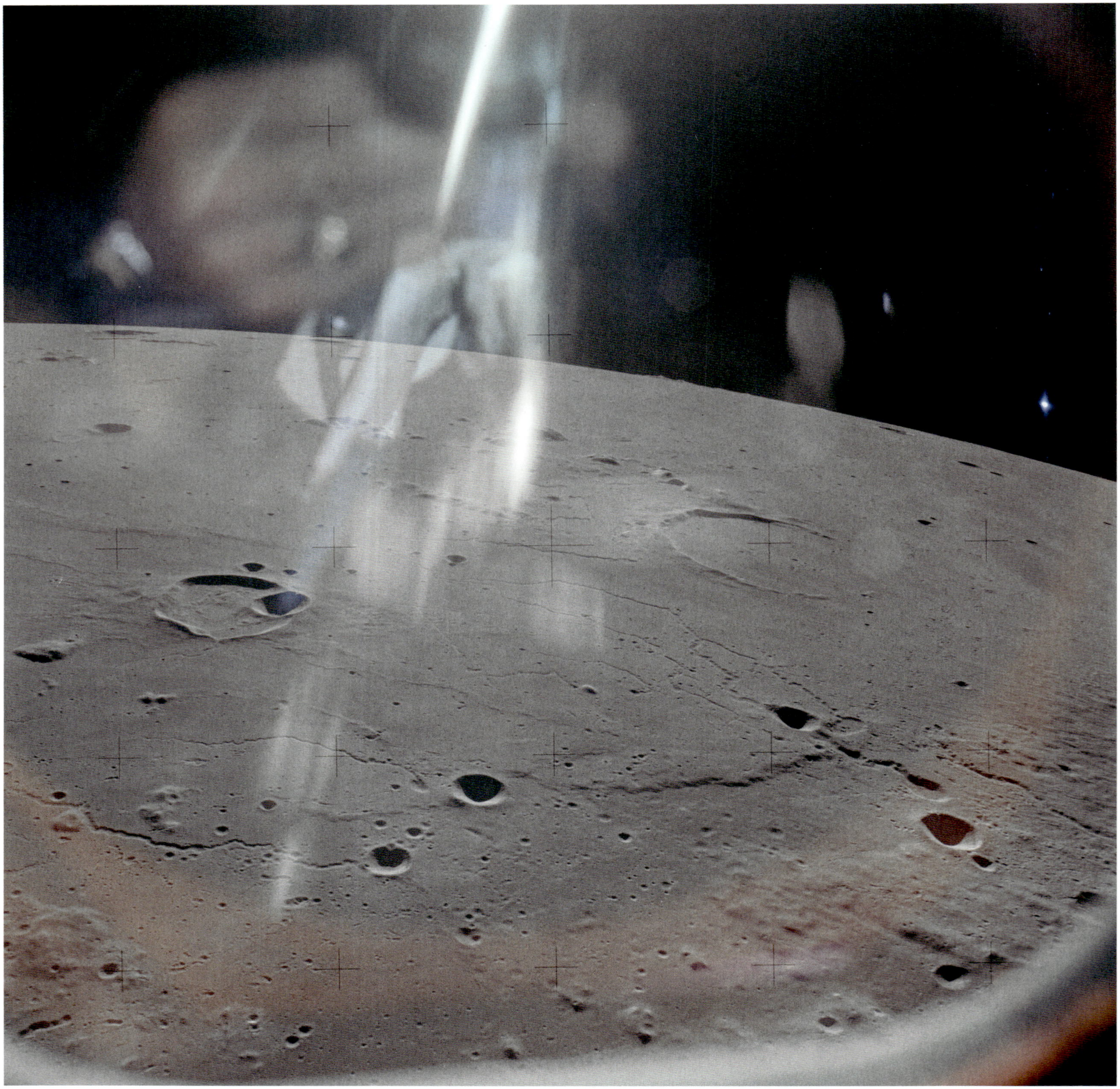

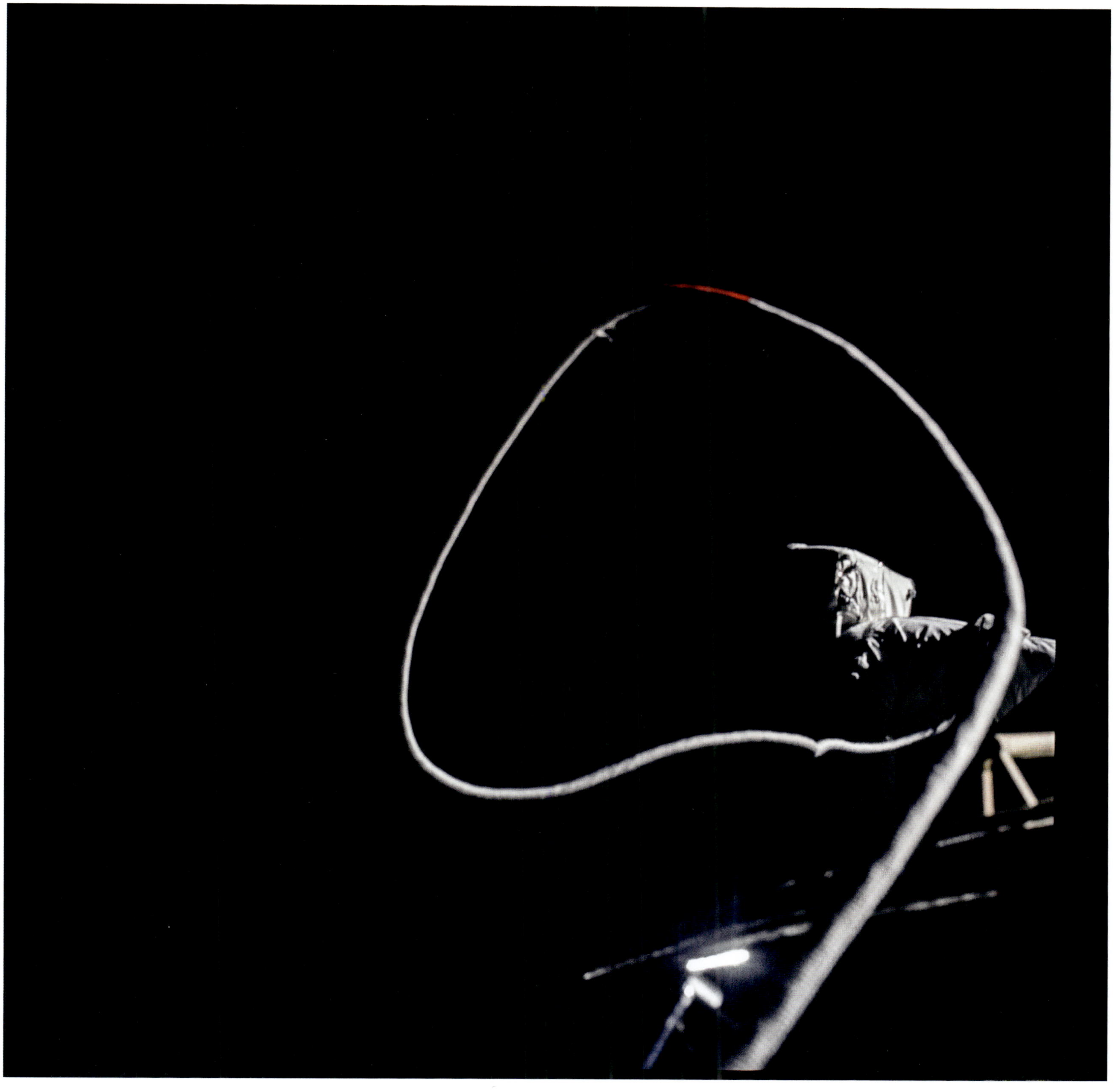

205 (AS15-81-10985)
A view of the lunar terrain west-southwest of Delisle Crater in Oceanus Procellarum. The crater's sharp edges suggest it is 'relatively' young, estimated to be about 1 billion years old.

206 (AS15-81-10920)
A crater on the northeast rim of Gibbs Crater, located on the far side of the Moon. This image contributed to studies of impact cratering and lunar geology.

207 (AS15-97-13190)
A detailed view of Sklodowska Crater on the far side of the Moon. Named after Marie Curie (née Maria Skłodowska) to distinguish it from Curie Crater, it remains a key site for studying lunar impact history due to its well-preserved features.

208-209
(AS15-92-12428–AS15-92-1231)
Lunar Module Pilot James B. Irwin's Extravehicular Activity (EVA) Apollo Lunar Surface Experiments Package (ALSEP) panorama, showing the Swann Range and the sunlit face of Mons Hadley. The Lunar Roving Vehicle (LRV) tracks and Irwin's bootprints are visible in the lunar dust. The Swann Range was named after geologist Gordon Swann, who trained Apollo astronauts in field geology.

210 (AS15-84-11318)
A close-up of the south ridge of Mons Hadley, a towering lunar mountain rising 15,100ft (4,600m) above the surface. The image captures steep, eroded slopes, loose regolith, and scattered boulders, evidence of the mountain's formation by the Imbrium Basin impact.

211 (AS15-84-11250)
A view of Silver Spur in the Hadley-Apennine region, where deep shadows and sunlit ridges highlight its rugged terrain. The sharp contrasts reveal steep slopes and linear formations shaped by tectonic forces and the Imbrium impact.

212 (AS15-85-11425)
A view of Hadley Rille, looking northwest. This deep, winding channel stretches 80mi (130km) at the base of the Apennine Mountains. With walls up to 1,200ft (370m) deep, it likely formed from ancient volcanic activity, possibly as a collapsed lava tube.

213 (AS15-85-11365)
A section of a panorama featuring Mons Hadley, one of the Moon's tallest peaks at 2.8mi (4.5km). The image captures the sharp contrast between sunlit ridges and deep shadows, showcasing the rugged terrain of the Hadley-Apennine region.

214 (AS15-88-11894)
A memorial on the lunar surface, featuring the 'Fallen Astronaut' figurine and a plaque listing 14 astronauts and cosmonauts who lost their lives in space exploration. Created by Belgian artist Paul Van Hoeydonck, the tribute was not publicly disclosed by the Apollo 15 crew until after the mission.

215 (AS15-88-11866)
This iconic composition captures Irwin saluting the American flag, with the Lunar Module (LM) *Falcon*, LRV, and the rugged Hadley-Apennine terrain in one frame. Commander David R. Scott is faintly visible in Irwin's visor reflection.

216-217
(AS15-87-11805–AS15-87-11824)
A panorama of the Apollo 15 landing site, showing the LM *Falcon* with Mons Hadley in the background. The LM sits at a tilt of 11°, caused by one of its legs settling in a small depression.

219 (AS15-86-11603)
Irwin stands beside the LRV at the Hadley-Apennine landing site. The low-angle sunlight casts deep shadows, highlighting the golden reflections of the LRV's antenna and creating a striking contrast against the stark lunar landscape.

220 (AS15-88-11984)
A view of Aristarchus Crater and Schröter's Valley, one of the Moon's most geologically intriguing regions. The 40mi (64km)-wide Aristarchus Crater stands out for its bright ejecta, while Schröter's Valley, the largest sinuous rille at 99mi (160km) long, was likely formed by ancient lava flows.

221 (AS15-91-12374)
A detailed view of the Sea of Ingenuity, one of the few lunar maria on the far side of the Moon. The image highlights its smooth basaltic plains and lunar swirls, features thought to be linked to magnetic anomalies and solar wind interactions.

222 (AS15-82-11146)
Part of a panorama at Station 9A, near the rim of Hadley Rille. The gnomon stands atop the sampled boulder, where Scott chipped fragments 15595 to 15598. He holds a sample bag in his left hand, while an EVA cuff checklist is visible on his right wrist.

223 (AS15-85-11514)
Scott using his camera on the slope of Hadley Delta, with the LRV visible in the background. The sharp shadows and textured terrain highlight the challenges of lunar exploration at this scientifically rich landing site.

224-225
(AS15-85-11448–AS15-85-11454)
A panorama of Hadley Rille, showing Scott and the LRV on its edge. Taken by Irwin from St. George Crater, the image looks northwestward at Hadley Rille's 80mi (130km) winding channel, shaped over time by erosion. Large talus blocks are visible along the rille's walls and floor.

226 (AS15-82-11097)
A panoramic view of a lunar scarp in the Hadley-Apennine region, showing a steep, cliff-like formation created by tectonic activity. The Apollo 15 crew used the LRV to explore and collect samples from this geologically significant site.

227 (AS15-89-12094)
A close-up black-and-white photograph of the east wall of Hadley Rille, showing distinct layers of lunar rock. These formations include ancient basalt from lava flows and brecciated material from meteorite impacts, offering insight into the Moon's volcanic and impact history. Captured during Apollo 15's geological survey.

228 (AS15-88-12004)
Photograph of lunar craters Krieger (left) and Prinz (right of center), taken from orbit during Revolutions 64-72. By this time, Scott and Irwin were on the surface, while Command Module Pilot Alfred M. Worden continued working alone in *Endeavour*. A reflection of Worden's hand can be seen in the window — a subtle, lonely reminder of his solo work in lunar orbit.

229 (AS15-96-13100)
Worden performs the first deep-space EVA, floating outside the Command/Service Module (CSM) *Endeavour*. Tethered by an umbilical cord, Worden retrieves film cassettes from the Scientific Instrument Module (SIM) bay, over 196,000mi (315,000km) from Earth — a record-breaking spacewalk in the vast emptiness of space.

XVI

Apollo 16

Launch: April 16th, 1972
Lunar landing: April 20th, 1972
Splashdown: April 27th, 1972

Command Module: Casper
Lunar Module: Orion

NASA's tenth crewed lunar mission — spent 71 hours on the surface of the Moon. Astronauts John Young and Charles Duke conducted three moonwalks, totaling over 20 hours of extravehicular activity.

The landing site in the Descartes Highlands had been chosen for its geological interest following analysis of Apollo 14's orbital photography. Scientists hoped the area would yield ancient volcanic rock. Ultimately, Descartes was found not to be volcanic in origin but shaped by meteoric impact — no volcanic material was recovered.

Duke, the youngest person to walk on the Moon, left behind a photo of his family — and photographed it lying on the surface as a keepsake and tribute. He later explained that during the long training period, he spent so much time away from his family that bringing their photo along felt like a way to include them in the adventure.

The mission proceeded largely as planned, completing most photographic targets and scientific experiments. One technical hiccup occurred when a fender over one of the Lunar Rover's wheels broke off during unloading. This caused lunar dust to kick up during driving, coating the astronauts and the rover — a detail visible in many photographs. Interestingly, similar fender damage occurred on Apollo 17.

In much of the live footage broadcast back to Earth, it was clear that Young and Duke felt very comfortable on the Lunar surface, moving around easily and often taking full advantage of the low gravity environment. This is beautifully shown on page 241, where Young can be seen jumping up while saluting the American flag. Video footage shows both Duke and Young performing the 'Lunar Olympics', jumping to see how high they could get while wearing their heavy space suits. Young's jump was successful, but Duke lost his balance and fell backwards onto the lunar surface.

John W. Young
Commander

Born on September 24th, 1930 in San Francisco, California. He was Pilot of Gemini 3, Backup Pilot of Gemini 6A, Command Pilot of Gemini 10, Backup Command Module Pilot of Apollo 7, Command Module Pilot of Apollo 10 and Backup Commander of Apollo 13 and 17. He also commanded the first and ninth Space Shuttle mission (STS-1 & STS-9).

Thomas K. Mattingly II
Command Module Pilot

Born on March 17th, 1936 in Chicago, Illinois. Scheduled to be Command Module Pilot of Apollo 13, he was replaced by his backup because he had been exposed to measles. After flying on Apollo 16, he headed the Astronaut Office ascent/entry group from December 1979 to April of 1981. He later became Spacecraft Commander of STS-4 and STS-51C.

Charles M. Duke, Jr.
Lunar Module Pilot

Born on October 3rd, 1935 in Charlotte, North Carolina. He was chosen with the fifth group of astronauts in 1966. He was Backup Lunar Module Pilot of Apollo 13 and Apollo 17 and was the tenth man to walk on the Moon. He resigned from NASA and the Air Force on January 1st, 1976.

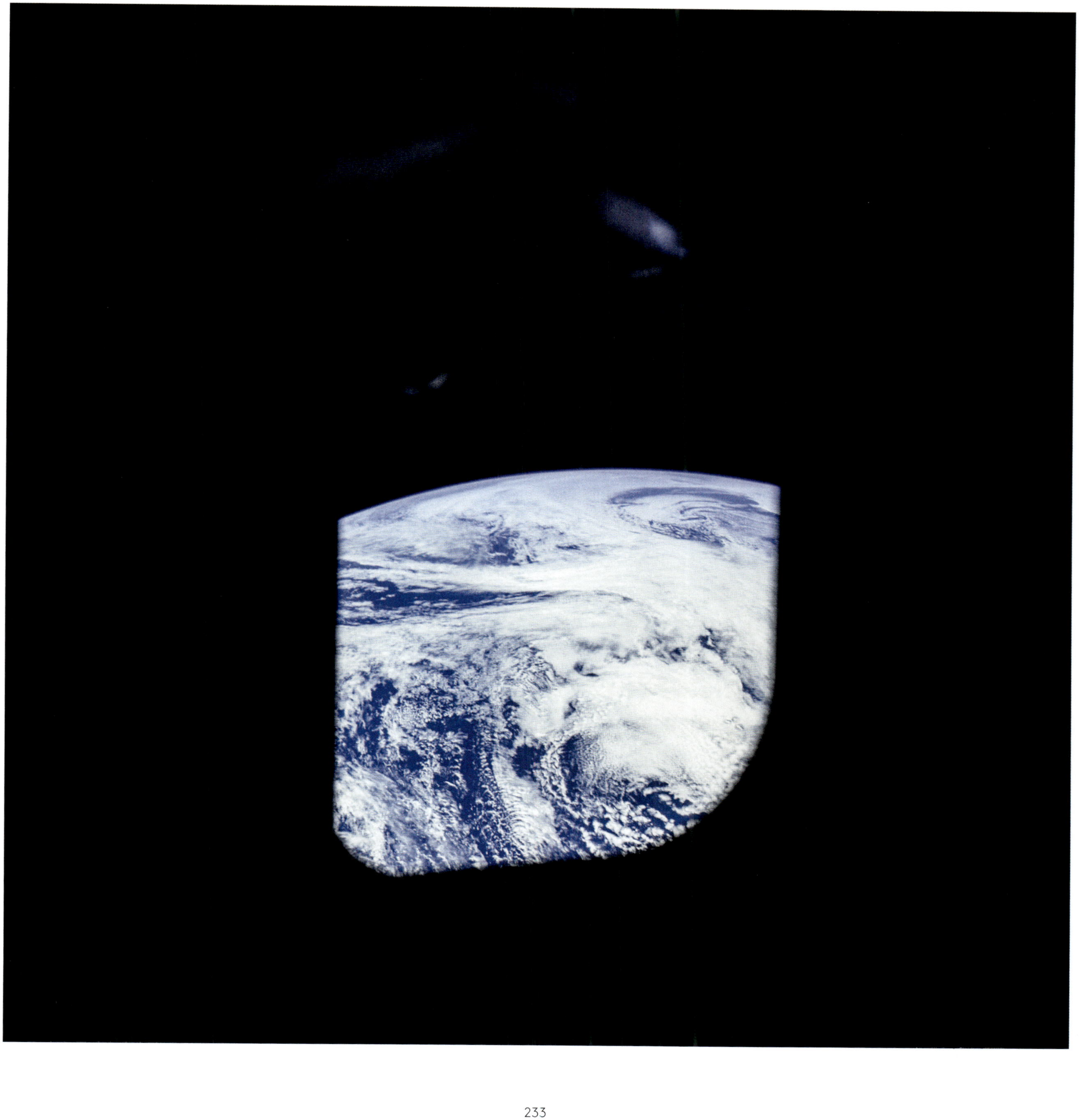

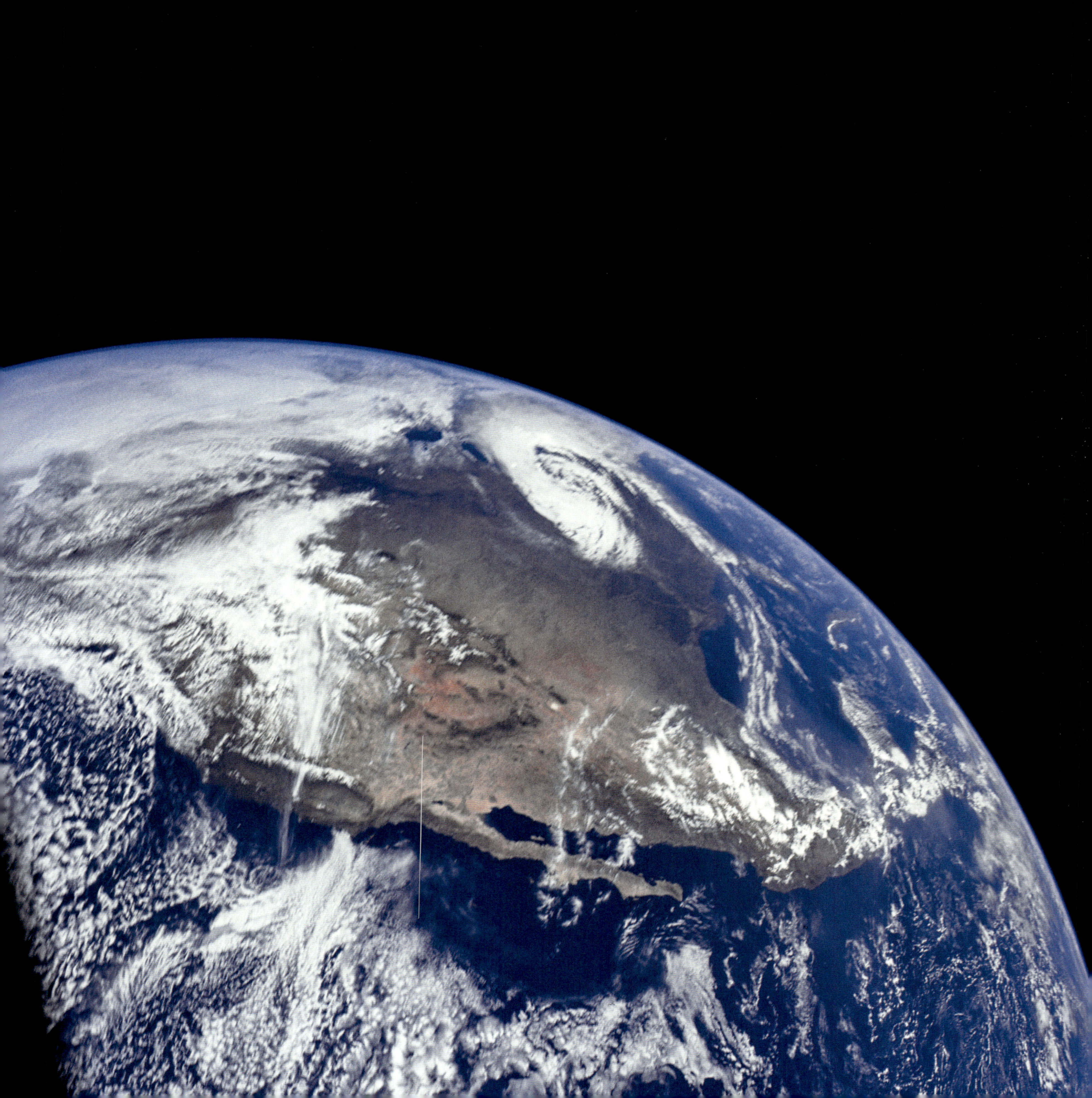

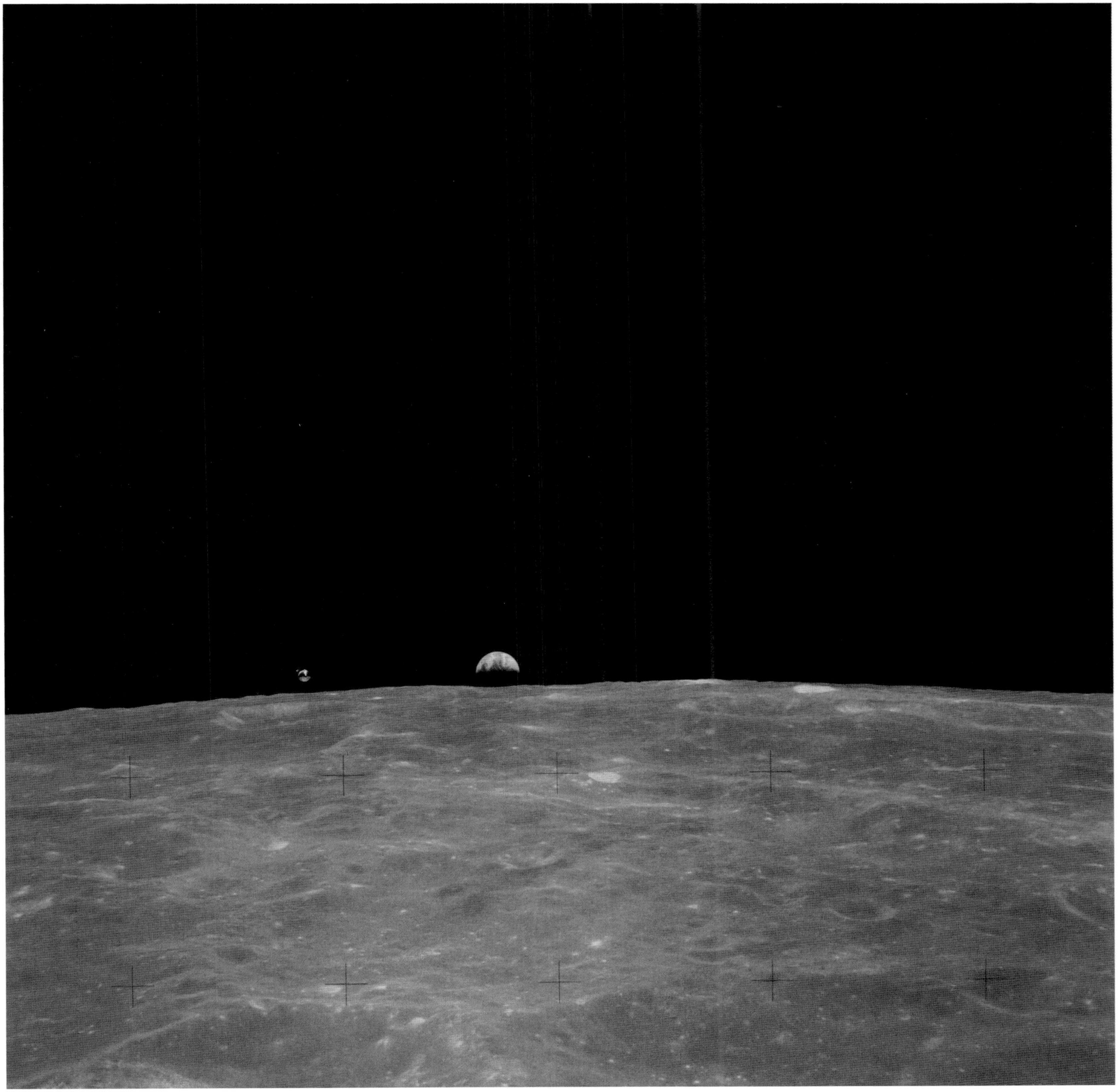

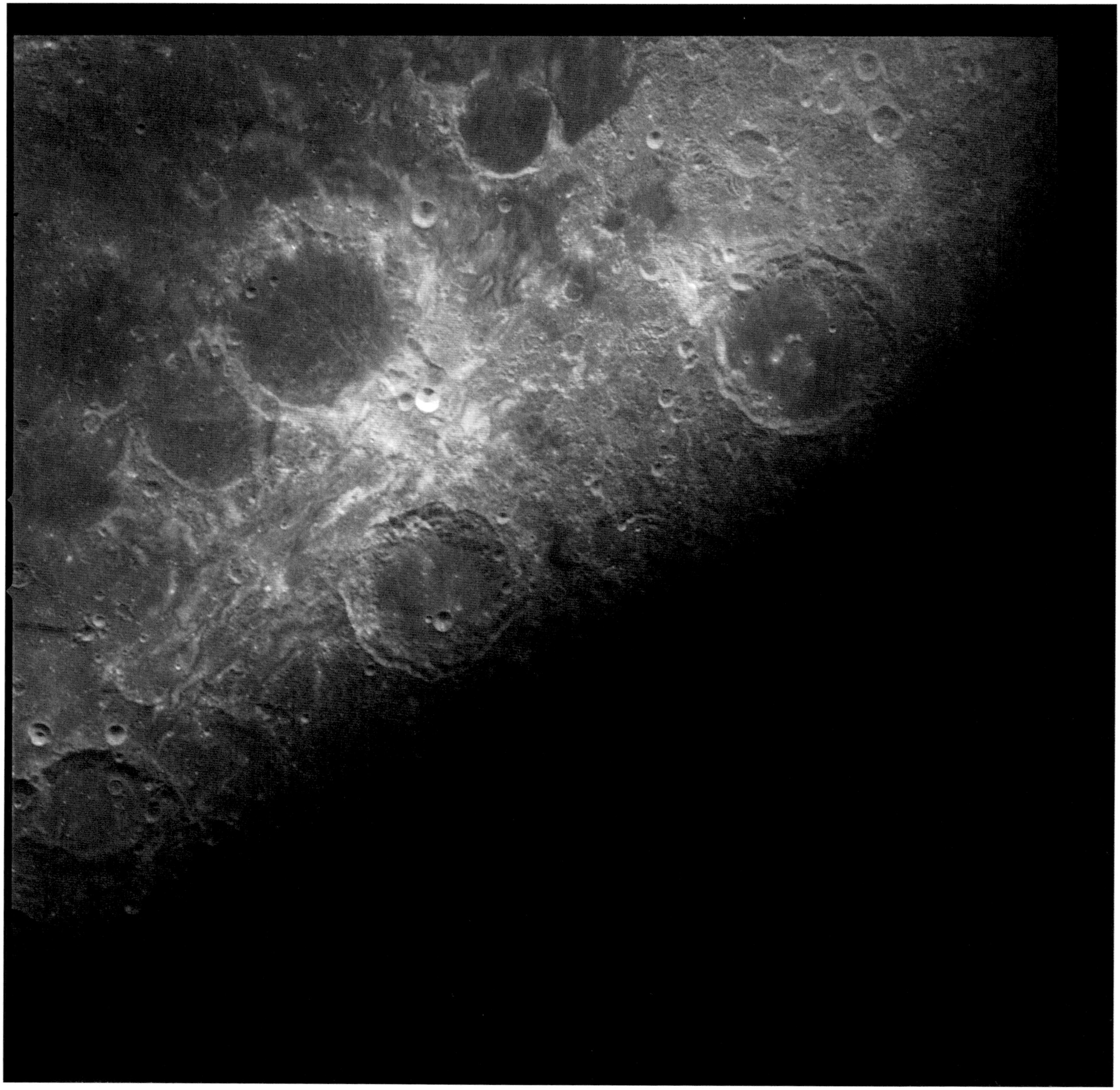

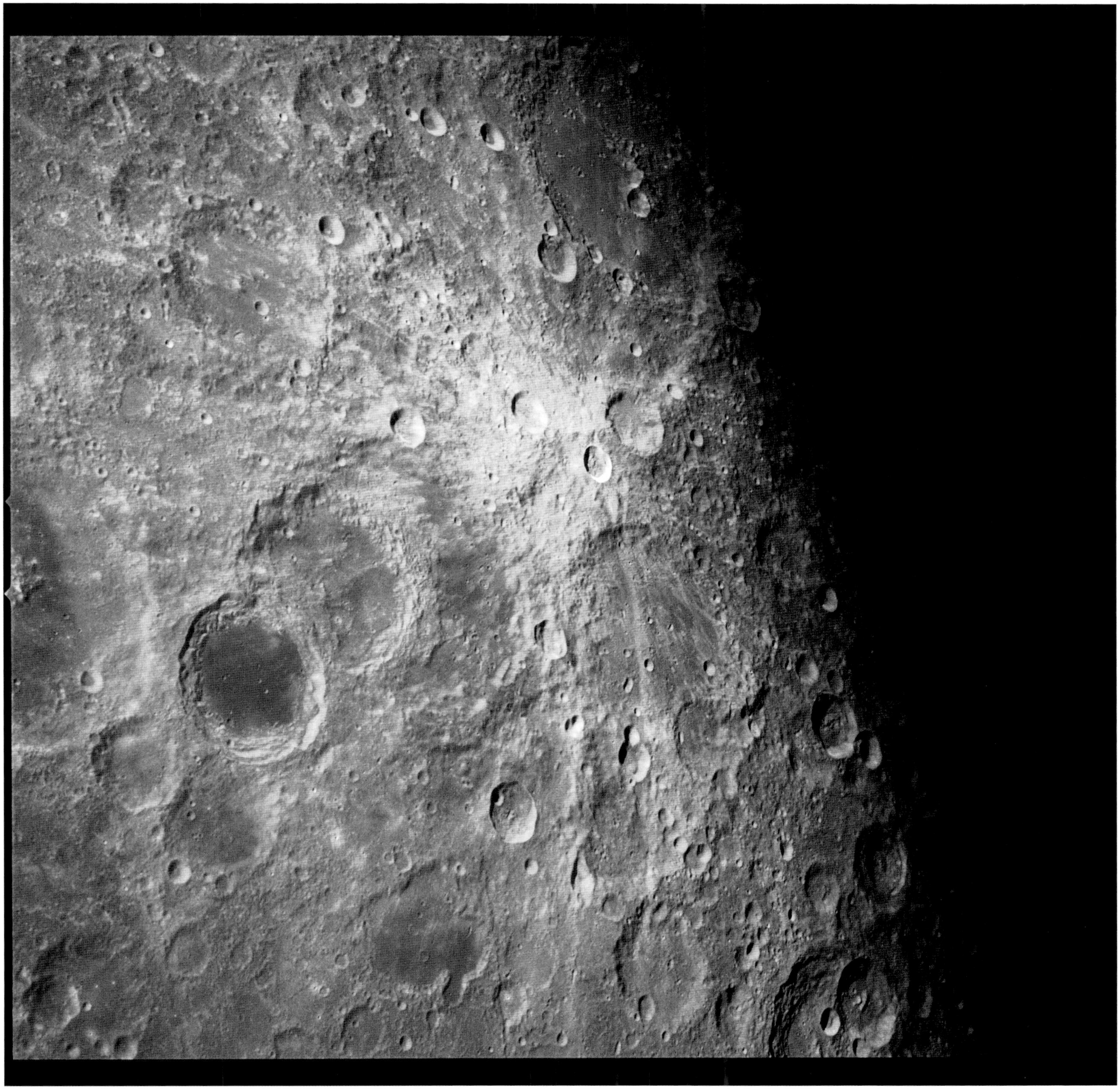

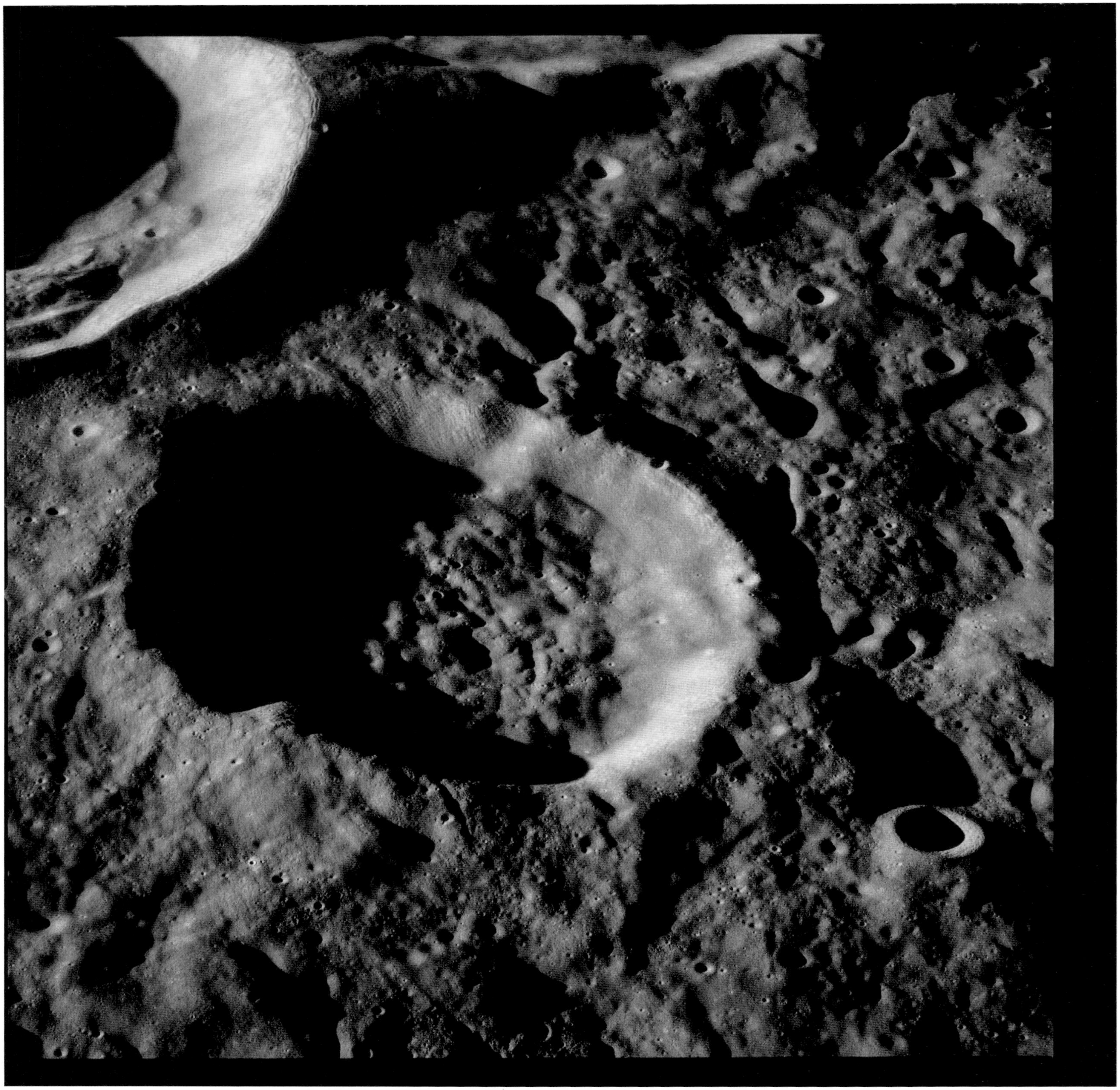

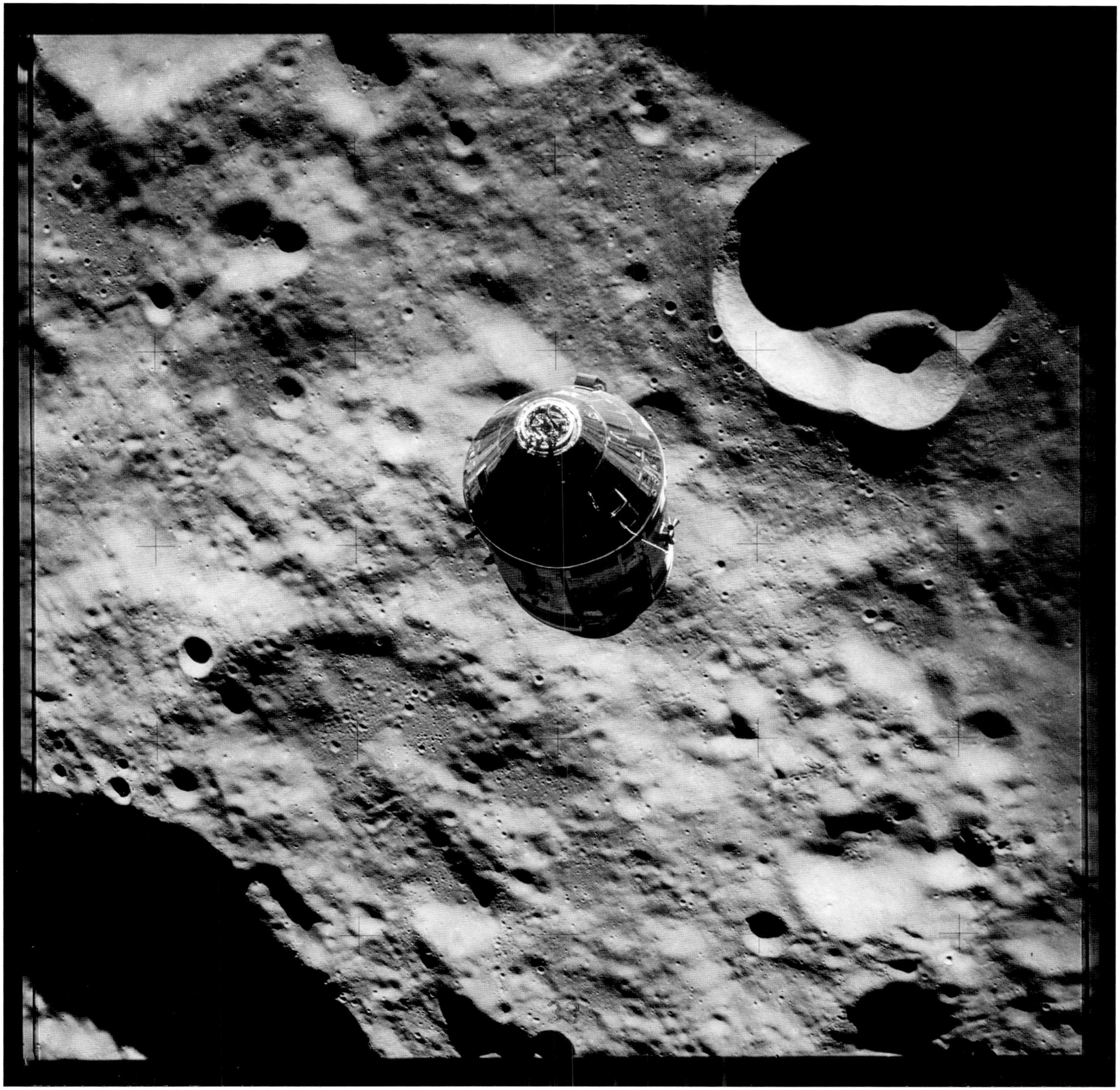

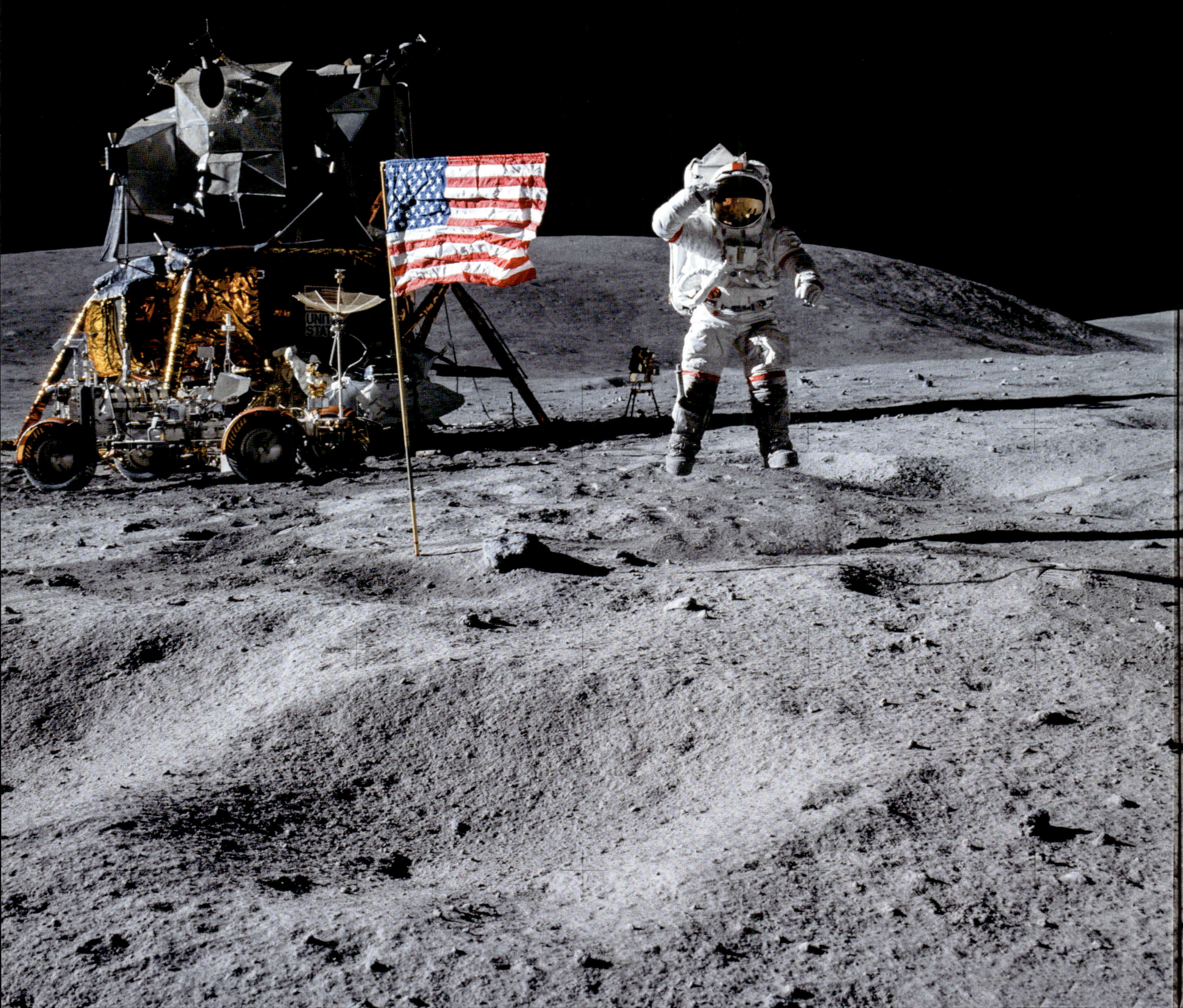

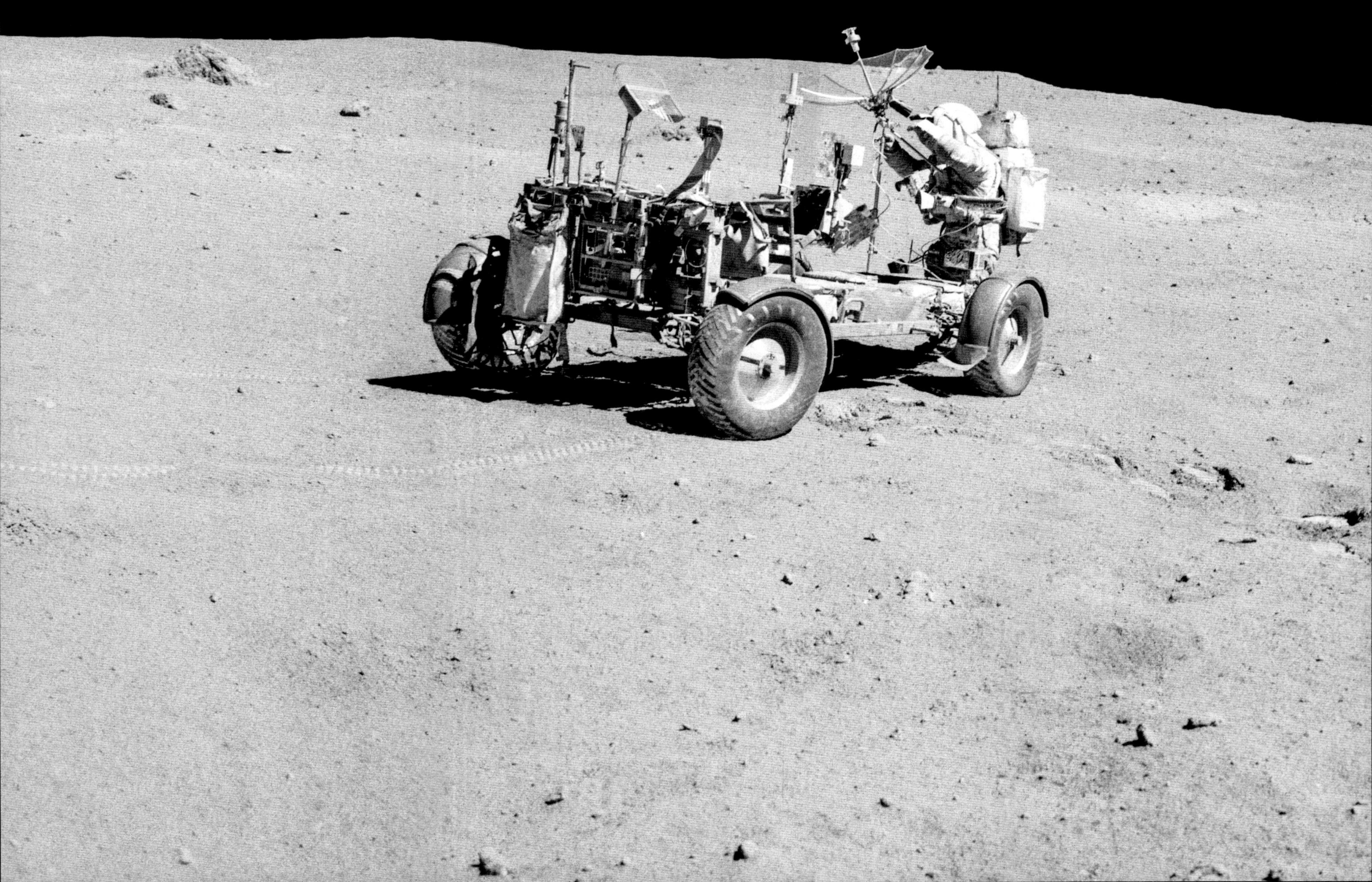

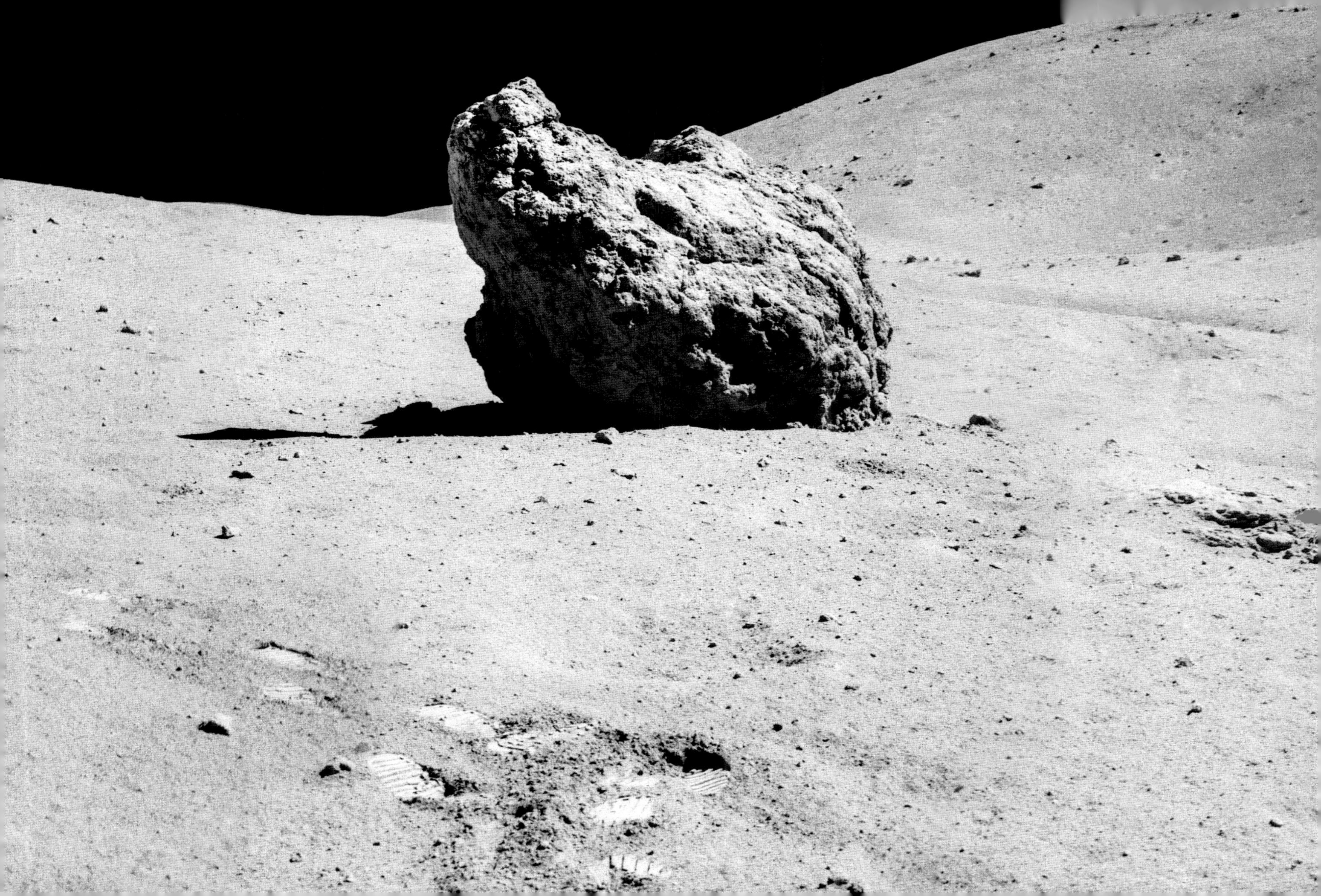

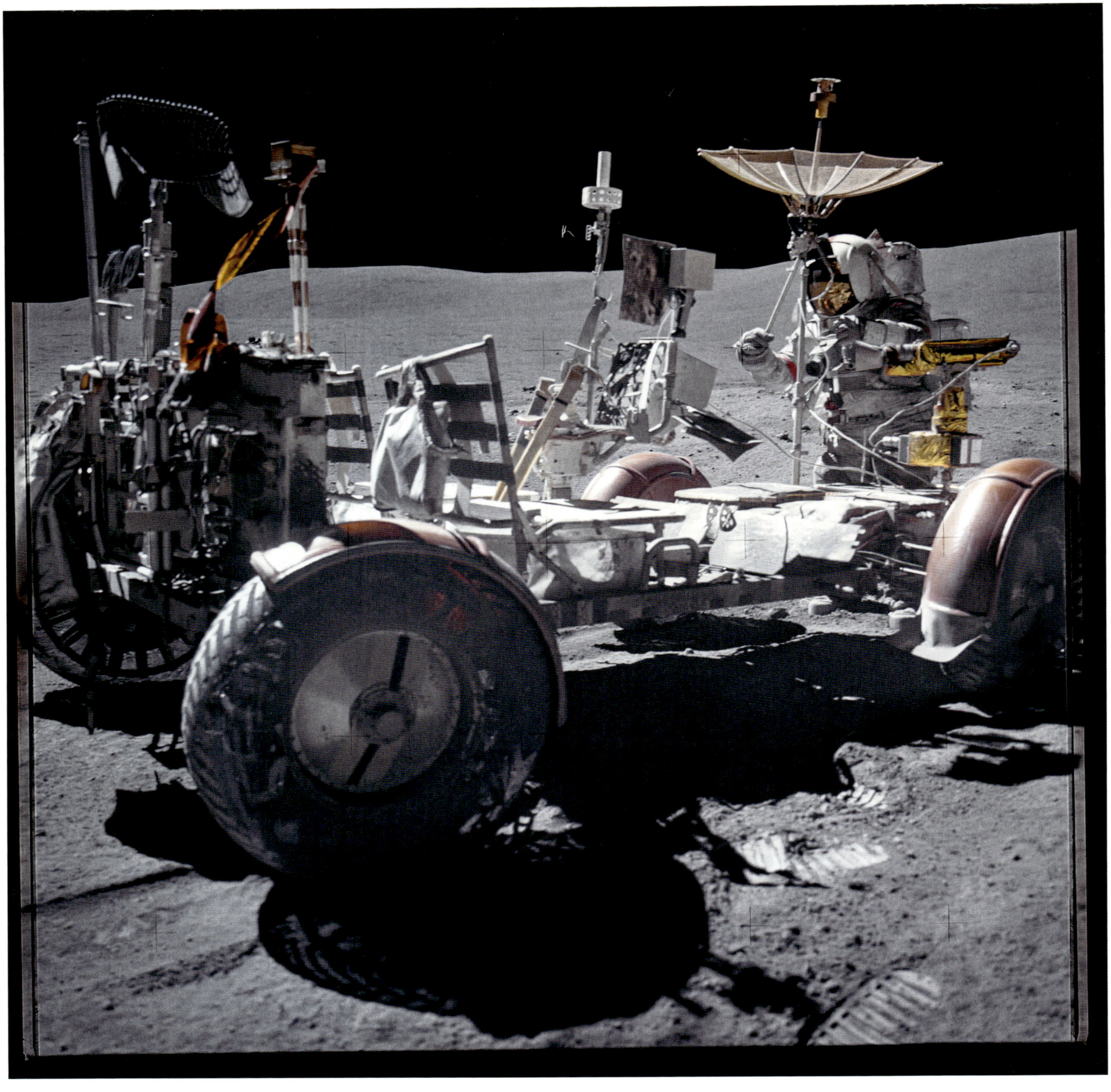

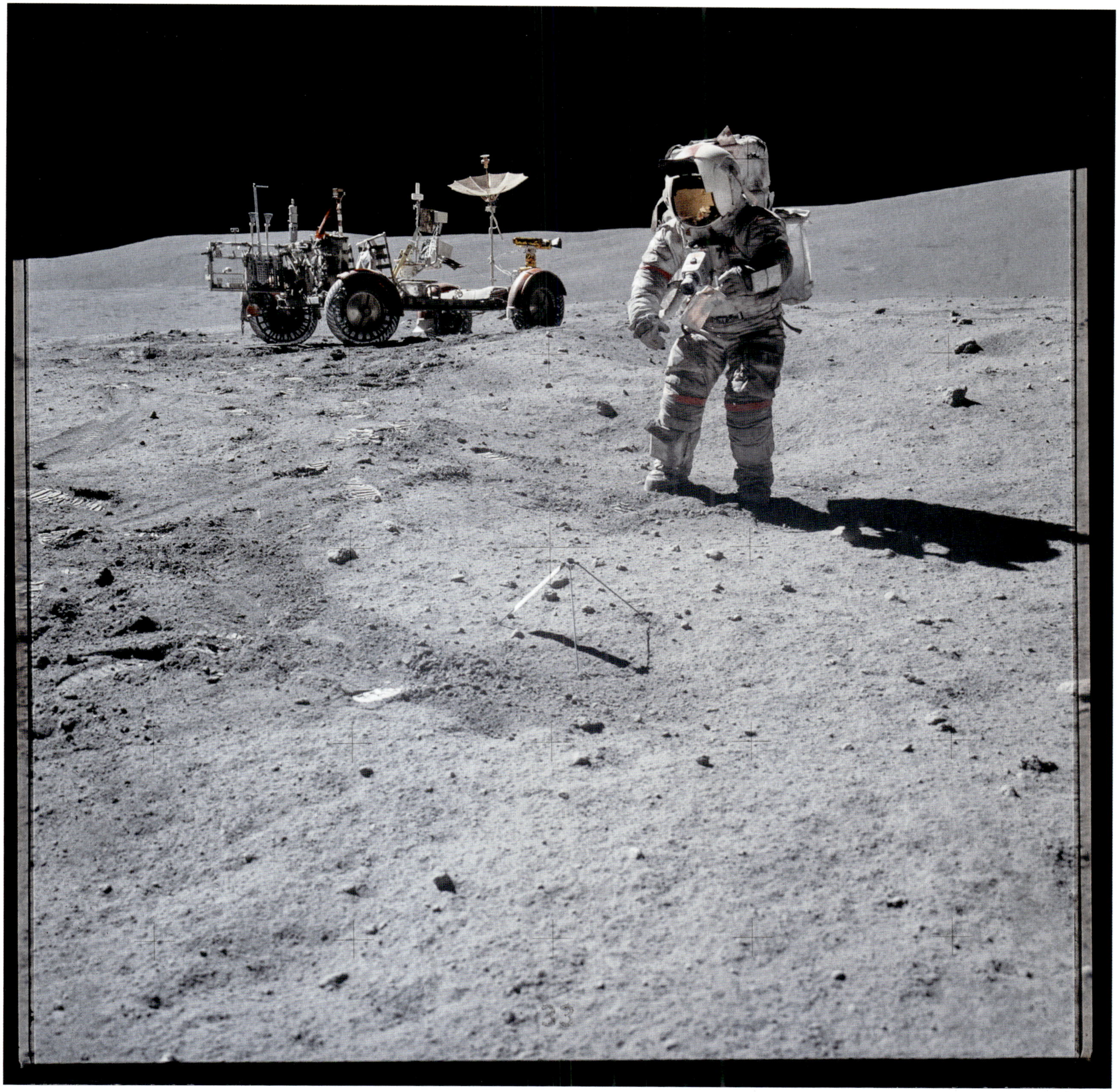
133

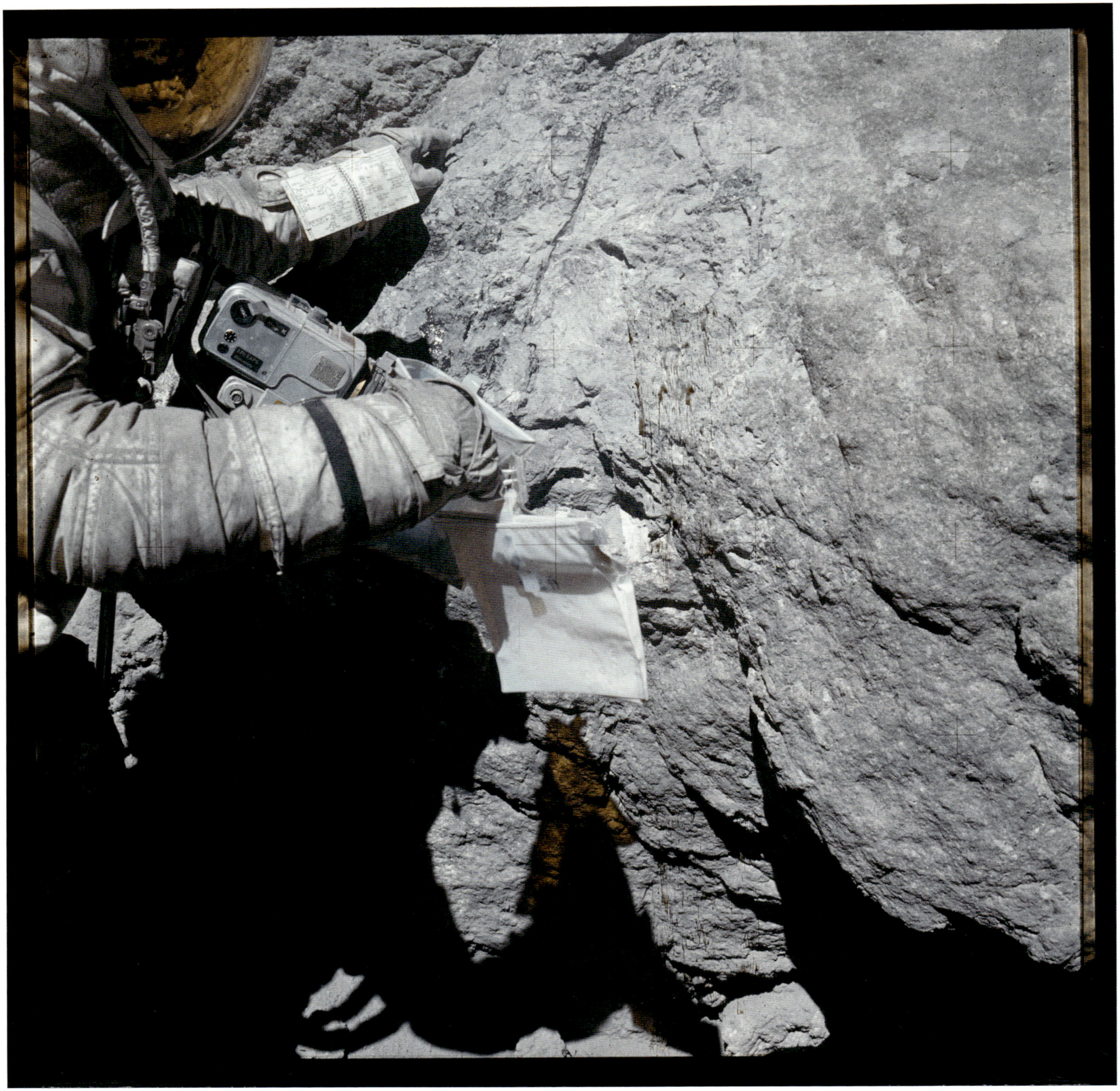

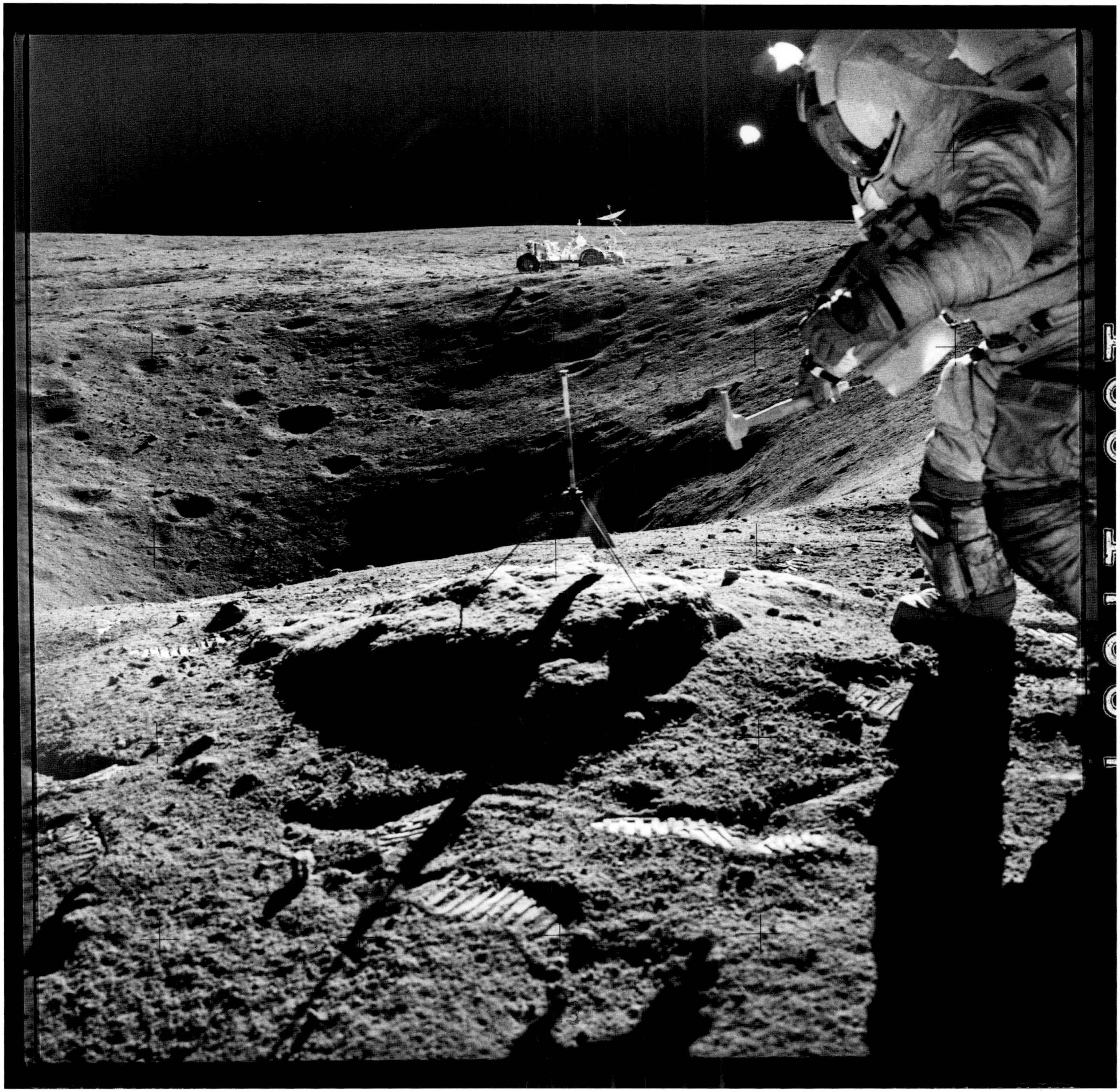

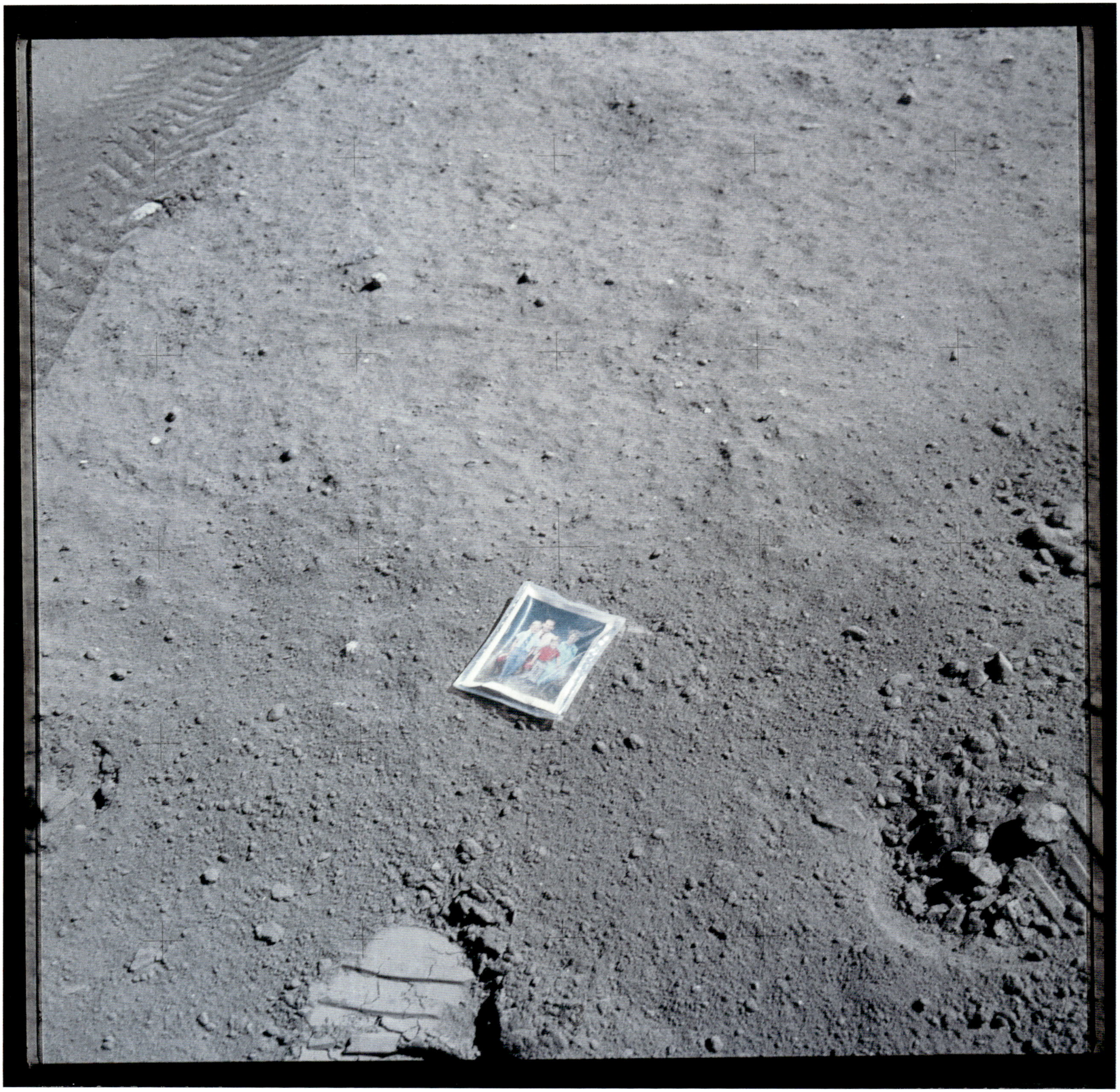

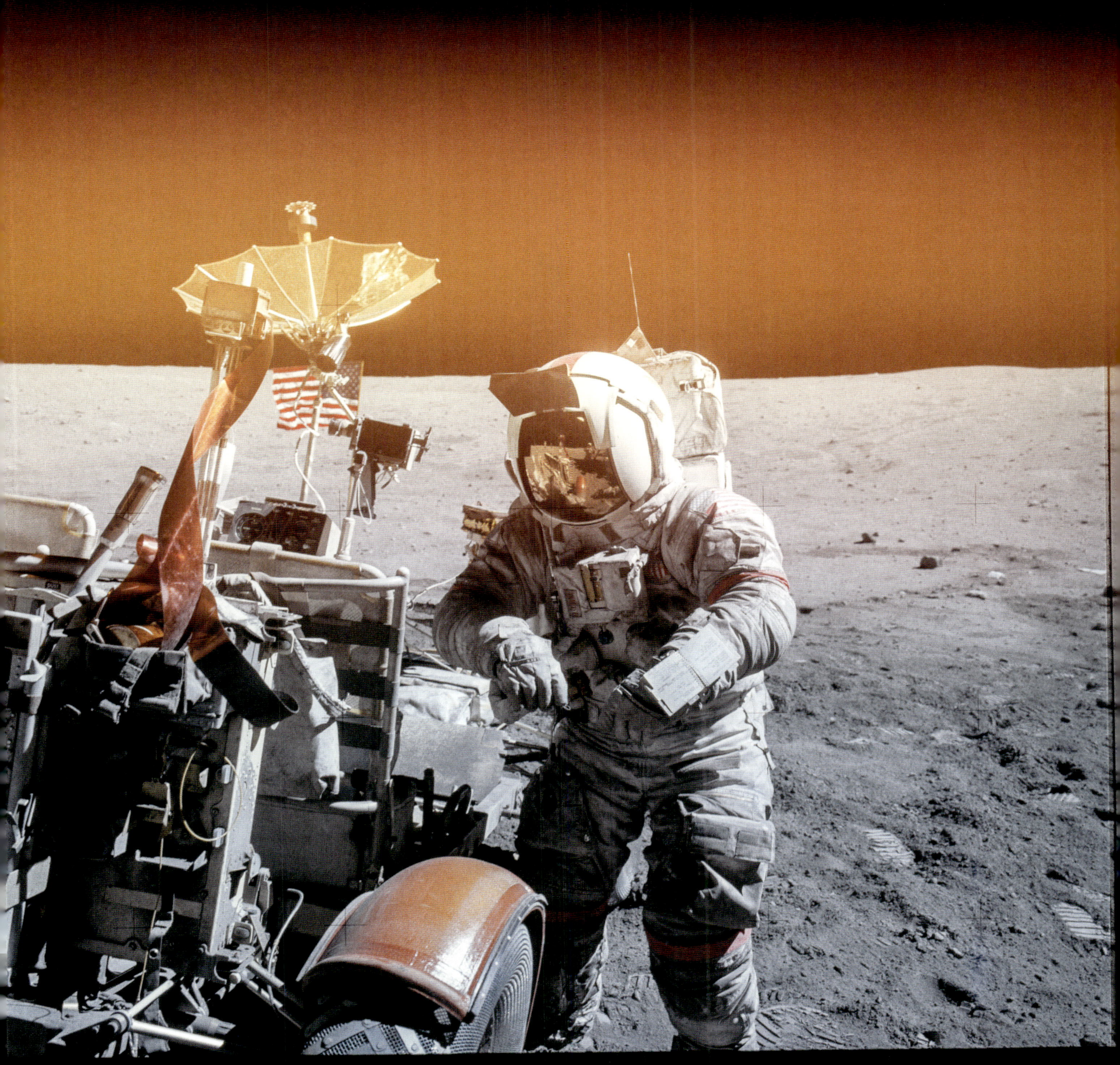

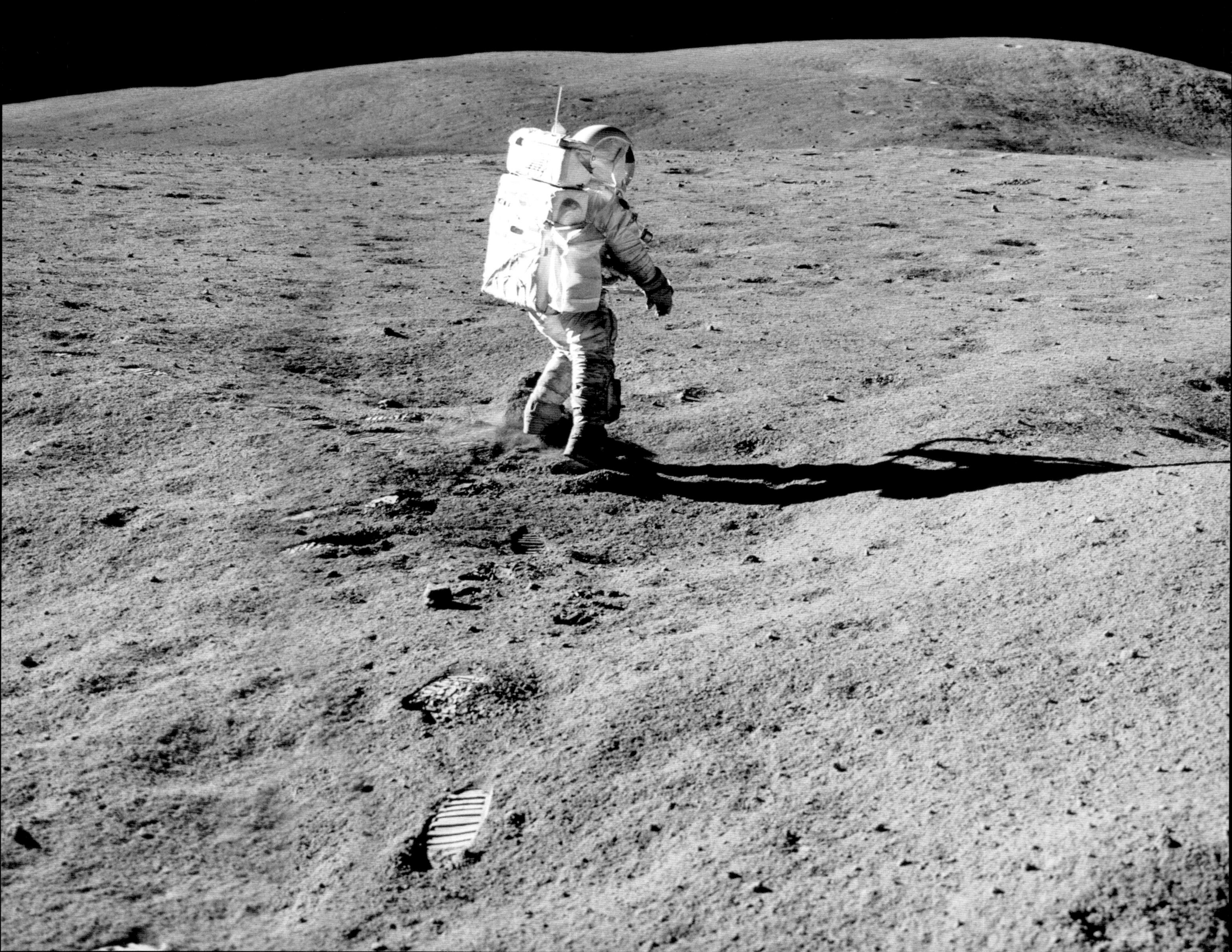

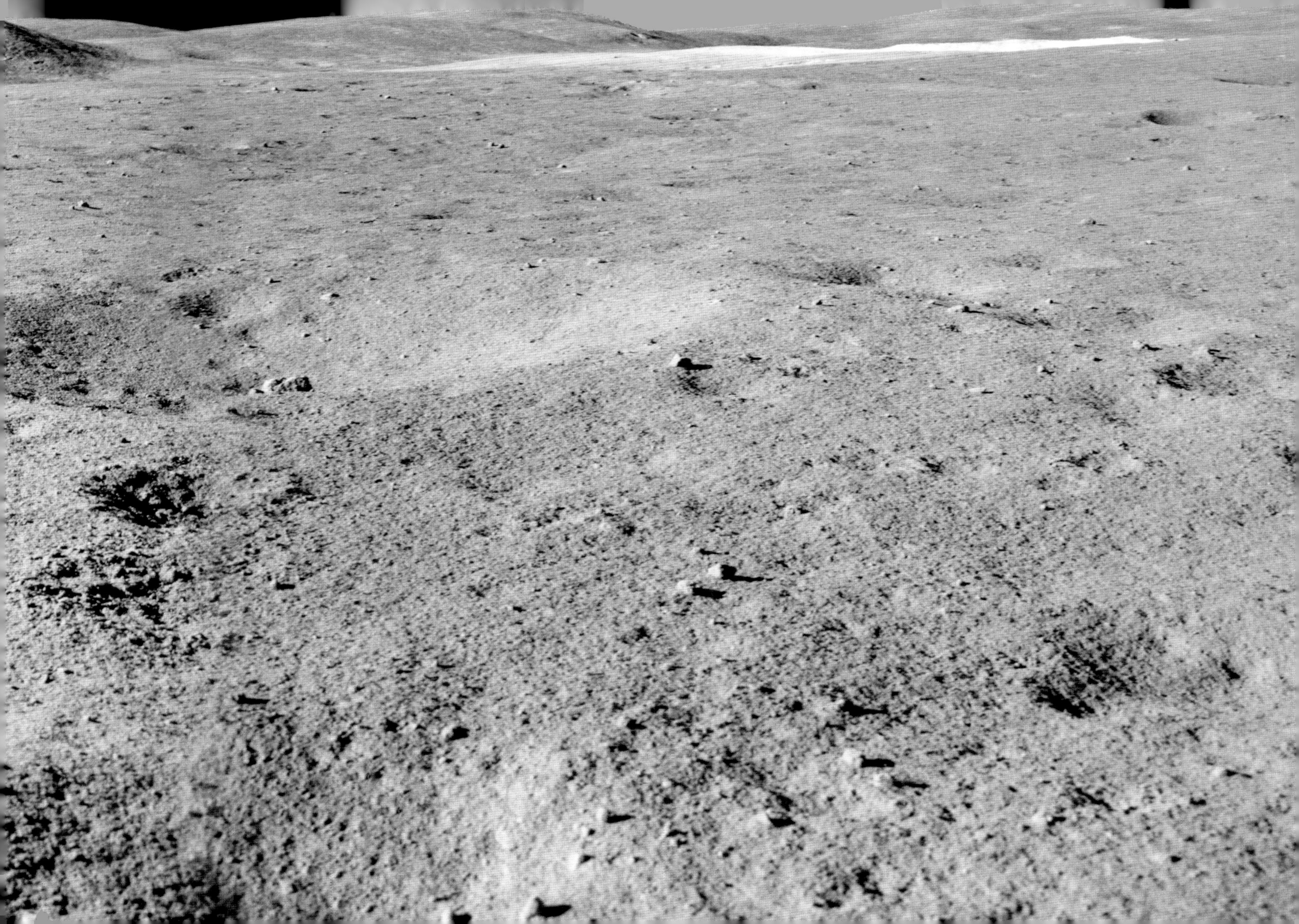

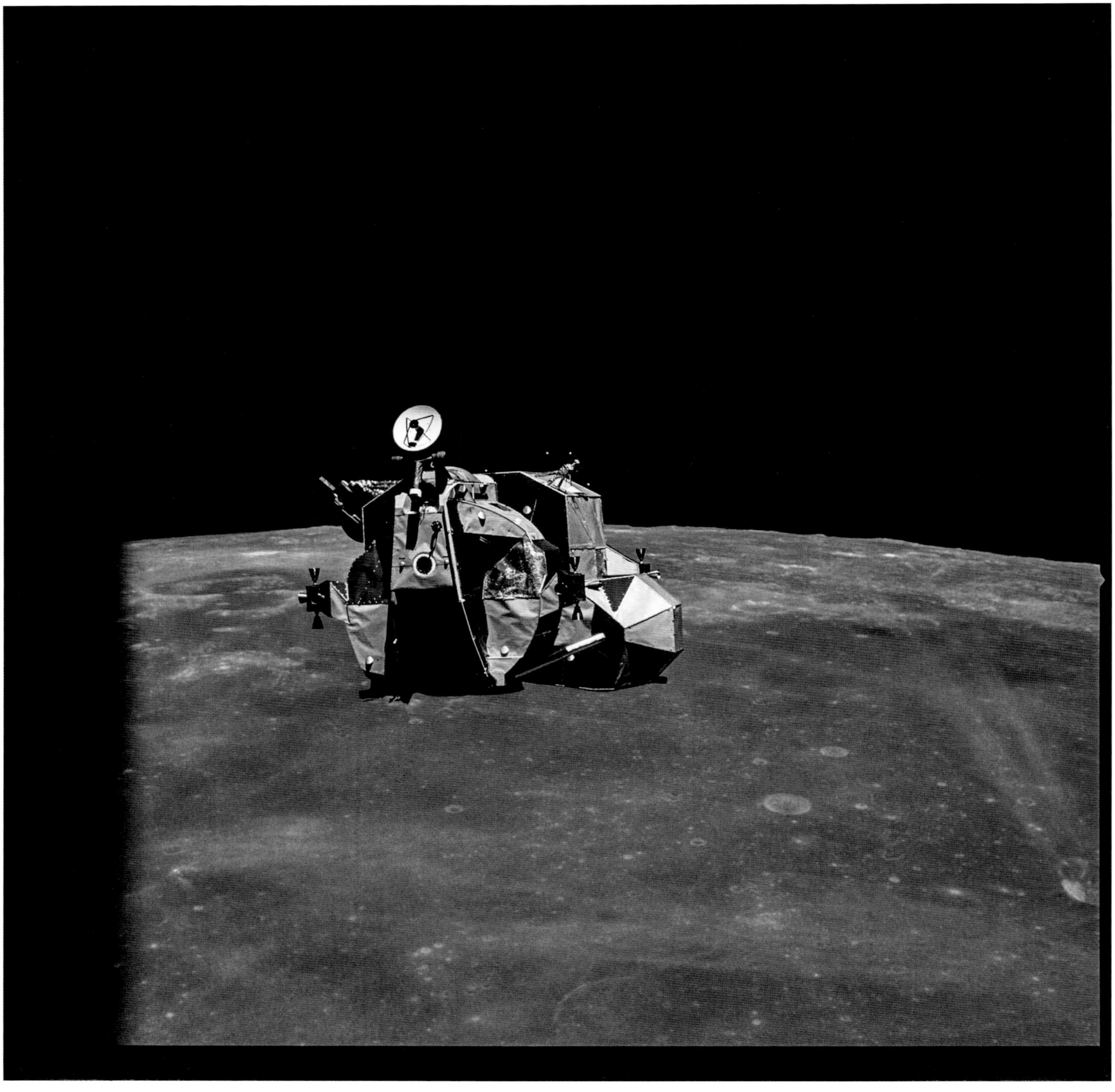

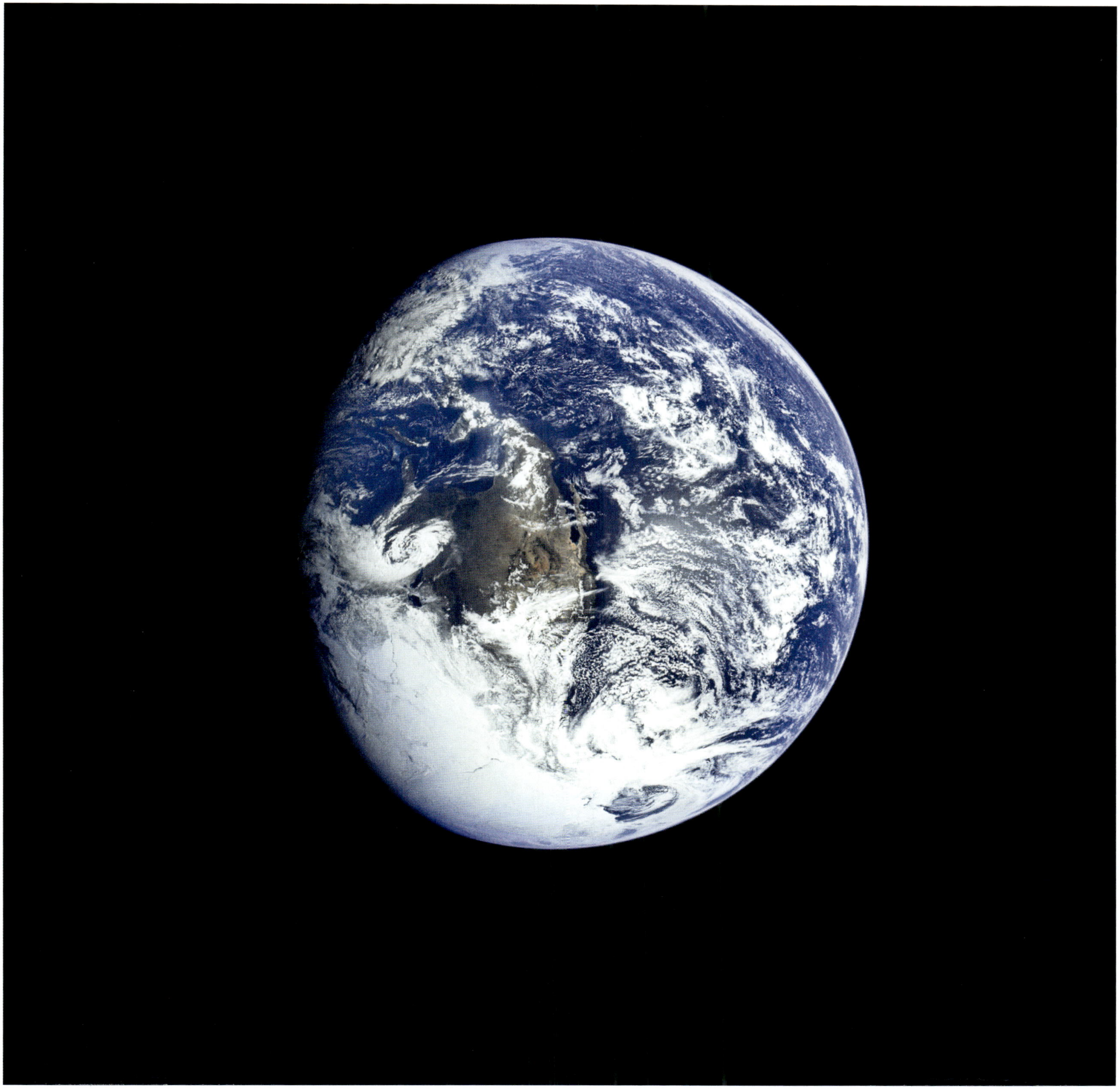

233 (AS16-118-18876)
Seen through the Command Module (CM) window, Earth appears as a glowing blue crescent against the void of space. Captured shortly after Trans-Lunar Injection (TLI).

234 (AS16-118-18873)
As Apollo 16 departs Earth, this view captures North and Central America, with the Rocky Mountains, Great Lakes, and Bahama Banks visible beneath scattered clouds. Taken shortly after TLI, the photograph shows Earth's landmasses in striking detail.

235 (AS16-113-18288)
Earth rises over the Moon's horizon, with the Command/Service Module (CSM) *Casper* visible to the left. Captured from the Lunar Module (LM) *Orion* at 57mi (92km) altitude, this image was taken during a hold period before landing in the Descartes Highlands.

236 (AS16-121-19431)
Goddard and Al-Biruni Craters on the Moon's far side. The rugged Goddard Crater, named after the father of modern rocketry, sits beside Al-Biruni Crater, known for its terraced walls and central peak — both shaped by ancient impacts.

237 (AS16-121-19434)
Lomonosov, Maxwell, and Szilard Craters on the Moon's far side. Lomonosov's eroded rim, Maxwell's central peak, and Szilard's smooth floor highlight the varied impact history of the lunar surface, shaped by billions of years of collisions and volcanic activity.

238 (AS16-120-19304)
Gassendi A and B Craters, located within Gassendi Crater, a lava-flooded impact site near Mare Humorum. The fractured crater floor hints at past volcanic activity.

239 (AS16-113-18294)
CSM *Casper*, drifts above the lunar farside after undocking from LM *Orion*. At this moment, both spacecraft were out of communication with Earth, regaining signal as they moved toward the Moon's Earth-facing side.

241 (AS16-113-18339)
Commander John W. Young leaps off the lunar surface, saluting in mid-air during Apollo 16's 'Lunar Olympics.' The Moon's reduced gravity allowed for higher jumps, though Lunar Module Pilot Charles M. Duke Jr.'s attempt ended with a fall.

242-243
(AS16-106-17390–AS16-106-17394)
Apollo 16 Station 13: Young adjusts the high-gain antenna on the Lunar Roving Vehicle (LRV), aligning it with Earth for TV transmission. Shadow Rock, a 13ft (4m) boulder, sits in a permanently shadowed area, key for studying lunar volatiles. Smoky Mountain rises in the background, 0.6mi (1km) away.

244 (AS16-117-18818)
Young carefully aligns the high-gain antenna on the LRV, ensuring a clear signal back to Earth. The antenna's golden mesh contrasts against the stark lunar surface.

245 (AS16-117-18825)
Young collects lunar samples near North Ray Crater for analysis. In the background, the LRV sits parked on the uneven terrain, its tracks imprinted in the lunar dust. Large boulders surrounding the site provide clues to the Moon's impact history.

246 (AS16-116-18649)
Duke examines Outhouse Rock, pointing to a sample site while holding a pack of sample bags. Attached to his left arm, an open checklist displays pages LMP-30 and LMP-31, outlining sampling procedures for this station. His Hasselblad camera, secured to his chest, is visible in the photo.

247 (AS16-109-17804)
Young stands on the west rim of Plum Crater, surrounded by a rugged landscape of large boulders and steep crater walls. Samples collected here were among the oldest lunar rocks, over 3.8 billion years old. Their analysis disproved earlier theories that the region was volcanic, instead confirming the highlands were shaped by ancient impacts, reshaping our understanding of lunar geology.

248 (AS16-117-18841)
One of the Apollo program's most well-known personal mementos, this family portrait was left on the Moon by Duke during Apollo 16. Encased in plastic, it shows Duke, his wife, and their two sons. Placed near the LM during Extravehicular Activity (EVA), it remains undisturbed in the lunar dust.

249 (AS16-117-18852)
Young works at the seat of the LRV. His visor reflects the rugged terrain, the rover, and mission equipment in striking detail.

250-251
(AS16-114-18415–AS16-113-18432)
Duke walks alone across the Descartes Highlands, his long shadow stretching over the lunar surface. Taken around the same time as the panorama on pages 242–243, this frame captures a different moment, showing Duke moving further into the vast lunar landscape.

252 (AS16-122-19537)
The ascent stage of LM *Orion* approaches CM *Casper* during rendezvous in lunar orbit. Captured after departing the Moon, the image shows the LM's thrusters and docking target, critical for rejoining Command Module Pilot Thomas K. Mattingly II aboard the CM before returning to Earth.

253 (AS16-118-18885)
Earth photographed about 1.5 hours after TLI on April 16th, 1972. Despite cloud cover, the United States, most of Mexico, and parts of Central America are visible. The Great Lakes (Michigan and Superior) and the Bahama Banks can be seen beneath the clouds.

XVII

Eugene A. Cernan

Ronald B. Evans

Apollo 17

Launch: December 7th, 1972
Lunar landing: December 11th, 1972
Splashdown: December 19th, 1972

Command Module: America
Lunar Module: Challenger

Apollo 17 concluded a highly successful series of crewed lunar landing missions, and remains the last time humans have traveled beyond low Earth orbit. As the final mission of the Apollo program, it included the longest moonwalks, the most time spent in lunar orbit, and the largest lunar sample return of any Apollo flight.

One of the most iconic photographs in space history — a full-disk image of Earth — was taken during Apollo 17. Richard W. Underwood of NASA's Photographic Technology Lab recalled: "They launched at night and went trans-lunar over Madagascar, giving them a fully sunlit Earth. I kept telling Jack Schmitt — who was a geologist — 'That's going to be the classic picture. Make sure you get it after you go trans-lunar.' He worked it into his schedule and got a beautiful series. That one's at 28,000mi (45,000km) — it's perfect." That image became known as *Blue Marble*, and appears as originally framed on page 315.

Another story from Apollo 17 is about a photograph that was never taken — and long regretted. Commander Eugene Cernan later described his final actions before climbing back into the Lunar Module: "I left my Hasselblad on the Moon, lens aimed at the zenith, thinking someone might one day come back and study the glass for cosmic radiation damage. But going up the ladder, I didn't take a photo of my last footprint. How dumb! I should've taken the picture, pulled off the film pack, and if weight was a concern, thrown the camera away."

Eugene A. Cernan
Commander

Born on March 14th, 1934 in Chicago, Illinois. He was the Pilot of Gemini 9A, Backup Pilot of Gemini 12, Backup Lunar Module Pilot of Apollo 7, Lunar Module Pilot of Apollo 10, and Backup Commander of Apollo 14. As Commander of Apollo 17, he was the eleventh, and currently last, man to walk on the Moon.

Ronald B. Evans
Command Module Pilot

Born on November 10th, 1933 in St. Francis, Kansas. He was in the fifth group of astronauts chosen in 1966. He was Backup Command Module Pilot of Apollo 14. After Apollo, he was the Backup Command Module Pilot for the Apollo-Soyuz Test Project. Evans retired from NASA on March 15th, 1977.

Harrison H. Schmitt
Lunar Module Pilot

Born on July 3rd, 1935 in Santa Rita, New Mexico. A geologist, he was Backup Lunar Module Pilot of Apollo 15 and the twelfth man on the Moon. In May 1974 he was named NASA's Assistant Administrator for Energy Programs, a position he held until he resigned from NASA on August 30th, 1975.

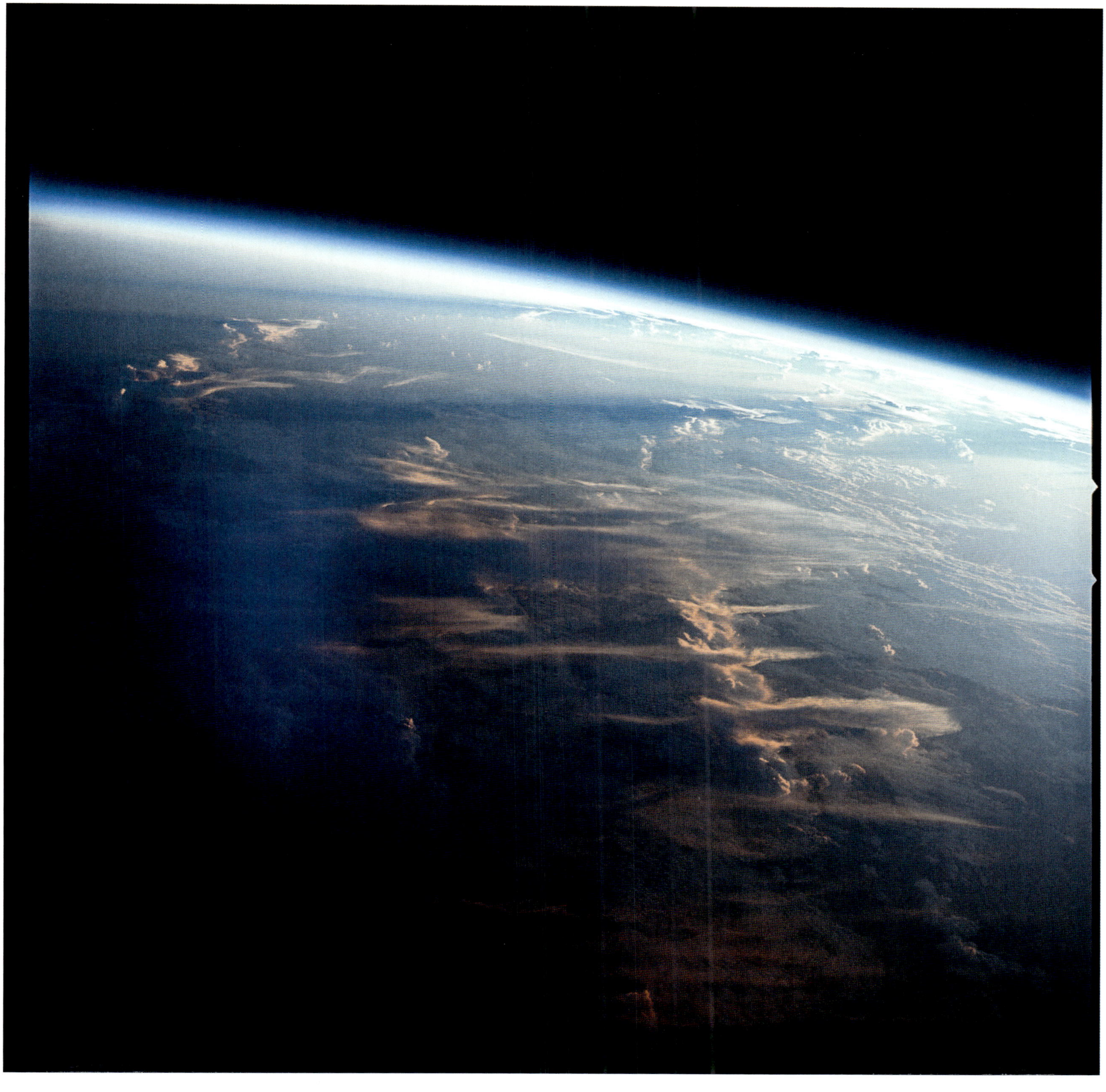

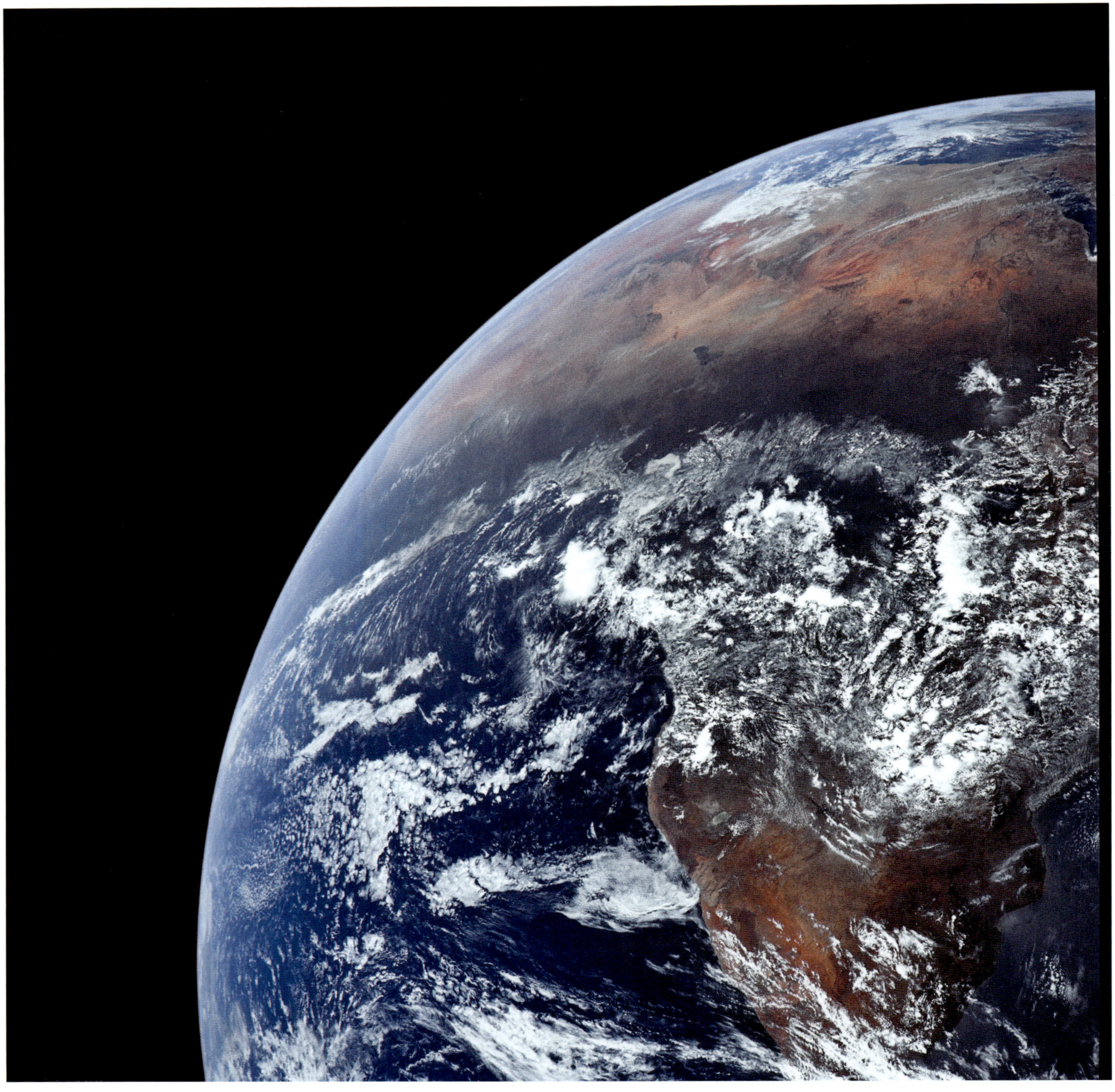

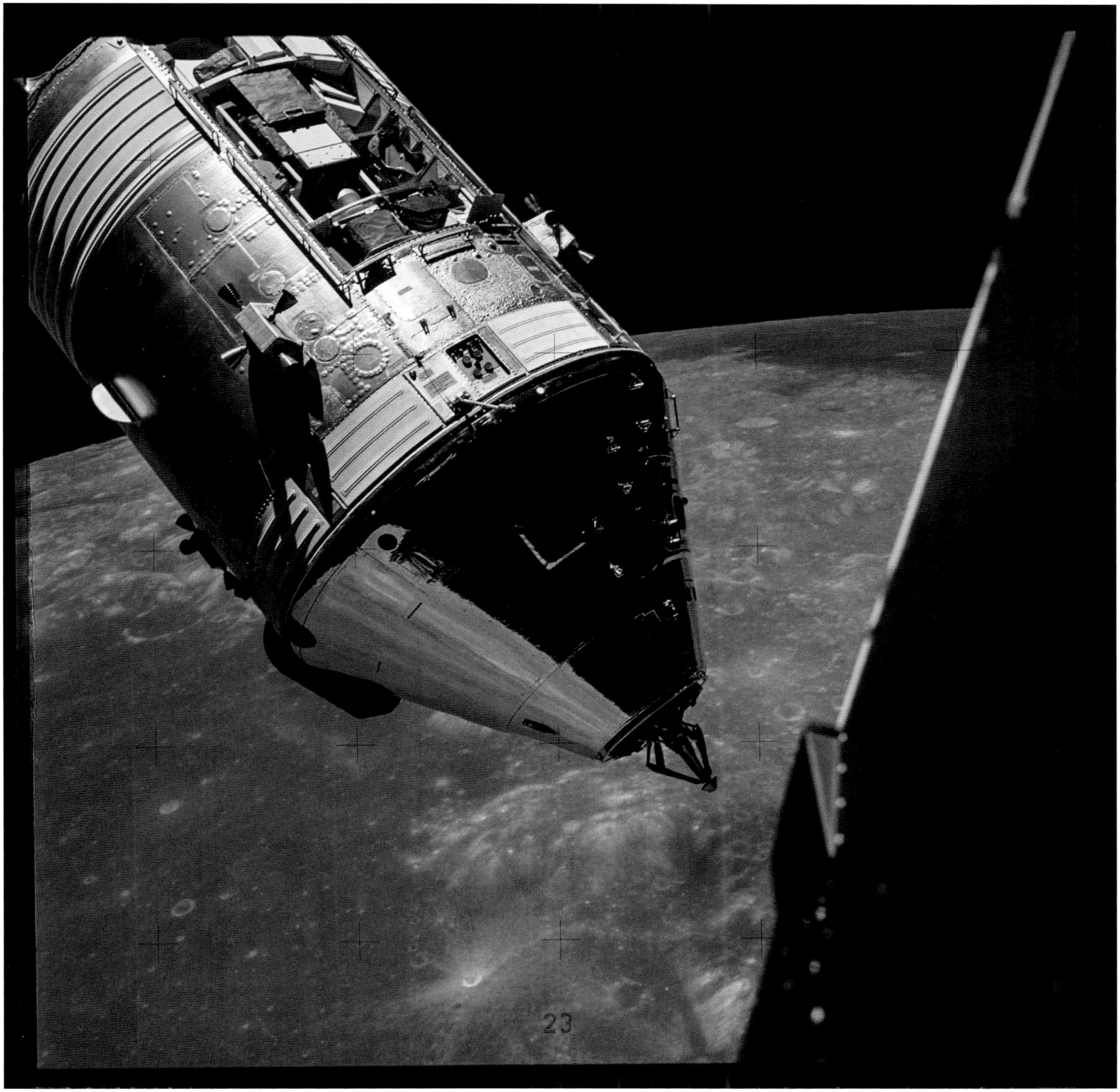
23

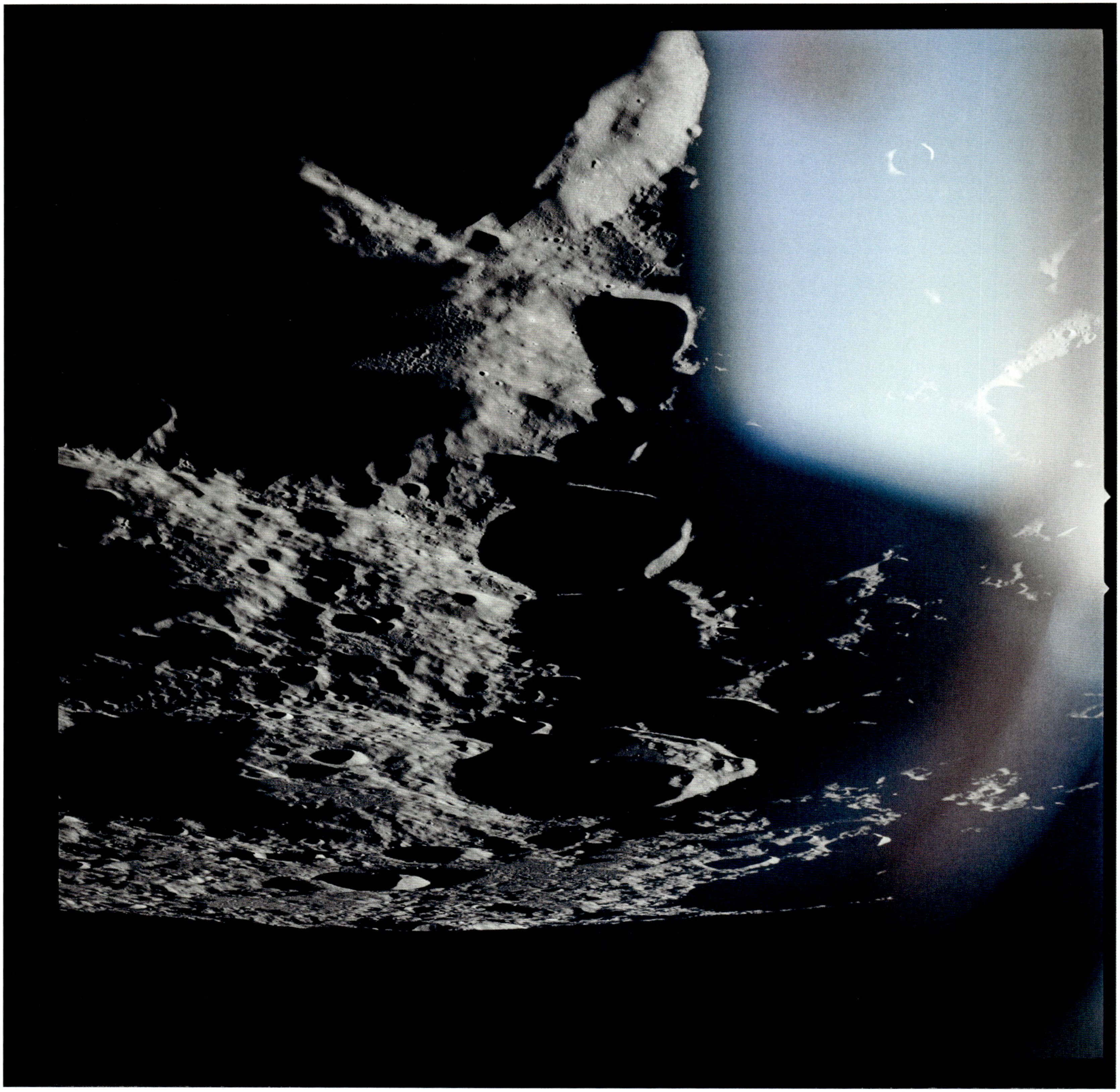

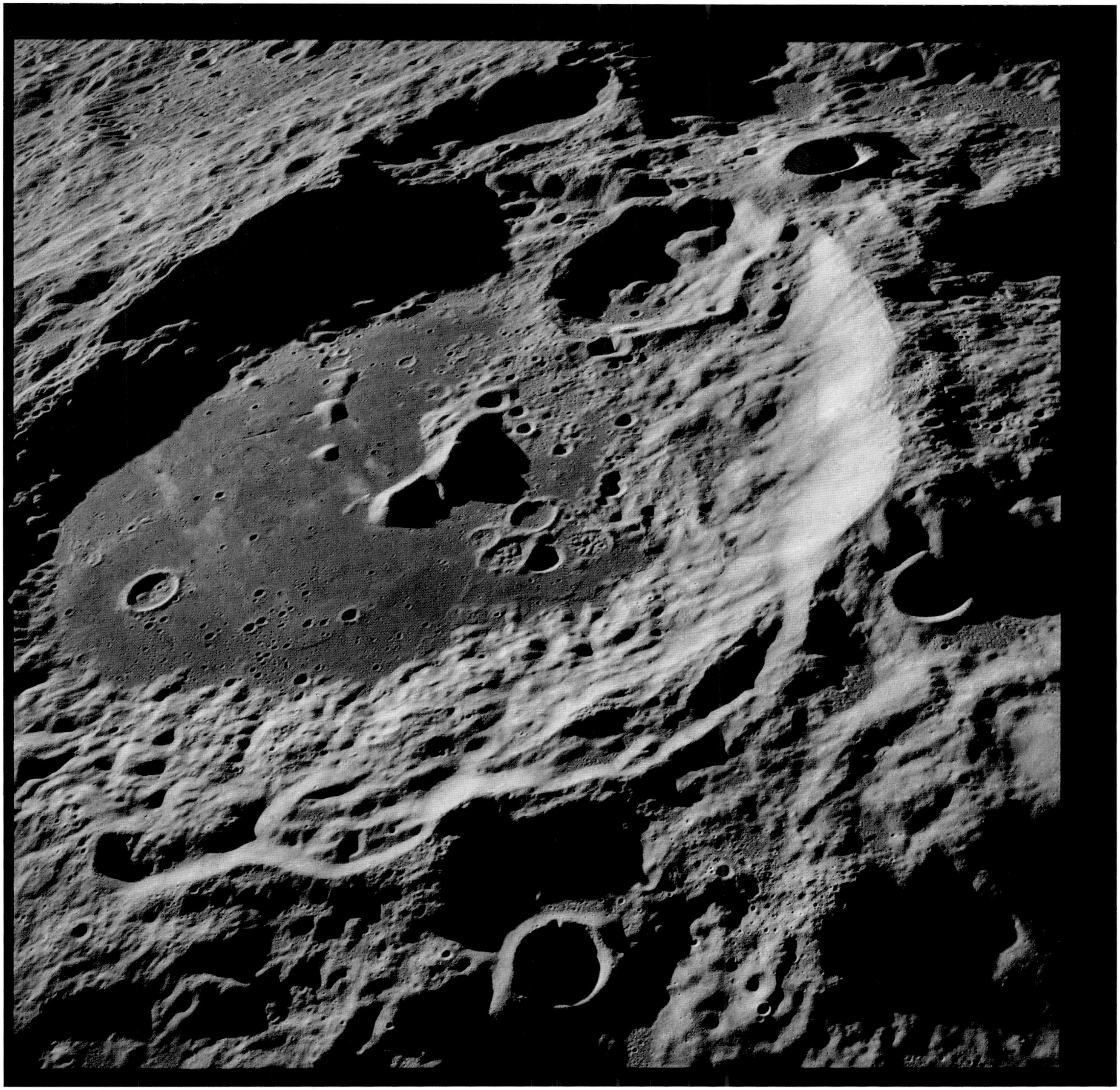

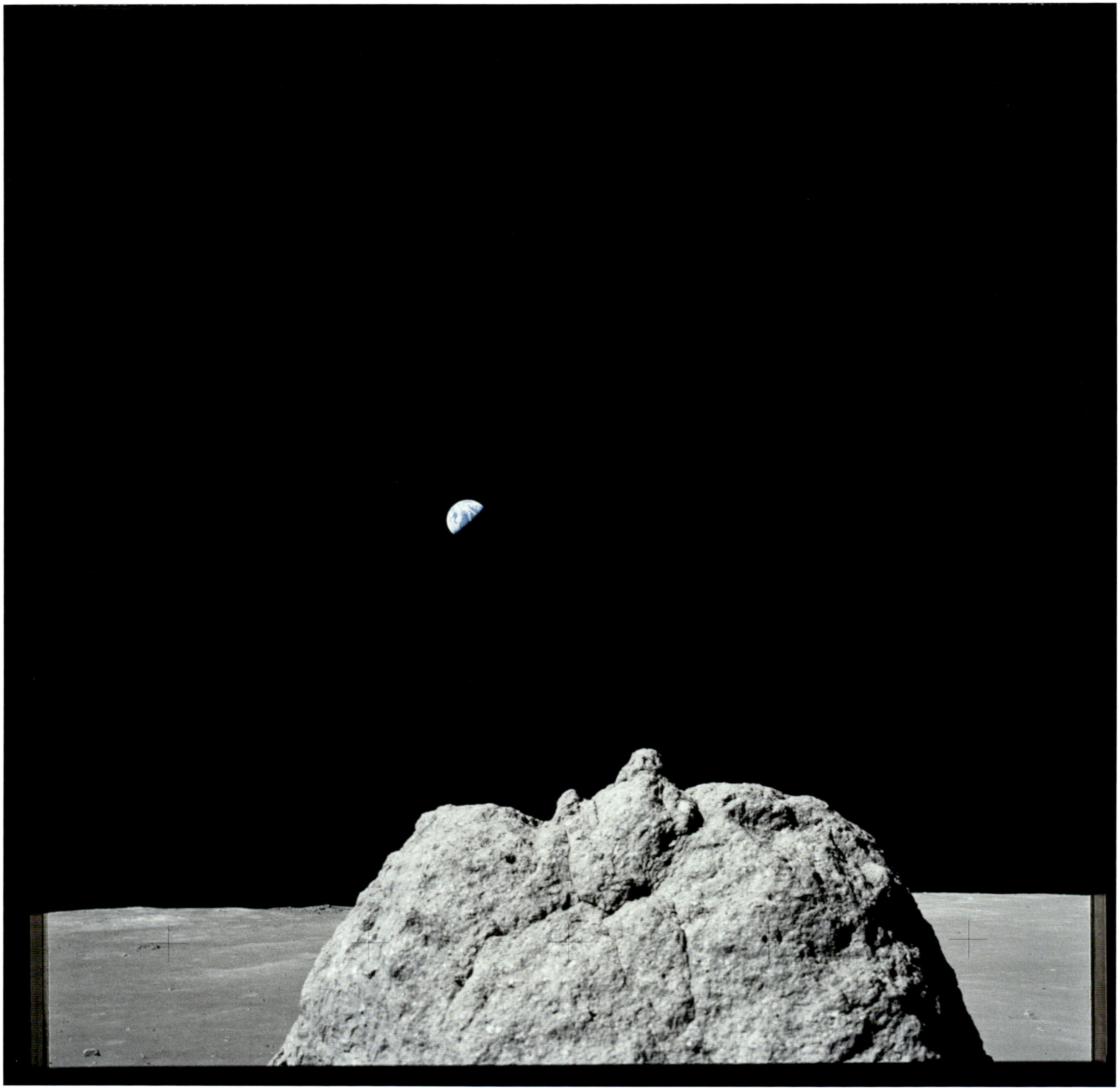

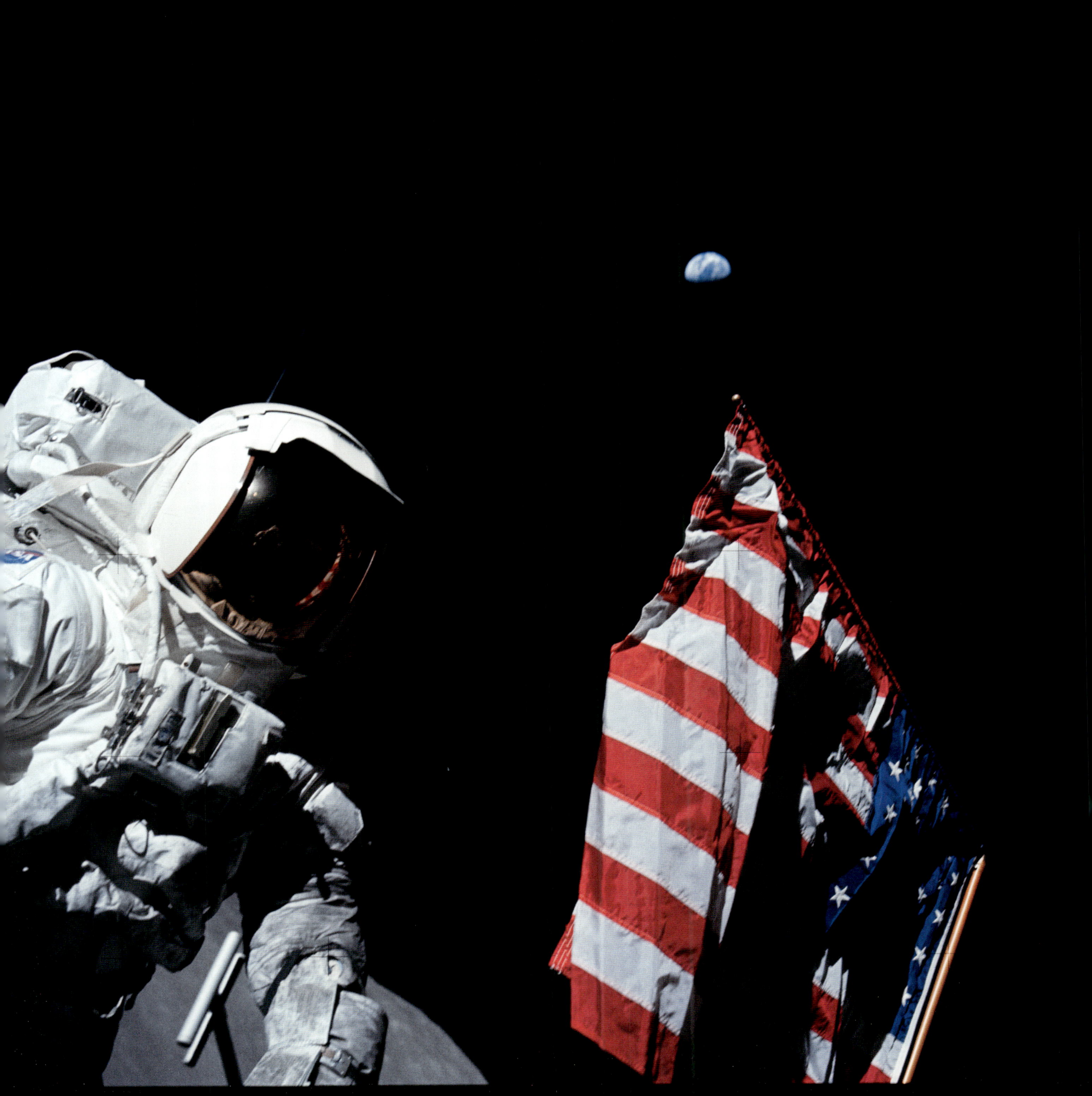

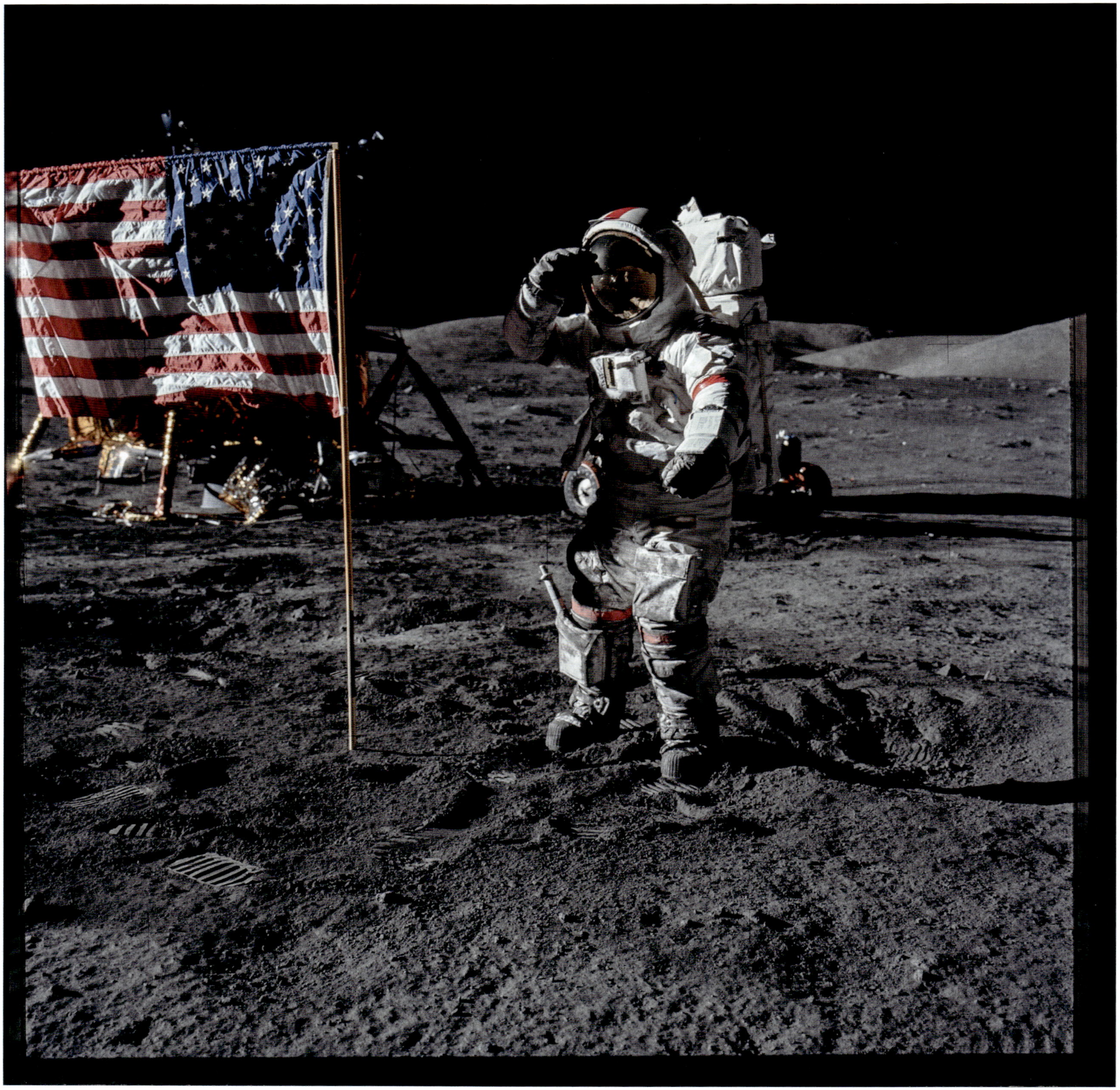

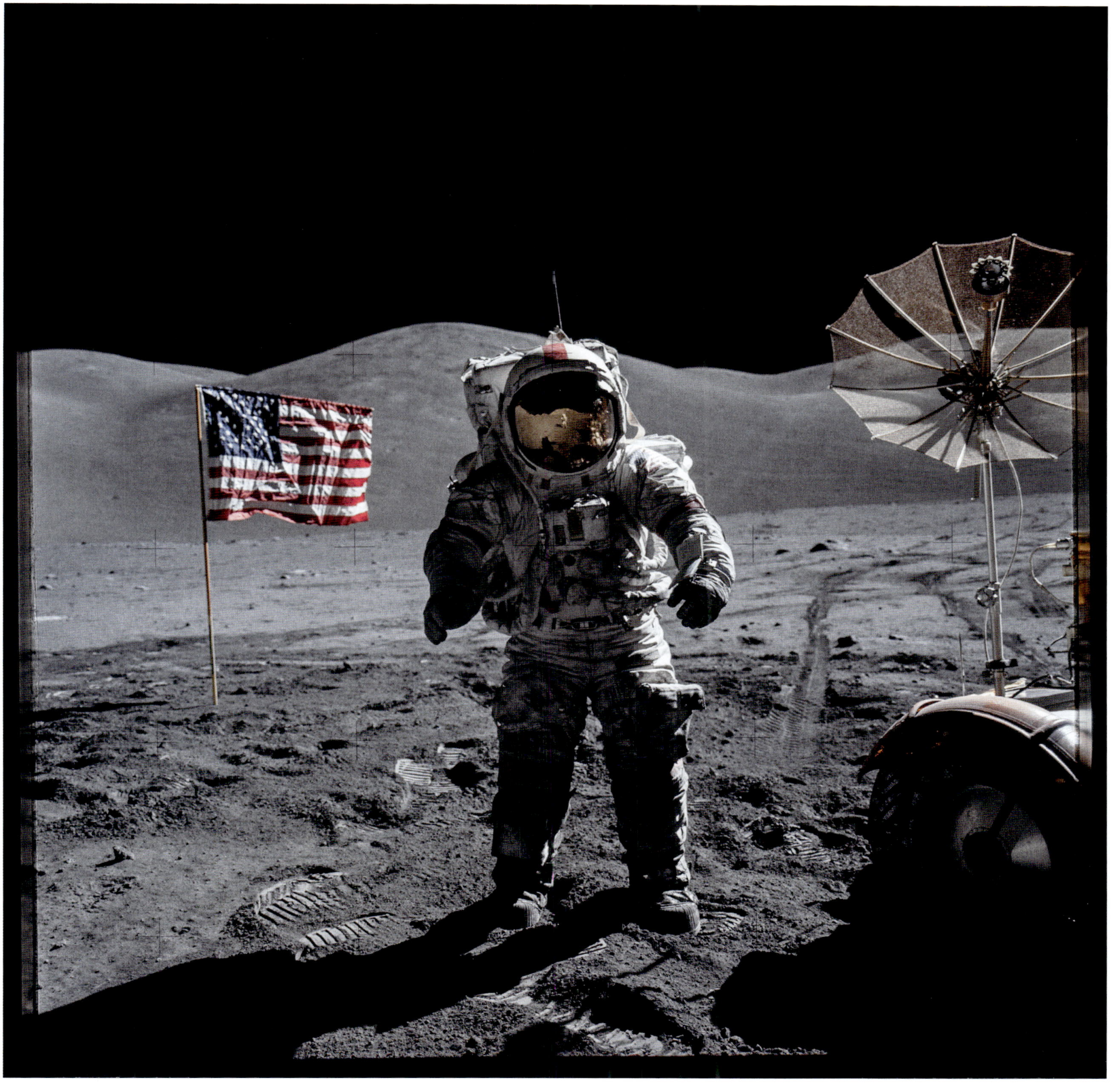

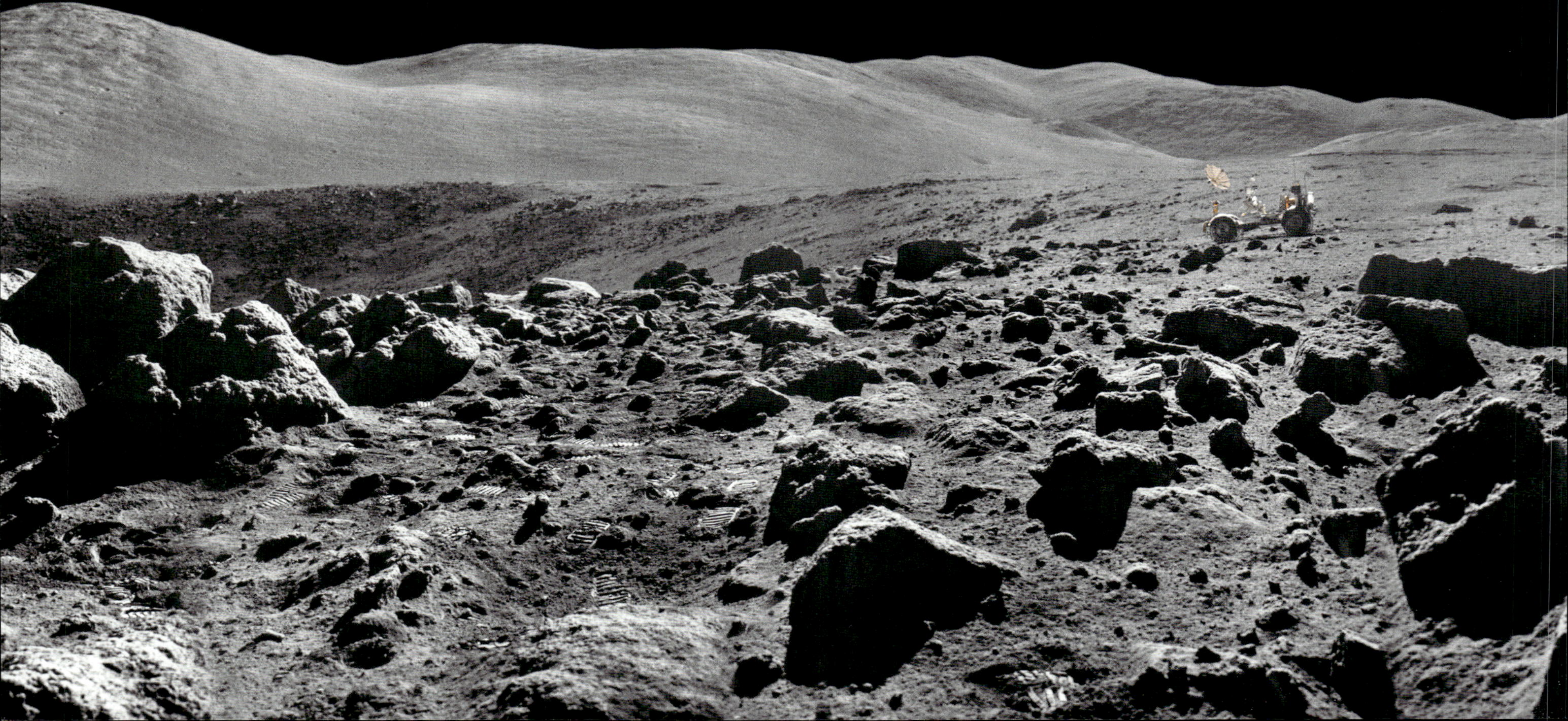

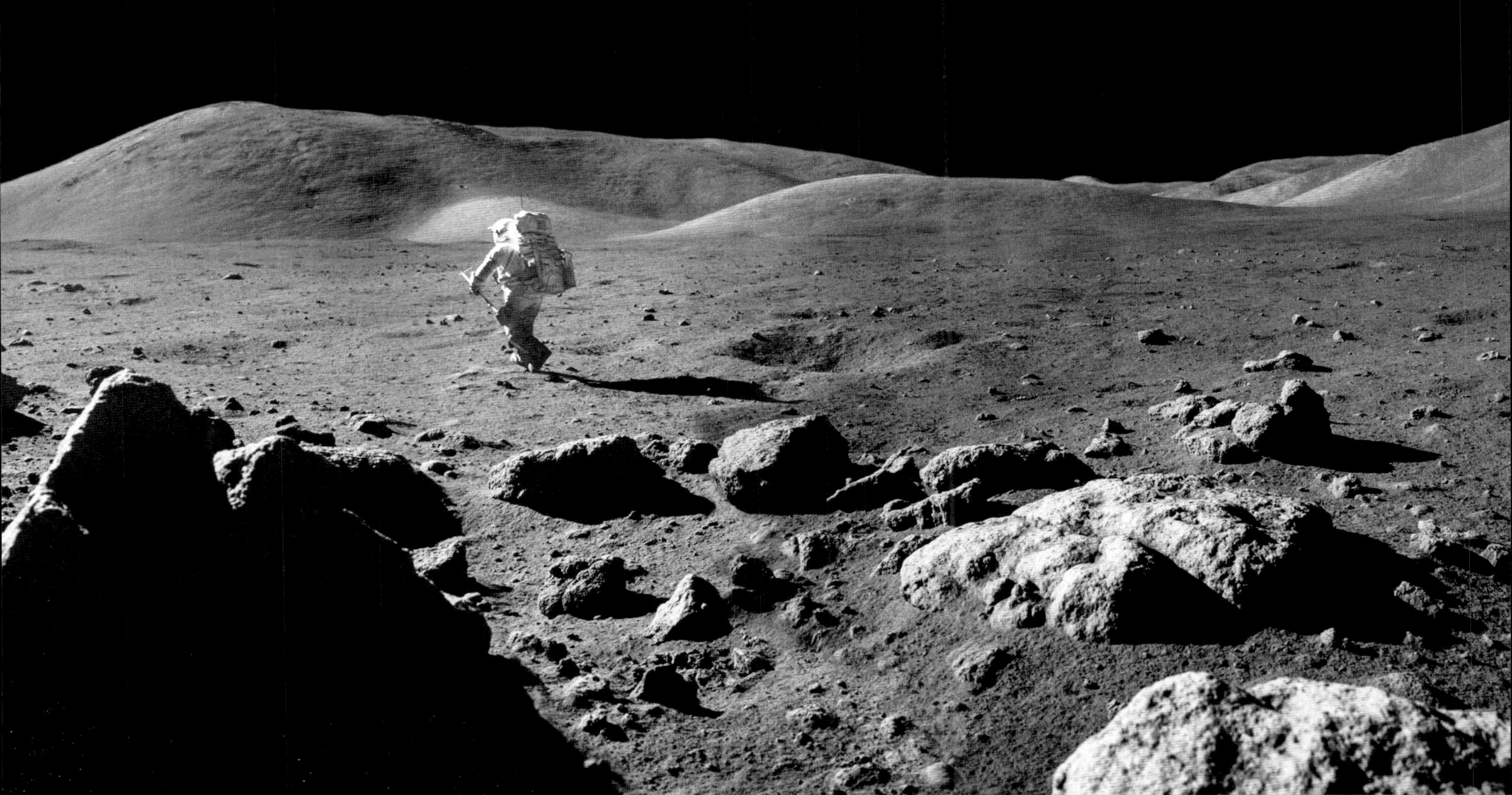

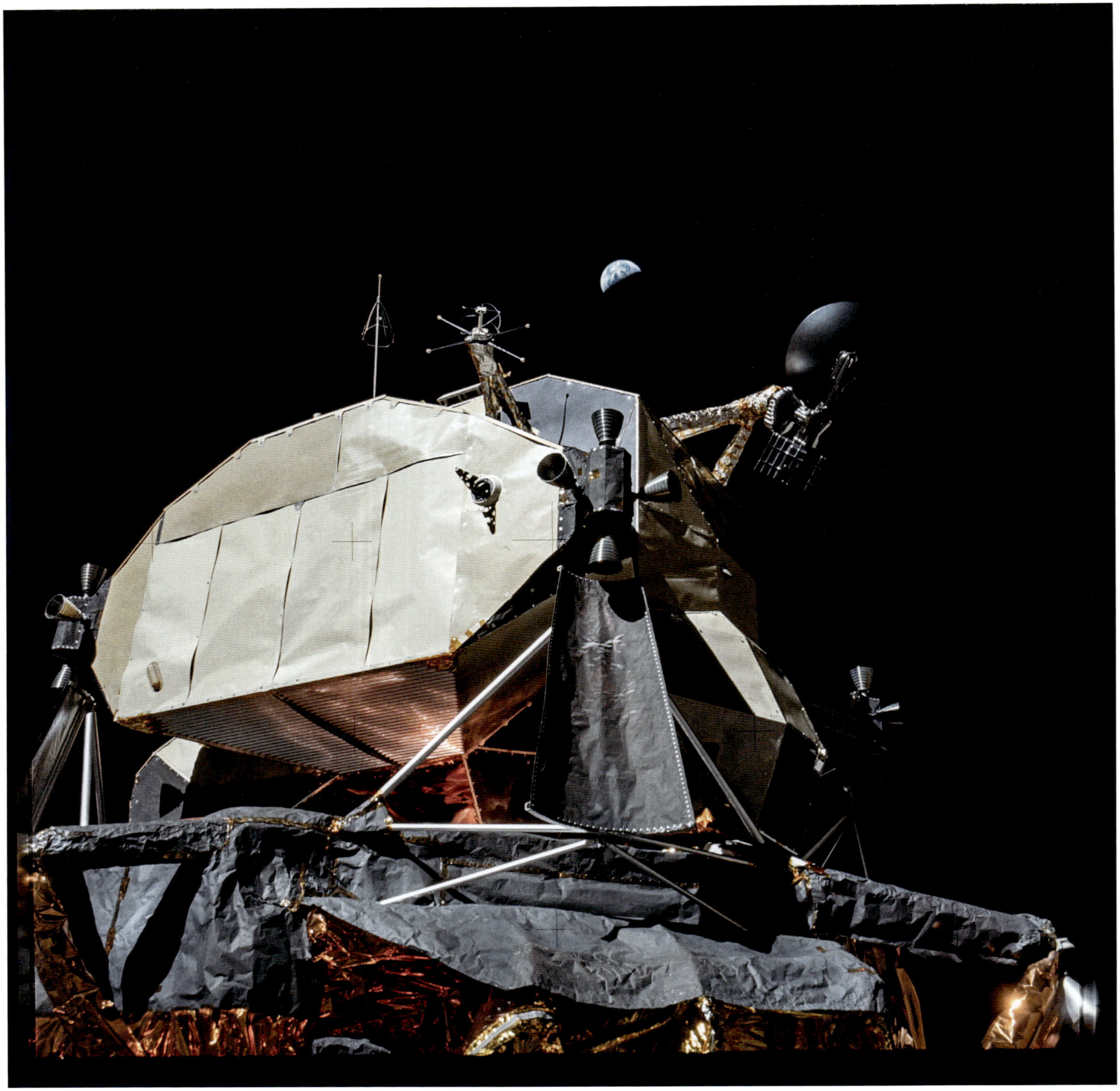

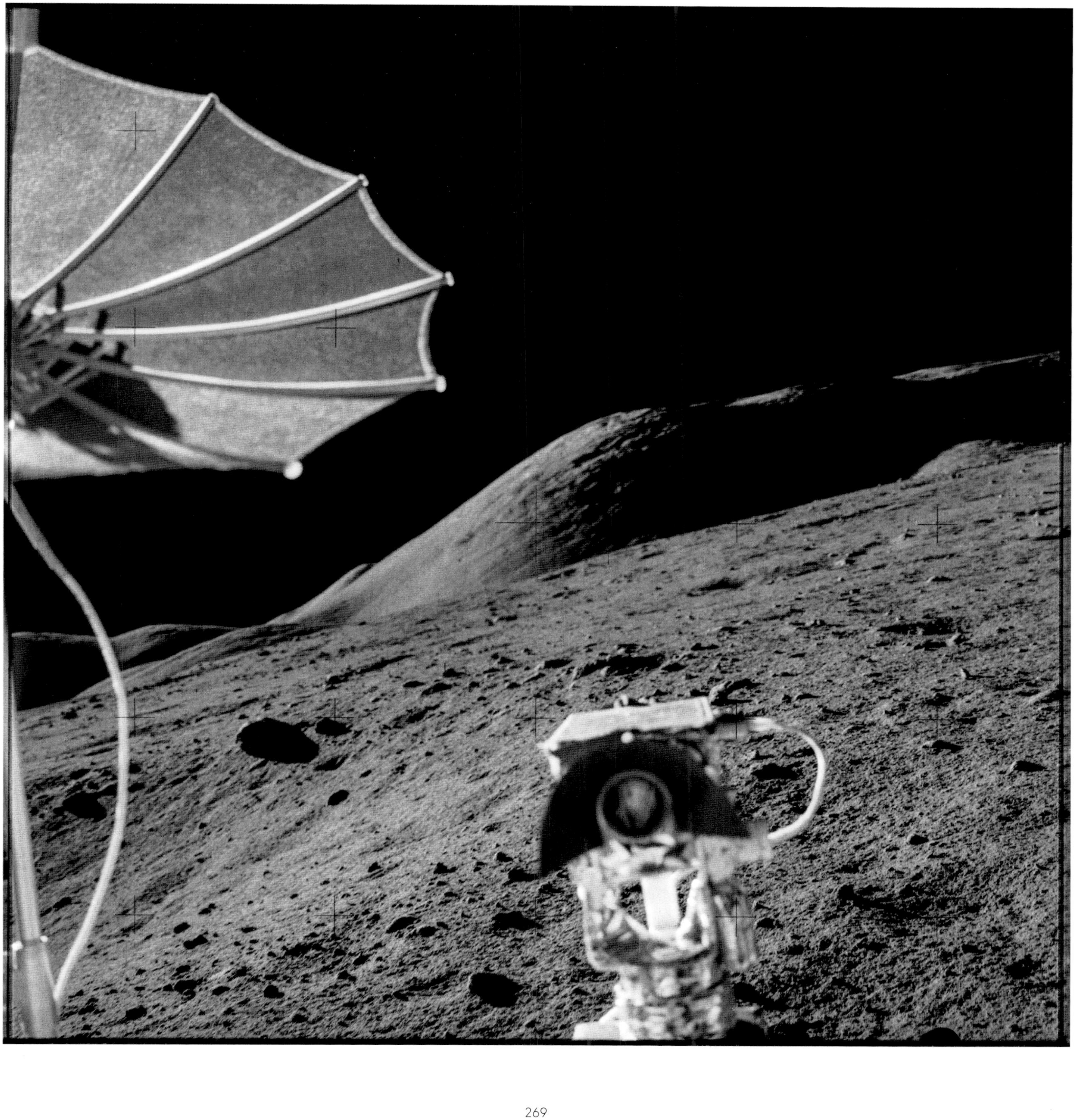

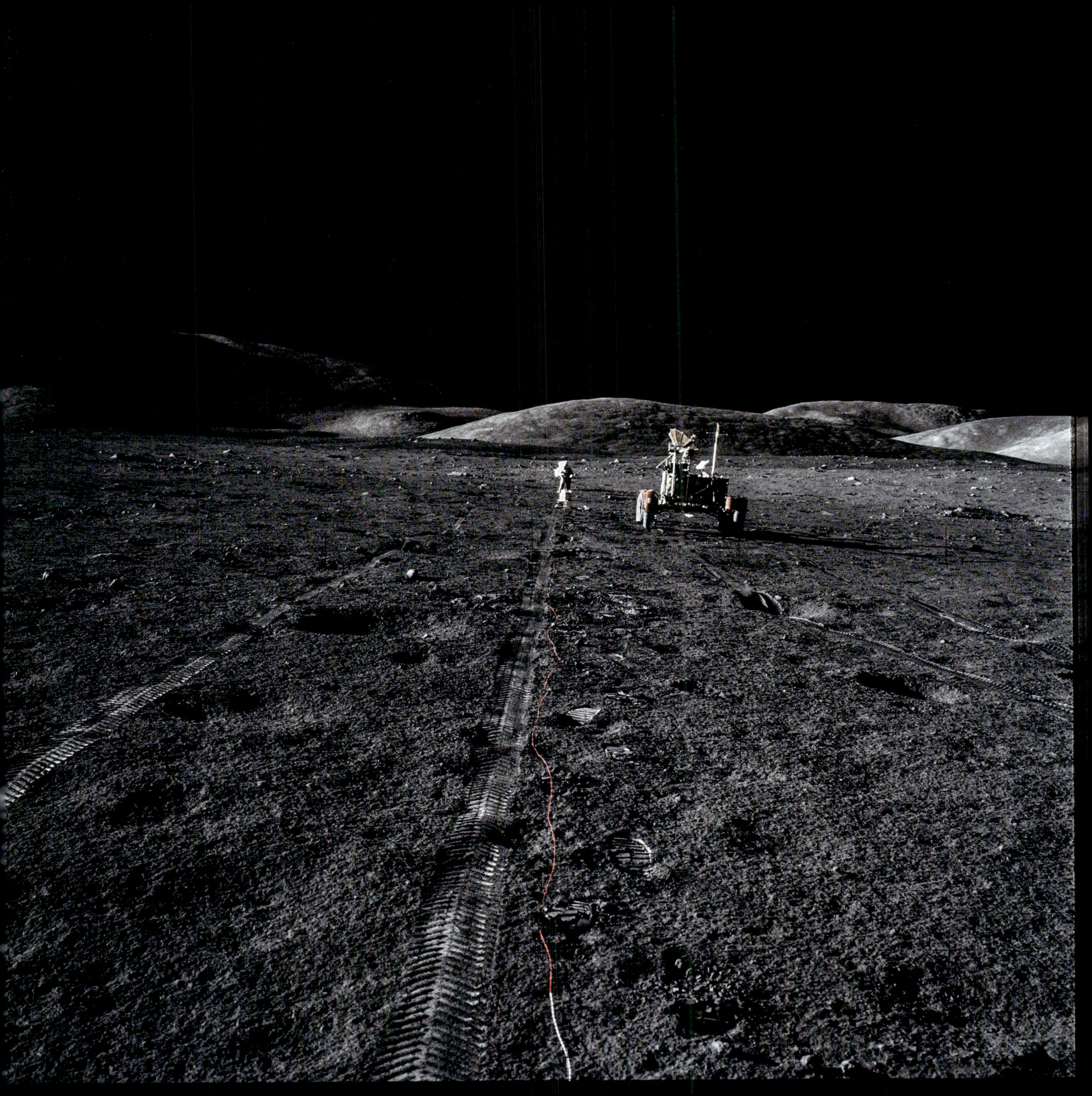

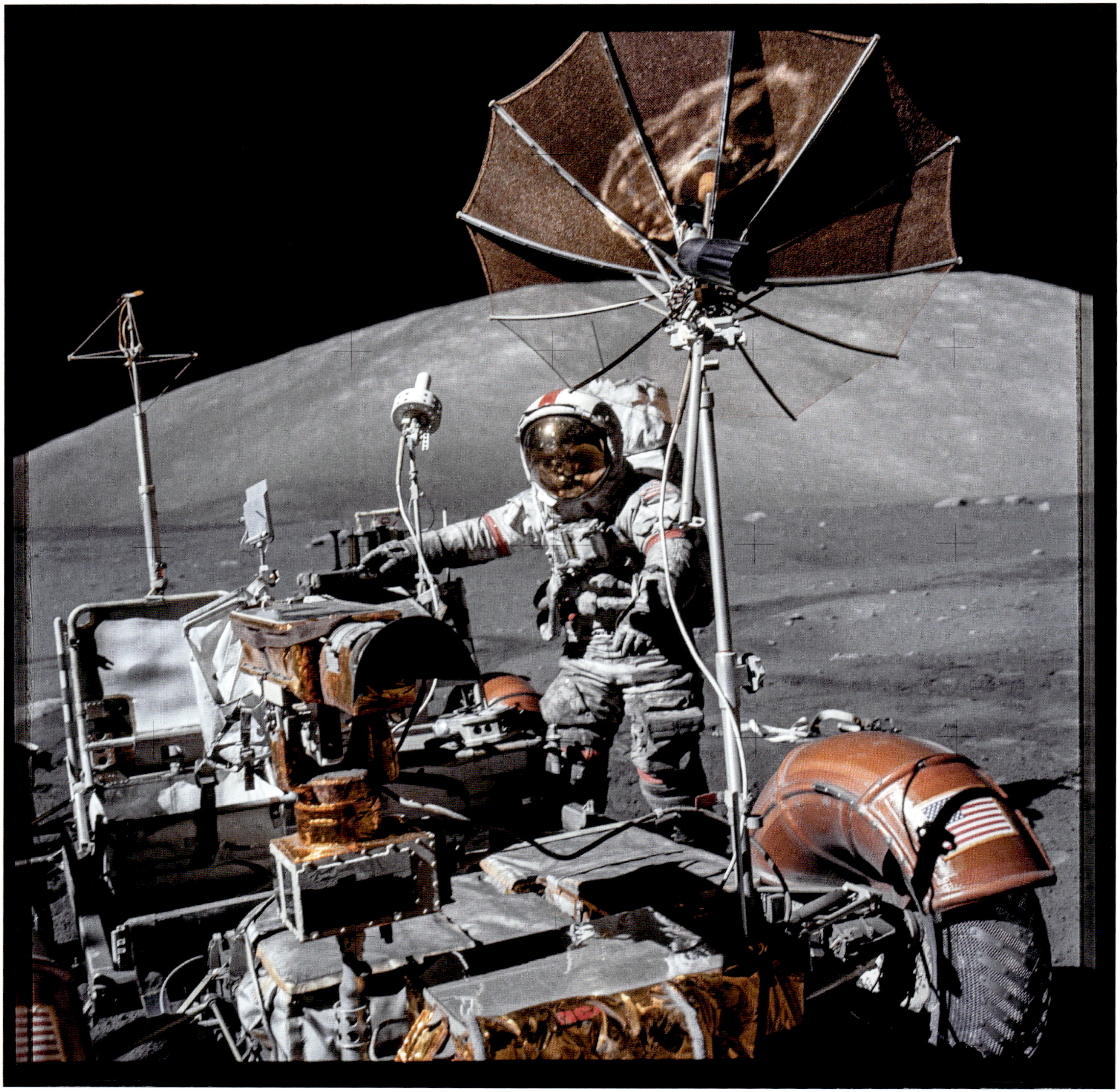

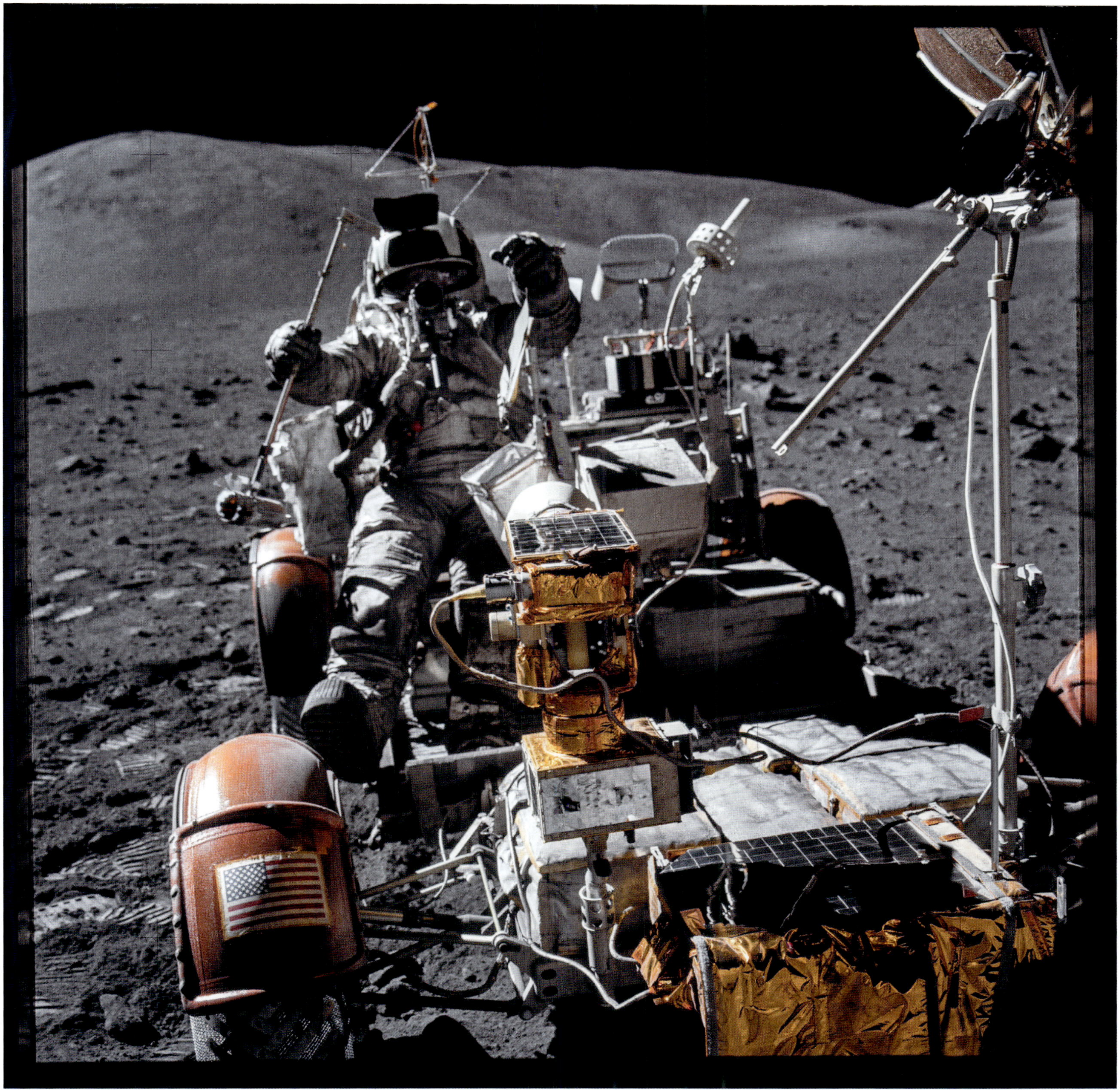

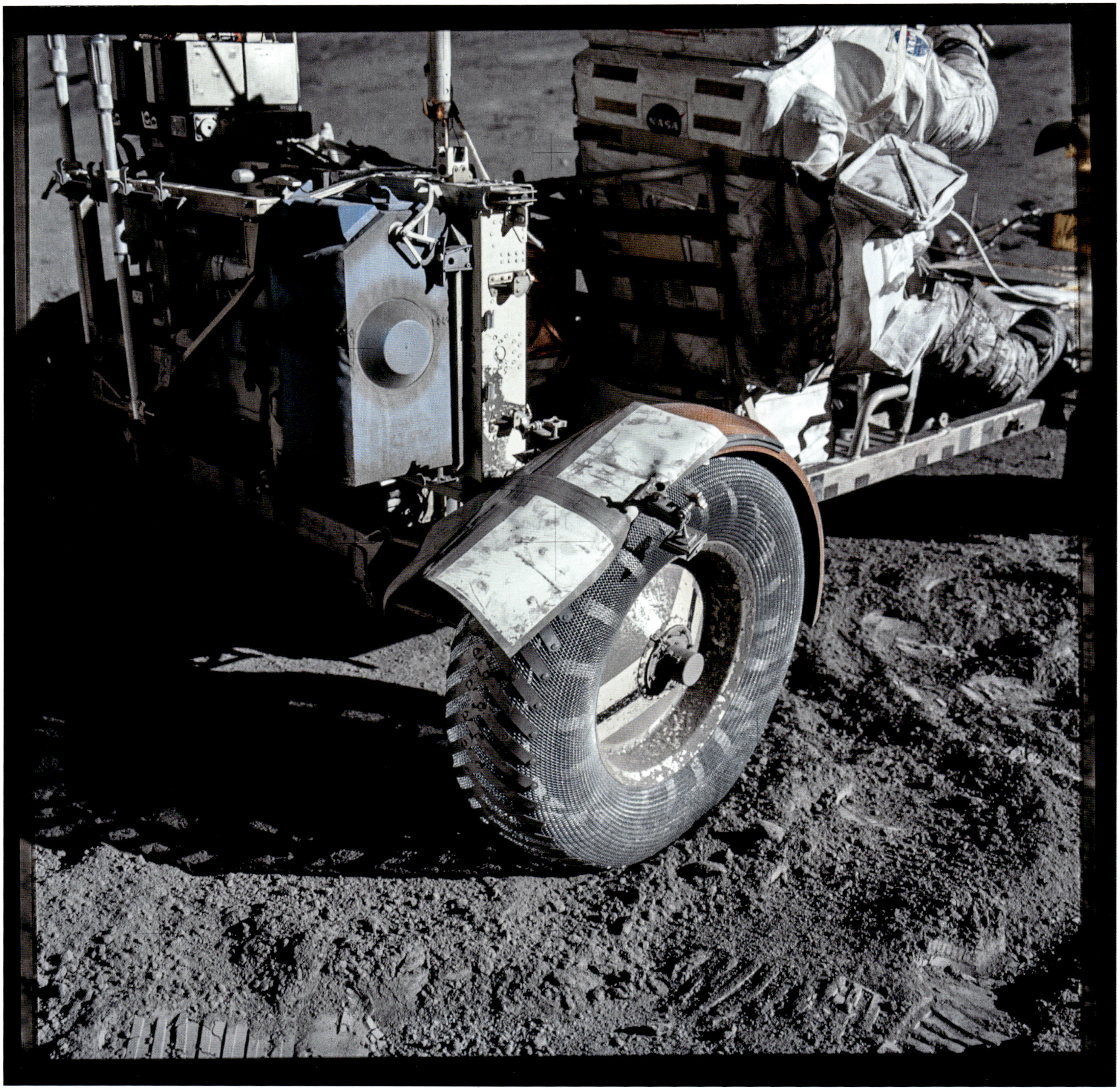

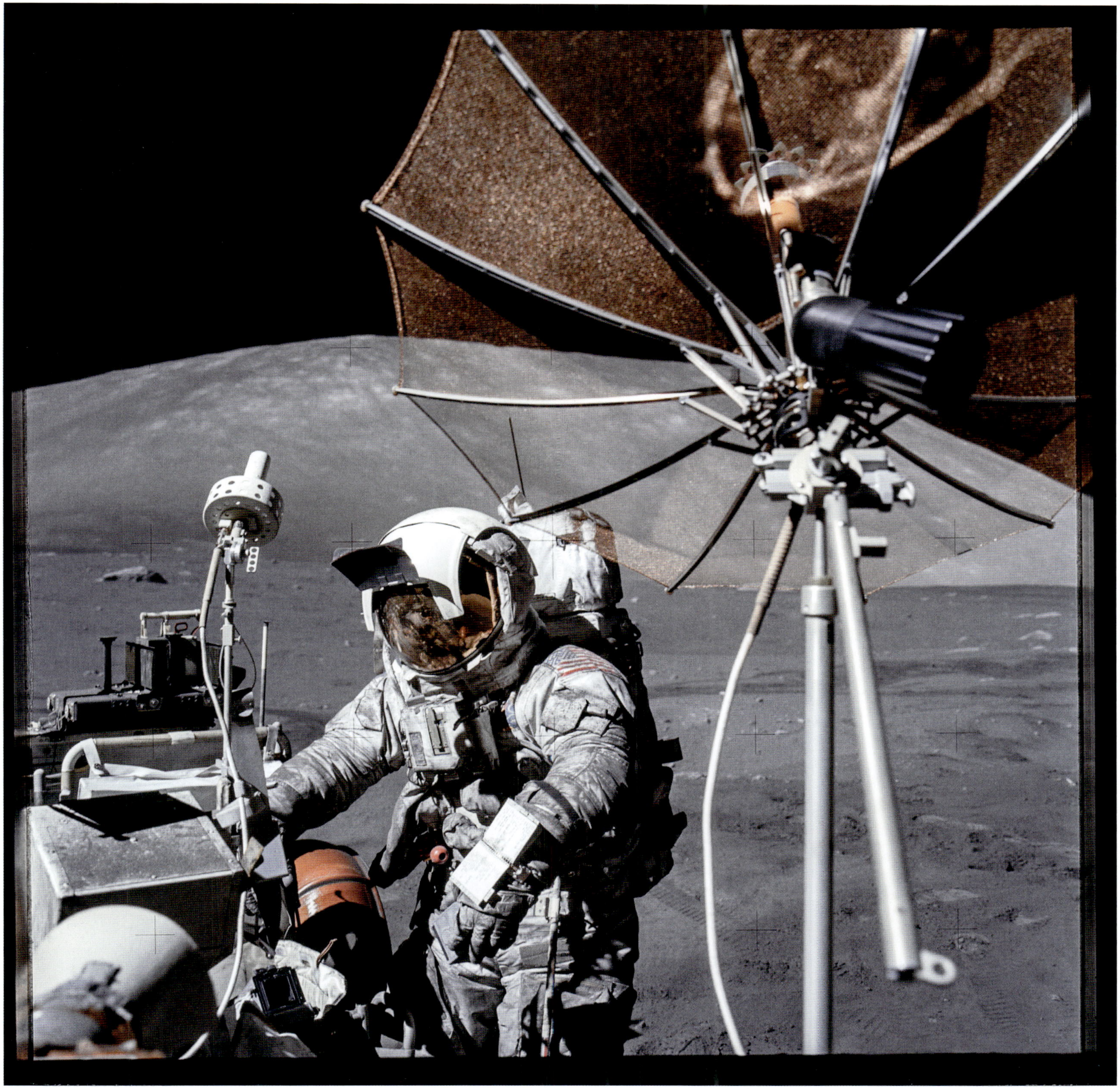

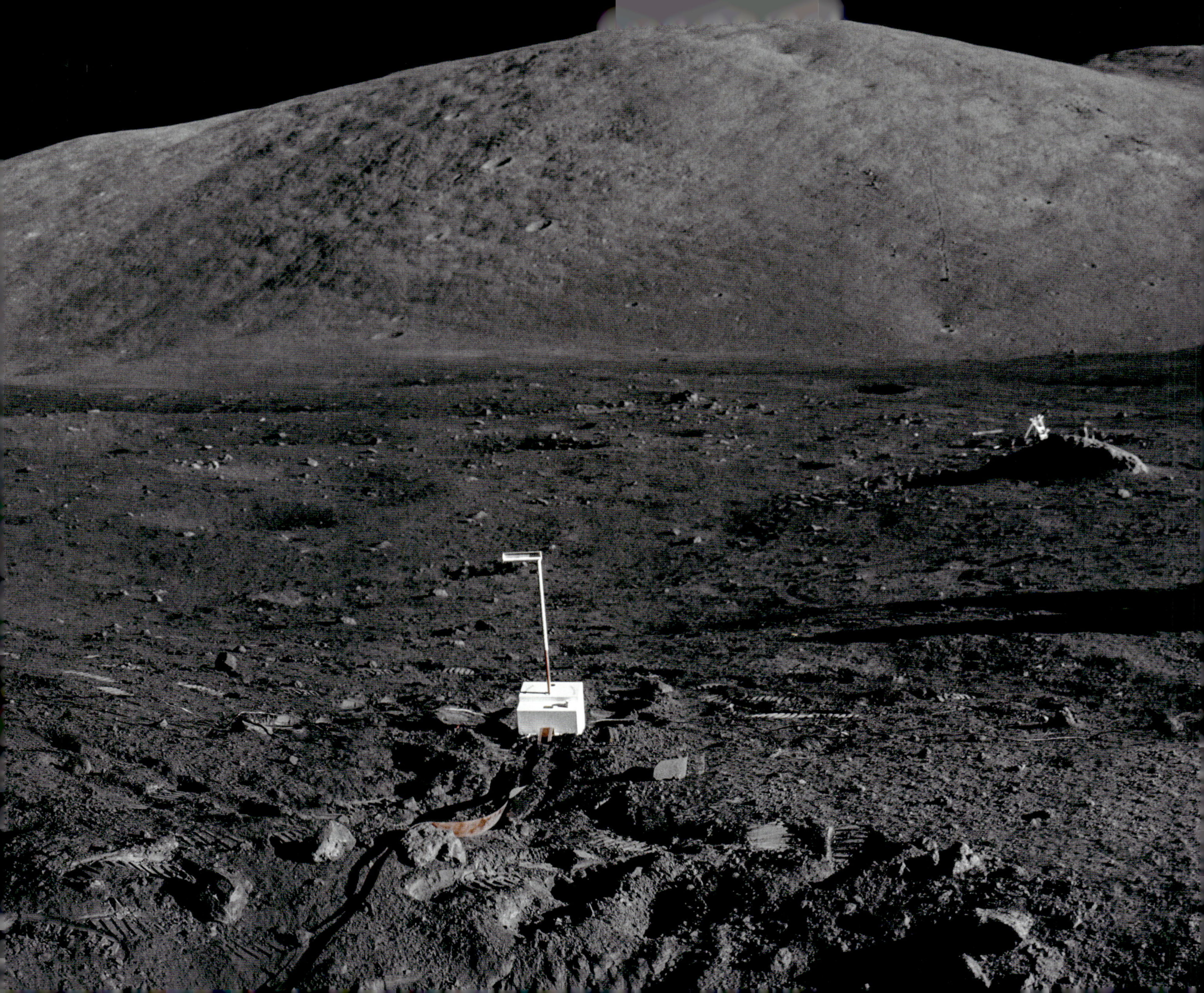

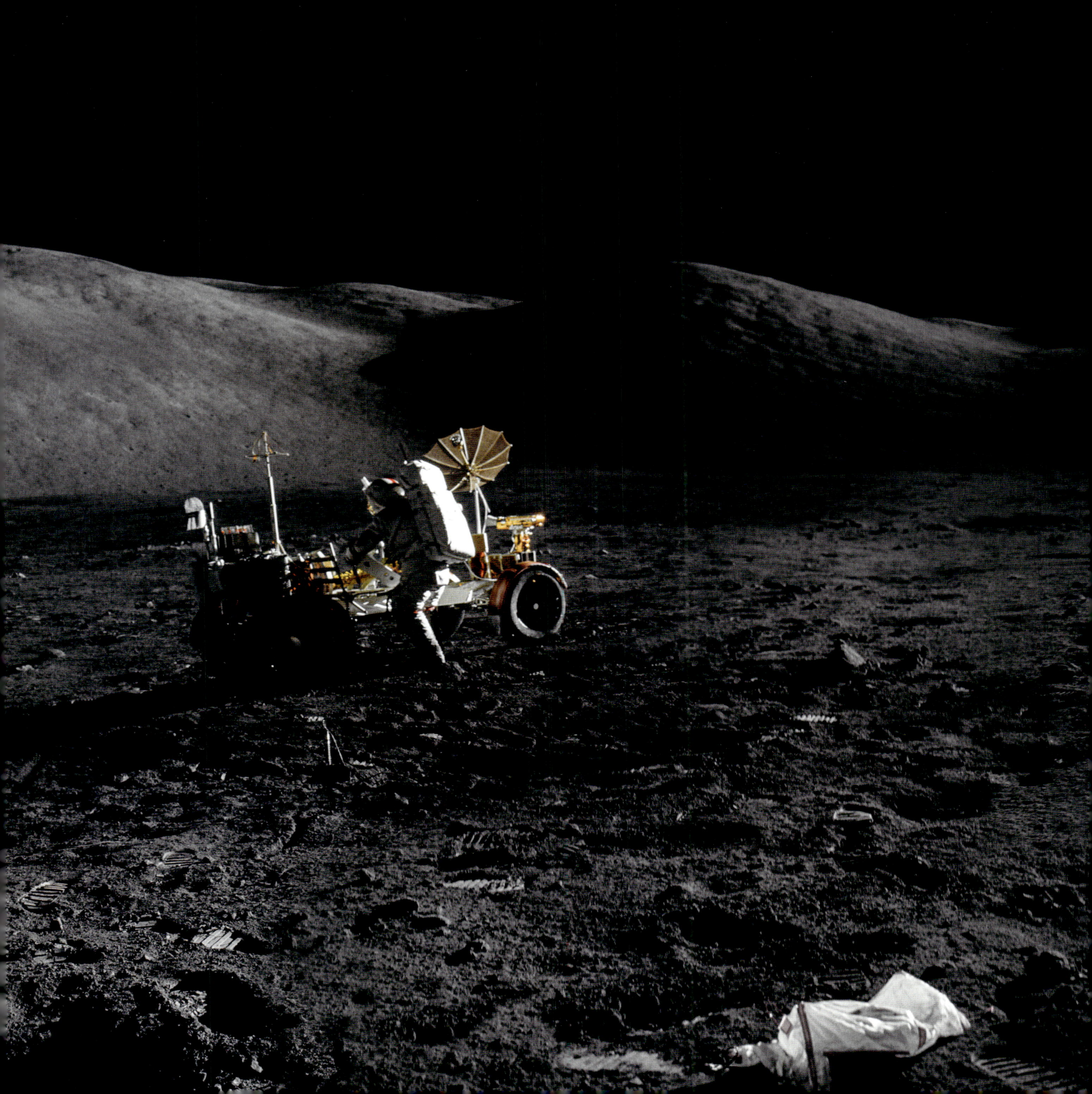

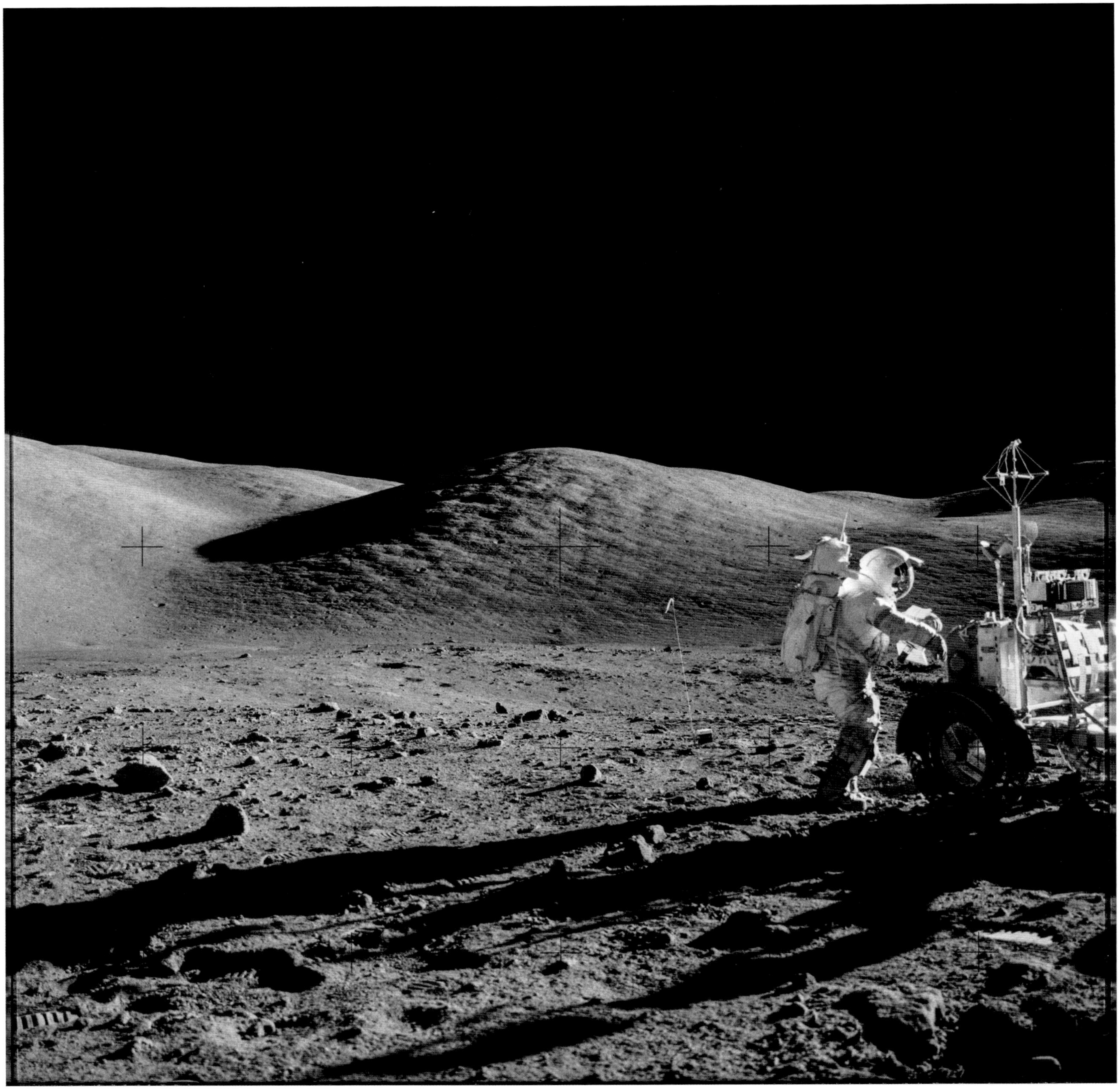

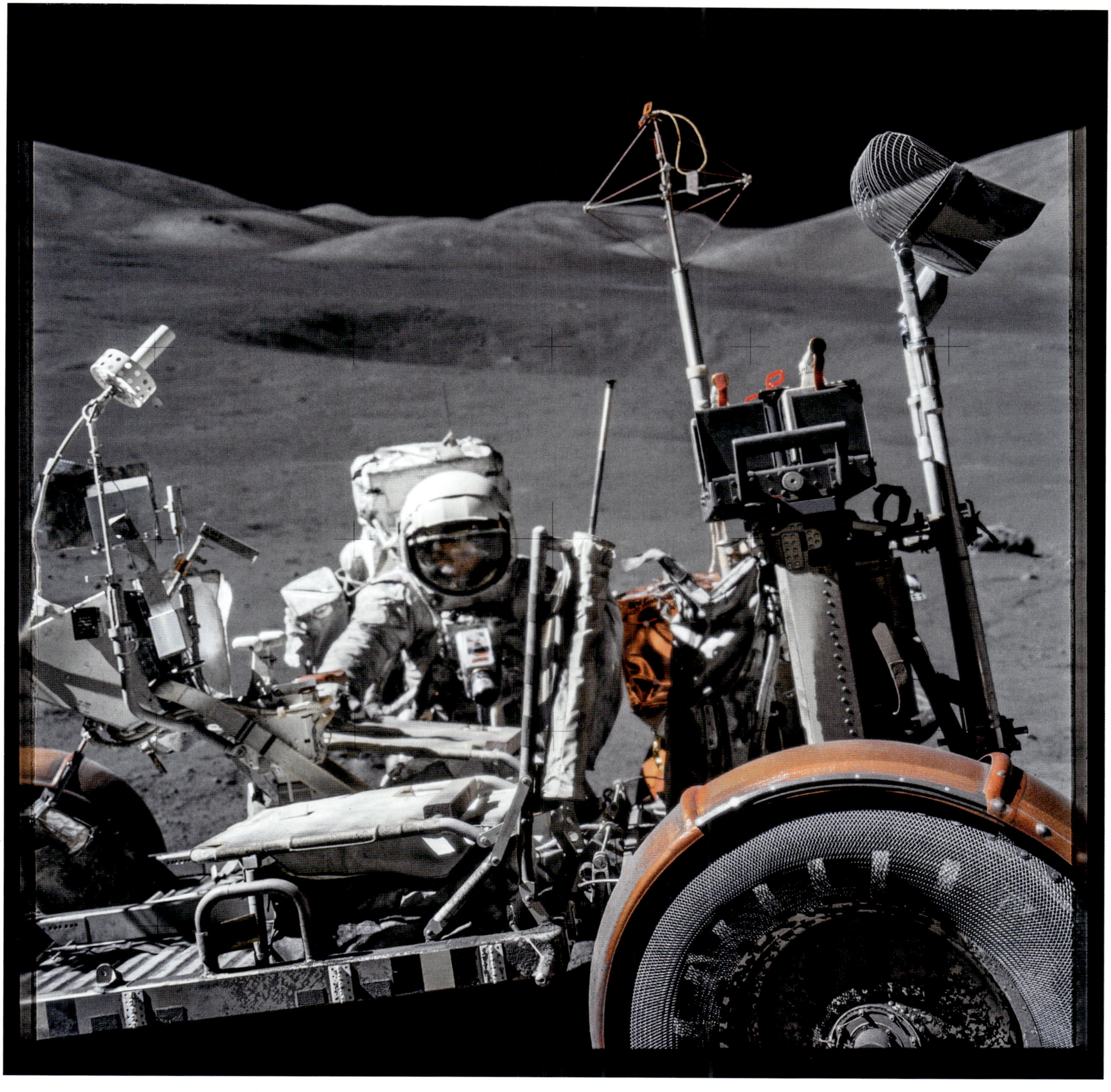

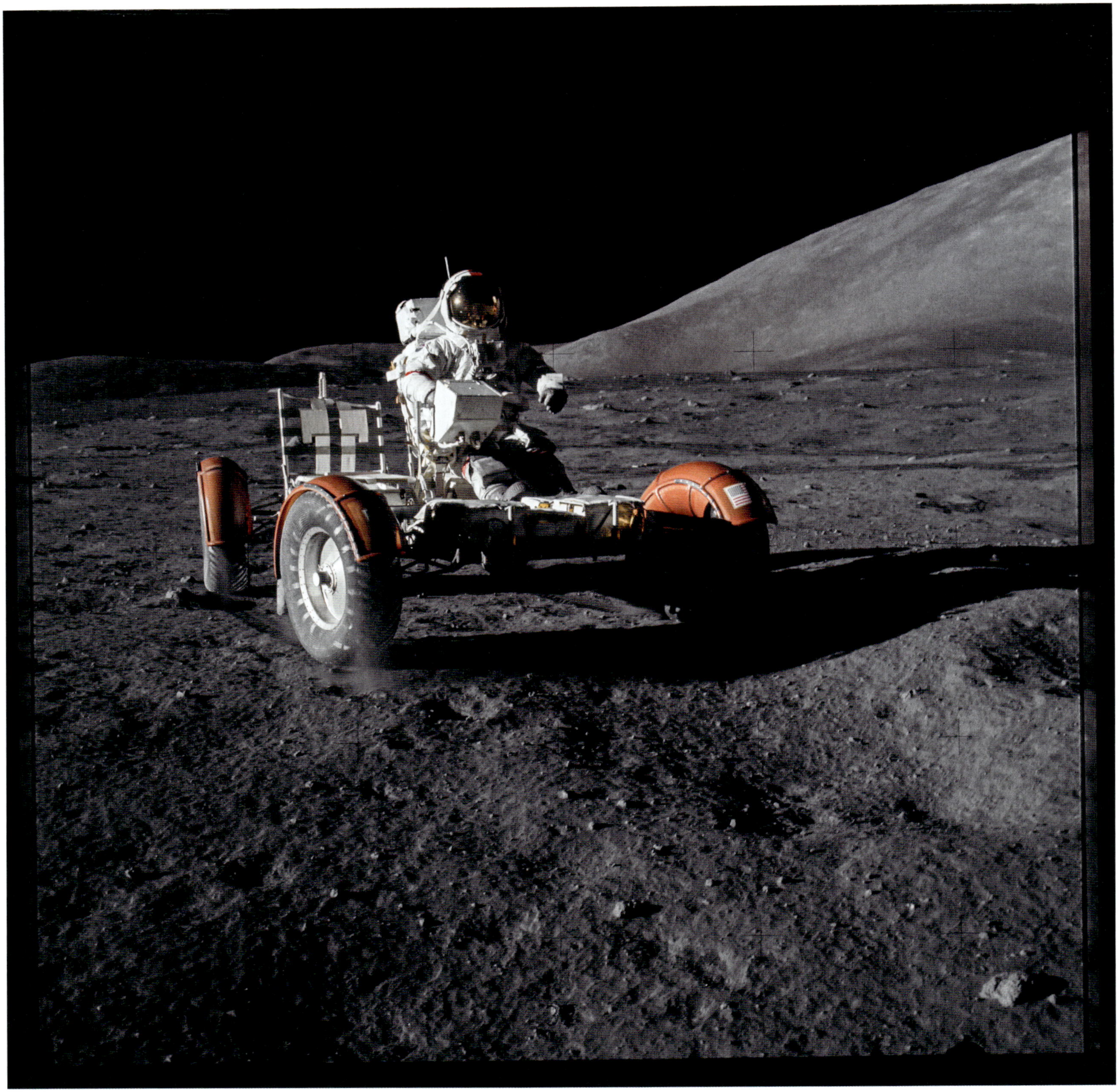

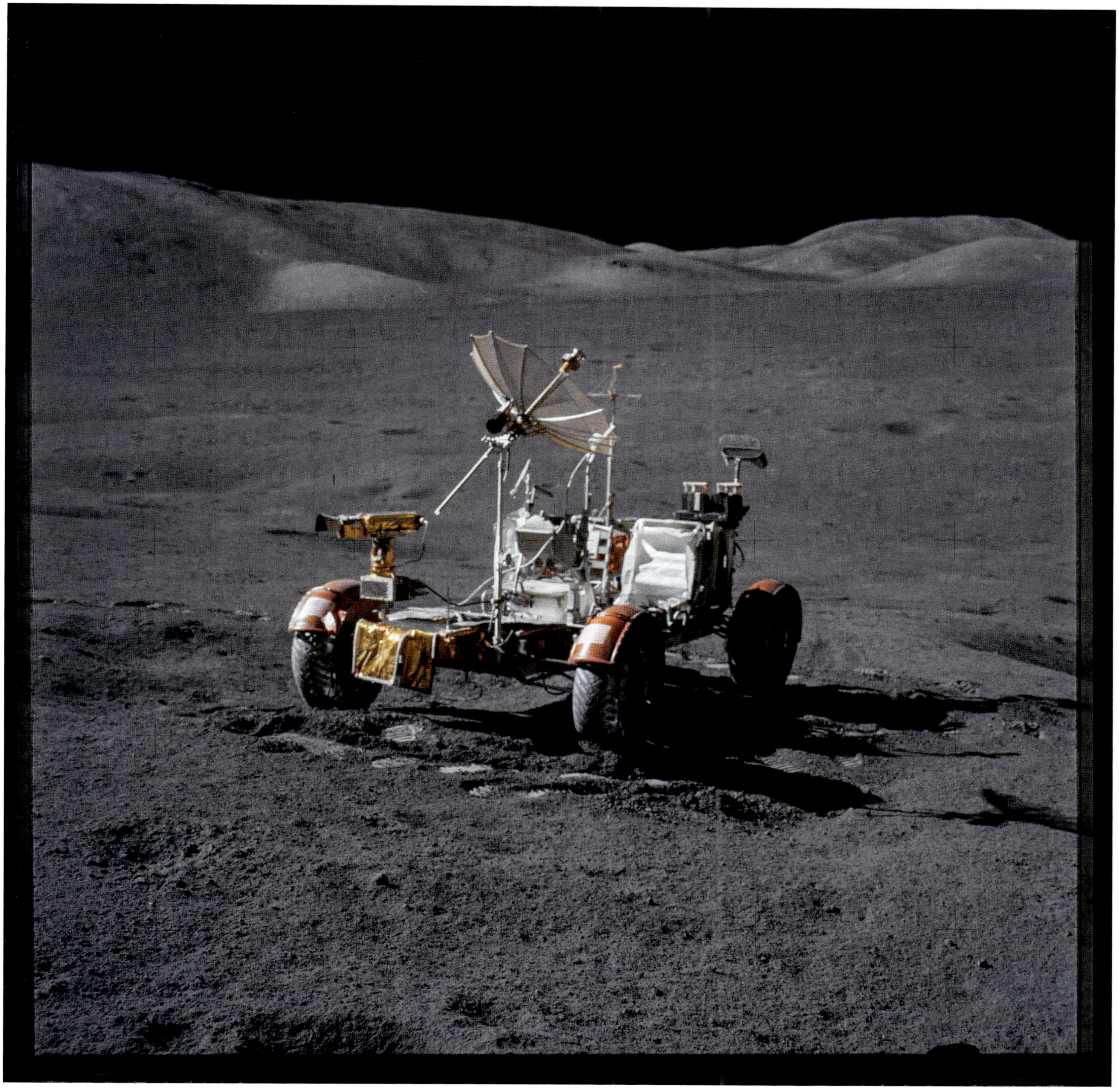

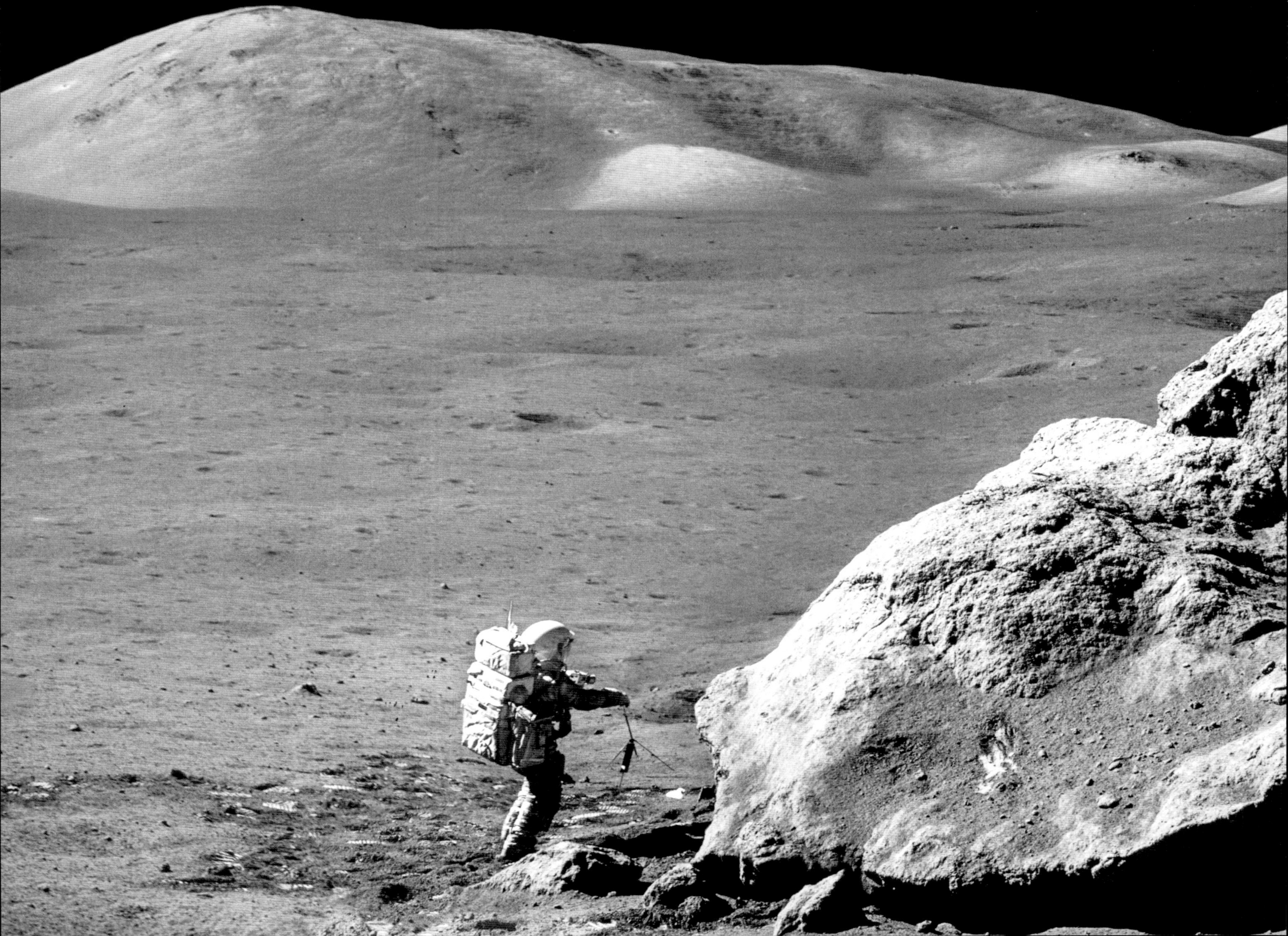

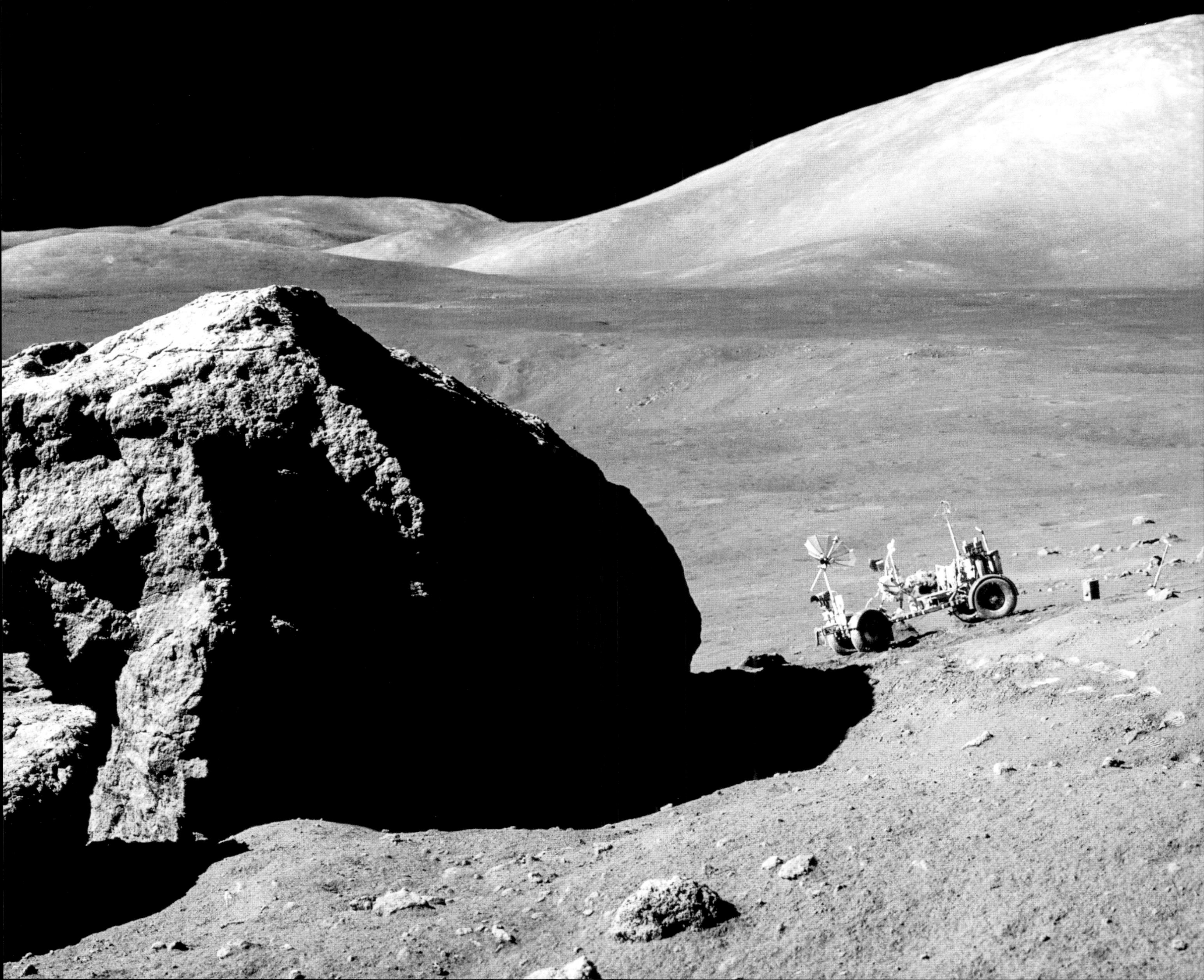

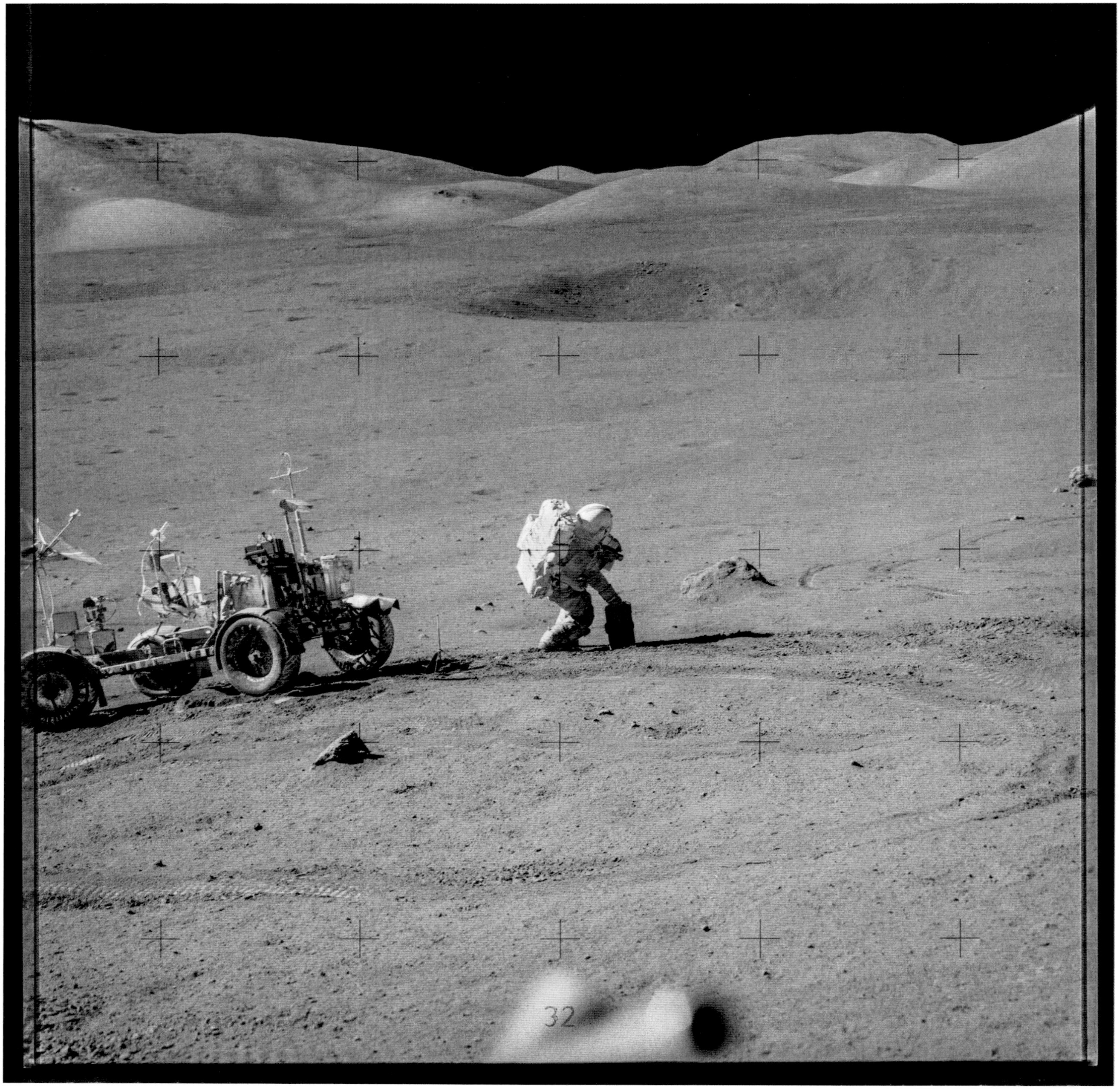

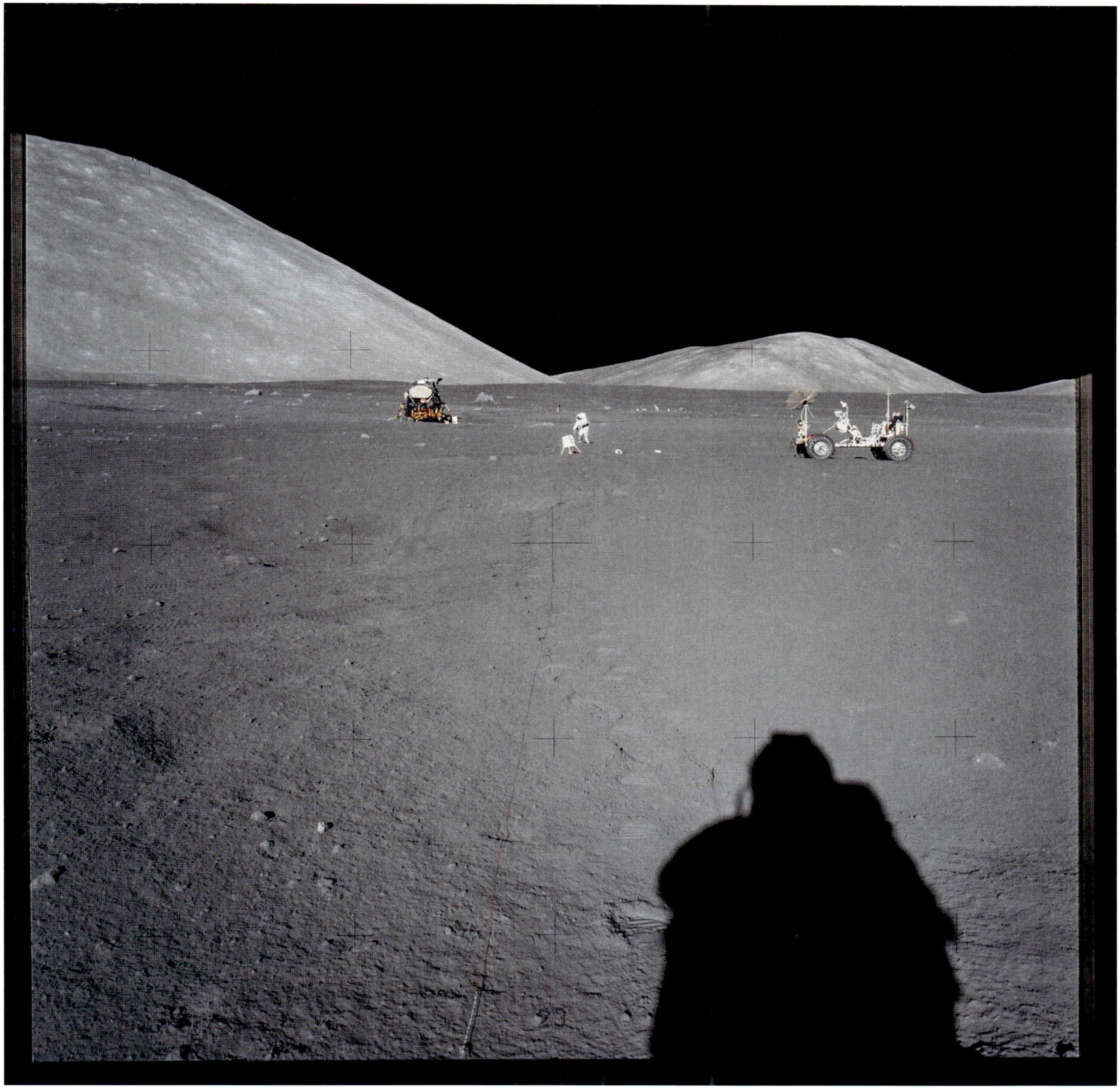

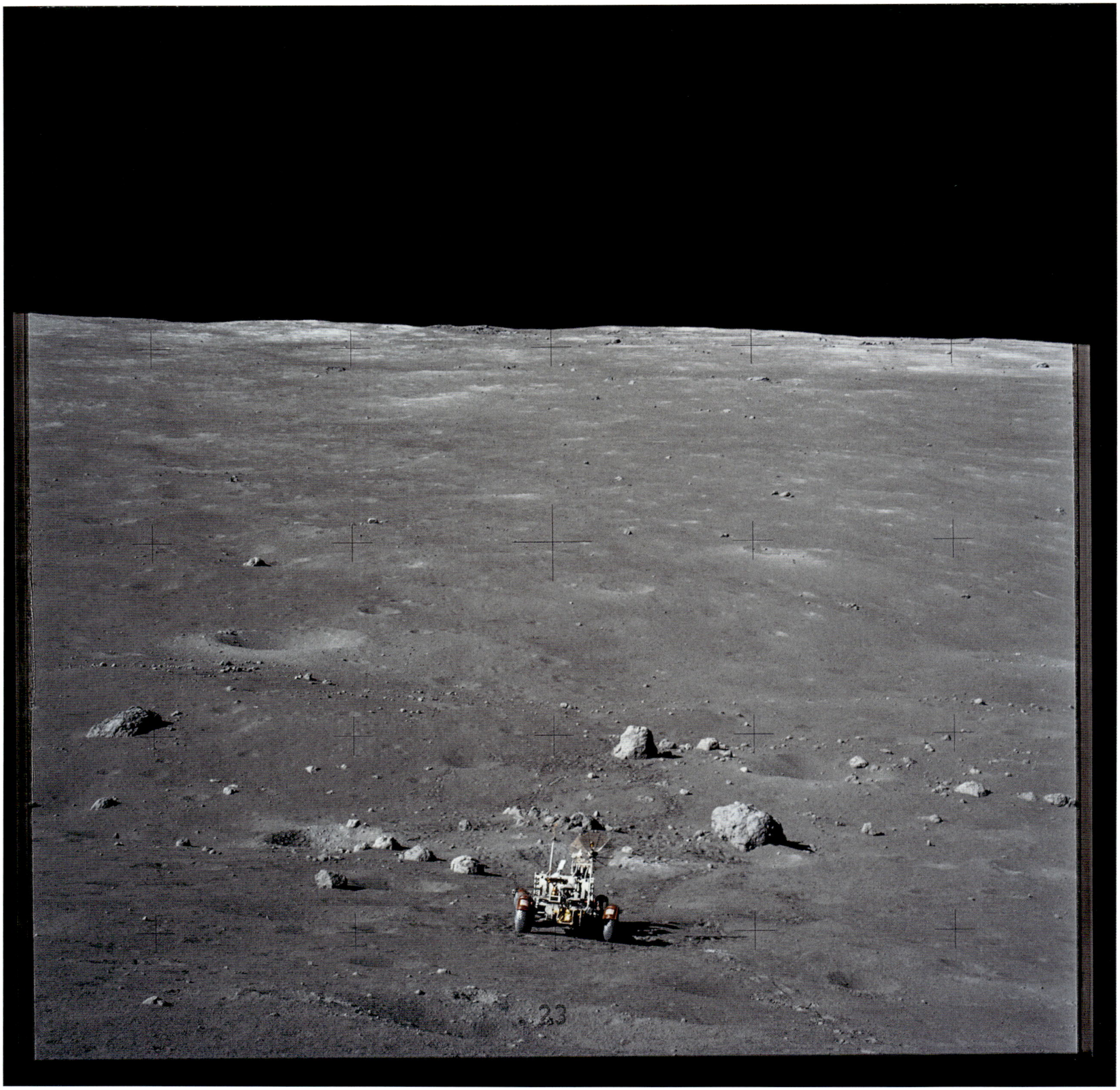

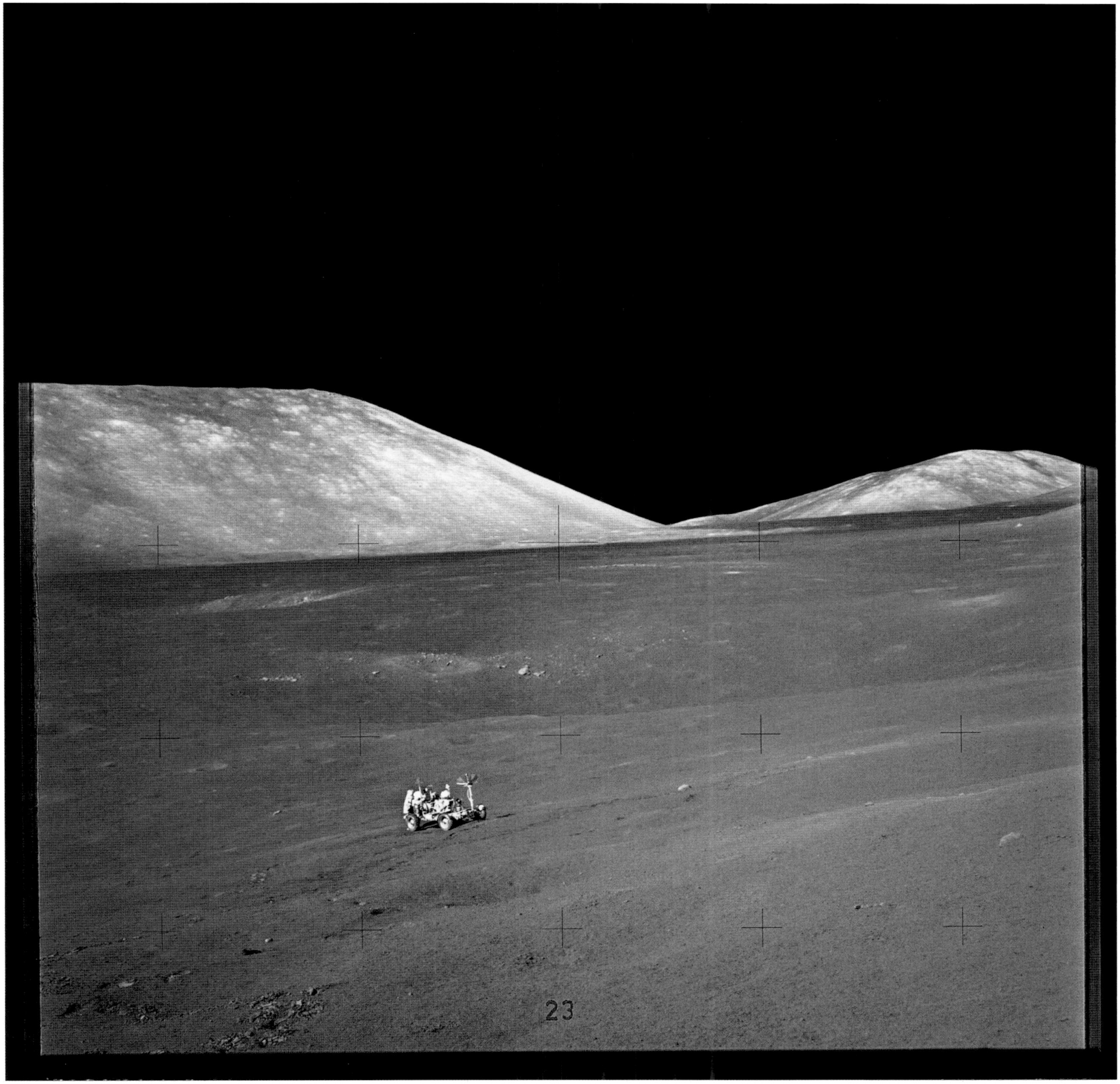
23

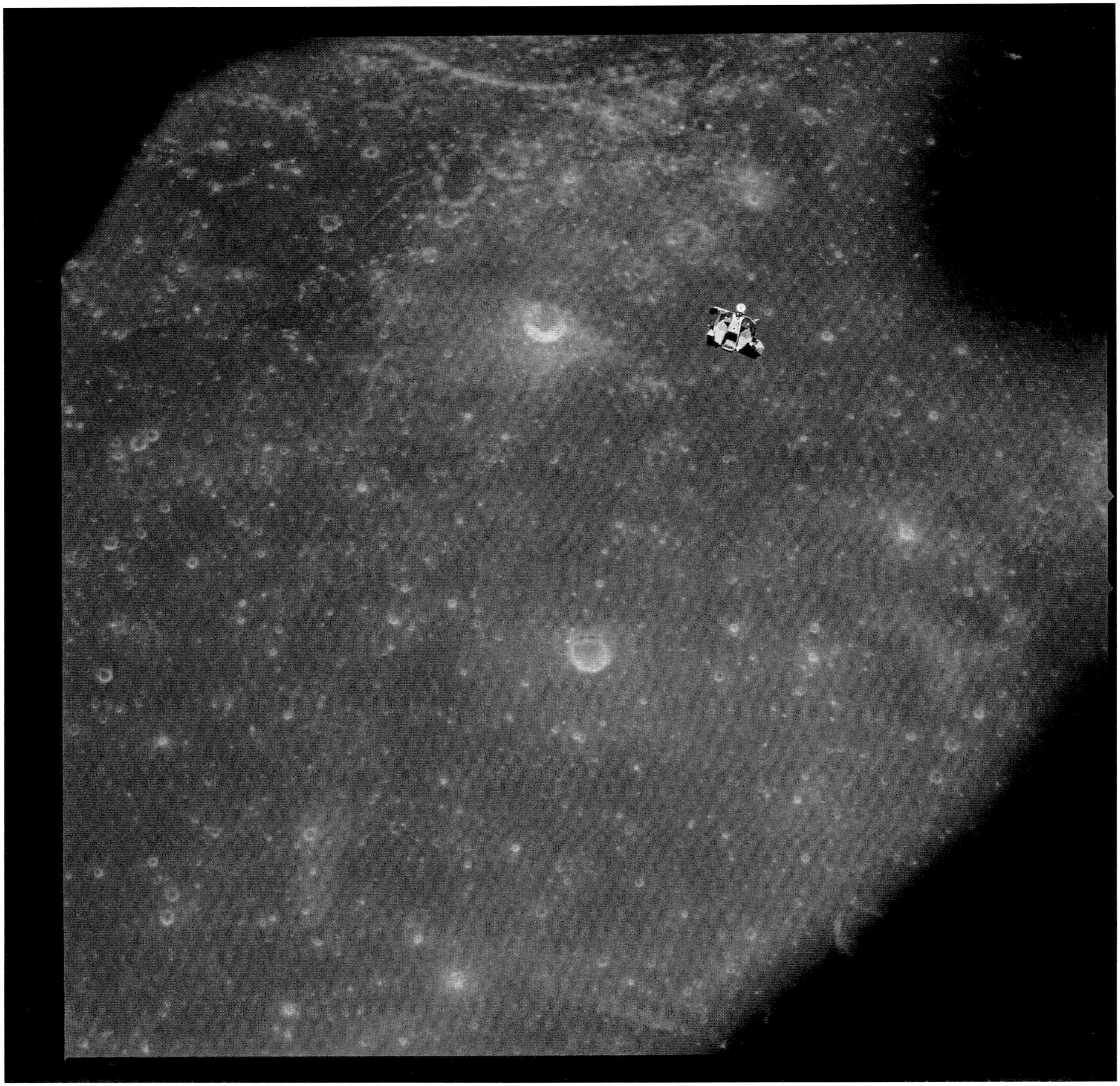

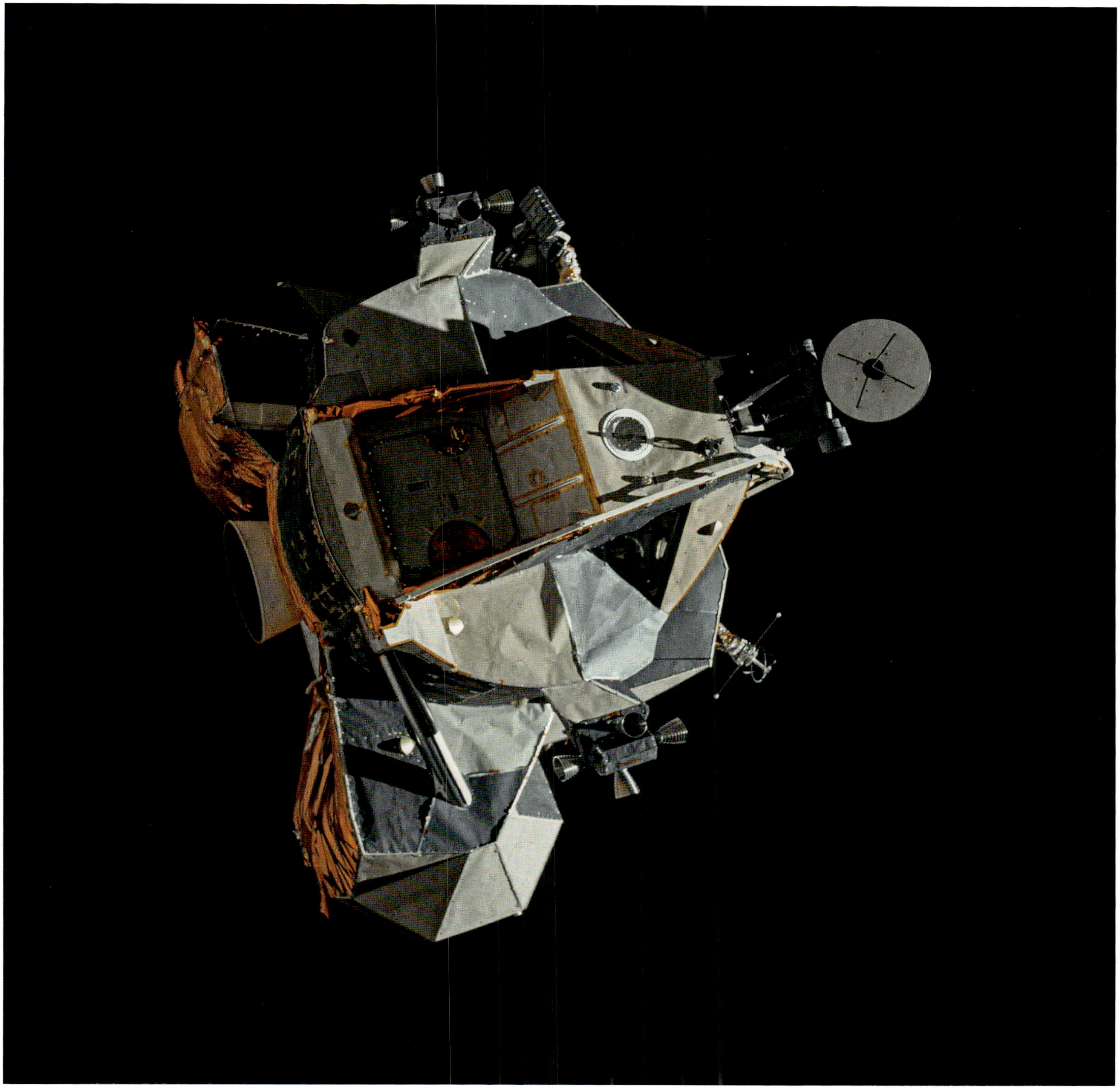

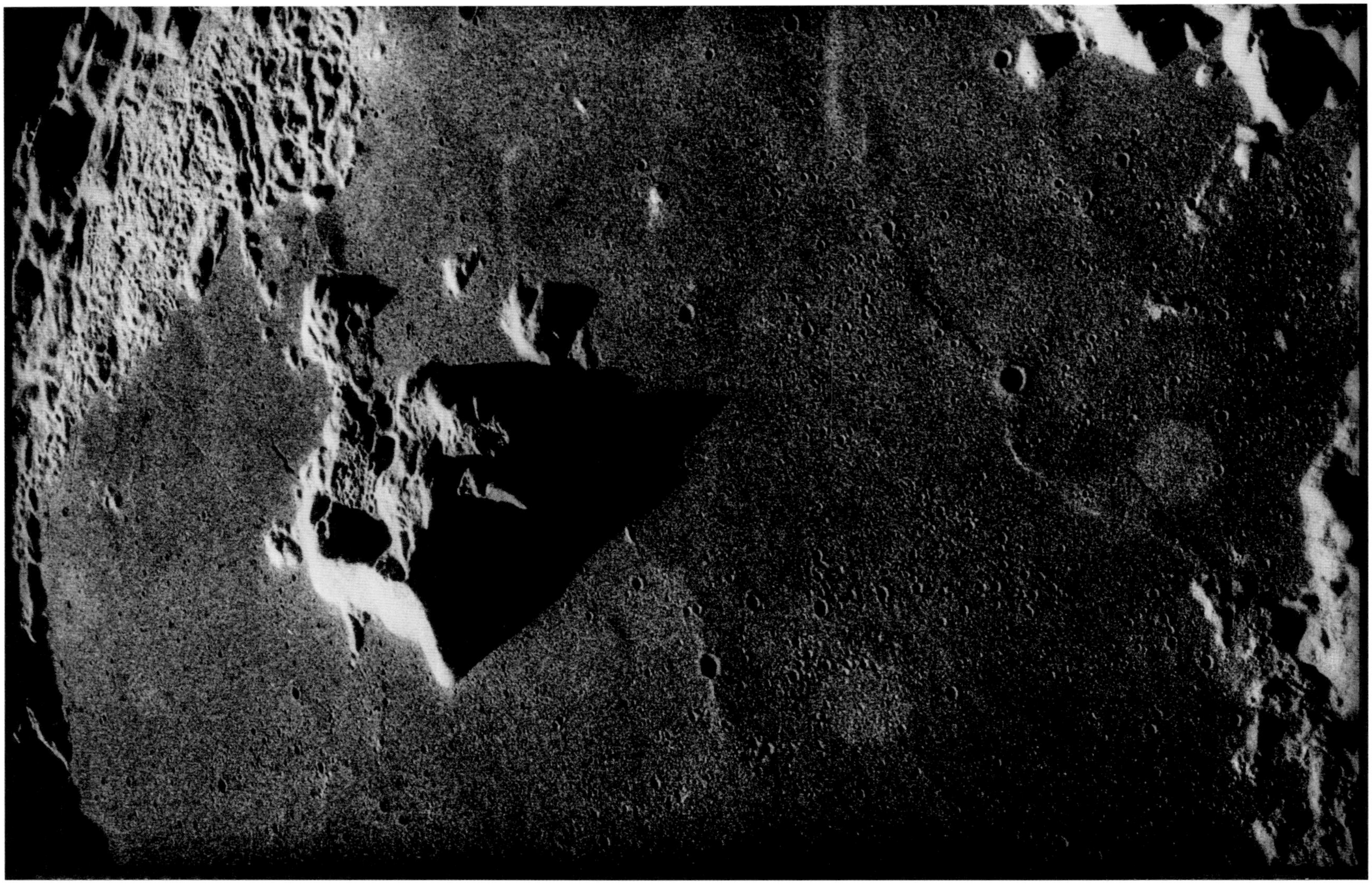

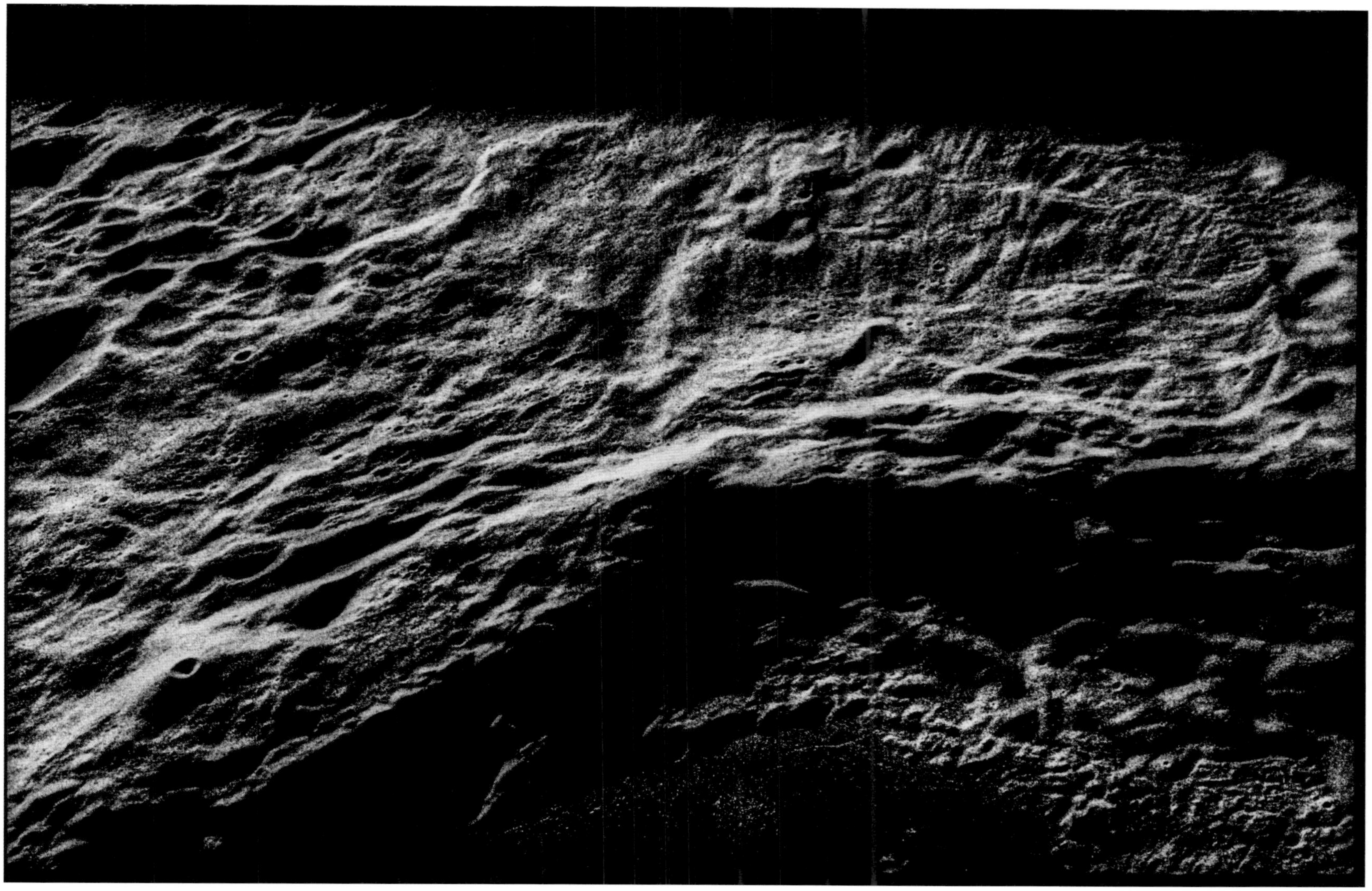

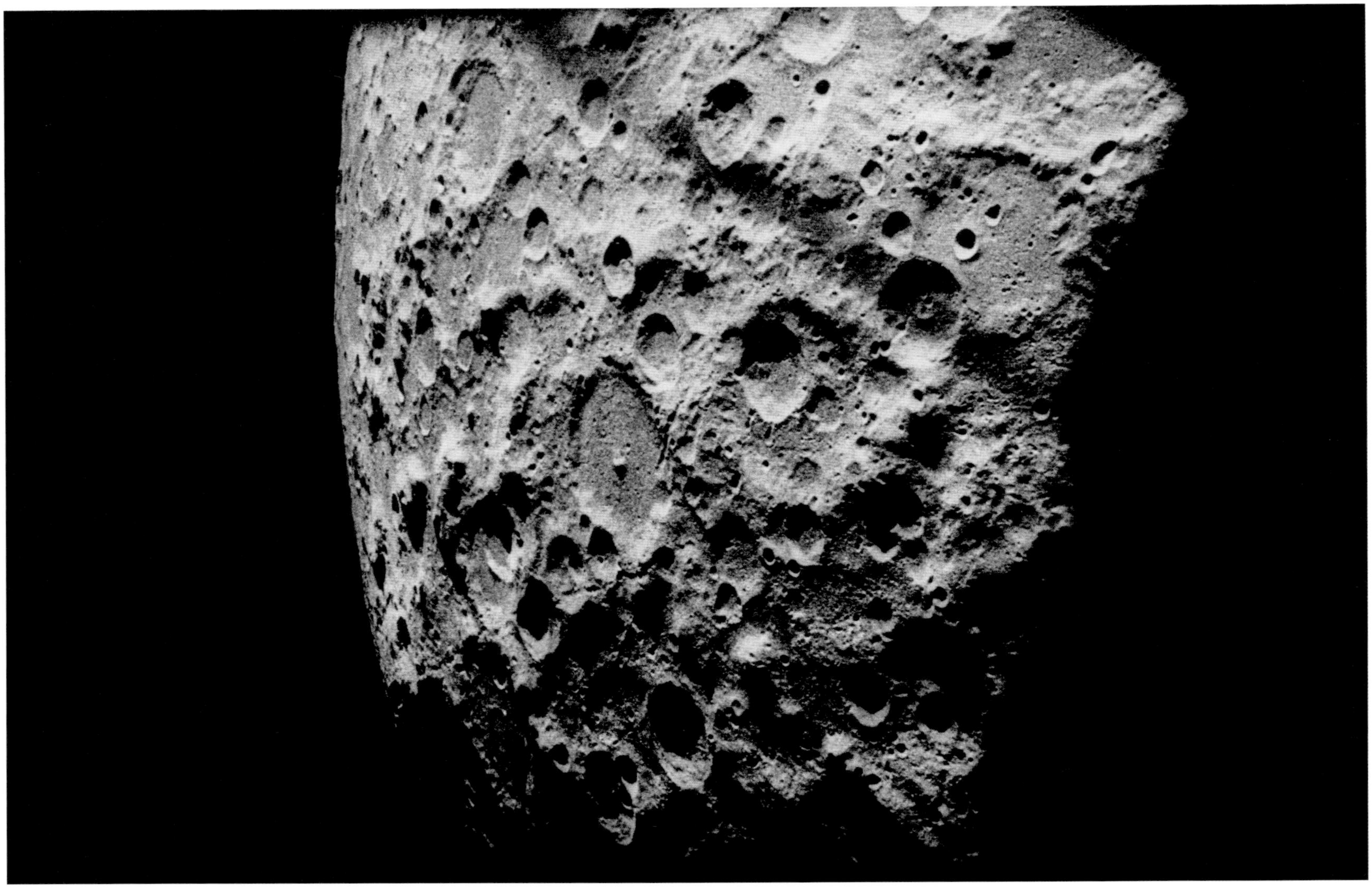

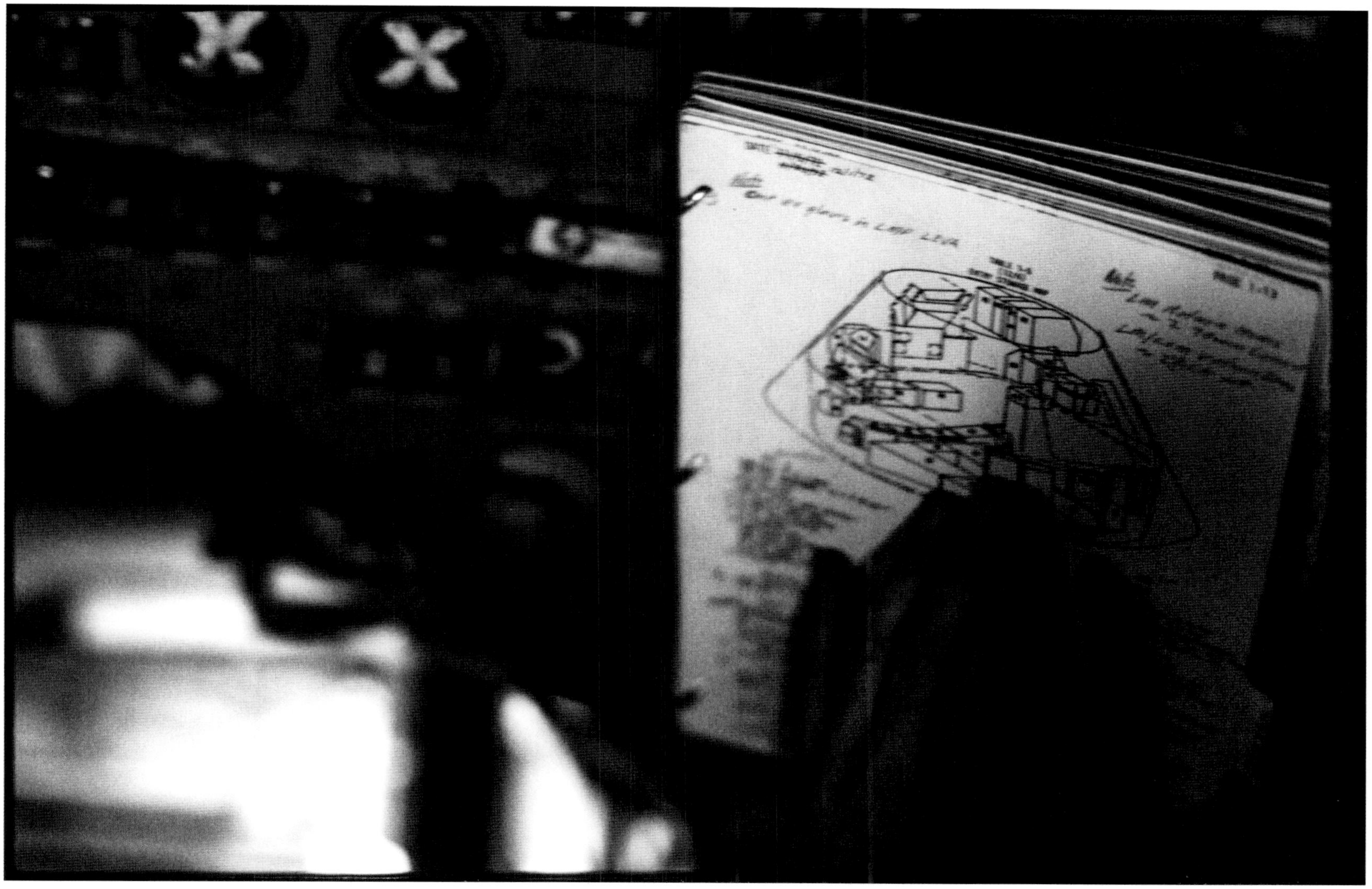

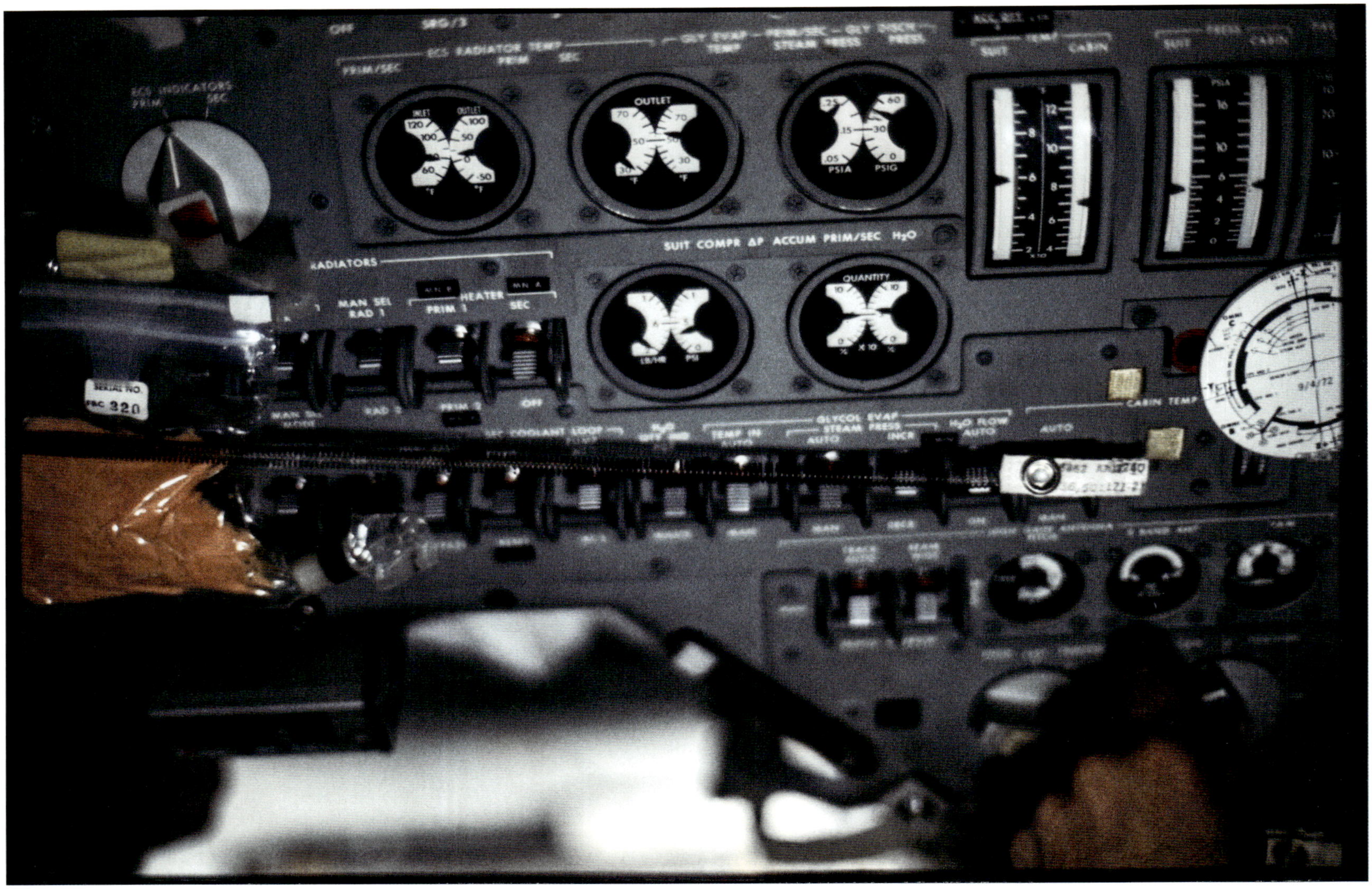
ECS RADIATOR TEMP
OUTLET
PSIA PSIG
SUIT COMPR ΔP ACCUM PRIM/SEC H2O
RADIATORS
MAN SEL
RAD 1
HEATER
PRIM 1
SEC
QUANTITY
GLYCOL EVAP
STEAM PRESS
CABIN TEMP
AUTO
SERIAL NO.
9/4/72

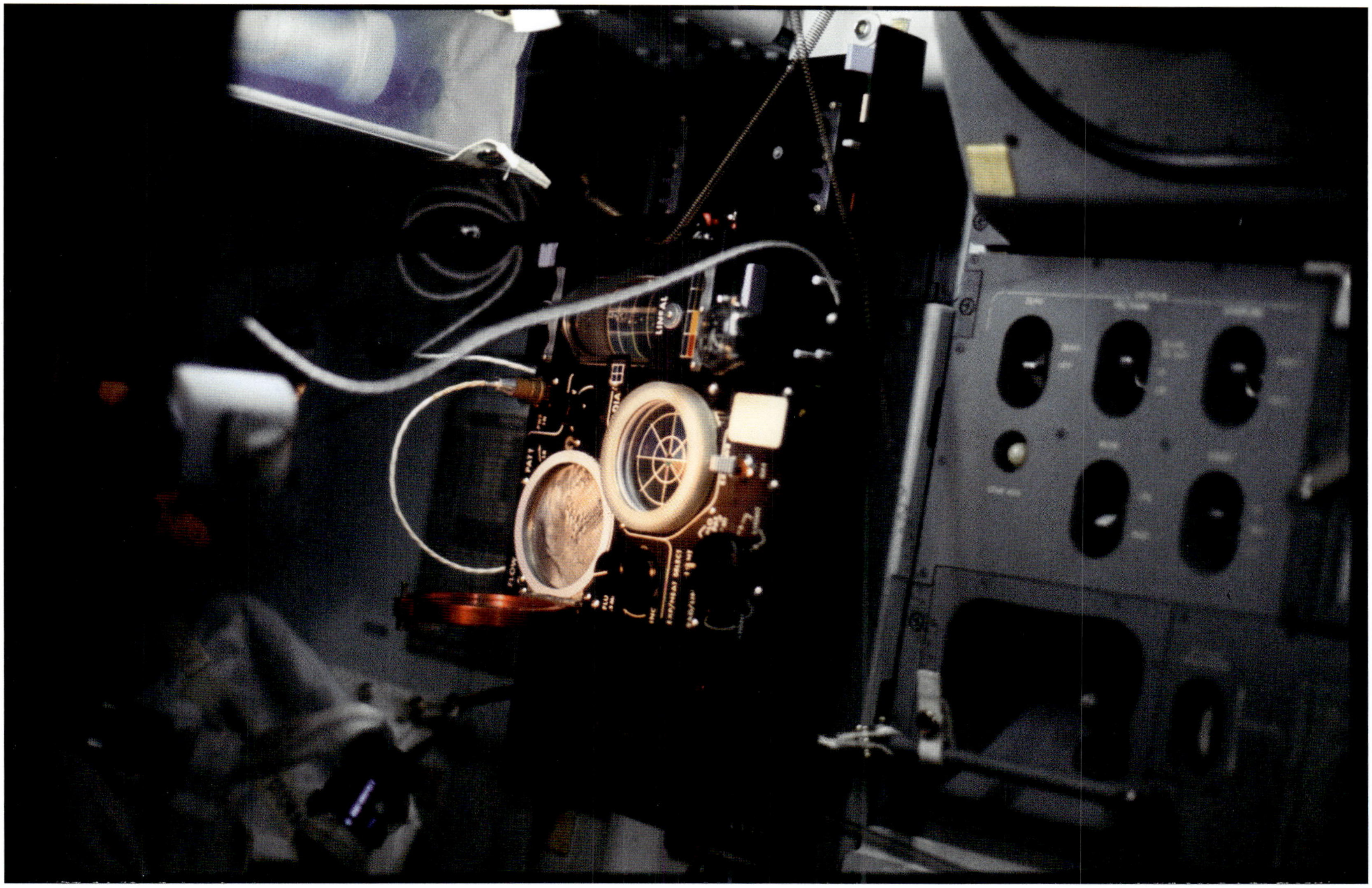

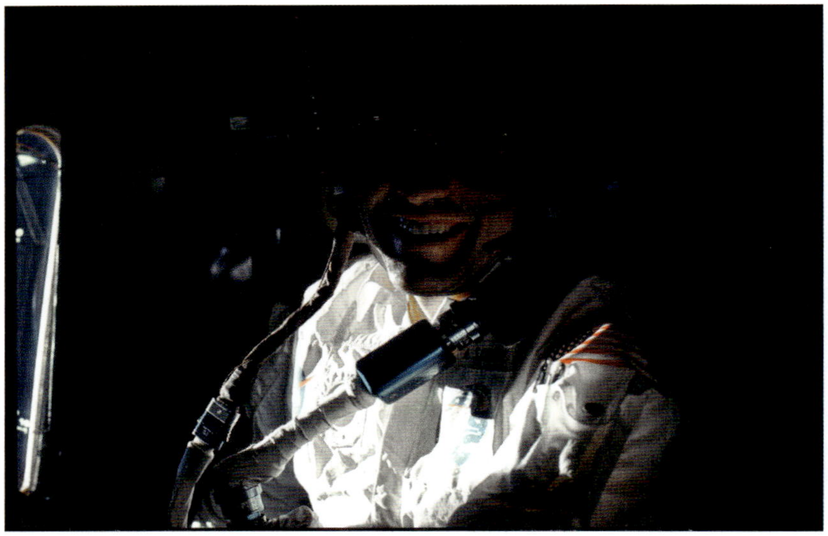

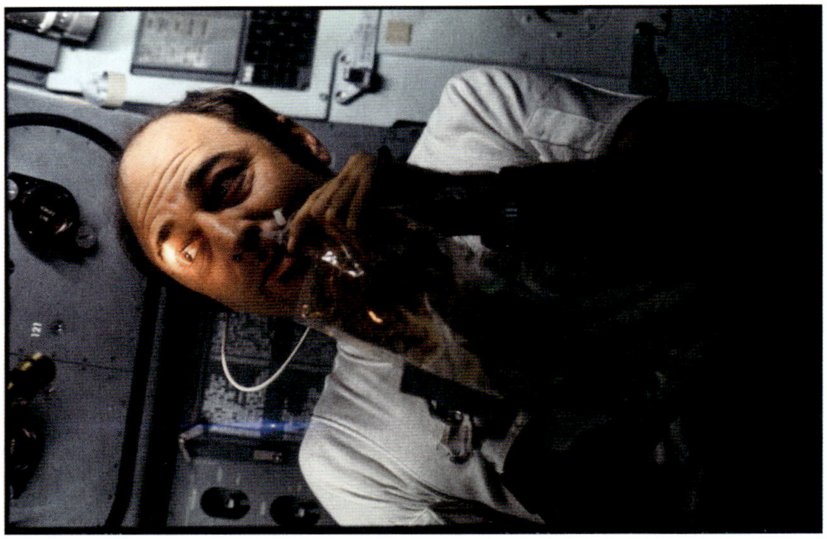

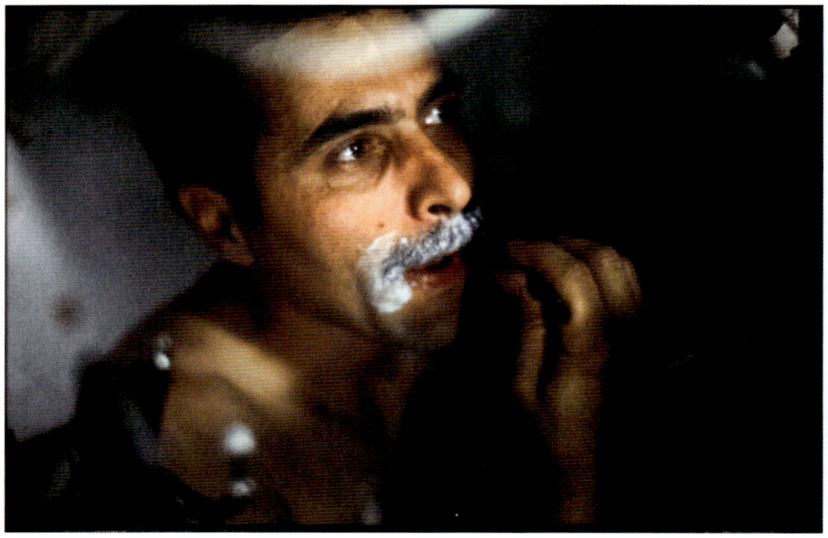

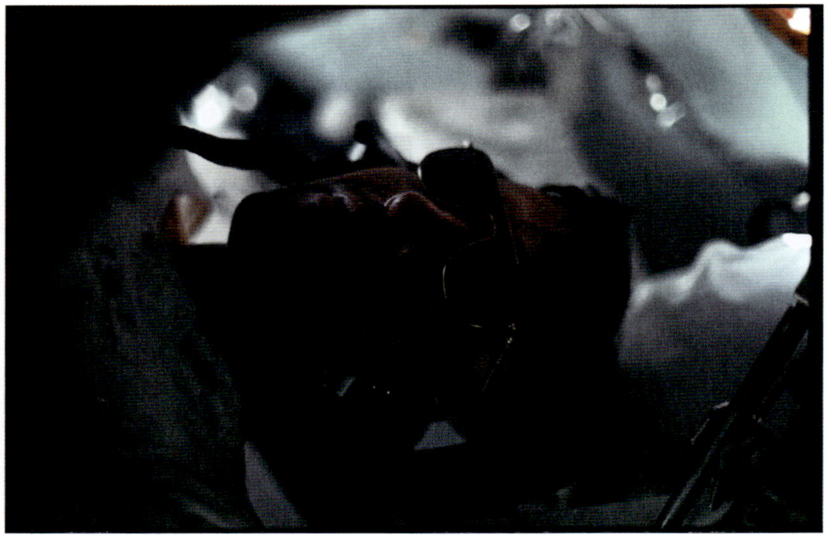

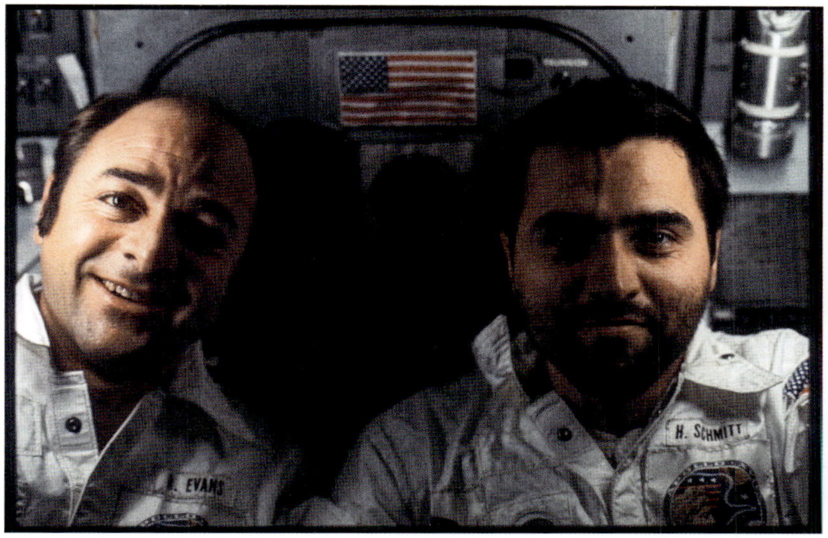
EVANS
H. SCHMITT

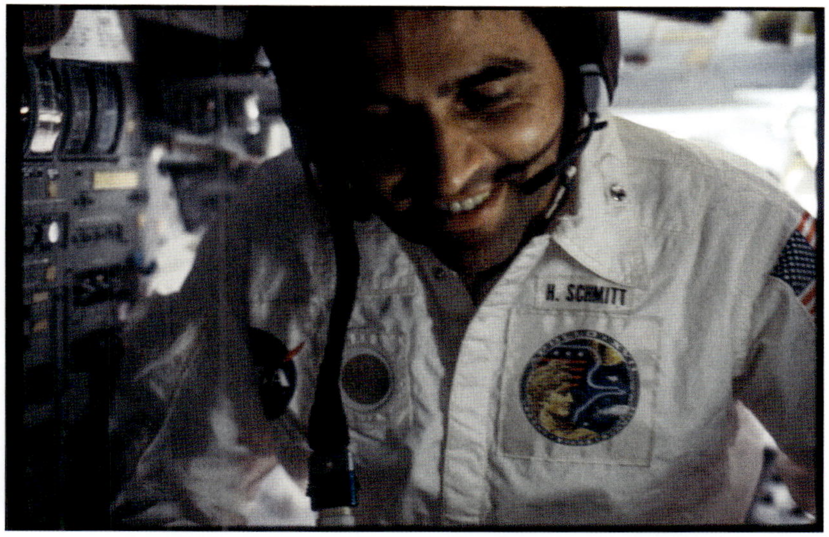
H. SCHMITT

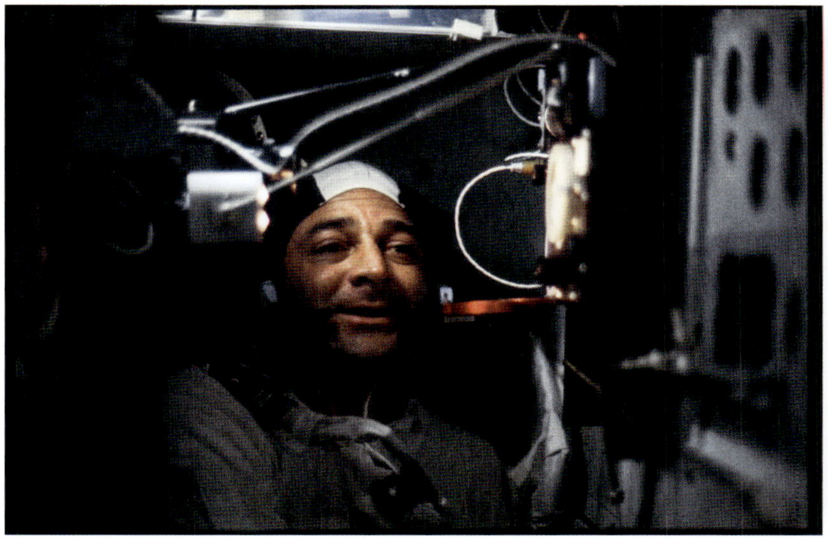

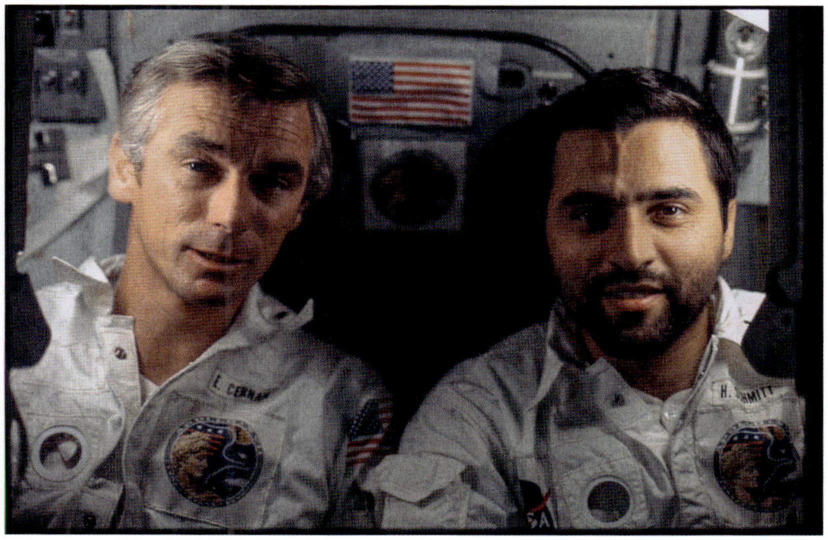
H. SCHMITT

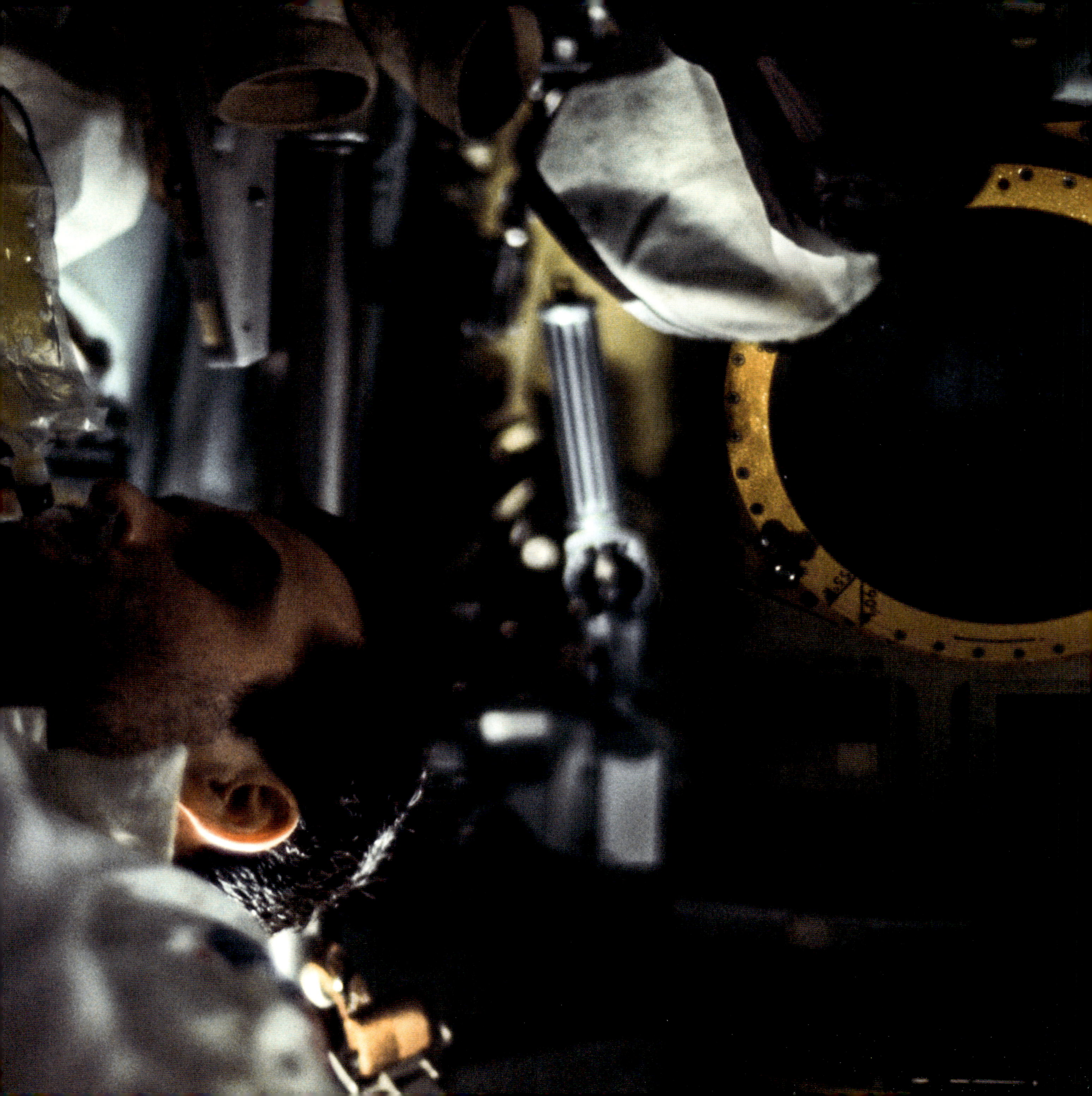

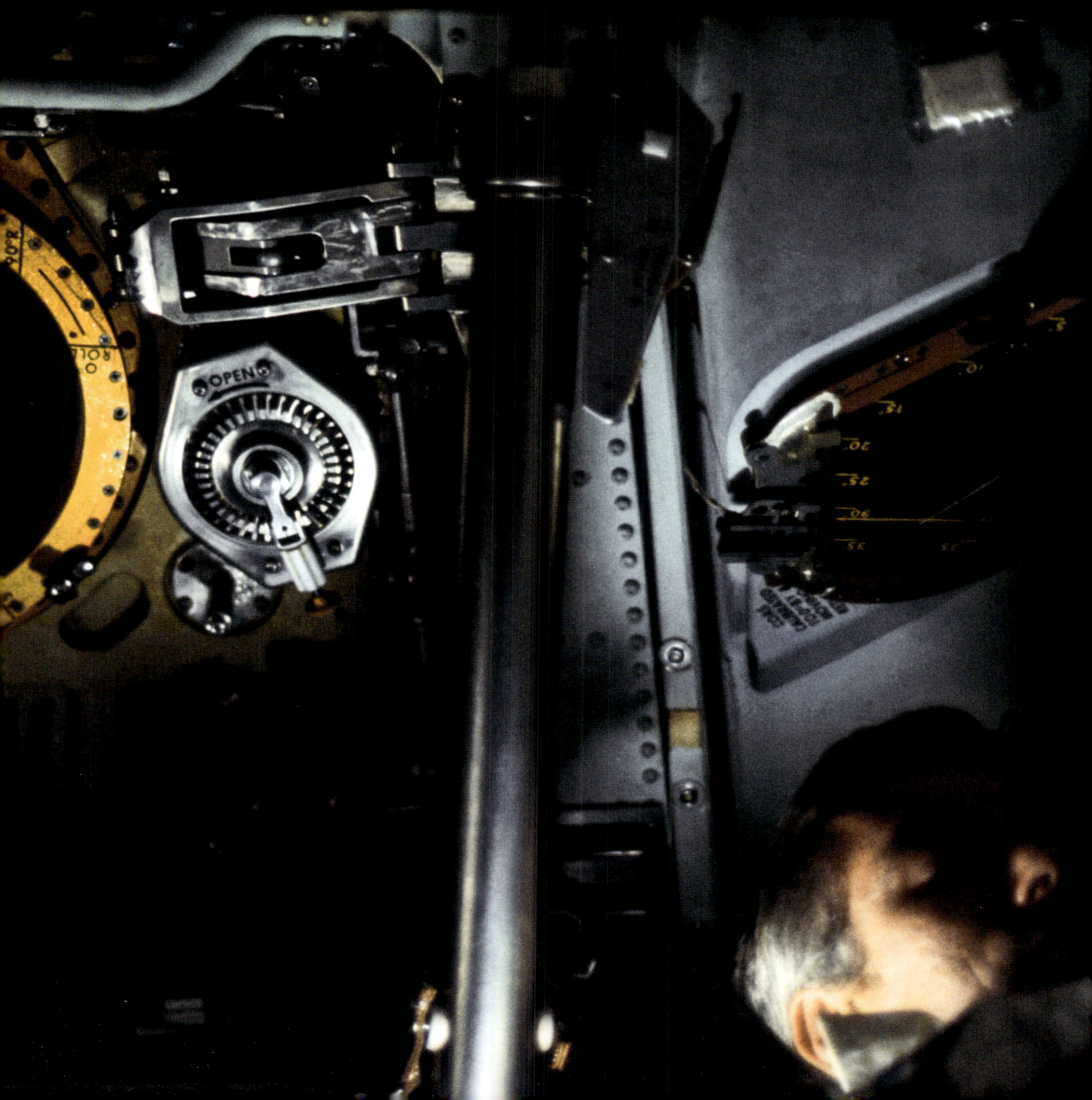
OPEN

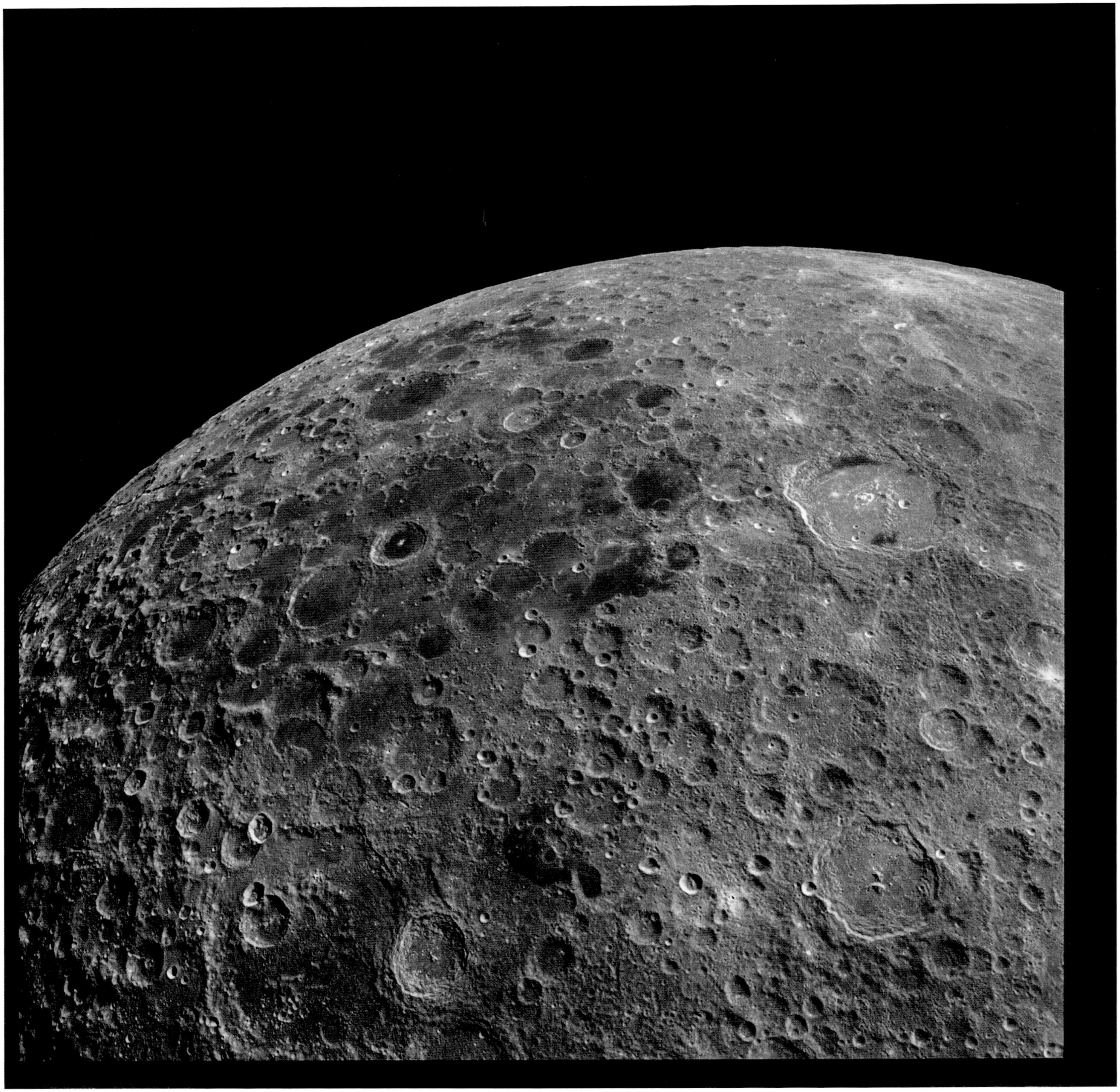

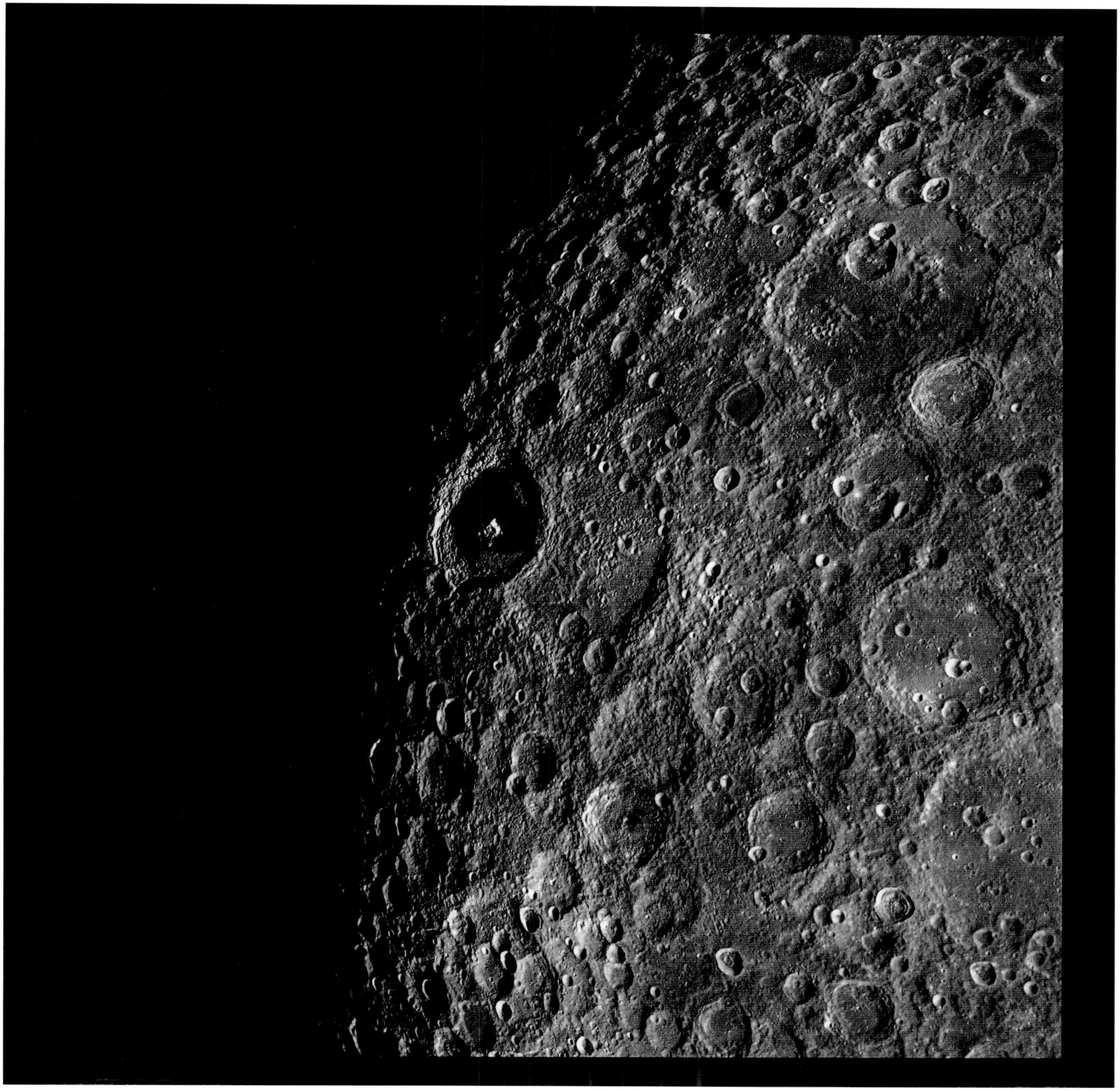

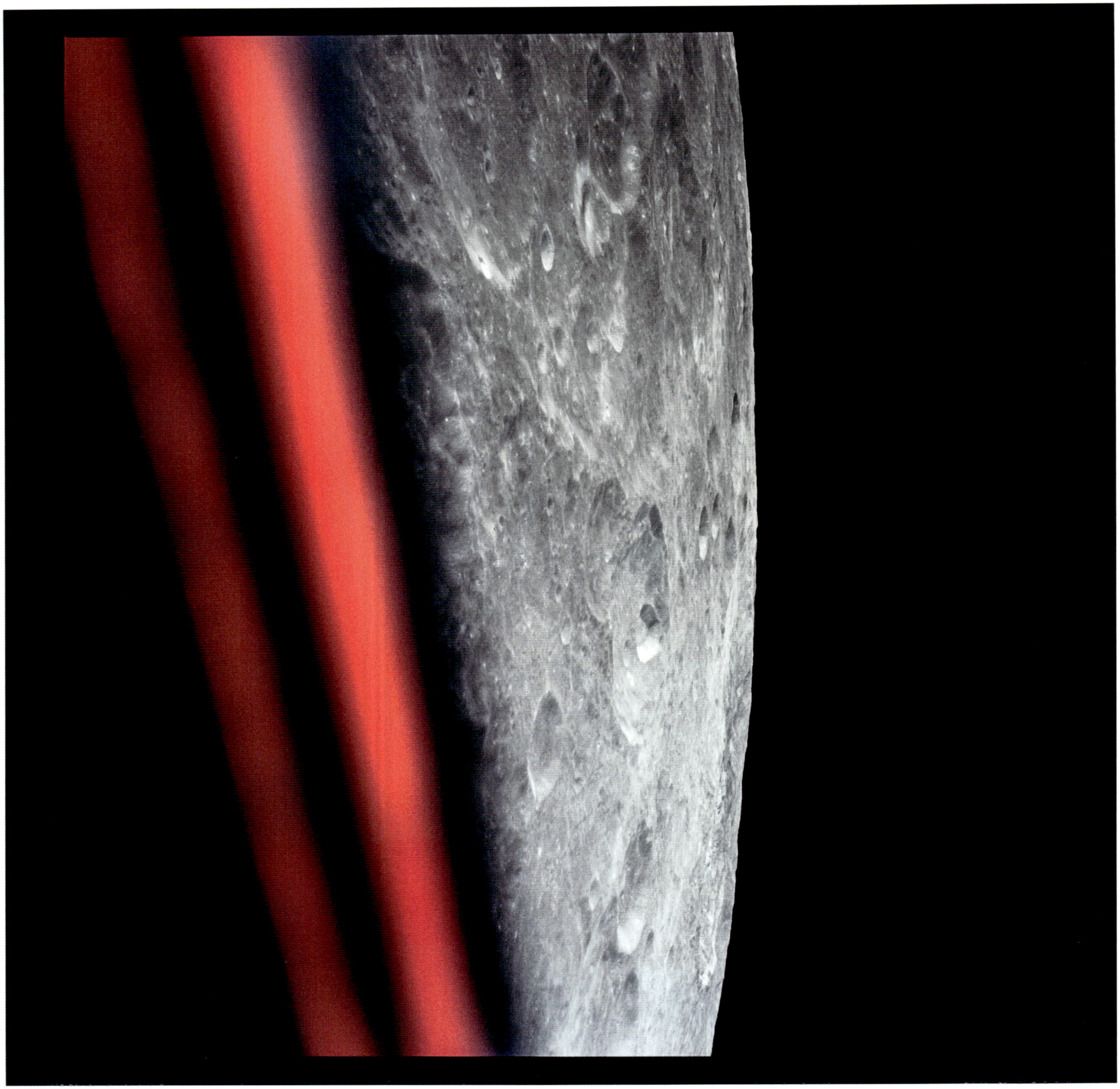

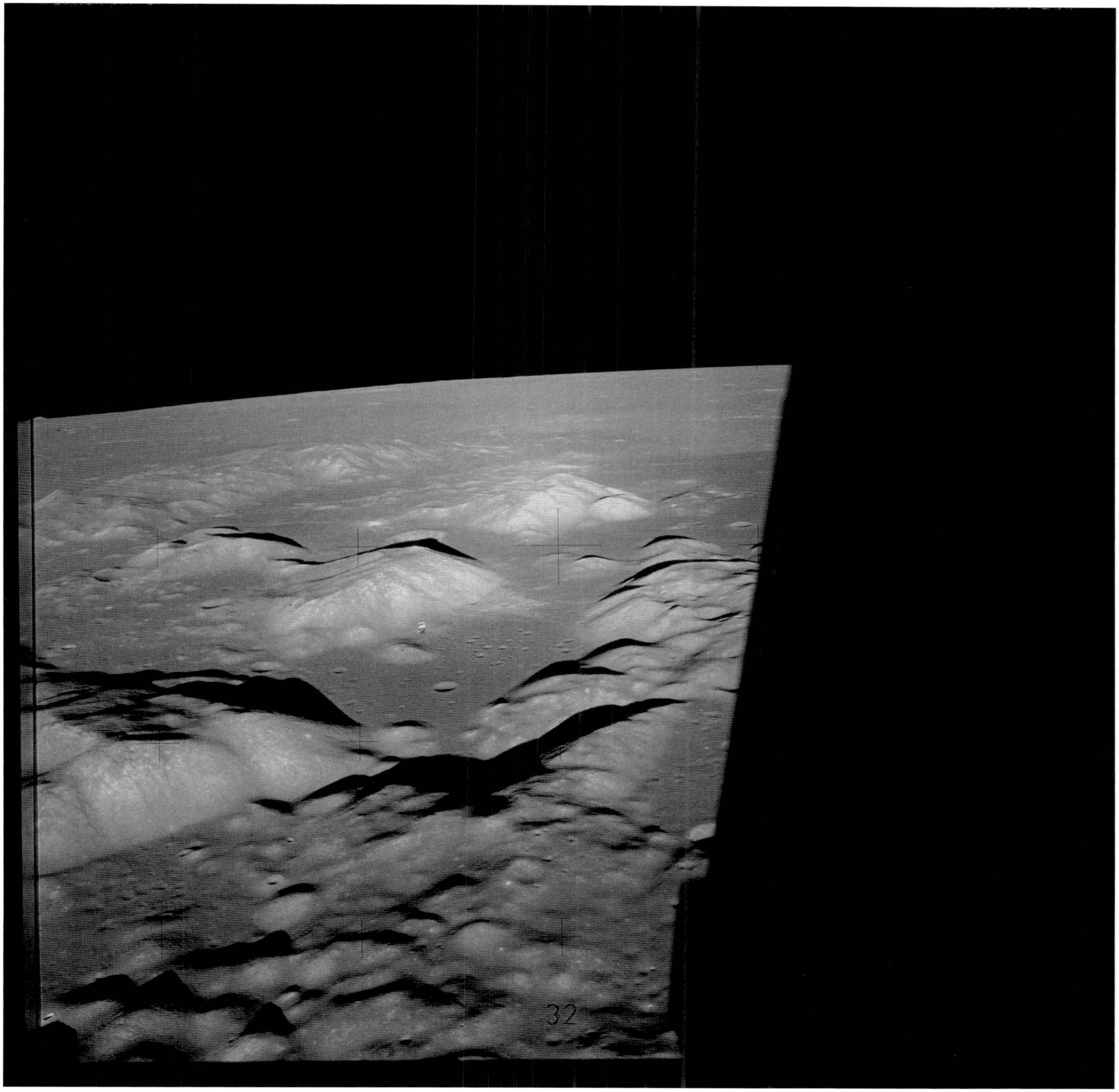
32

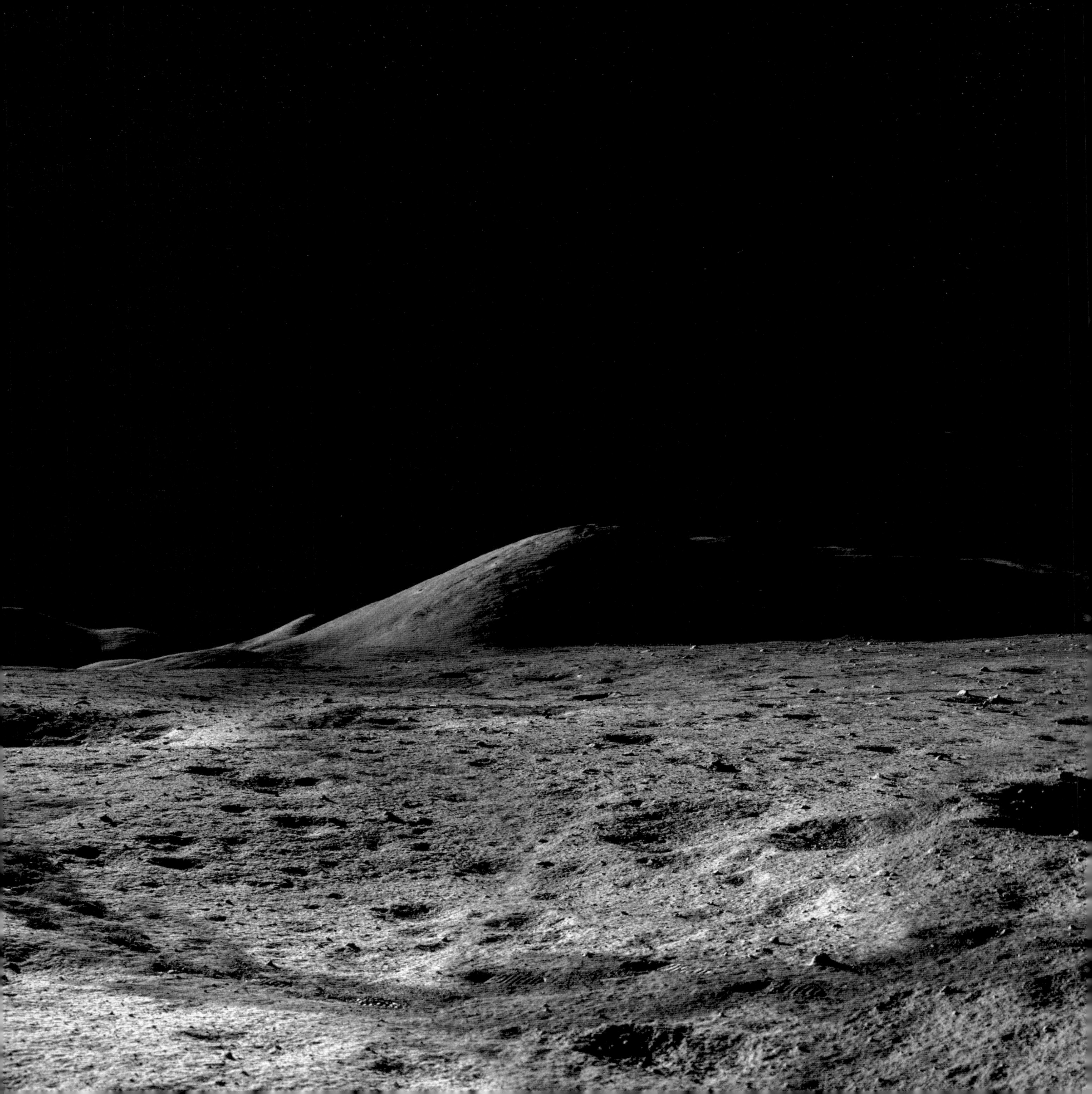

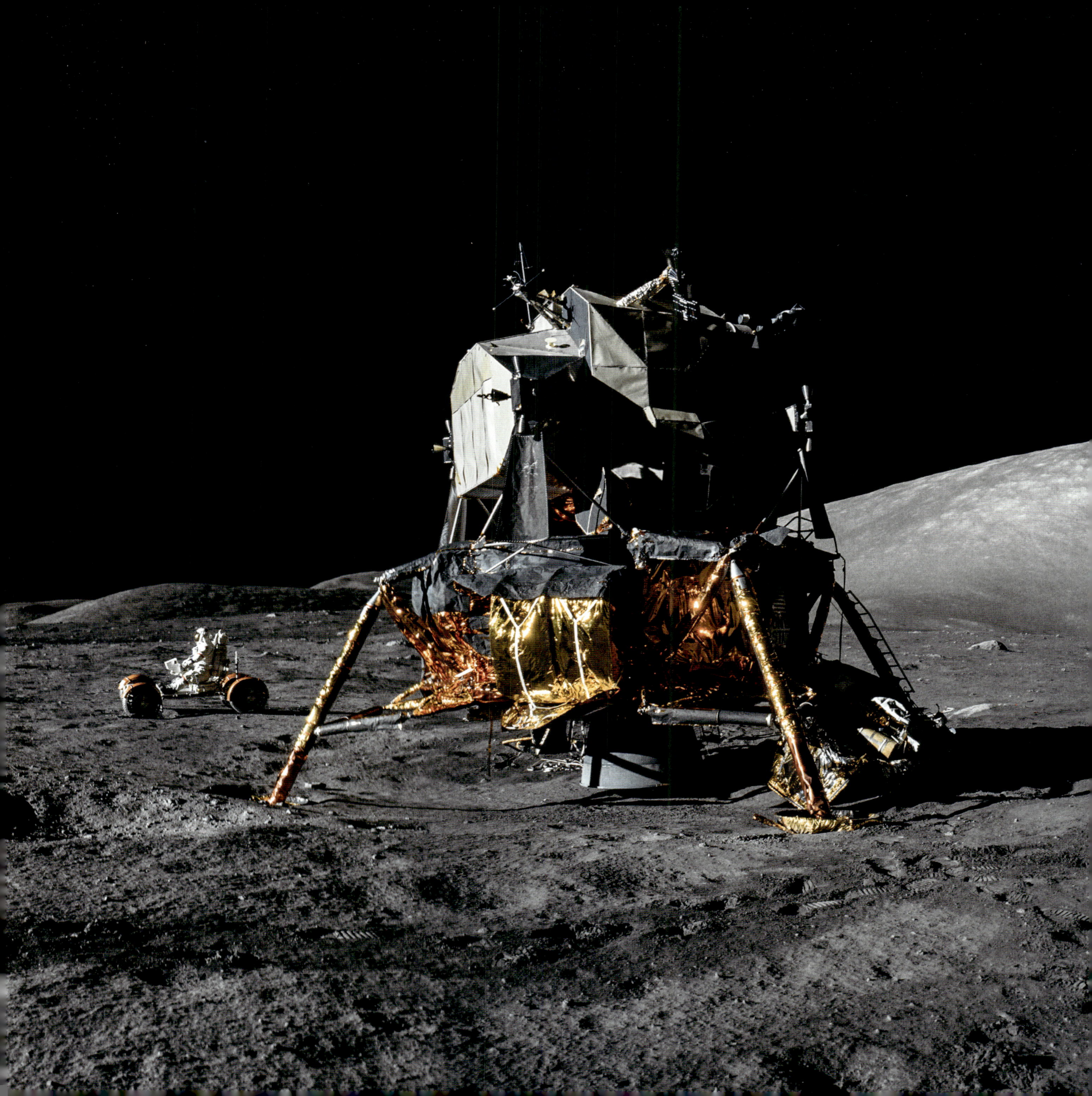

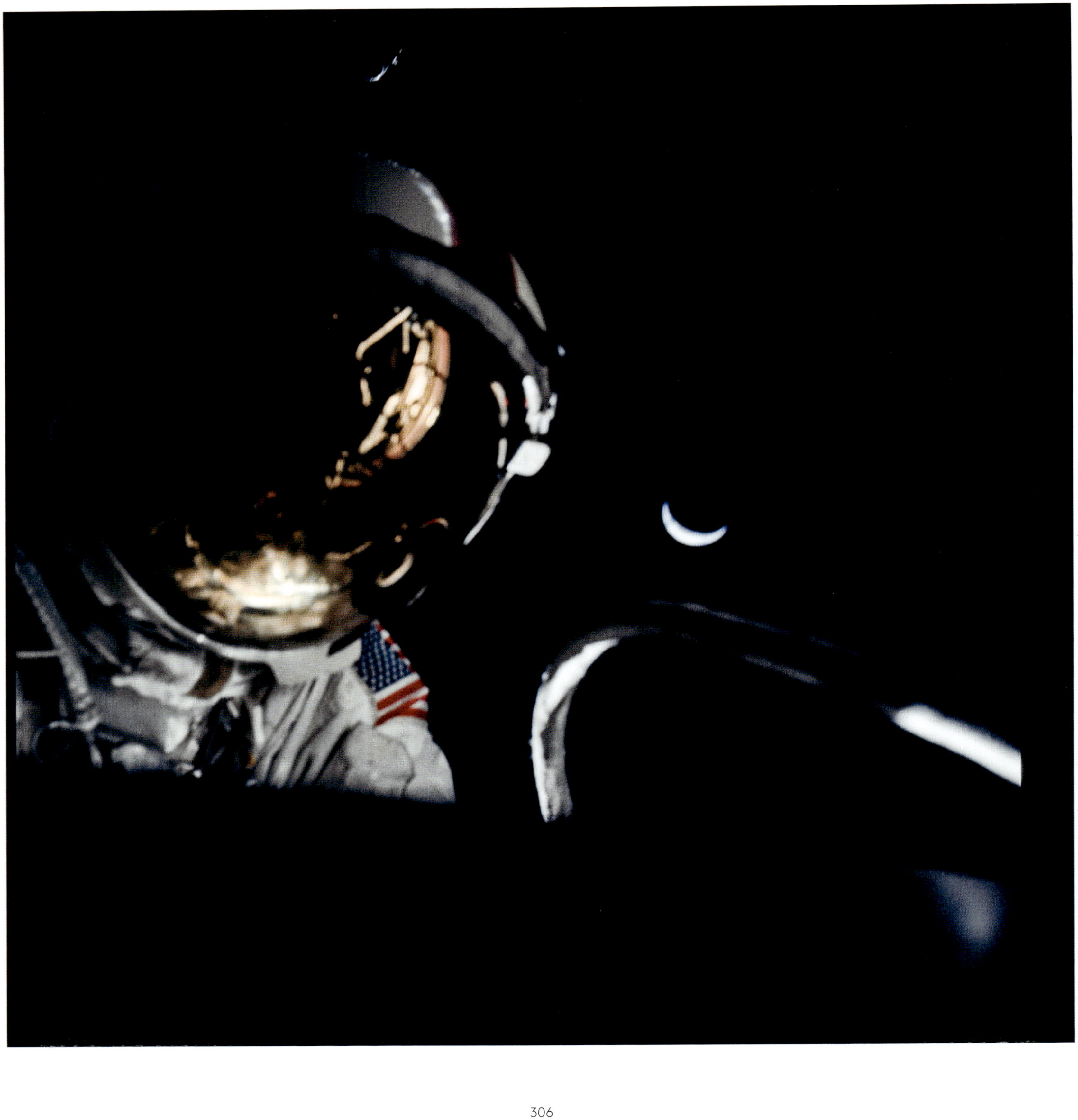

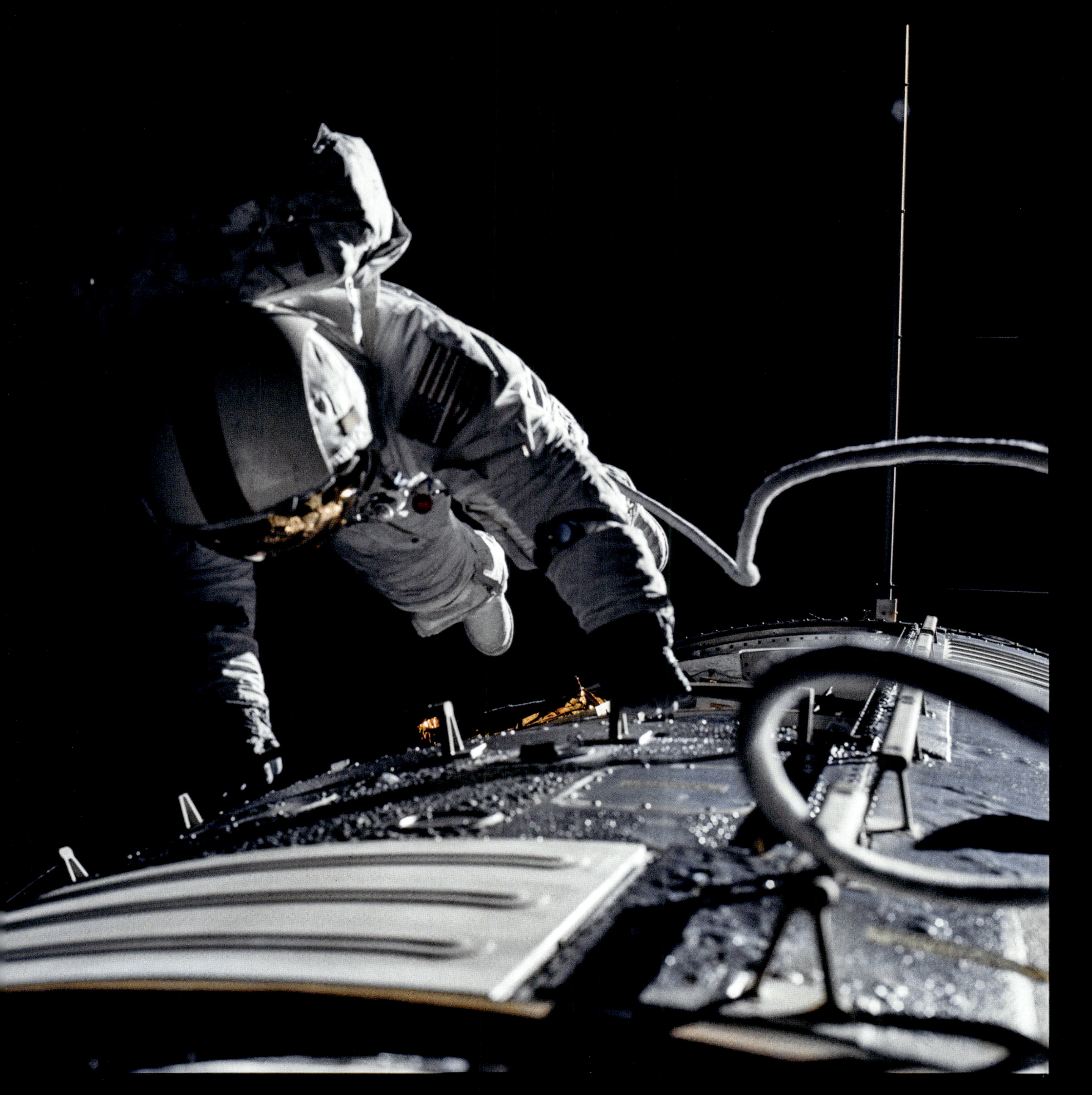

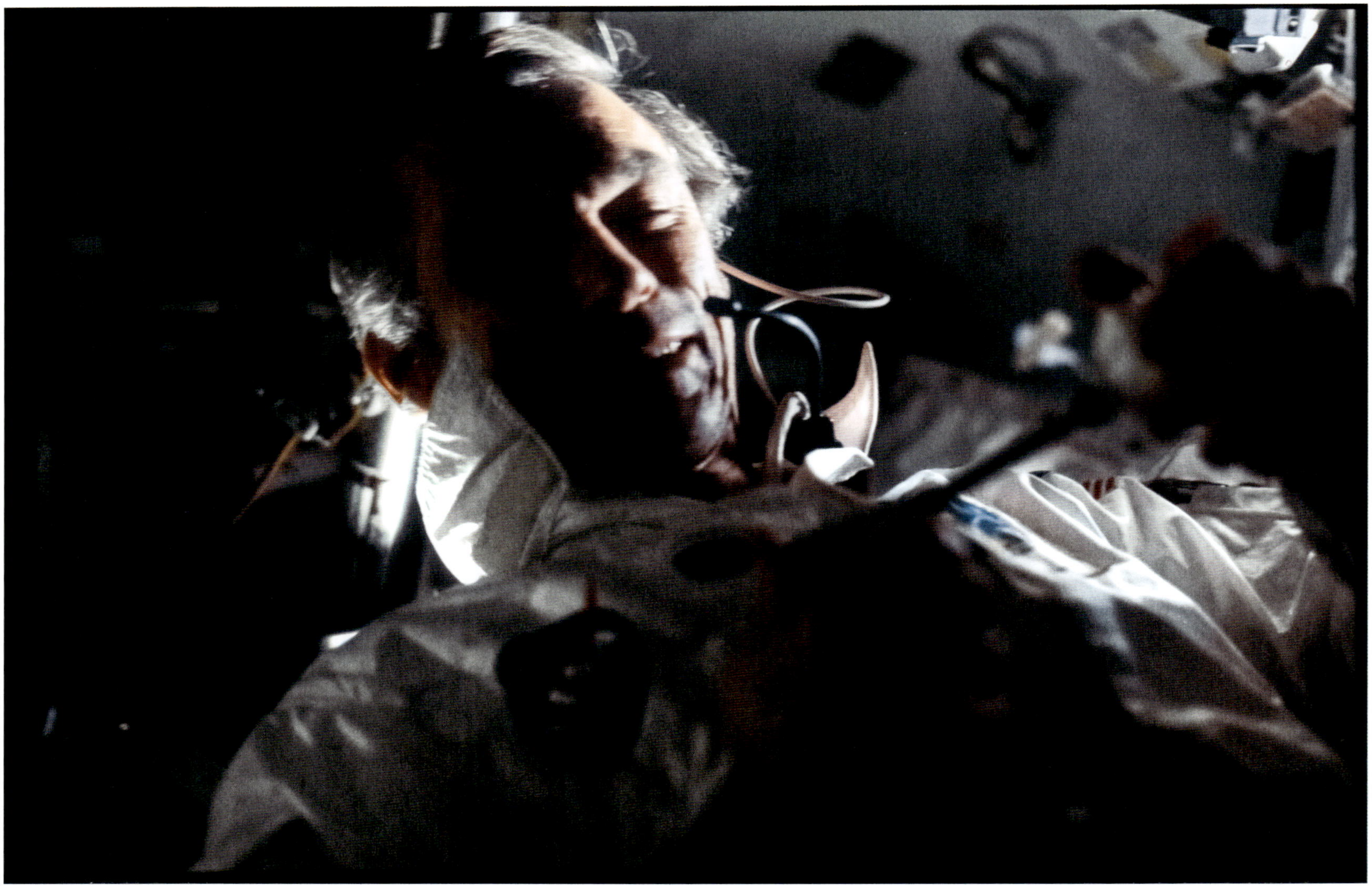

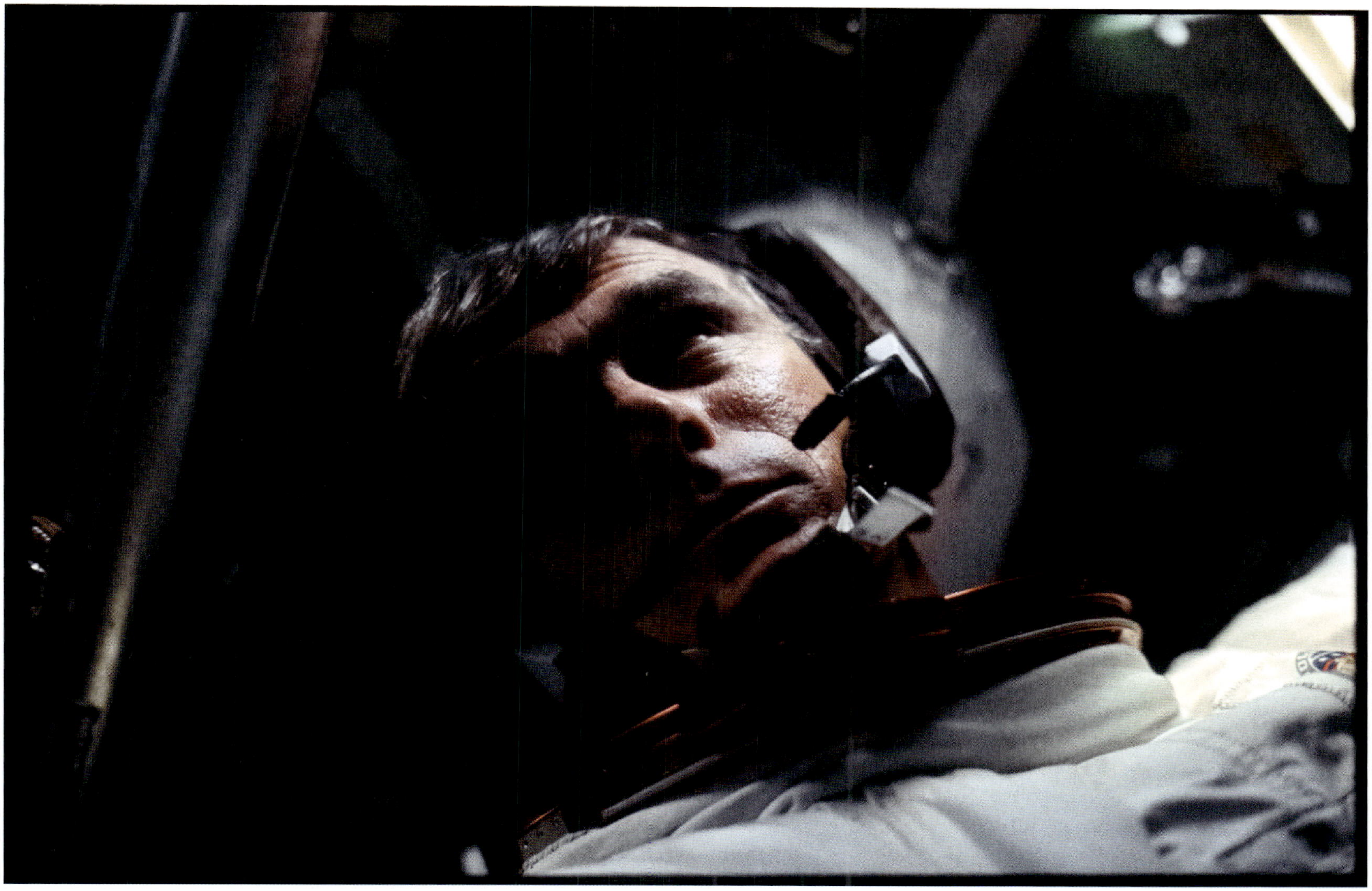

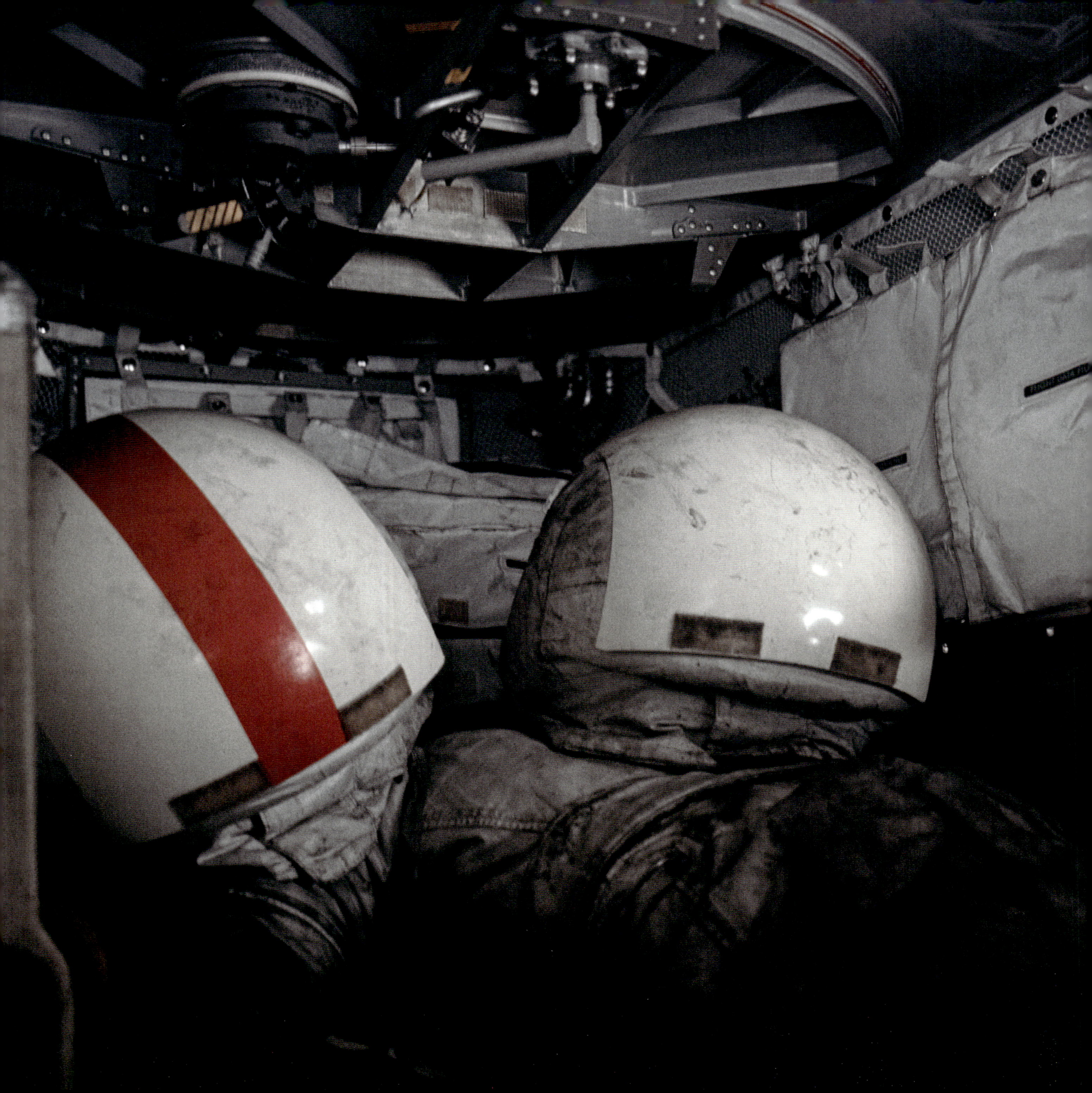

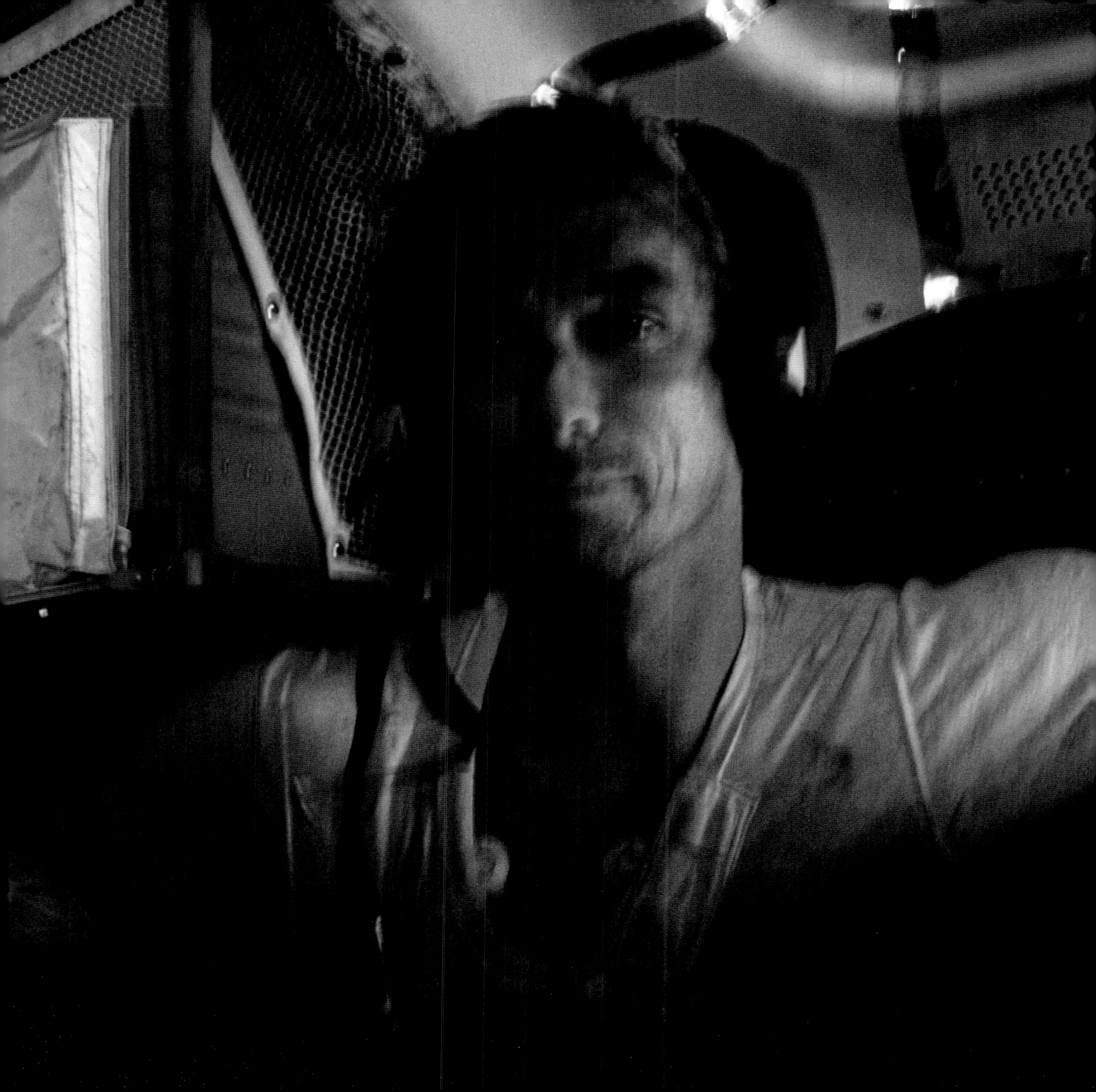

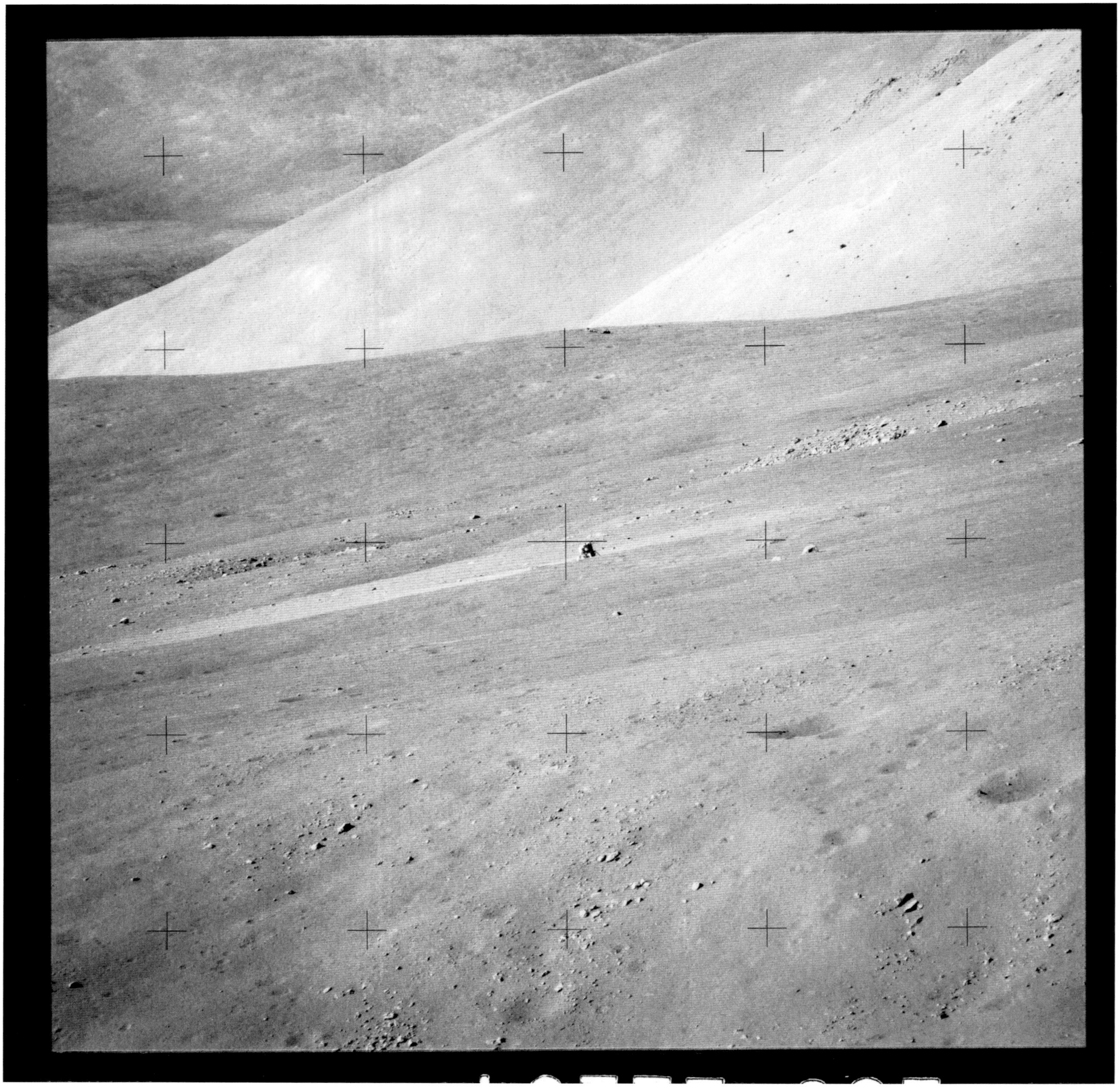

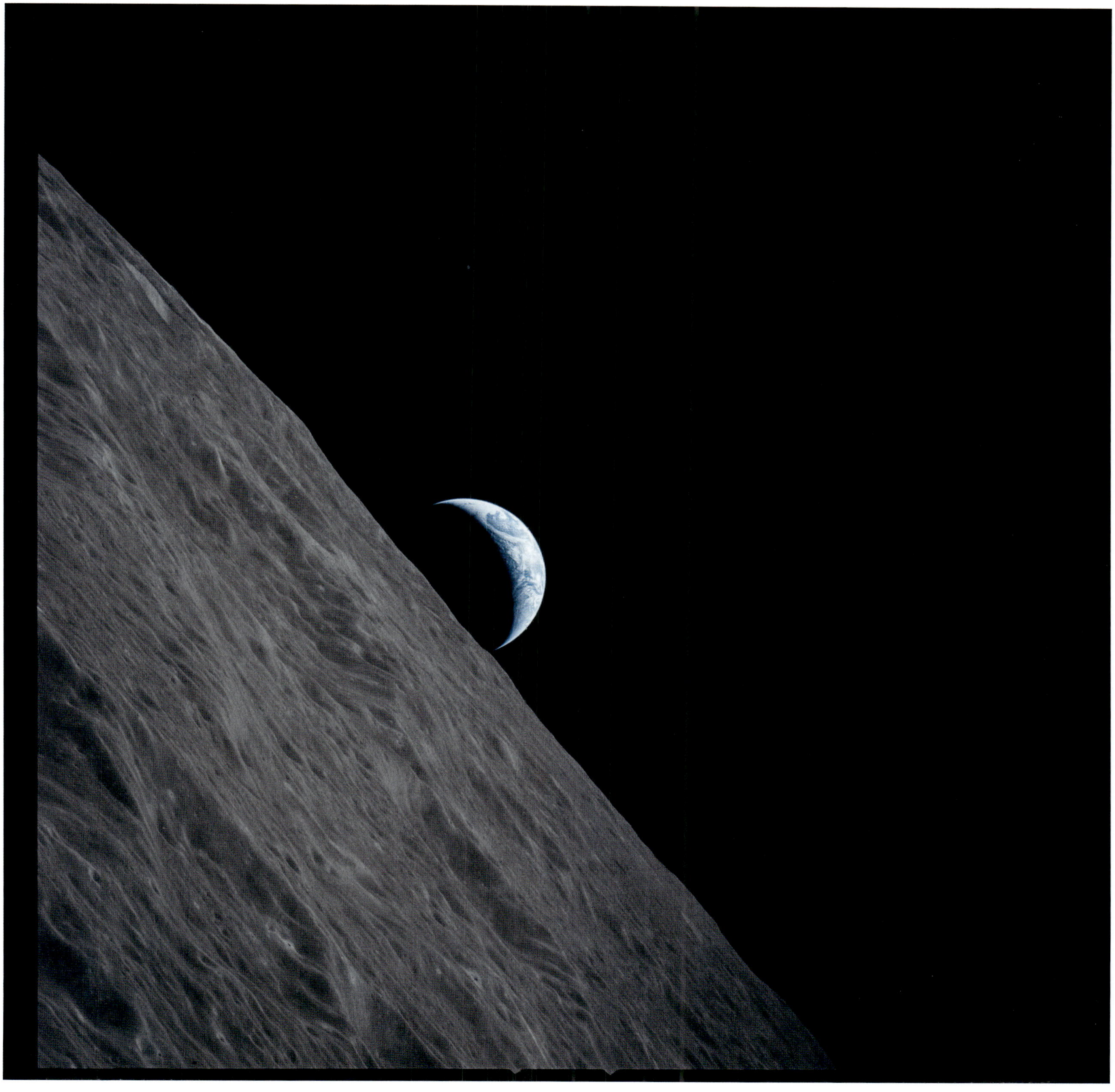

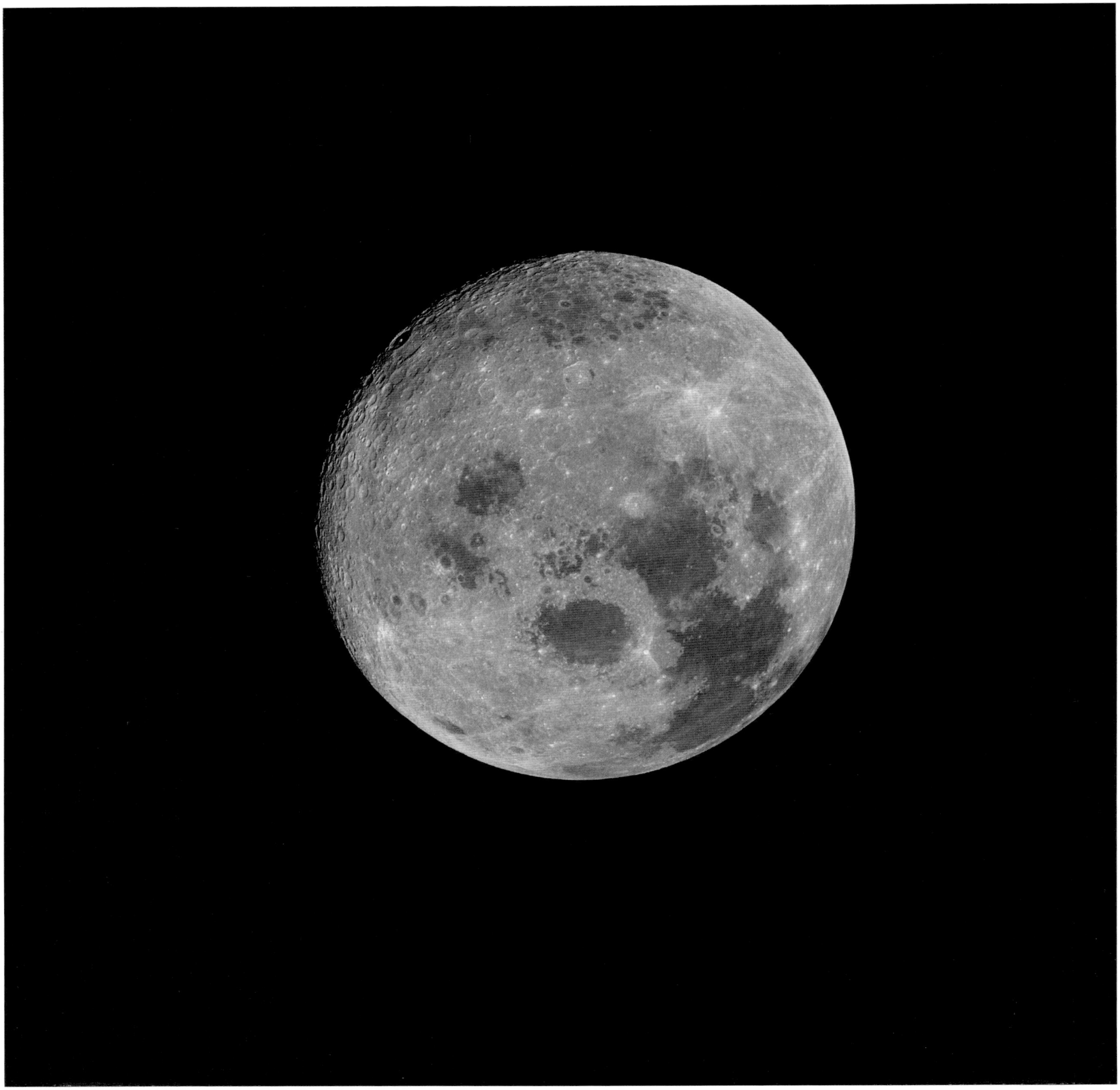

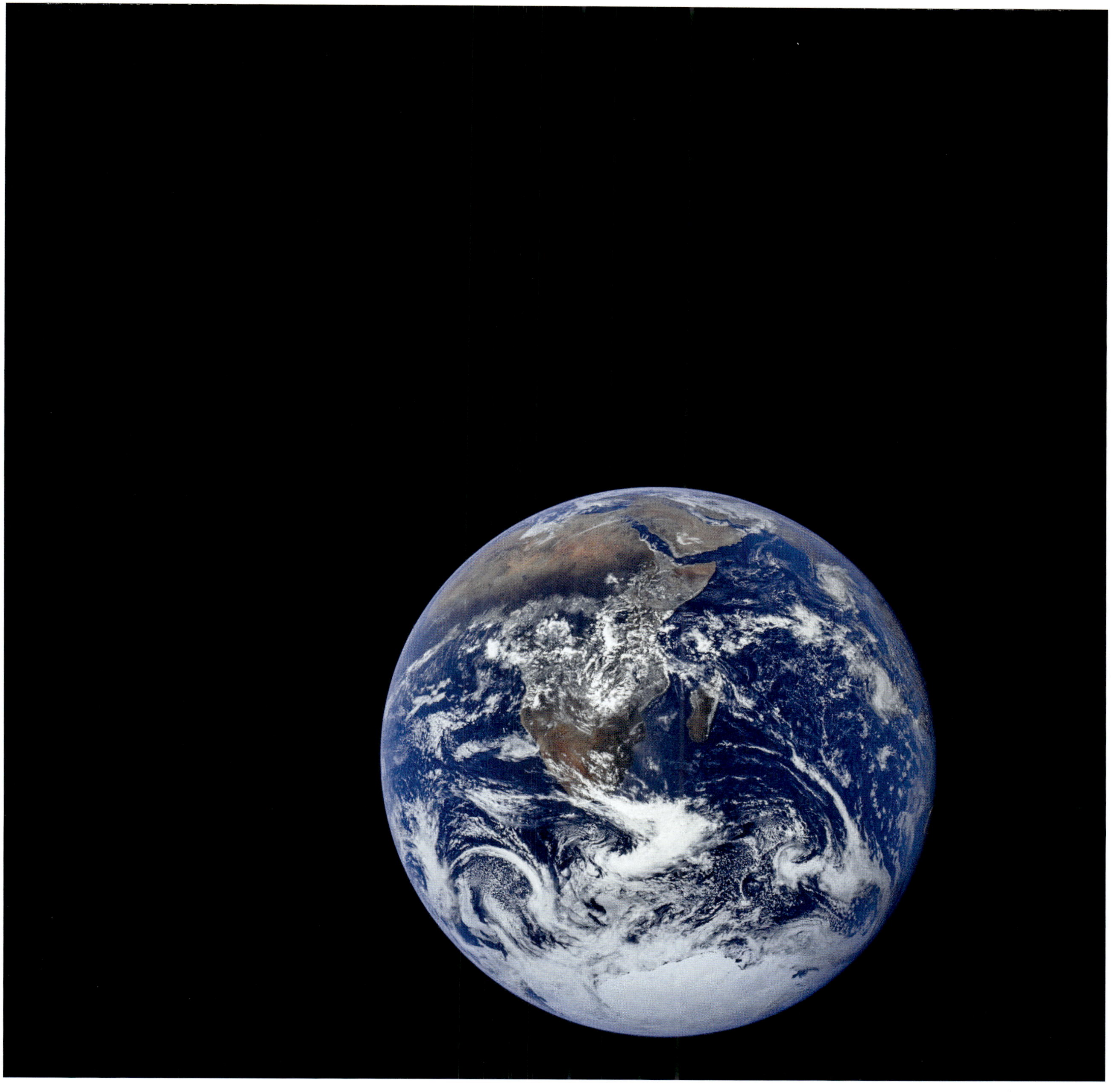

257 (AS17-148-22615)
A unique angled photograph of Earth shows it as a glowing crescent against the blackness of space.

258 (AS17-148-22719)
Africa's terrain with the Sahara Desert, Nile River, and contrasting lush regions.

259 (AS17-145-22257)
The Command/Service Module (CSM) *America* nearing rendezvous with the Lunar Module (LM) *Challenger* after lunar exploration.

260 (AS17-150-22942)
The rugged, impact-scarred terrain of Rumford and Sniadecki Craters.

261 (AS17-151-23210)
Aitken Crater, one of the largest impact structures in the solar system.

262 (AS17-137-20961)
A massive boulder at Station 2, with Earth shining as a small sphere above the rugged Taurus-Littrow horizon.

263 (AS17-134-20384)
Lunar Module Pilot Harrison H. Schmitt stands by the flag with Earth shining in the lunar sky.

264 (AS17-134-20380)
Commander Eugene A. Cernan salutes the American flag, marking Apollo's final lunar mission.

265 (AS17-140-21390)
Cernan, framed between the Lunar Roving Vehicle (LRV) and the 'Stars and Stripes,' a defining moment from the final Moon landing.

266-267
(AS17-145-22159 - AS17-145-22171)
Station 5 panorama, with the LRV centered and Schmitt collecting samples in Taurus-Littrow.

268 (AS17-134-20463)
Earth rises above *Challenger*, a reminder of home.

269 (AS17-136-20731)
Abstract shot capturing part of the LRV, leaving the Surface Electrical Properties (SEP) site.

271 (AS17-134-20436)
Schmitt on the South Arm of the SEP Transmitter Array, conducting experiments.

272 (AS17-134-20476)
Cernan works beside the LRV, with a vast mountain in the background.

273 (AS17-134-20453)
A lively photograph of Schmitt jumping into the LRV, illustrating the challenges of movement in lunar gravity.

274 (AS17-137-20979)
Cernan's makeshift fender replaces the original, which broke off when his hammer accidentally struck the LRV.

275 (AS17-134-20472)
Schmitt working at the seat of the LRV.

276-277
(AS17-147-22572 - AS17-147-22590)
A panoramic view of Apollo 17's Extravehicular Activity (EVA) site, with Cernan working near the LRV, with Wessex Cleft in the background.

278 (AS17-136-20758)
Cernan working by the LRV, with the rugged Taurus-Littrow Valley behind him.

279 (AS17-146-22296)
A photo of Schmitt, highlighting the challenges of working on the Moon's sloped terrain.

280 (AS17-147-22526)
Cernan in action, driving the LRV during their first EVA.

281 (AS17-146-22367)
A beautifully framed shot of the LRV in its final parking spot, with brilliant light, shadow, and bootprints visible.

282-283
(AS17-140-21488 - AS17-140-21507)
Schmitt stands by the Split Boulder, with the LRV parked on the valley slopes.

284 (AS17-141-21599)
Cernan places a gravimeter on the lunar surface, with the vast Taurus-Littrow Valley in the background.

285 (AS17-134-20435)
Schmitt stands at the end of the SEP array, showing distances between the astronauts, rover, and lander.

286 (AS17-137-20977)
The LRV sits against the towering South Massif summit in Taurus-Littrow.

287 (AS17-146-22388)
Schmitt stands with the LRV close to Shorty Crater, where he discovered the famous orange soil.

288 (AS17-149-22842)
Challenger ascends above Mare Smythii, leaving the Moon for the final time.

289 (AS17-149-22857)
Challenger approaches *America* for the final rendezvous at the Moon.

290 (AS17-157-23854)
The 115mi (185km)-wide Tsiolkovskiy Crater features a central peak rising from its dark basalt floor.

291 (AS17-157-23851)
This photo of the west rim of Tsiolkovskiy Crater offers a rare view of the Moon's far side.

292 (AS17-157-23826)
Clark, Pizzetti, and van der Waals Craters, captured by the crew during lunar orbit.

293 (AS17-163-24156)
Cernan reviews a manual inside *America*.

294 (AS17-163-24159)
A close-up of the control panels and instruments inside the CM.

295 (AS17-162-24056)
The Heat Flow Experiment (HFE) in the CM studied thermal transfer in space.

296 (AS17-162-24051)
Cernan inside the CM, reflecting on the mission. One of eight intimate photos on this spread, captured with the compact Nikon F 35mm camera, introduced for its ease of use in the cabin. Its smaller size allowed for a more candid, personal style of photography, offering a rare glimpse into the astronaut experience.

296 (AS17-163-24165)
Evans consuming a food packet inside the CM during Trans-Earth Coast (TEC).

296 (AS17-163-24170)
Schmitt shaving inside the CM, a rare look at daily life in space.

296 (AS17-162-24038)
Schmitt wearing sunglasses inside the CM, shielding his eyes from bright sunlight.

297 (AS17-163-24141)
Schmitt and Evans pose for a 'selfie'.

297 (AS17-162-24045)
A candid shot of Schmitt inside the CM, taken during Trans-Lunar Coast (TLC).

297 (AS17-162-24057)
The HFE panel mounted in the CM, with Evans observing in the background.

297 (AS17-163-24148)
This time, Cernan poses with Schmitt for a photo inside the CM.

298-299
(AS17-163-24133 - AS17-163-24134)
A unique stitched panorama showing Cernan and Schmitt, with Schmitt consuming a food packet in the CM.

300 (AS17-152-23288)
Mare Australe, Humboldt, and Milne Craters, showcasing the Moon's rugged, lava-flooded terrain.

301 (AS17-152-23292)
Tsiolkovskiy Crater's dark floor contrasts with the cratered terrain of Hilbert and Milne Craters.

302 (AS17-151-23241)
Abul Wafa and Buisson craters on the Moon's rugged far side.

303 (AS17-147-22465)
America can just be seen crossing in front of the Taurus-Littrow Valley.

304-305
(AS17-147-22493 - AS17-147-22521)
A panoramic view of the Apollo 17 landing site, with the LRV in operation.

306 (AS17-152-23401)
Evans retrieves film cassettes during a deep-space EVA on the return to Earth.

307 (AS17-152-23374)
Evans spent over an hour outside *America*, retrieving film cassettes from the Scientific Instrument Module (SIM) bay.

308 (AS17-163-24111)
A close-up of Cernan, illuminated by the glow of instrument panels.

309 (AS17-162-24035)
A portrait of Cernan, wearing part of his A7LB spacesuit.

310-311
(AS17-134-20522 - AS17-134-20523)
A poignant moment — the last moonwalker, Cernan, sits covered in lunar dust, next to his helmet and suit.

312 (AS17-139-21204)
A lonely view of *Challenger*, shining in Taurus-Littrow, captured from 2mi (3.2km) away with the 500mm lens.

313 (AS17-152-23275)
A crescent Earth rises above the lunar horizon, overlooking Ritz Crater.

314 (AS17-152-23342)
A stunning full-disc view of the Moon, captured from space.

315 (AS17-148-22727)
The iconic *Blue Marble* Earth photo, captured after Apollo 17 separated from the S-IVB, offering a final glimpse of home on the journey to the Moon.

"With Mercury, space photography was born. With Gemini, it struggled toward maturity so that Apollo space photography would give you and me, indeed the whole world, an opportunity to reach out and practically touch the Moon."

Richard W. Underwood
NASA Chief of Photography for Mercury, Gemini and Apollo

Above:
Underwood gives his evaluation of the progression of photography during space exploration in the Mercury, Gemini, and Apollo programs.

Richard W. Underwood Interviewed by Summer Chick Bergen
Houston, Texas – October 17th, 2000

Behind the photographs

The Apollo astronauts are hailed as being some of the greatest scientists, adventurers, explorers and heroes, but rarely are they portrayed as some of the most influential photographers of all time.

Their photographs are undoubtedly some of the most iconic and well known ever taken. Images of man standing on the Moon have traveled to almost every part of the globe. It is even quite well known that Hasselblad cameras and Kodak film were used to document the missions for the world to see.

However, little thought has been given as to how astronauts managed to capture these images while operating in low gravity, wearing pressurized suits, bulky gloves, and helmets. Bringing these photographs back to Earth required the coordinated efforts of hundreds of engineers, scientists, and technicians.

NASA organized a complex operation in order to achieve their goal of capturing scientific and historical imagery. It required huge resources, extensive research and engineering advancements in several fields. Major advancements in photographic technology can be attributed to the Apollo program. Without the collaboration between NASA and companies like Hasselblad, Kodak, Carl Zeiss and Nikon, the astronauts could not have brought back the images that will be preserved for future generations to see.

The next few pages are devoted to the role of photography during the Apollo program, detailing the requirements and modifications that were necessary to ensure that the equipment functioned in the harsh environment of dust, vacuum and low gravity, as well as the extensive training given to the astronauts, mostly test pilots, to transform them into competent and well-practised photographers.

Opposite:
John Young is bagging a sample from a small crater behind the Rover at the ALSEP site. (AS16-114-18387)

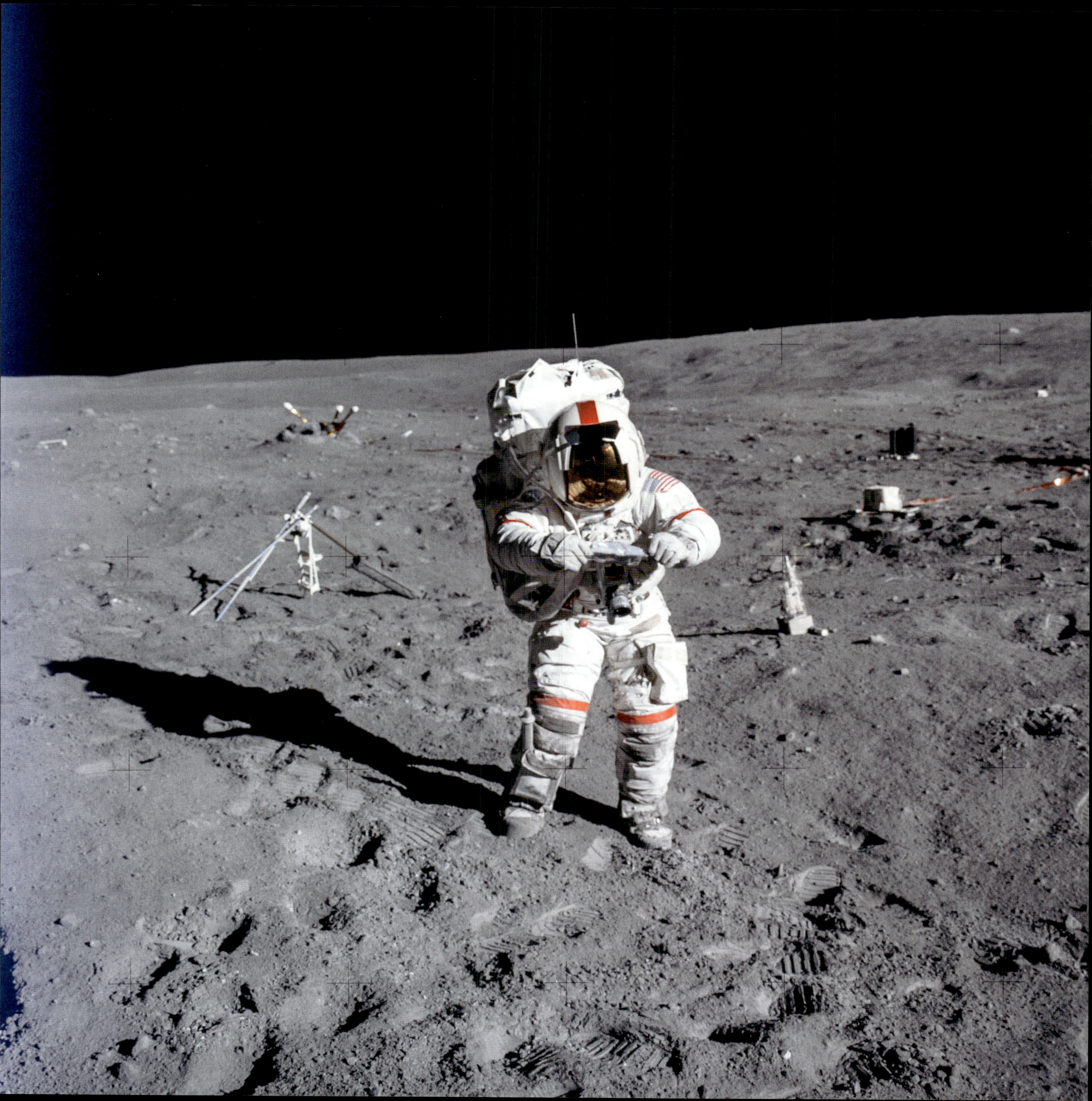

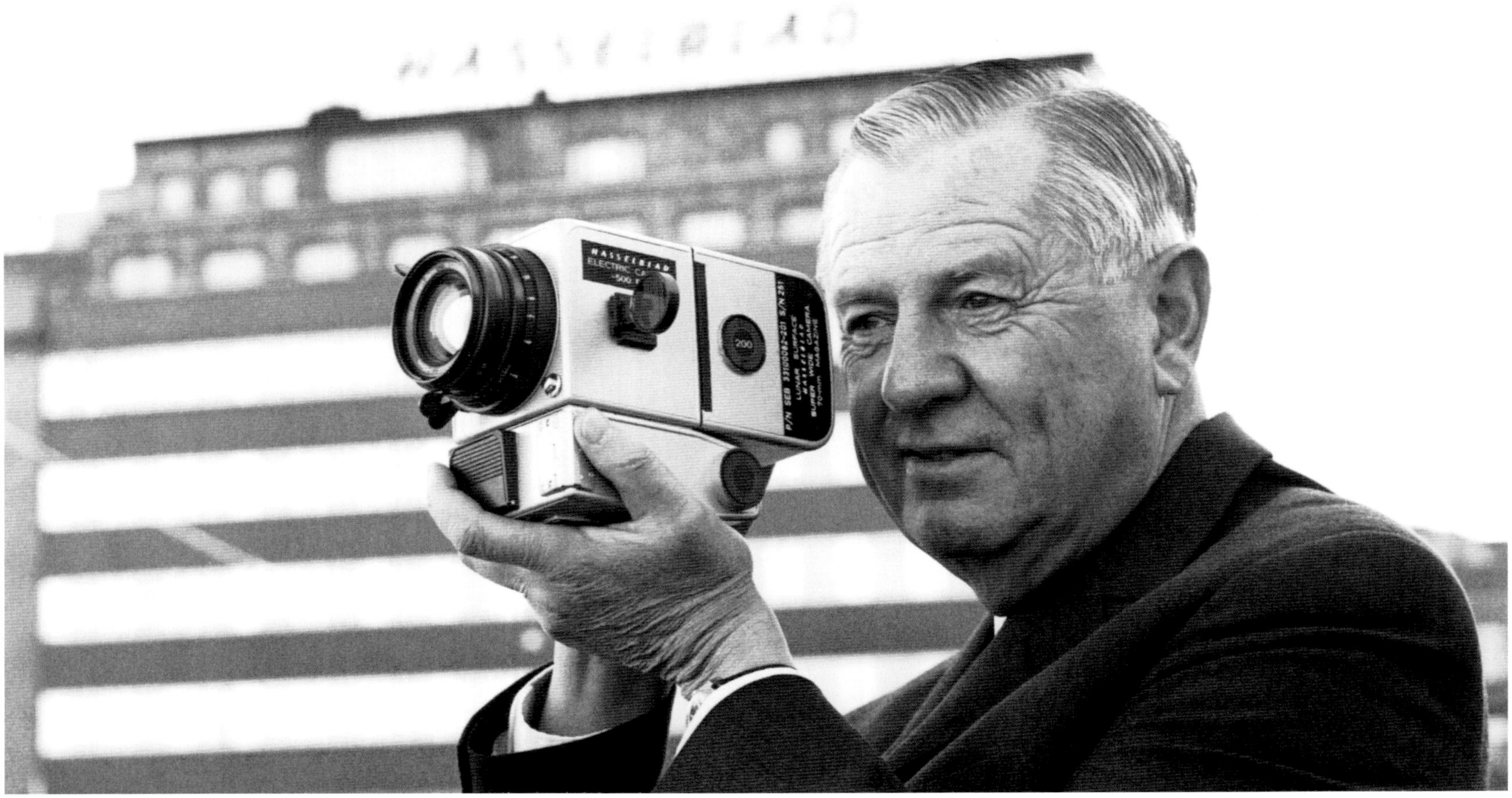

Victor Hasselblad posing with a NASA modified 500 EL (Hasselblad Electronic Data Camera).

The role of photography during the Apollo program

In the early days of NASA, the organization lacked a solid space photography program. The astronauts, piloting one-man spacecraft, were fully focused on the actual flying which left little or no time for photographs to be taken. It was not until Walter Schirra's mission, piloting Sigma 7 on the Mercury-Atlas 8, that NASA's photography program started to take form. Schirra brought up the idea with NASA engineers about the possibility of using his own Hasselblad camera during his Mercury mission. As a keen photographer he knew he could take beautiful pictures from space. Eventually, after some long discussions, he was allowed to fly with a heavily modified Hasselblad 500C.

Schirra's photos were well received in the end and, as the ventures into space progressed, the importance of photography became more and more obvious and, at the same time, the role of space photography also changed.

At the beginning of the Apollo flights, the initial photographic objectives included scientific research in fields such as oceanography, meteorology, and geology. As the missions progressed, the priority for the astronauts became photography of the lunar surface from orbit. From Apollo 8 until Apollo 10, their 'target of opportunity' lists focused on exploring suitable landing sites and areas of significant geological interest.

Other roles of photography involved documenting the spacecraft's condition and durability during docking and undocking maneuvers, providing detailed feedback for the engineers back on Earth. The missions that finally landed on the Moon were, of course, mainly focused on documenting topographic and geological data.

These various roles of photography required a number of different cameras, lenses and film types for each mission. Each camera had a set purpose and list of 'targets' that were defined during the mission planning. In addition the astronauts were instructed to photograph anything that 'appeared interesting' and to communicate with Mission Control whenever an image was taken. These messages documented the time, location and key information of the shot and were later passed on to the film engineers who processed the rolls.

Hasselblad in space

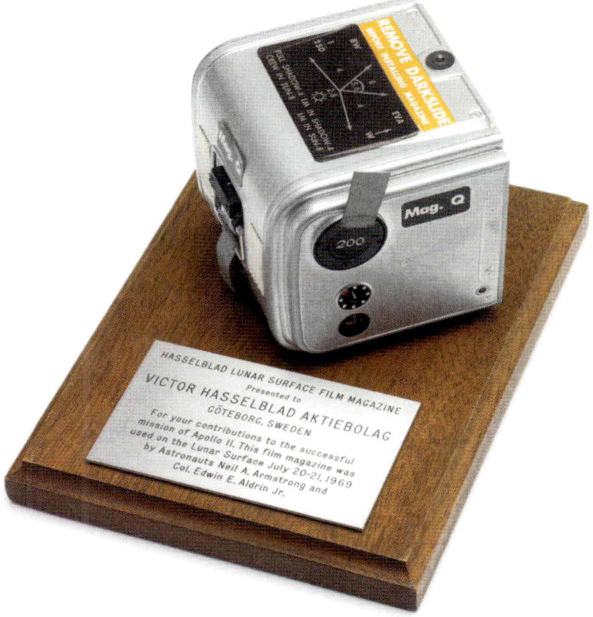

A lunar surface film magazine that was flown on the Apollo 11 mission, presented to Victor Hasselblad AB.

In the search for a camera that was fit for use both on board the spacecraft and on the lunar surface, NASA drew up various requirements such as interchangeable film magazines, the ability to hold black-and-white and color films of different sensitivities and the use of different lenses.

Astronaut and amateur photographer Walter Schirra first brought to the attention of the engineers responsible for the flight cameras the fact that his personal camera, a Hasselblad 500C, seemed to meet NASA's requirements. Beyond the obvious quality of its photographs, further inspection of the camera by the engineers confirmed that its build, reliability, configuration, and ease of modification complied with all demands.

In 1962, Schirra purchased another Hasselblad 500C from a Houston camera store for NASA engineers to adapt for use in space. He took it with him on the Mercury-Atlas 8 mission just a few months later. The stunning images that he took of the Earth from orbit marked a turning point for spaceflight photography and led to a strong and cooperative relationship between NASA and Hasselblad.

This collaboration benefited both parties. NASA's expectations of beautiful, high-quality imagery were easily fulfilled and the worldwide publication of the images became a real breakthrough for Victor Hasselblad and his company. Hasselblad gained brand recognition within the photography circles at the time as a company whose equipment could produce photographs of the finest quality in the harshest of conditions.

When NASA approached Victor Hasselblad with the possibility that his cameras could go to the Moon, it became his personal challenge to modify them for use in space. It formed a relationship between Hasselblad and NASA which lasted for decades.

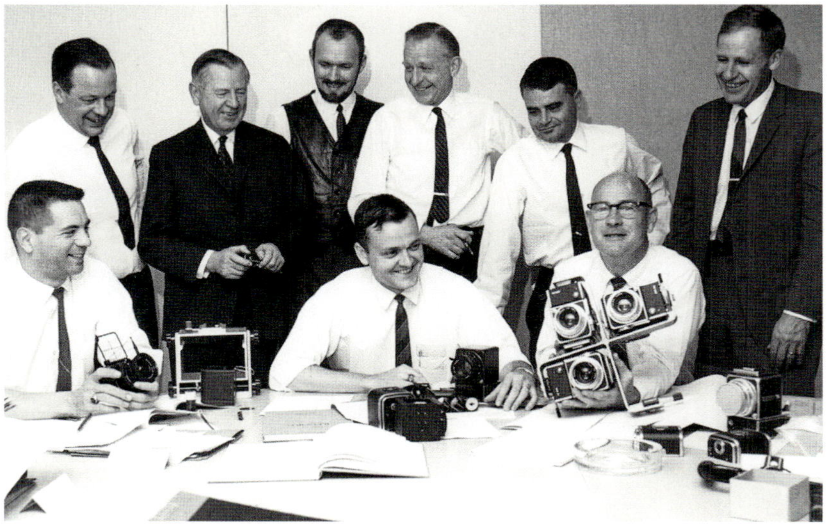

The 'NASA-Group', a mix of NASA and Hasselblad employees together with Victor Hasselblad pose together with the range of different Hasselblad modified especially for the Apollo program.

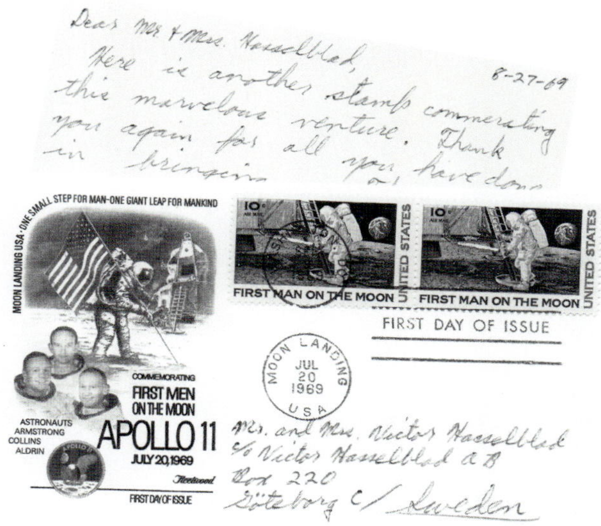

A 'first day cover' card to Victor and Erna Hasselblad sent by Richard W. Underwood of the NASA Photographic Technology Laboratory, thanking them for the help and service to the Apollo program. It read: *"Dear Mr. and Mrs. Hasselblad, Here is another stamp commemorating this marvelous venture. Thank you again for all you have done in bringing us 'The Hasselblad'. Very truly yours, Dick".*

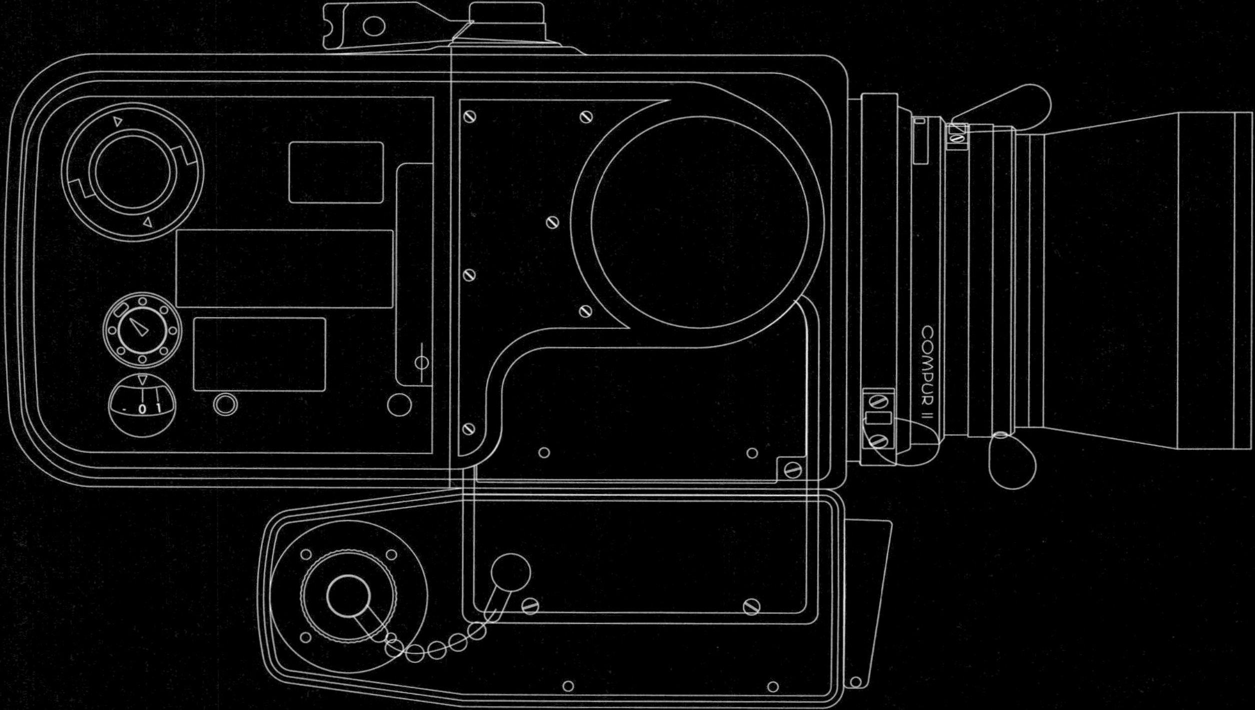

Technical drawing of the modified Hasselblad 500EL Camera

The cameras and equipment

Unmodified commercial cameras posed several problems for the Apollo astronauts. Their bulky space suit gloves made it impossible to adjust small and complicated buttons, and a traditional viewfinder was of little use to astronauts when wearing a space helmet.

Many modifications and specialist pieces were manufactured to meet these very unique operational requirements. The equipment list varied for each mission, but generally a number of different Hasselblad camera bodies, grips and mounts were carried on board along with a variety of lenses and different 70mm color and black-and-white films with varying sensitivities and speeds. The following pages show an overview of key photographic equipment used during the Apollo program.

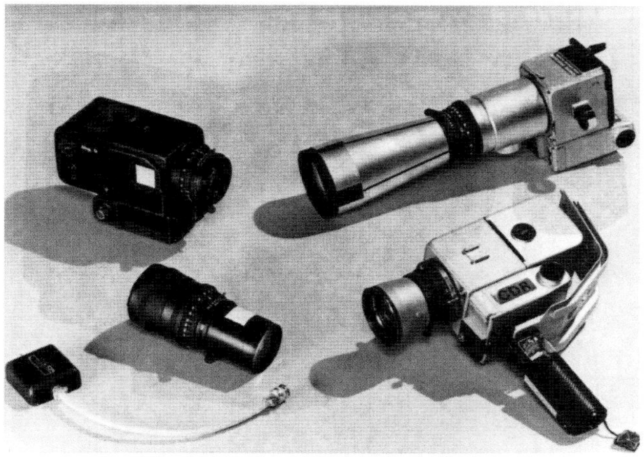

Overview of the camera bodies and lenses that were carried during the Apollo 14 mission.

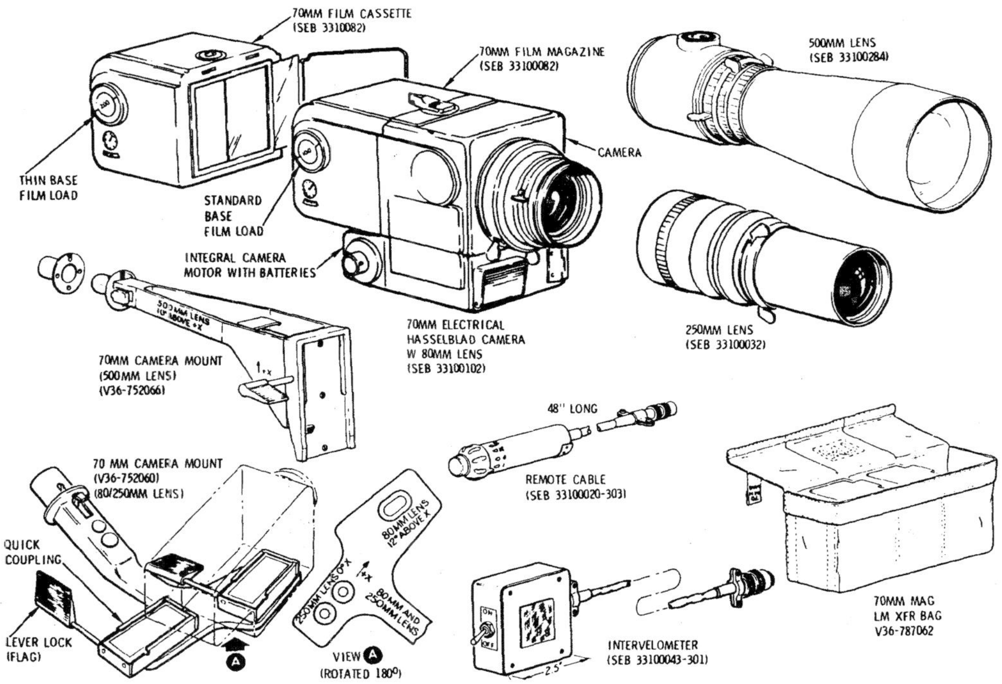

Illustration of the camera equipment taken on board the Apollo 16 mission:

- 70mm Hasselblad Electric Camera (HEC)
- 80mm f/2.8 Planar lens
- 250mm f/5.6 Sonnar telephoto lens
- 500mm f/8 telephoto lens
- remote control cable
- 70mm film magazines
- lunar surface 70mm film magazines
- magazine Lunar Module (LM) transfer bag
- 70mm camera mount for 80 & 250mm lenses
- intervalometer
- HEC timing cable

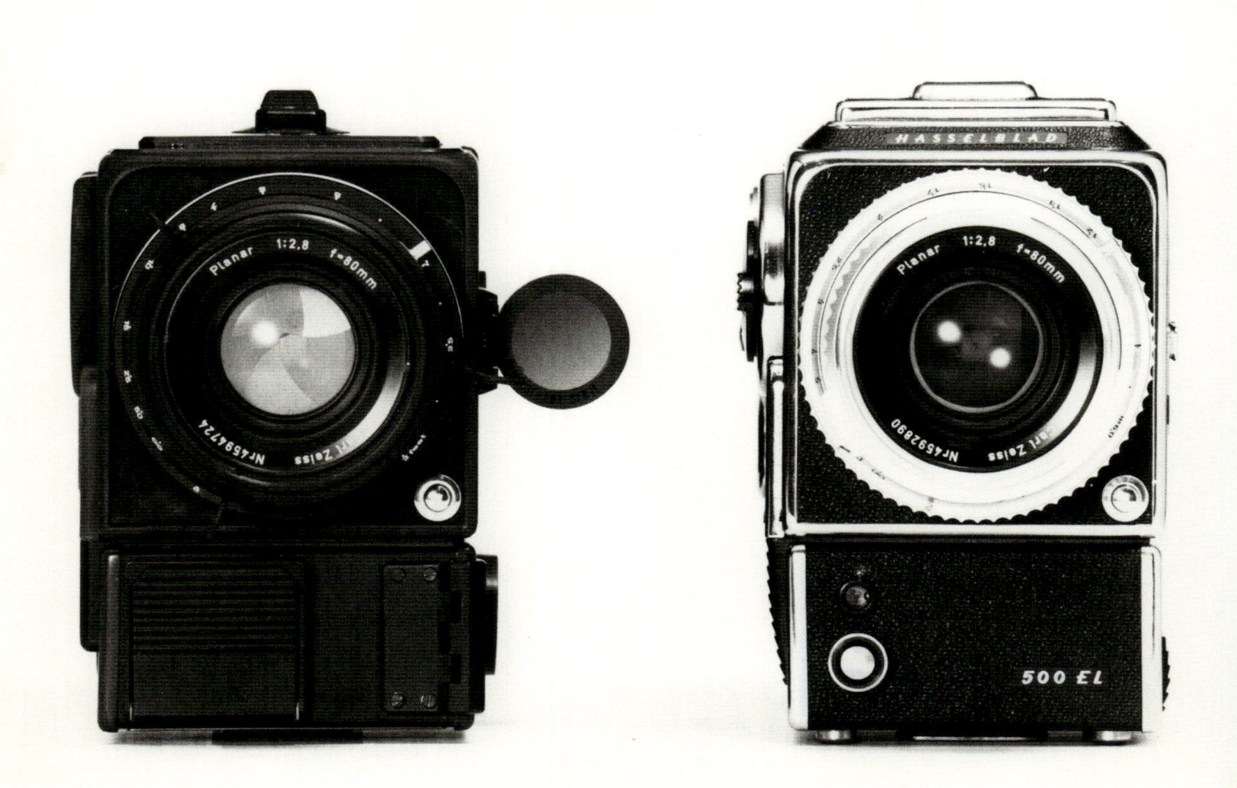

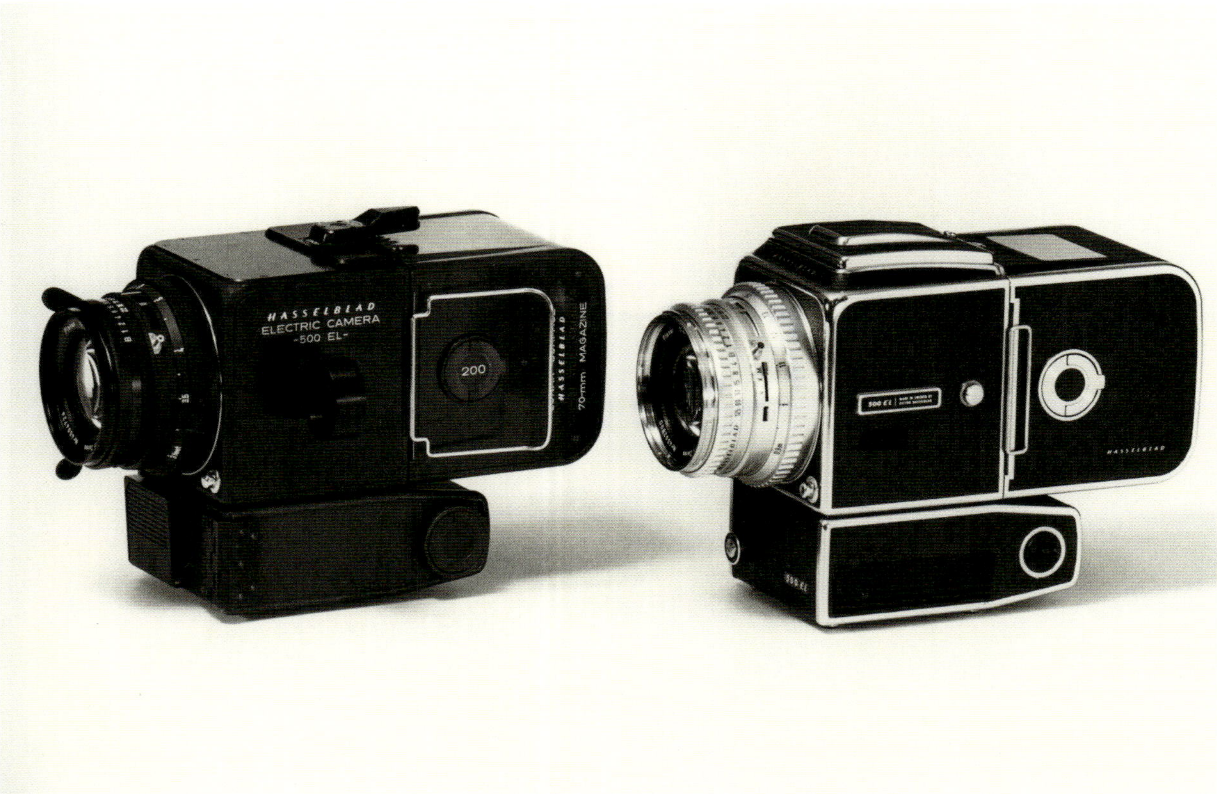

An original commercial Hasselblad 500EL (right) photographed next to the modified NASA Hasselblad Electrical Camera (HEC).

Some of the many modifications can be seen. The viewfinder on the top of the camera body has been removed and a basic sight placed on the side of the HEC. The leather wrap has been removed from the HEC, exposing the painted matte black metal.

The special cameras made for the Apollo program

Hasselblad 500C

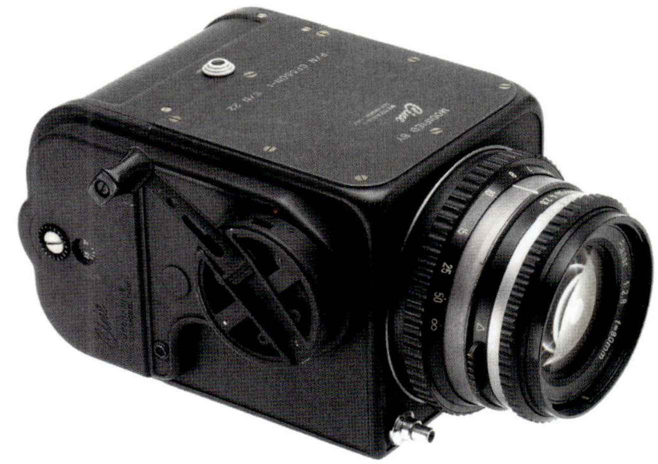

The 500C first flew with Walter Schirra on Sigma 7 during the Mercury-Atlas 8 mission in 1962. The photographs he captured were highly satisfactory, which led to its continued use on subsequent Mercury and Gemini missions. This marked the beginning of Hasselblad becoming NASA's preferred supplier of spaceflight cameras. During the Apollo missions, the 500C was used on Apollo 7 and Apollo 9, before being replaced by the Hasselblad 500EL, which featured automatic electric film winding.

NASA technicians adapted a commercially available 500C body by modifying numerous components to make it safe and functional for spaceflight. They removed the mirror and auxiliary shutter, along with the top-mounted viewfinder and protective glass — effectively turning the camera into a simplified 'point-and-shoot' system. A rudimentary sight was added to the side of the camera body to help astronauts frame their images.

The camera's buttons and levers were also simplified and enlarged for use in low gravity and while wearing pressurized gloves. Its leather covering was removed, exposing the bare metal body, which was then anodized matte black to prevent reflections from interfering with instruments or visibility.

The 500C was the model that first brought Hasselblad and NASA together — laying the foundation for a decade of collaborative innovation that would ultimately help carry photographic technology to the surface of the Moon.

The Hasselblad Electric Camera

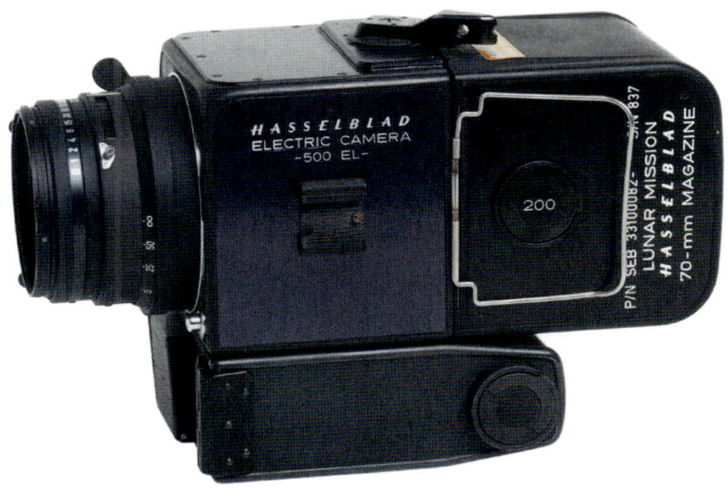

The NASA Hasselblad Electric Camera (HEC) originated from the standard 500EL model. More technically advanced than the earlier 500C, it featured an electric motor that automatically advanced the film and primed the shutter — reducing the number of manual steps astronauts had to perform when taking photographs.

NASA intended the HEC for use inside both the Lunar Module and the Command/Service Module. It underwent similar modifications to the 500C: the mirror and auxiliary shutter were removed, and the body was anodized in a matte black finish to eliminate potentially blinding solar reflections.

The standard lens for the HEC was an 80mm f/2.8 Carl Zeiss Planar. A 250mm f/4 Carl Zeiss Sonnar telephoto lens was also frequently used for photographing distant objects — including in the famous *Earthrise* image.

The Hasselblad Electric Data Camera

The Hasselblad Electric Data Camera (HEDC) was developed specifically for use on the surface of the Moon. In late 1968, nearly a year before the first lunar landing, NASA approached Hasselblad to further upgrade the existing HEC. The new camera would need to withstand the Moon's extreme conditions while accurately and reliably documenting the lunar surface.

The HEDC incorporated many of the same modifications as the HEC, with one major addition: the Réseau plate. A new shutter mechanism was also developed, using specialized lubricants — as standard lubricants would boil in the vacuum of space. Unlike other NASA Hasselblad models, the HEDC's exterior was painted silver, not matte black. This reflective finish helped disperse heat and prevent the camera from overheating under direct solar exposure. On the Moon, surface temperatures can swing from 248°F (120°C) in sunlight to -292°F (-180°C) in shadow — extreme enough to damage both hardware and film.

The lens used on the lunar surface was also purpose-built. The 60mm f/5.6 Biogon Carl Zeiss lens was designed to deliver a wide field of view with minimal distortion. Its development posed a significant challenge, as it had to align with the precise calibration of the Réseau plate's grid markings.

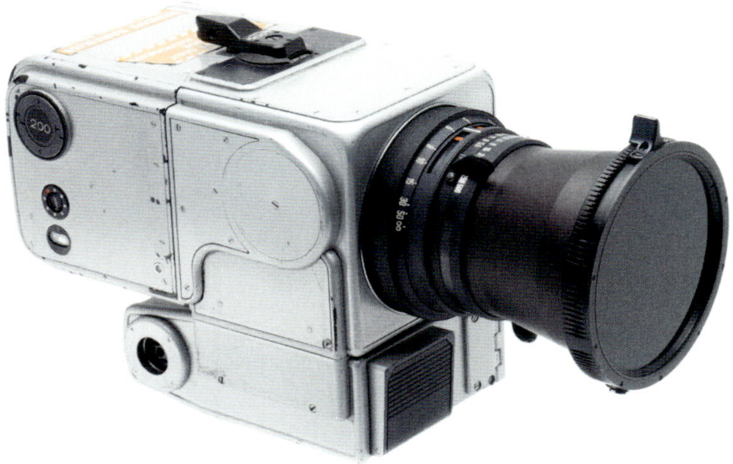

Réseau plate

The lunar surface cameras were fitted with a transparent glass Réseau plate, mounted inside the camera body near the film plane. Fiducial markers — a grid of tiny crosshairs — were precisely engraved onto the plate and are visible in many Apollo photographs. These markings were calibrated to within micrometers, with a margin of error as small as 0.002 mm. When combined with known camera geometry, they allowed scientists and engineers to accurately calculate the distance, scale, and elevation of objects captured in the image.

While cameras with Réseau plates were not uncommon — the technique had been widely used in aerial photography for decades — those systems were typically large, stationary, and not suited to handheld operation. The challenge for NASA was to integrate this level of precision into a compact, astronaut-friendly system. Working closely with Carl Zeiss and Hasselblad, NASA engineers managed to adapt the technology into a much smaller format, compensating for potential lens distortion common in wide-angle optics.

This compact implementation of the Réseau plate became a remarkable technological achievement — not only critical for lunar surface studies, but also highly influential in advancing scientific photography in other fields, particularly aerial and remote sensing applications.

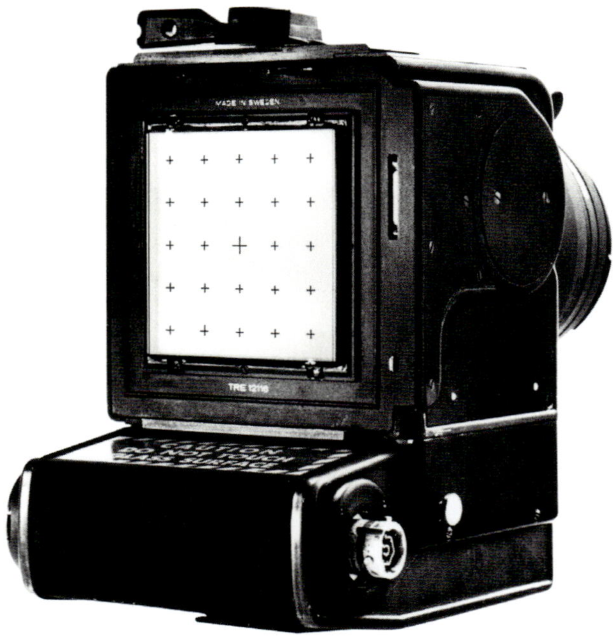

The Hasselblad Super Wide Camera

Used frequently during the Gemini missions, the Hasselblad Super Wide Camera (SWC) was a special-purpose model designed for wide-angle photography. It delivered images with exceptionally broad coverage, making it particularly well-suited for lunar surface documentation.

In early 1967, with this potential in mind, NASA assembled a team of engineers to adapt the SWC for use on the Moon. Using the existing model as a foundation, they enlarged the top-mounted viewfinder and redesigned the film winder and shutter release to improve usability. The resulting prototype produced higher-quality images than the 500EL models then in use.

However, the development of the Lunar Super Wide Camera was ultimately discontinued. During testing and training, astronauts overwhelmingly preferred the ease of the automatic film winding found in the 500EL — which became the basis for the Hasselblad Electric Data Camera (HEDC) used during Apollo surface missions.

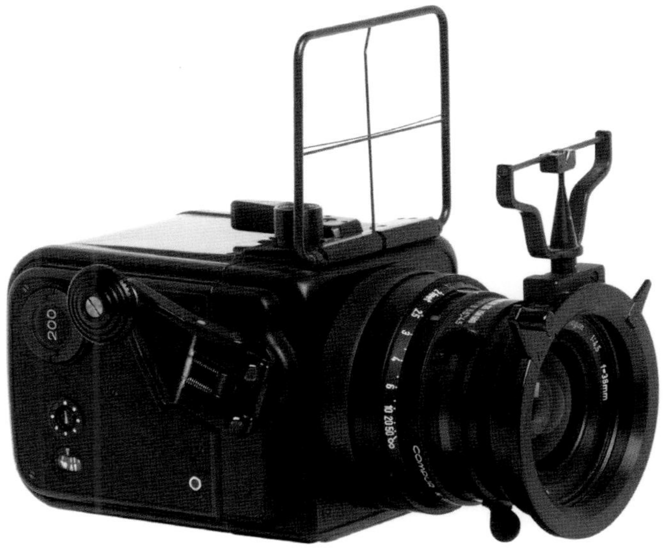

Nikon 35mm Photomic FTn NASA Modified

As the Apollo program progressed, the need for a smaller, more portable camera for use inside the Command/Service Module became apparent. NASA wanted to capture more dynamic and informal moments of the mission — particularly personal and candid interactions among the crew.

To meet this need, NASA approached Nikon to modify their 35mm Photomic FTn camera and flash system. The camera underwent several adjustments similar to those made to the Hasselblad models, ensuring it could withstand the intense gravitational forces and environmental conditions of spaceflight. Electrical components were rigorously tested to ensure they would not interfere with onboard systems in either the Lunar Module or Command/Service Module. As with other flight hardware, the camera body was painted matte black to minimize reflections. The grip and control buttons were enlarged for ease of use while wearing gloves.

The Nikon 35mm camera was introduced on Apollo 15 and remained in use on subsequent missions. It offered a new visual perspective — capturing more personal, candid, and often playful glimpses of life aboard the spacecraft.

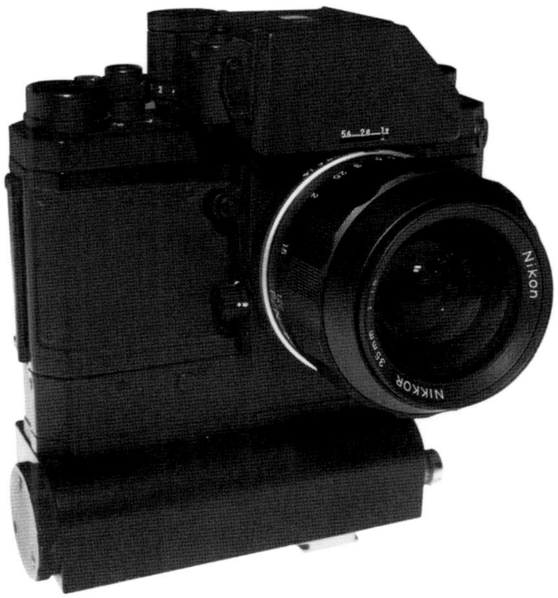

Carl Zeiss

German lens manufacturer Carl Zeiss played a vital role in developing the optical technology used during the Apollo program.

Zeiss worked closely with NASA and Hasselblad to design and supply a range of lenses to meet the program's demanding and varied requirements — from the ultra-fast 50mm f/0.7 to the large 500mm f/8 telephoto.

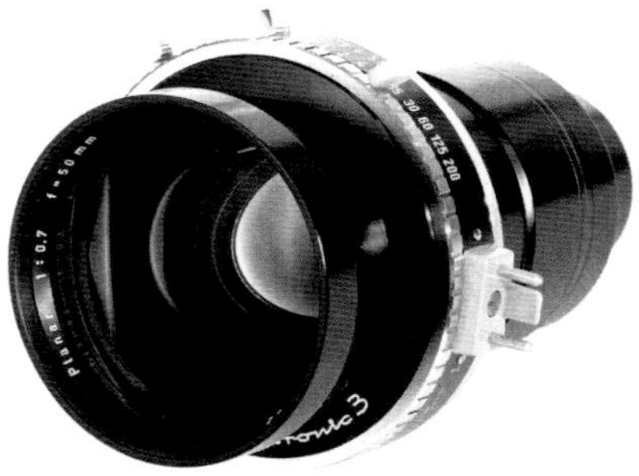

50mm f/0.7 Planar
This lens was unique and specially developed for NASA as one of the fastest camera lenses ever made. It was designed to allow crews to photograph low-light scenes on the dark side of the Moon. Only ten of these lenses were ever produced. One of them later found fame in Hollywood — used in the Oscar-winning production of Stanley Kubrick's film Barry Lyndon to shoot ultra low-light, candlelit scenes.

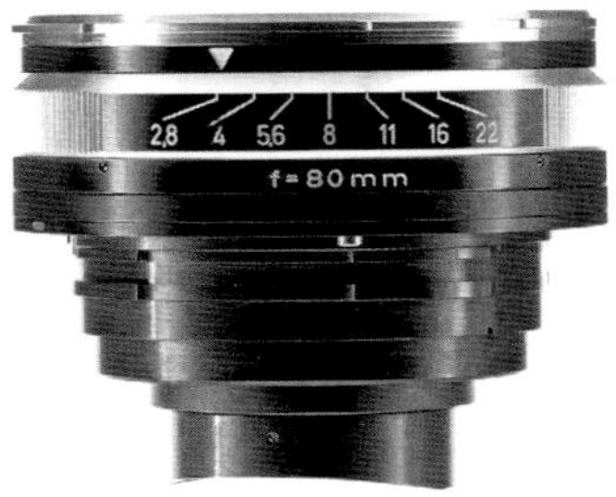

80mm f/2.8 Planar
This standard lens was carried onboard for general photography. The body of the lens was painted matte black, and the focus ring was simplified to offer three preset focus positions: near, medium, and far.

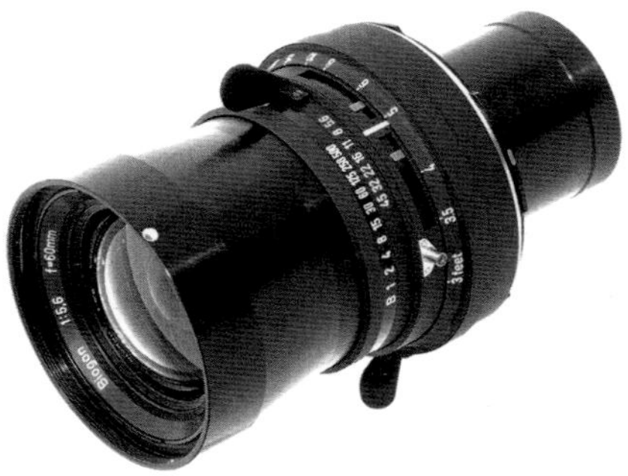

60mm f/5.6 Biogon
Another specially designed lens which was later released commercially. Designed specifically for use on the lunar surface, this wide-angle lens offered a high-quality field of view with extremely low distortion.
It was precisely calibrated to work in conjunction with the Réseau plate, ensuring scientific accuracy and allowing for the measurement and calculation of objects in the photographs.

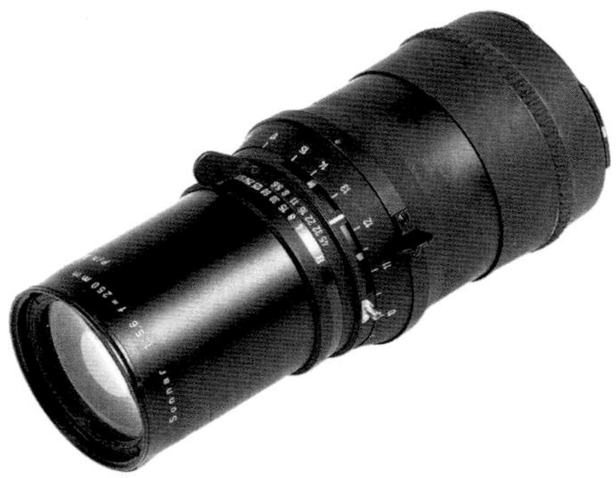

250mm f/5.6 Sonnar telephoto
This lens was used to capture distant subjects, including Earth from lunar orbit and features on the Moon's surface. It was also the lens that captured the iconic *Earthrise* photograph.

Eastman Kodak film

One of the biggest photographic challenges NASA faced was how to bring back as many images as possible from each mission, working within the tight space and weight restrictions of the Command Module.

The traditional 70mm Hasselblad camera held an average of 12 6×6cm square-format exposures per roll. During the Mercury missions, NASA approached Cinematics, a California-based company that modified cameras for Hollywood film productions. They were asked to enlarge the Hasselblad film magazines. With these changes, they could fit 18ft (5.5m) of film — enough for around 70 exposures.

During the Apollo program, however, a larger film magazine alone would not meet the new requirements. NASA asked Eastman Kodak to develop a film system that could hold even more exposures per roll. Kodak's research team created a thinner polyester-based film, which could still carry the standard emulsions. As a result, the same magazine could now hold up to 42ft (12.8m) of film, or around 200 exposures. NASA also stimulated the development of faster color reversal film — up to 50 times more sensitive than previous versions.

To adapt to the vastly different lighting conditions encountered during missions, Apollo flights used several film types: Kodak Panatomic-X, a fine-grained 80 ASA black-and-white film; Kodak Ektachrome SO-68 and SO-121 for color photography; and Kodak 2485, a black-and-white film rated at ASA 16,000, used for low-light conditions.

Camera setting instructions

Recommended camera settings were printed on the film magazines to assist astronauts during operation. These guides helped reduce the number of poorly exposed shots caused by incorrect settings — especially under pressure or in challenging lighting conditions.

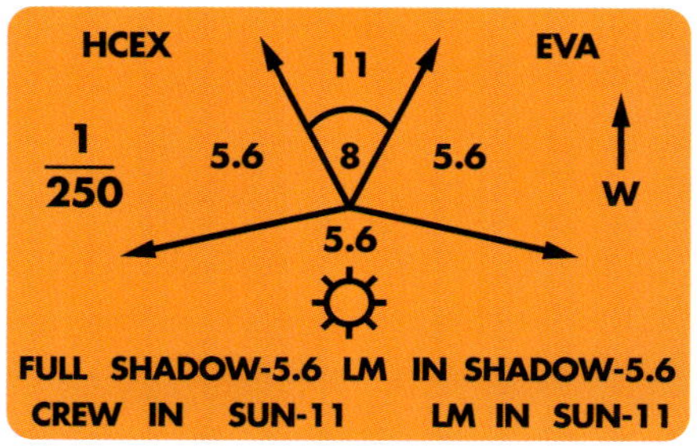

Operational sticker found on magazine S (top right). High-speed Color Exterior; the film type. 1/250 is the required camera shutter speed. The diagram with arrows gives the recommended aperture settings for the camera. It suggests using f/5.6 for either crew (astronauts) or Lunar Module (LM) in shadow (poorly lit areas), f/8 for objects near the camera and f/11 for distant objects in sunlight.

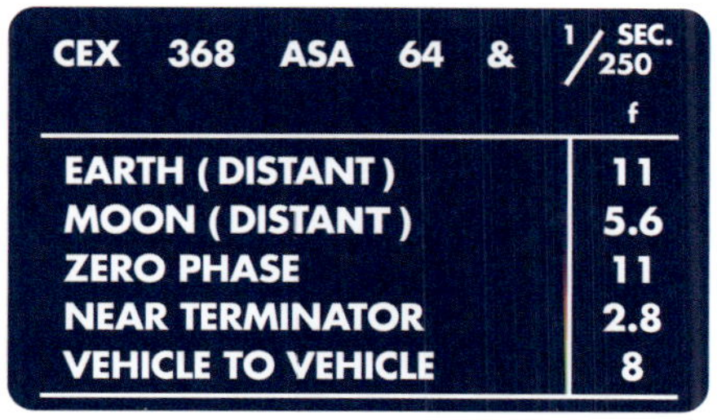

The blue sticker on Magazine N (right), intended for onboard use, shows the film type (CEX) and ASA rating (ASA 64). The recommended shutter speed is 1/250 second. The sticker also lists aperture settings for different subjects the astronaut might photograph. Zero phase refers to when the sun is directly behind the camera, producing flat light with minimal shadows. Near terminator refers to when you see both night and day on either the Earth or the Moon. Half the subject is well lit, the other half in darkness.

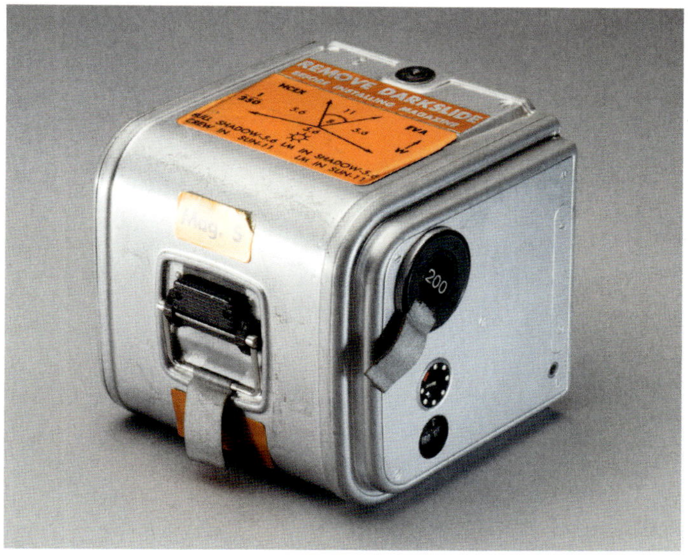

Magazine S
This magazine was used by Neil Armstrong during his time on the lunar surface. Its silver-coated exterior was designed to match the 'HEDC'. Loaded with High-speed Color Exterior (HCEX) Ektachrome film, it captured approximately 130 exposures. Several of the most widely recognized photographs from the Apollo 11 mission — including those on pages 115 and 131 — were taken using this specific magazine.

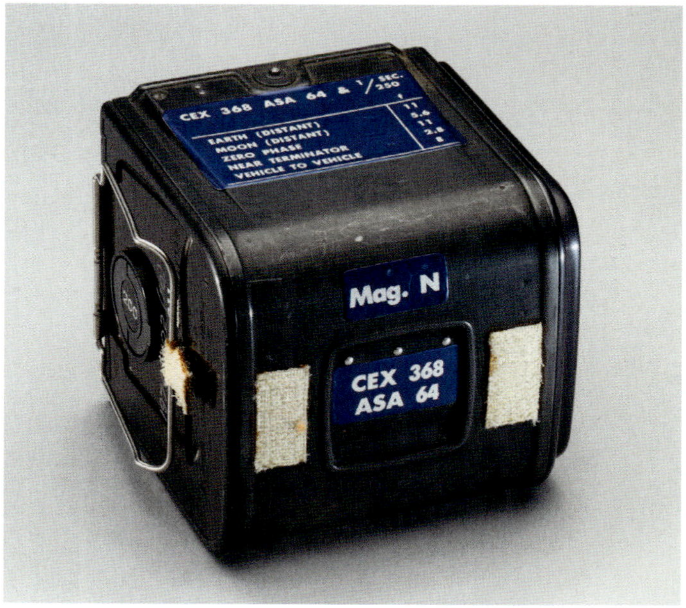

Magazine N
This magazine flew onboard the Apollo 11 Command/Service Module. It was loaded with a roll of Eastman Kodak SO-368 Color Exterior (CEX) film, specially balanced for exterior lighting conditions. Photographs on pages 116 and 130 were taken from this film magazine.

The Training

When NASA recognized that photography would be a key part of the Apollo program, they ensured the astronauts became highly proficient with the modified equipment during training. Astronauts were encouraged to carry their training cameras at all times — even on personal trips — to build familiarity with the gear and develop their photographic eye.

They also received tutorials on the cameras and their operation, and frequently trained in locations with geological features similar to those of the lunar surface — such as Nevada, Arizona, and Hawaii. On these field exercises, the crews reviewed their photographs and gave feedback on technique to improve image quality. As missions progressed, each returning crew passed along tips, insights, and practical experience to the next, leading to a noticeable improvement in image quality over the course of the program.

Since the cameras were largely automatic, astronauts mainly had to consider three things: framing the shot, selecting the correct exposure setting, and estimating the subject distance to set focus accurately.

NASA also provided in-mission photographic support. A specialist in Mission Control was always available to answer photography-related questions from the crew during flight.

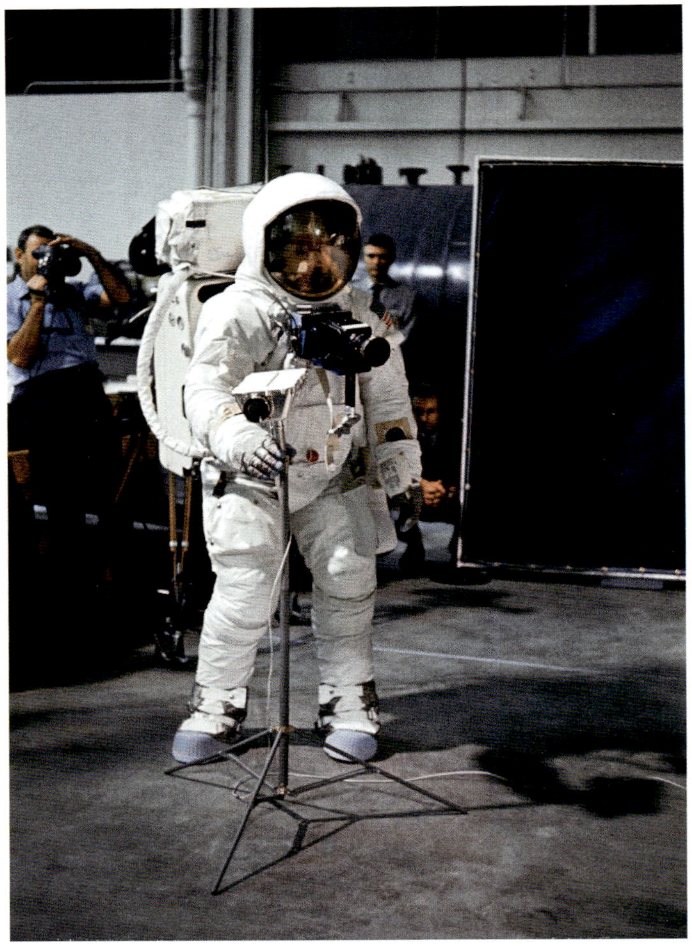

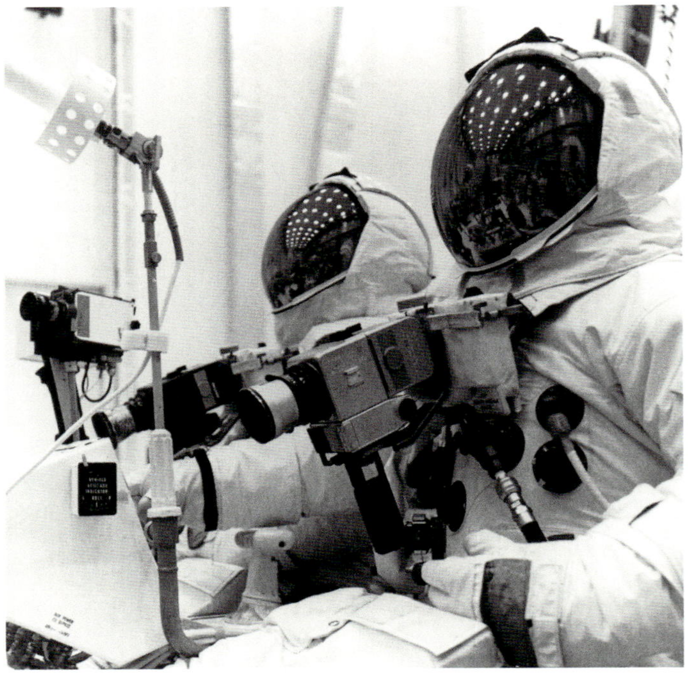

Top right:
Neil Armstrong trains with the Hasselblad still camera while standing beside the Westinghouse Lunar Television Camera. This black-and-white TV camera was flown on Apollo 11 and used to transmit the live broadcast of Armstrong's first steps on the Moon.

Bottom right:
Astronauts John W. Young (right) and Charles M. Duke Jr. participate in simulation training with the Lunar Roving Vehicle (LRV) during Apollo 16.

Opposite:
Aldrin (left) and Armstrong (right) during their Apollo 11 Extravehicular Activity (EVA) training exercises.

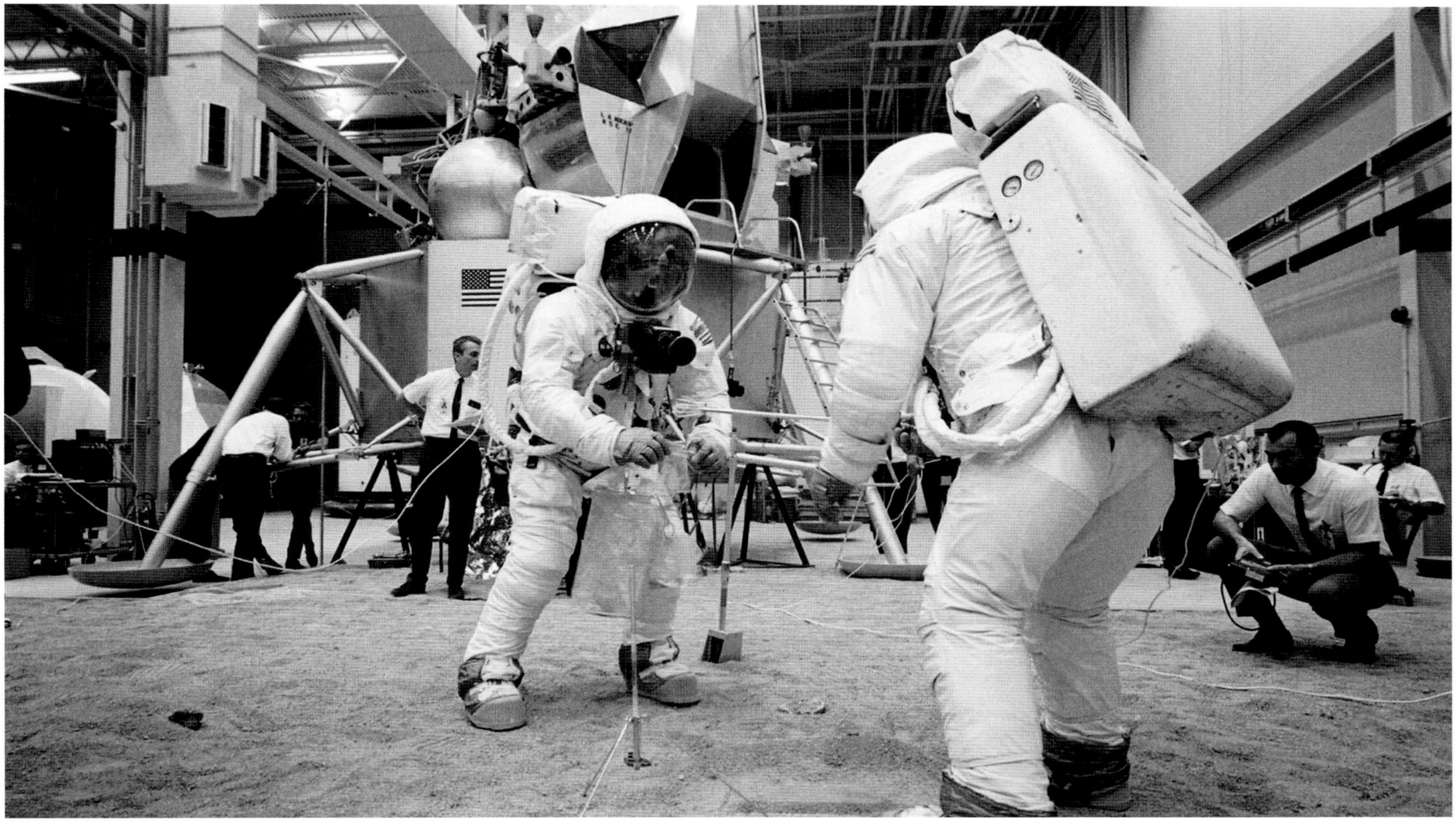

Framing
The lack of mobility in a spacesuit, and in particular the restrictive view of the helmet, meant that the Hasselblad camera's normal viewfinder was replaced with a simple sight mounted on the side of the camera. The astronauts were also taught to shoot from the hip when they were in the Lunar Module or Command/Service Module.

When on the lunar surface, the cameras were mounted on the chest of the spacesuit. The astronauts were trained to simply aim their body towards the point of interest. A generally inaccurate method but, with the intense training the astronauts received, they became very adept at this peculiar framing method.

Exposure
The exposure settings on the camera were simplified in advance as engineers had been able to calculate the lighting conditions beforehand. Simple guidelines were printed onto the film magazines. The shutter speed for the lunar cameras for example was set at 1/250 and the aperture recommendations were f/5.6 for objects in shadow or f/11 for objects in direct sunlight. When taking photographs of high interest or importance, the astronauts were instructed to use an exposure bracketing technique of one f-stop up and down from the suggested setting to guarantee a well exposed shot.

Focus
After significant trial and error during the training period, focusing became a relatively simplified procedure. The combination of a high f-stop and the limited amount of lenses taken on board meant that the image generally had a large depth of field. The astronaut only really had to estimate the distances of the main subject in order get the right focus. The focus ring was further simplified to offer three presets; near, medium and far to further make the camera operation as simple as possible.

Expert advice always on hand (CAPCOM)

Expert photographic advice was always available to the astronauts. An experienced photographer — familiar with space-to-ground photography — was stationed at a Mission Control desk whenever the crew was awake. This specialist remained on hand to offer real-time guidance on camera settings, especially when astronauts wanted to take a shot they hadn't specifically trained for, or one that required adjustments beyond standard procedures.

The radio exchange below, between the Apollo 11 crew and Mission Control, offers a glimpse into how astronauts consulted with photography experts to ensure they captured the perfect image.

Commander Neil Armstrong (CDR)
Houston, do you read Apollo 11?

Capsule Communicator, Mission Control, Houston (CAPCOM)
Roger, 11. We're reading you loud and clear now. We were down in the noise as we switched antennas a minute or so ago. Over.

Command Module Pilot Michael Collins (CMP)
Roger. What sort of (aperture) settings could you recommend for the solar corona? We've got the Sun right behind the edge of the Moon now.

Lunar Module Pilot Buzz Aldrin (LMP)
It's quite an eerie sight. There is a very marked three-dimensional aspect of having the Sun's corona coming from behind the Moon the way it is.

CAPCOM
Roger.

LMP
And it looks as though — I guess what's giving it that three-dimensional effect is the earthshine. I can see Tycho fairly clearly — at least if I'm right side-up, I believe it's Tycho, in moonshine — I mean, in earthshine. And, of course, I can see the sky is lit all the way around the Moon, even on the limb of it where there's no earthshine or sunshine.

CAPCOM
Apollo 11, this is Houston. Over.

LMP
Go ahead.

CAPCOM
Roger. If you'd like to take some pictures, we recommend using magazine Uniform which is loaded with high speed black-and-white film, interior lights off, electric Hasselblad with the 80mm lens. And you're going to have to hand-hold us, I guess. We're recommending an f-stop of 2.8, and we'd like to get a sequence of time exposures. Over.

LMP
Okay. You want magazine Uniform instead of magazine Tango? Over.

CAPCOM
Roger. We're not trying to get you all wrapped up in a procedure here. This is on a not-to-interfere basis, of course. Over.

LMP
Okay.

CAPCOM
And on the exposures we're looking for an eighth of a second, a half a second. And, if you think you can steady the camera against anything to get longer exposures, 2 seconds, 4 seconds, and 8 seconds. Over.

LMP
Roger. We copy.

Opposite:
The photograph of the solar corona that was being discussed between the crew of Apollo 11 and the CAPCOM on the July 19th, 1969. (AS11-42-6179)

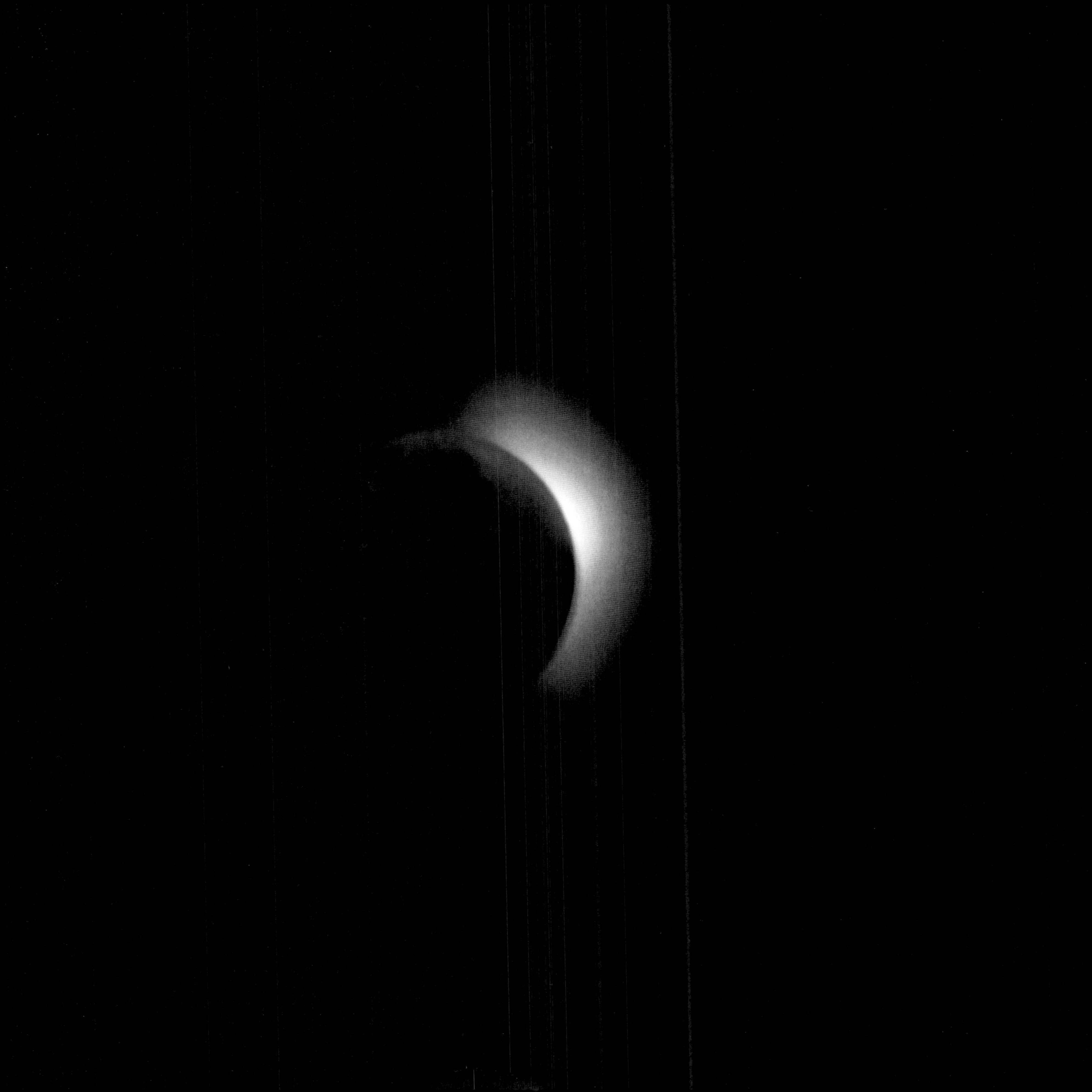

References

Apollo program. (Online) https://en.wikipedia.org/wiki/Apollo_program (Accessed Jun. 2016)

Derr, A. Photography Equipment and Techniques — A Survey Of Nasa Developments. (Online) http://history.nasa.gov/alsj/apollo.photechnqs1.pdf (Accessed Jun. 2016)

Earth Science and Remote Sensing Unit, NASA Johnson Space Center. Gateway to Astronaut Photography of Earth. (Online) https://eol.jsc.nasa.gov/FAQ/ (Accessed Jun. 2016)

Gunnarsson, S. (2016). Victor Hasselblad The Man Behind the Camera. Gothenburg: Journal, p.264–274

Hartwell, E. Apollo Cameras. (Online) http://www.ehartwell.com/Apollo17/BlueMarblePhotography_Cameras.htm (Accessed Jun. 2016)

Hasselblad AB, (02.07.1998). Hasselblad 30-year Lunar Mission Anniversary Press Release. Gothenburg.

Hasselblad and NASA. (Online) Available at: http://www.clubhasselblad.com/hasselblad-and-nasa (Accessed Jun. 2016)

Horgmo, Ø. The Apollo 11 Hasselblad Cameras. (Online) https://sterileeye.com/2009/07/23/the-apollo-11-hasselblad-cameras/ (Accessed Jun. 2016)

Jones, E & Glover, K. Apollo Lunar Surface Journal. (Online) https://www.hq.nasa.gov/alsj/ (Accessed Jun. 2016)

JSC PAO Web Team. Apollo Gallery. (Online) http://spaceflight.nasa.gov/gallery/images/apollo/ (Accessed Jun. 2016)

Kitmacher, G. Astronaut Still Photography During Apollo. (Online) http://history.nasa.gov/apollo_photo.html (Accessed Jun. 2016)

Lunar Planetary Institute. (Online) http://www.lpi.usra.edu/lunar/missions/apollo/ (Accessed Jun. 2016)

Multiple Authors. Apollo Flight Journal. (Online) http://history.nasa.gov/afj/index.htm (Accessed Jun. 2016)

Nordin, R. (1998). Hasselblad System Compendium. Small Dole: Hove Books, p.120–126 & p.160–162)

Parker, P. Apollo-11 Hasselblad Cameras. (Online) http://history.nasa.gov/alsj/a11/a11-hass.html (Accessed Jun. 2016)

Ringo. Historic Photography Questions: The Apollo XI Crew Wants to Take a Picture in Space. (Online) www.lomography.com/magazine/28327-historic-photography-questions-the-apollo-xi-crew-wants-to-take-a-picture-in-space (Accessed Jun. 2016)

Teague, K. Apollo Archive. (Online) http://www.apolloarchive.com/apollo_gallery.html (Accessed Jun. 2016)

The Truth Behind The Moon Landings: Stranger Than Fiction. Hollywood, United States of America: Virginia Quinn, 2013. DVD.

Underwood, R. (2000) Richard W. Underwood Oral History. Interviewed by Summer Chick Bergen for Johnson Space Center Oral History Project. (Online) http://www.jsc.nasa.gov/history/oral_histories/UnderwoodRW/UnderwoodRW_10-17-00.htm (Accessed Jun. 2016)

Photo & Image credits

All Images are courtesy of The National Aeronautics and Space Administration (NASA) photographic archives unless stated below.

https://www.flickr.com/photos/projectapolloarchive/

Page 11:
Launch of Apollo 11.
Courtesy of The Project Apollo Archive
https://www.flickr.com/photos/projectapolloarchive/27736779953/in/dateposted/

Page 12:
Launch of Apollo 7.
Photograph from NASA Apollo Flight Journal.
www.history.nasa.gov/ap07fj/a7images.htm

Photograph of Walt Cunningham and T-38.
Courtesy of Walt Cunningham

Page 14:
Walt Cunningham preparing for launch.
Photograph from NASA Apollo Flight Journal
www.history.nasa.gov/ap07fj/a7images.htm

Page 17:
Apollo 7 orbital chart.
Courtesy of Walt Cunningham

Page 18:
CSM Window viewing area diagram.
Courtesy of Walt Cunningham

Page 320:
Photograph of Victor Hasselblad
©Hasselblad Foundation

Page 321:
NASA Group & Postcard
Courtesy of Gösta Flemming at Journal-media and Sören Gunnarsson, author of Victor Hasselblad, The Man Behind the Lens.

Film magazine plaque
©Hasselblad Foundation

Page 322:
Hasselblad Camera illustration
Drawing by Max K. Reice

Page 323:
Photograph of Apollo 14 cameras
Courtesy of the Lunar Planetary Institute
www.lpi.usra.edu/lunar/missions/apollo/apollo_12/photography/

Illustration of Apollo 16 camera equipment
Photograph from NASA Apollo Flight Journal
http://history.nasa.gov/ap16fj/02photoequip.htm

Page 324:
Hasselblad Camera modification images
©Victor Hasselblad AB

Page 325:
Hasselblad 500c
©Hasselblad Foundation, Photo Jens Karlsson

Hasselblad HEC Camera
©Victor Hasselblad AB

Page 326:
Hasselblad HEDC
©Hasselblad Foundation

Hasselblad Réseau plate
©Hasselblad Foundation

Page 327:
Hasselblad SWC
©Hasselblad Foundation

Nikon 35mm Photomic FTn
Photograph by Morio, www.commons.wikimedia.org/wiki/File:1971_Nikon_Photomic_FTn_NASA_Modified_2012_CP%2B.jpg

Page 328:
Carl Zeiss Lens photographs
Courtesy of the 'ZEISS Archives'
www.zeiss.com/corporate/int/history/archives.html

Page 329:
Film magazine photographs
Smithsonian National Air and Space Museum (TMS A19980006000_PS03). Smithsonian National Air and Space Museum (TMS A19980009000_PS03).

Pages 330–331:
Astronaut training images
Courtesy of NASA

The process behind making this book

As you can imagine, it was incredibly difficult to select just a few images from the thousands of brilliant photographs taken during the Apollo program. We went through multiple rounds of selection, narrowing over 27,000 photographs down to the 255 you see in this book. Our decision was based on photographic quality, not simply fame or historical importance. With that in mind, we realize that some of your favorite images might not have made the cut — and the same goes for ours.

From the beginning, our intention was to create a beautifully crafted book that celebrates these extraordinary images. Each photograph was meticulously retouched to enhance tone, color, and depth, always with deep respect for the original material. Early on, we defined a few guiding principles: to keep the images true and honest to what was captured. This wasn't about using the latest technology to sharpen, de-noise, or add color that wasn't there. It was about honoring the photograph and the photographer, captured as-is. The human aspect mattered — the slightly out-of-focus frames, the flares, the strange crops. These weren't flaws to be corrected, but part of the experience, part of the story. Our aim wasn't technical perfection, but emotional presence.

One ongoing debate in reproducing these images is the color of the Moon's surface — and how accurately it should be printed. Opinions vary widely on what shade of gray the lunar surface truly is. After much research, we made a conscious decision: not to pursue absolute accuracy in coloring, but to highlight the visual and aesthetic strength of each image.

Beyond color correction and the careful removal of dust and scratches, we left the scans largely untouched — keeping frame edges, imperfect film cuts, handwritten notes, and visible markings. We believe these artifacts add character and authenticity, enhancing the photographic story behind each image.

We gave careful thought to how the photographs are presented — not just as documents, but as fine art. Each image sits cleanly on the page, framed in white, with no captions or technical distractions. This approach, inspired by fine art photography, gives each image space, clarity, and respect for its visual power.

With Gratitude

Rusty Schweickart, for his generosity and thoughtfulness in recounting his experiences from the Apollo 9 mission, and for helping us reflect on the meaning of these photographs in the context of both their time and what lies ahead with Artemis.

Walt Cunningham for his incredible energy and enthusiasm, and for taking time from his busy schedule to recount and write about his experiences with taking photographs from the Apollo 7 mission he flew on.

Neil Pearson, our editor, who has taken countless hours editing and revizing the many versions of the book that we put before him.

Sinc Agency, the advertizing agency in Amsterdam where the four of us worked when creating the book. Thanks to everyone at Sinc, especially to Fanny and Olivier, for your support and guidance, and of course the use of the workspace and coffee! www.sinc.agency

Elsa Modin, who warmly welcomed us to the Hasselblad Foundation Archive in Gothenburg, Sweden, and then who patiently spent many hours searching for many of our obscure picture requests.

Søren Dalsgaard for his hours of reading and advice.

Ida Gustafsson and Ove Bengtson at Hasselblad AB, for the introduction to such a wonderful and extensive archive that they have in Gothenburg.

Benoit Faiveley for his active involvement and tips throughout the making process.

Dr. Wolfgang Wimmer at the ZEISS Archive.

A note from the authors

A heartfelt thank you to our original Kickstarter backers — your early support over a decade ago made this book possible. What began as a crowdfunded idea has since grown into a published reality, and we remain deeply grateful for the belief you placed in us from the very start.

Floris is grateful for the opportunity to create something lasting with friends, for teNeues supporting us in our process, and for his family and loved ones for their support.

Joel would like to thank his friends, family and especially Bianca for their love, support and inspiration throughout this wild and wonderful ride. It's been an unforgettable journey, now shared with Rosie and June —to whom he dedicates this book, with all his love.

Simon is thankful to his friends and family, and particularly to Pía and Noemí, for their love, support, and endless encouragement over the years.

Delano wants to thank his friends and family for their support throughout this journey. A special thanks to Esmee for her patience and understanding. To my wonderful kids, Joas and Lilou, you mean the world to me.

Additional thanks to Karim Jobe & Esmee Friederichs at Doogle Productions, Bauke Bakker, Peter Bas Mensink, Andre Kuipers, Michael Light, Kees Dorresteijn & Daan Musters at NOS.

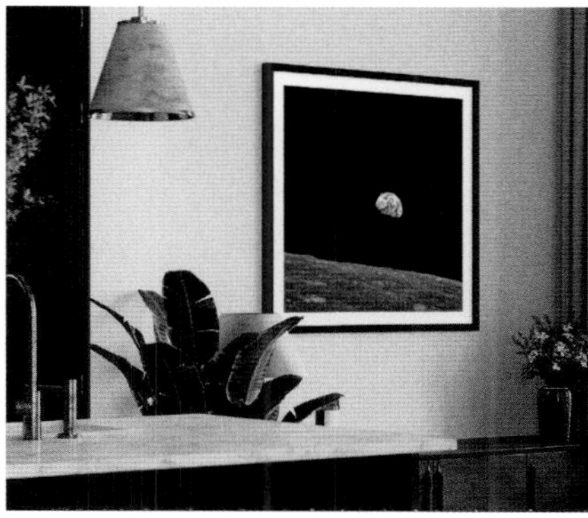

First Printing

Forwords by Walt Cunningham & Rusty Schweickart
Edited by Neil Pearson
Editorial coordination by Roman Korn
Production by Alwine Krebber
Color separation by Jens Grundei

Published by gestalten, Berlin 2025
ISBN 978-3-96171-706-4

Printed in Slovakia by Neografia a.s.

Bibliographic information published by the Deutsche Nationalbibliothek: The Deutsche Nationalbibliothek lists this publication in the Deutsche Nationalbibliografie; detailed bibliographic data are available on the Internet at dnb.dnb.de.

FSC
www.fsc.org
MIX
Paper | Supporting responsible forestry
FSC® C020353

For more information, and to order books, please visit
www.teneues.com and www.gestalten.com

Die Gestalten Verlag GmbH & Co. KG
Mariannenstrasse 9–10
10999 Berlin, Germany
hello@gestalten.com

Düsseldorf Office
Waldenburger Straße 13
41564 Kaarst, Germany
verlag@teneues.com

teNeues Press Department
press@gestalten.com

www.teneues.com

https://instagram.com/teneuespublishing

www.teneues.com